# UNITEXT for Physics

**Series Editors**

Michele Cini, University of Rome Tor Vergata, Roma, Italy

Stefano Forte, University of Milan, Milan, Italy

Guido Montagna, University of Pavia, Pavia, Italy

Oreste Nicrosini, University of Pavia, Pavia, Italy

Luca Peliti, University of Napoli, Naples, Italy

Alberto Rotondi, Pavia, Italy

Paolo Biscari, Politecnico di Milano, Milan, Italy

Nicola Manini, University of Milan, Milan, Italy

Morten Hjorth-Jensen, Department of Physics and Astronomy, University of Oslo, Oslo, Norway

Alessandro De Angelis , Physics and Astronomy, INFN Sezione di Padova, Padova, Italy

UNITEXT for Physics series publishes textbooks in physics and astronomy, characterized by a didactic style and comprehensiveness. The books are addressed to upper-undergraduate and graduate students, but also to scientists and researchers as important resources for their education, knowledge, and teaching.

Giulia Ricciardi

# Introduction to Neutrino and Particle Physics

From Quantum Field Theory to the Standard Model and Beyond

Giulia Ricciardi
Università di Napoli Federico II
Naples, Italy

ISSN 2198-7882   ISSN 2198-7890 (electronic)
UNITEXT for Physics
ISBN 978-3-031-65095-6   ISBN 978-3-031-65096-3 (eBook)
https://doi.org/10.1007/978-3-031-65096-3

© Springer Nature Switzerland AG 2024

This work is subject to copyright. All rights are solely and exclusively licensed by the Publisher, whether the whole or part of the material is concerned, specifically the rights of translation, reprinting, reuse of illustrations, recitation, broadcasting, reproduction on microfilms or in any other physical way, and transmission or information storage and retrieval, electronic adaptation, computer software, or by similar or dissimilar methodology now known or hereafter developed.
The use of general descriptive names, registered names, trademarks, service marks, etc. in this publication does not imply, even in the absence of a specific statement, that such names are exempt from the relevant protective laws and regulations and therefore free for general use.
The publisher, the authors and the editors are safe to assume that the advice and information in this book are believed to be true and accurate at the date of publication. Neither the publisher nor the authors or the editors give a warranty, expressed or implied, with respect to the material contained herein or for any errors or omissions that may have been made. The publisher remains neutral with regard to jurisdictional claims in published maps and institutional affiliations.

This Springer imprint is published by the registered company Springer Nature Switzerland AG
The registered company address is: Gewerbestrasse 11, 6330 Cham, Switzerland

If disposing of this product, please recycle the paper.

# Preface

The field of neutrino physics remains remarkably dynamic in contemporary research. Advancements in technology over recent decades have significantly enhanced our capacity to explore the properties of neutrinos, sustaining considerable interest in this captivating particle.

Neutrinos are presently under investigation across various domains of physics, spanning from high-energy physics to quantum field theory, cosmology, astrophysics, nuclear physics, and geophysics. In particle physics, the Standard Model faces a challenge in accommodating neutrinos with mass, and their precise nature is still unknown. This makes neutrino physics a compelling avenue for probing physics which extends the Standard Model. In astrophysics and cosmology, the importance of neutrinos cannot be overstated. As an example, just consider that by observing neutrinos emitted by stars, one gains significant understanding of the internal mechanisms of stellar evolution, including processes like nuclear fusion and supernova explosions.

The array of topics where neutrinos provide valuable insights could be expanded much further. However, this breadth can pose a challenge for a student who is just beginning to delve into this subject. Many textbooks offer an overview of the diverse facets of neutrino physics, assuming readers possess the foundational knowledge required for a thorough comprehension. This background is typically covered in quantum field theory or particle physics books, which, however, can prove overwhelming for those with a specific interest in neutrinos. This book aims to be a middle ground.

Its purpose is not just to provide a comprehensive examination of both theoretical and experimental advancements in neutrino physics but also to supply the solid background necessary for their thorough understanding. The emphasis is on clear explanations of the foundations, and concepts are discussed in the most possible self-contained way. Importantly, this book caters to both theorists and experimentalists. In the experimental sections, the most significant physical aspects and contexts are explored to avoid simply listing technical details that may become outdated in a few years. The relationship between theoretical framework and experimental probing is consistently emphasised and clearly articulated.

To enhance overall understanding, each chapter in the book begins with a paragraph summarising its content. The opening chapter is an historical survey, providing essential context, followed by an ideal division of the book into two parts.

The first part endeavours to provide the foundation in field theory and particle physics necessary for a comprehensive understanding of neutrino physics. It includes elements of group theory, global symmetries, gauge theories, and the description of the Standard Model. The treatment is pedagogical and introductory. Specialised jargon and traditionally ambiguous concepts are carefully explained. Great care has been taken to ensure that the simplicity of the explanations does not compromise their scientific rigour. This part stands as a comprehensive primer in quantum field theory and particle physics in its own right. It can also be used as a resource for courses in quantum field theory and particle physics.

The second part focuses on neutrino physics. It can also function as a reference text for advanced readers or researchers specialised in other fields and merely interested to acquire the essential notions to stay abreast of developments in neutrino physics. It begins with the subjects of neutrino masses and their mixing, followed by a chapter addressing astrophysical, cosmic neutrinos, and geoneutrinos. Another chapter is dedicated to the thriving subject of neutrino oscillations at reactors and accelerators. Standard neutrino textbooks sometimes neglect topics such as neutrino-nucleus interactions and cross-sections, which are given a dedicated chapter. Their importance has grown in recent years because of the improved precision of experiments and the resulting need for precisely calculated cross-sections. The final chapter provides an overview of theoretical and experimental prospects, such as the searches for sterile neutrinos or neutrinoless double beta decay. The hope is for readers to contribute to some of these forthcoming endeavours.

I wish to express my gratitude to my research collaborators, whose valuable interactions have greatly enriched my knowledge and broadened my perspectives. In particular I would like to thank for their support and comments my colleagues and friends Ilkaros Bigi, Gianfranca De Rosa, Speranza Falciano, José W. F. Valle, Eleonora Fusco, Marco Pallavicini, Francesco Vissani. I also thank the students who attended my courses, particularly those at the Department of Physics E. Pancini, University of Naples Federico II, as well as the students whom I advised, for their interest and insightful questions. I extend my sincere appreciation to Marina Forlizzi and Barbara Amorese at Springer for their kindness, professionalism, and patience throughout all stages of this book's development.

Naples, Italy                                                                                       Giulia Ricciardi
May 2024

# Contents

1 **Historical Survey** .................................................. 1
   1.1 The Dirac Field .................................................. 2
       1.1.1 The Fermion in the Electromagnetic Field ............... 4
   1.2 The Idea of the Neutrino ......................................... 6
   1.3 The $\beta$ Decay ................................................ 9
   1.4 Neutrino Observations ........................................... 12
   1.5 The Pursuit of Neutrino Properties .............................. 16

2 **Global Symmetries** ................................................ 19
   2.1 Elements of Group Theory ........................................ 21
   2.2 Classical Physics: Parity and Time Reversal ..................... 25
   2.3 Magnetic and Electric Dipole Moments ............................ 28
   2.4 Lorentz Transformations ......................................... 29
       2.4.1 The Proper Lorentz Group ................................. 32
   2.5 Poincaré Transformations ........................................ 34
   2.6 Flavour Quantum Numbers ......................................... 36
   2.7 Charge Conjugation .............................................. 38
       2.7.1 $G$-Parity ............................................... 39
   2.8 Parity Eigenstates .............................................. 39
       2.8.1 Observations of Parity Violation ......................... 44
   2.9 Charge Conjugation Eigenstates .................................. 46
       2.9.1 Charge Symmetry and G-Parity ............................. 48
   2.10 $CP$ Invariance ................................................ 51
   2.11 Time Reversal .................................................. 54
       2.11.1 Observations of Time Reversal Violation ................. 55
   2.12 Lorentz Transformations for Spin 1/2 Fields .................... 59
   2.13 Field Transformations Under $C$, $P$ and $T$ ................... 61
   2.14 Helicity and Chirality ......................................... 69
   2.15 The CPT Theorem ................................................ 75

## 3 Gauge Theories ........... 79
- 3.1 Electromagnetism ........... 81
- 3.2 Gauge Theories and the Covariant Derivative ........... 83
- 3.3 Spontaneous Symmetry Breaking ........... 87
  - 3.3.1 Internal Discrete Symmetry ........... 90
  - 3.3.2 Continuous Global Symmetry ........... 93
  - 3.3.3 Linear Sigma Model ........... 96
  - 3.3.4 The Higgs Mechanism ........... 98

## 4 The Standard Model ........... 101
- 4.1 The Electroweak Gauge Symmetry ........... 102
- 4.2 The EW Lagrangian for Massless Fermions ........... 107
  - 4.2.1 Isospin and CVC Hypothesis ........... 111
- 4.3 Charged and Neutral Currents ........... 112
- 4.4 The Higgs Mechanism in the SM ........... 114
  - 4.4.1 Massive Gauge Fields and Higgs Interactions ........... 116
- 4.5 The Yukawa Lagrangian ........... 120
- 4.6 Fermion Mass Eigenstates ........... 122
- 4.7 The CKM Matrix ........... 126
  - 4.7.1 Two Fermion Generations ........... 127
  - 4.7.2 Three Generations ........... 129
- 4.8 $C$, $P$ and $CP$ Symmetries in the EW Lagrangian ........... 133
  - 4.8.1 Matrix Elements of Hadronic Currents ........... 134
- 4.9 Accidental Symmetries ........... 135
- 4.10 Lepton Universality ........... 138

## 5 The Mass of Neutrinos ........... 143
- 5.1 The Neutrino in the Standard Model ........... 144
- 5.2 Dirac Mass Terms ........... 146
- 5.3 Majorana Neutrinos ........... 149
- 5.4 Majorana Mass Terms ........... 152
- 5.5 Mass Terms and Gauge Symmetry ........... 157
- 5.6 Dirac-Majorana Mass Term ........... 159
  - 5.6.1 One-Generation Case ........... 160
    - 5.6.1.1 Real Dirac-Majorana Mass Matrix ........... 162
  - 5.6.2 More Generations ........... 164
- 5.7 The Seesaw Mechanism ........... 164
  - 5.7.1 More Generations ........... 166
- 5.8 The Effective Lagrangian ........... 169
- 5.9 Opening Up the Weinberg Operator at Tree Level ........... 172
- 5.10 Type II See-Saw Mechanism ........... 175
- 5.11 Type III See-Saw Mechanism ........... 179

## 6 The Mixing of Neutrinos ........... 183
- 6.1 The Lepton Mixing Matrix ........... 184
  - 6.1.1 Parameter Counting ........... 186
- 6.2 Neutrino Oscillations ........... 188

|   |     |       |                                         |       |
|---|-----|-------|-----------------------------------------|-------|
|   | 6.3 |       | Two-Generations Mixing                  | 194   |
|   | 6.4 |       | CP Violation                            | 197   |
|   | 6.5 |       | Hierarchy and Simplified Mixing         | 202   |
|   |     | 6.5.1 | Two-Neutrino Formalism                  | 204   |
|   |     | 6.5.2 | Electron Neutrino Disappearance         | 206   |
|   | 6.6 |       | Neutrino Oscillations in Matter         | 207   |
|   |     | 6.6.1 | Uniform Density                         | 212   |
|   |     | 6.6.2 | Variable Density                        | 214   |

# 7  Natural Neutrino Sources .................................. 219

|   |     |       |                                              |     |
|---|-----|-------|----------------------------------------------|-----|
|   | 7.1 |       | The Standard Solar Model                     | 219 |
|   | 7.2 |       | Solar Neutrino Experiments                   | 225 |
|   |     | 7.2.1 | The Homestake Experiment                     | 227 |
|   |     | 7.2.2 | The Cherenkov Detector                       | 228 |
|   |     | 7.2.3 | The Kamiokande Experiment                    | 231 |
|   |     | 7.2.4 | The GALLEX and SAGE Experiments              | 232 |
|   |     | 7.2.5 | The SNO Experiment                           | 233 |
|   |     | 7.2.6 | The KamLAND Experiment                       | 234 |
|   |     | 7.2.7 | The Borexino Experiment                      | 235 |
|   | 7.3 |       | Supernova Neutrinos                          | 236 |
|   |     | 7.3.1 | Formation of a Supernova                     | 238 |
|   |     | 7.3.2 | Core-Collapse Supernovae                     | 240 |
|   |     | 7.3.3 | Detection                                    | 244 |
|   | 7.4 |       | Atmospheric Neutrinos                        | 246 |
|   |     | 7.4.1 | Production                                   | 247 |
|   |     | 7.4.2 | Detection                                    | 250 |
|   | 7.5 |       | Cosmic Neutrinos                             | 252 |
|   |     | 7.5.1 | Cosmic Neutrino Background                   | 252 |
|   | 7.6 |       | Neutrinos at High Energy                     | 256 |
|   | 7.7 |       | Cubic-Kilometer Detectors                    | 260 |
|   |     | 7.7.1 | The IceCube Experiment                       | 261 |
|   | 7.8 |       | Geoneutrinos                                 | 263 |

# 8  Oscillations at Reactors and Accelerators ............... 267

|   |     |       |           |                                              |     |
|---|-----|-------|-----------|----------------------------------------------|-----|
|   | 8.1 |       |           | Neutrinos at Reactors                        | 268 |
|   |     | 8.1.1 |           | The First Direct Neutrino Detection          | 271 |
|   |     | 8.1.2 |           | Early Oscillation Experiments                | 273 |
|   |     | 8.1.3 |           | Oscillation Experiments Since the Turn of the Century | 275 |
|   | 8.2 |       |           | Accelerator Experiments                      | 277 |
|   |     | 8.2.1 |           | Non Conventional Neutrino Beams              | 281 |
|   |     | 8.2.2 |           | First Generation LBL Experiments             | 283 |
|   |     |       | 8.2.2.1   | K2K                                          | 283 |
|   |     |       | 8.2.2.2   | MINOS                                        | 284 |
|   |     |       | 8.2.2.3   | ICARUS and OPERA                             | 286 |
|   |     | 8.2.3 |           | Off-Axis Technique                           | 288 |

|  |  | 8.2.4 | New Generation LBL Experiments | 292 |
|---|---|---|---|---|
|  |  |  | 8.2.4.1 T2K | 293 |
|  |  |  | 8.2.4.2 NOνA | 294 |
|  |  |  | 8.2.4.3 HK, T2HK and DUNE | 295 |
|  |  | 8.2.5 | SBL Accelerator Experiments | 297 |
|  | 8.3 | Oscillation Parameters Determination | | 299 |

## 9 Neutrino Cross Sections ... 303
- 9.1 Neutrino-Electron Scattering ... 305
  - 9.1.1 Elastic and QE Amplitudes ... 306
  - 9.1.2 QE Scattering Cross Sections ... 310
- 9.2 Neutrino-Nucleon Interactions ... 315
  - 9.2.1 Charged Current QE Scattering ... 317
  - 9.2.2 Single Pion Production ... 321
  - 9.2.3 Deep Inelastic Scattering ... 323
  - 9.2.4 The Bjorken Scaling and the Parton Model ... 327
- 9.3 The Gargamelle Neutrino Experiment ... 332
- 9.4 Neutrino Interactions with Nuclei ... 334
  - 9.4.1 Coherent Nuclear Scattering ... 336
  - 9.4.2 Nuclear Effects ... 340
  - 9.4.3 The Fermi Gas Model ... 342

## 10 Theoretical and Experimental Prospects ... 345
- 10.1 Absolute Values of Neutrino Masses ... 345
  - 10.1.1 Direct Neutrino Mass Measurements ... 348
- 10.2 Double $\beta$ Decay ... 352
- 10.3 Neutrinoless Double $\beta$ Decay ... 354
  - 10.3.1 Total Decay Rate and Effective Majorana Mass ... 356
  - 10.3.2 Experimental Searches ... 359
  - 10.3.3 Direct Detection Experiments ... 362
  - 10.3.4 Majorons ... 366
- 10.4 Neutrino Magnetic Moments ... 370
- 10.5 Sterile Neutrinos ... 374
  - 10.5.1 Mixing in the Dirac-Majorana Case ... 378
  - 10.5.2 In Search of eV-Scale Sterile Neutrino Oscillations ... 381
    - 10.5.2.1 LSND ... 382
    - 10.5.2.2 KARMEN ... 382
    - 10.5.2.3 MiniBooNE ... 383
    - 10.5.2.4 Short-Baseline Neutrino Program at Fermilab ... 384
    - 10.5.2.5 JSNS$^2$ ... 386
    - 10.5.2.6 Gallium and Reactor Anti-neutrino Anomalies ... 386
    - 10.5.2.7 Unstable Sterile Neutrinos ... 387

**References** ... 389

**Index** ... 413

# Chapter 1
# Historical Survey

At the end of the nineteenth century, an active field of research was the study of radiation originating in nuclear processes. In 1895 Wilhelm Conrad Roentgen had discovered X-rays [1], and one year later Henri Becquerel had witnessed for the first time natural radioactivity [2]. In 1897, experiments by Joseph John Thomson had led to the discovery of the electron [3].[1] In a few years it was realised that two kinds of radiative emission, named alpha ($\alpha$) and beta ($\beta$), accompanied the transmutation of an initial nucleus into another. They were distinct by their capacity of penetration, ionisation and behaviour under electric and magnetic fields. The beta rays, which were identified as streams of electrons, presented an unusual behaviour that led to hypothesise a yet undiscovered particle, the neutrino. This chapter starts reminding the main features of the Dirac equation (Sect. 1.1), the state of the art at the time of the Pauli proposal of neutrino in 1930 (Sect. 1.2). After discussing $\beta$ decay (Sect. 1.3), we continue to follow the main stages of the neutrino history along an ideal timeline, encompassing the actual discovery in 1956 and the assessment of its basic properties, the discovery of the muon neutrino (1962), the observation of neutrinos from the Sun and the atmosphere, the first direct observation of the tau neutrino (2000) (Sect. 1.4), ending with the observations of neutrino oscillations and other results of the last two decades (Sect. 1.5).

---

[1] J.J. Thomson was awarded the 1906 Nobel prize in Physics; his son George Paget Thomson was awarded the 1937 Nobel prize in Physics, together with Clinton Joseph Davisson, for the experimental discovery of the diffraction of electrons by crystals. One could say that the father showed that the electron is a particle, and the son that it also behaves like a wave.

## 1.1 The Dirac Field

In 1928, the debate about a relativistic extension of the Schrödinger equation had already lead to the formulation of the so-called Klein-Gordon equation

$$(\partial_\mu \partial^\mu + m^2)\phi = 0. \tag{1.1}$$

This Lorentz invariant equation was not felt as satisfactory at the time. The main reason was that in the related continuity equation there is the possibility of negative probabilities, which prevents a statistical interpretation of the wave function. In order to overcome this problem, Dirac proposed an equation linear in $\partial/\partial t$, rather than second order. To treat space and time on equal footing, linearity in $\partial/\partial x^i$ ($i = 1, 2, 3$) was also requested. Hence the Dirac equation was formulated as

$$i\frac{\partial \psi}{\partial t} = \left(-i\alpha^i \frac{\partial}{\partial x^i} + \beta m\right)\psi \tag{1.2}$$

where $\alpha^i$ and $\beta$ are arbitrary coefficients, and $m$ is the mass of the free particle described by the equation. By the quantum mechanics correspondence

$$i\frac{\partial}{\partial t} \to H \qquad -i\frac{\partial}{\partial x^i} \to P^i \tag{1.3}$$

where $H$ is the Hamiltonian and $\mathbf{P}$ is the momentum, we have

$$H\psi = (\boldsymbol{\alpha} \cdot \mathbf{P} + \beta m)\psi \tag{1.4}$$

The coefficients $\alpha^i$ and $\beta$ were determined by the requirement that a free particle must satisfy the relativistic energy-momentum relation

$$H^2 \psi = (\mathbf{P}^2 + m^2)\psi \tag{1.5}$$

It was demonstrated that this condition cannot be satisfied if $\alpha^i$ and $\beta$ are simply numbers. We are led to consider matrices operating on a wave function $\psi(x)$, which becomes a multi-component column vector. The lowest dimensionality of the matrices satisfying all these requirements is four, but the choice of the four matrices is not unique. One representation frequently used is the so-called Dirac-Pauli representation:

$$\alpha_i = \begin{pmatrix} 0 & \sigma_i \\ \sigma_i & 0 \end{pmatrix} \qquad \beta = \begin{pmatrix} I & 0 \\ 0 & -I \end{pmatrix} \tag{1.6}$$

## 1.1 The Dirac Field

where $I$ is the unit $2 \times 2$ matrix and $\sigma_i$ are the three $2 \times 2$ Pauli matrices

$$\sigma_1 = \begin{pmatrix} 0 & 1 \\ 1 & 0 \end{pmatrix} \quad \sigma_2 = \begin{pmatrix} 0 & -i \\ i & 0 \end{pmatrix} \quad \sigma_3 = \begin{pmatrix} 1 & 0 \\ 0 & -1 \end{pmatrix}. \tag{1.7}$$

The Pauli matrices have trace zero and determinant $-1$. Their commutation and anti-commutation relations are, for $i, j \in (1, 2, 3)$

$$[\sigma_i, \sigma_j] = 2i\,\varepsilon_{ijk}\sigma_k$$
$$\{\sigma_i, \sigma_j\} = 2\delta_{ij}I \tag{1.8}$$

where $\varepsilon^{ijk}$ is the completely antisymmetric tensor (or Levi-Civita tensor). Thus it follows

$$\sigma_i \sigma_j = i\varepsilon_{ijk}\sigma_k + \delta_{ij}I$$
$$\sigma_i^2 = I$$
$$\sigma_2 \sigma_i^* \sigma_2 = -\sigma_i. \tag{1.9}$$

By introducing the four Dirac $\gamma$ matrices

$$\gamma^\mu = (\beta, \beta\alpha_1, \beta\alpha_2, \beta\alpha_3) = (\gamma^0, \gamma^1, \gamma^2, \gamma^3) \tag{1.10}$$

we can express the Dirac equation in what is called its covariant form

$$\left(i\gamma^\mu \frac{\partial}{\partial x^\mu} - m\right)\psi(x) = (i\slashed{\partial} - m)\,\psi(x) = 0 \tag{1.11}$$

where $\partial_\mu = (\partial/\partial x^0, \vec{\nabla}) = (\partial/\partial x^0, \partial/\partial x^1, \partial/\partial x^2, \partial/\partial x^3)$ and $\slashed{\partial} = \gamma^\mu \partial_\mu$. The corresponding Lagrangian density is

$$\mathcal{L} = \overline{\psi}(x)\left(i\slashed{\partial} - m\right)\psi(x) \tag{1.12}$$

where $\overline{\psi} = \psi^\dagger \gamma_0$. The equation of Dirac for the electron, historically, has led to the concept of particles and antiparticles and to the definition of charge conjugation. The solution of the Dirac equation is a Dirac spinor with four independent complex components. Of the four independent solutions the two positive energy solutions describe a fermion with momentum $p^\mu$, and the two negative energy solutions are associated with the anti-particle. There is an extra twofold degeneracy, which means there must be another observable that commutes with the Dirac Hamiltonian and momentum, whose eigenvalues can be taken to distinguish the states and represent a good quantum number. This observable is the helicity, which will be discussed in detail in Sect. 2.14. One can interpret the negative energy solution (the anti-particle solution) as describing a state with opposite momentum and spin polarization.[2] The general solutions of the free Dirac equations can be expressed as a linear combinations of plane waves

$$\psi(x) = u(p,s)e^{-ip\cdot x} \qquad \psi(x) = v(p,s)e^{ip\cdot x} \qquad (1.13)$$

where the spinors $u(p,s)$ and $v(p,s)$ refer to particle and antiparticle solutions, respectively. The index $s = 1, 2$ distinguishes among the independent solutions and $p^\mu$ is the four-momentum of the particle. Note that $p^0 > 0$ and $p^2 = m^2$.

From the Dirac equation, and for plane waves solution, it follows that

$$\left(\slashed{p} - m\right) u(p,s) = \bar{u}(p,s) \left(\slashed{p} - m\right) = 0$$
$$\left(\slashed{p} + m\right) v(p,s) = \bar{v}(p,s) \left(\slashed{p} + m\right) = 0. \qquad (1.14)$$

We have used the notation $\slashed{a} \equiv \gamma^\mu a_\mu$, which holds for any four-vector $a_\mu$. There are four independent solutions of the energy eigenvalue equation $H\psi = E\psi$. If the four-momentum dependence of the plane wave function is $e^{-ip\cdot x}$, the Hamiltonian $H$ in (1.4) reads

$$H = \begin{pmatrix} m & \boldsymbol{\sigma} \cdot \boldsymbol{p} \\ \boldsymbol{\sigma} \cdot \boldsymbol{p} & -m \end{pmatrix}. \qquad (1.15)$$

Although the Dirac equation was derived as a single-particle quantum-mechanical wave equation, in modern quantum field theory it still occupies a significant role for the description of the free Dirac field.

### 1.1.1 The Fermion in the Electromagnetic Field

In quantum mechanics, even relativistic one, only problems in which the number of particles do not change can be formulated. The interaction can be introduced by means of the concept of an external field. Dirac generalised his equation to the case of a fermion $\psi$ of charge $q$ in an external electromagnetic field with vector potential $A^\mu$. He introduced a minimal coupling between the fermion and the electromagnetic field by slightly modifying the free equation. We derive it in modern terms, by using the covariant derivative $D^\mu = \partial^\mu + iqA^\mu$, which will be discussed in Sect. 3.2. The replacement $\partial_\mu \to D_\mu$ in the free Dirac equation (1.11) yields

$$\left[i\slashed{D} - m\right]\psi(x) = \left[i\slashed{\partial} - q\slashed{A} - m\right]\psi(x) = 0. \qquad (1.16)$$

---

[2] This can be confirmed by applying the charge conjugation operator, as we will do in Sect. 2.12.

## 1.1 The Dirac Field

The Dirac equation (1.16) describes the minimal coupling between a fermion $\psi$ of charge $q$ and the electromagnetic field $A^\mu$.

Applying the operator $[i\slashed{D} + m]$ on the left of Eq. (1.16), we obtain

$$\left[\slashed{D}^2 + m^2\right]\psi(x) = \left[D_\mu D^\mu + \frac{q}{2}F_{\mu\nu}\sigma^{\mu\nu} + m^2\right]\psi(x) = 0 \tag{1.17}$$

where $F^{\mu\nu} = \partial^\mu A^\nu - \partial^\nu A^\mu$ is the electromagnetic tensor. We have used the identity $\slashed{D}^2 = D_\mu D^\mu + \frac{q}{2}F_{\mu\nu}\sigma^{\mu\nu}$, where $\sigma^{\mu\nu} \equiv -\frac{i}{2}[\gamma^\mu, \gamma^\nu]$.

We can compare to the Klein-Gordon equation for a scalar field coupled to $A^\mu$, namely Eq. (1.1) when $\partial_\mu \to D_\mu$. The Dirac equation contains the extra term $\sigma_{\mu\nu}F^{\mu\nu}$, that can be modified as

$$\sigma^{\mu\nu}F_{\mu\nu} = 2\sigma^{0i}F_{0i} + \sigma^{ij}F_{ij}$$
$$= -2\sigma^{0i}E^i + \epsilon_{ijk}\sigma^{ij}B^k$$
$$= -2\sigma^{0i}E^i - 4s^k B^k \tag{1.18}$$

where $i, j, k \in \{1, 2, 3\}$. In the second passage we have used the relations, well known in classical electromagnetism, between $F^{\mu\nu}$, the electric field $E^i$ and the magnetic field $B^i$, namely $E^i = -F_{0i}$ and $F_{ij} = \epsilon_{ijk}B^k$. In the third passage we have defined $s^k \equiv \epsilon^{kij}\sigma^{ij}/4$. It is easy to demonstrate that $\sigma^{ij} = \epsilon_{ijn}\sigma^n$, where $\sigma^n$ ($n \in \{1, 2, 3\}$) is a Pauli matrix. Since $\epsilon^{kij}\epsilon_{ijn} = 2g_n^k$, where $g_n^k$ is the metric tensor, we find that $s^k = \sigma^k/2$ and can be identified with the spin vector in non-relativistic quantum mechanics.

We would like to estimate the contribute of this additional term to the energy in the non-relativistic approximation.[3] As a first step, we can go back to Eqs. (1.17), but do not use the natural units. We have to reintroduce factors of $\hbar$ and the speed of light $c$, and we can write

$$m^2c^4 + \frac{\hbar c^2 q}{2}\sigma^{\mu\nu}F_{\mu\nu} = m^2c^4\left[1 + \frac{1}{c^2}\left(\frac{\hbar q}{2m^2}\sigma^{\mu\nu}F_{\mu\nu}\right)\right]. \tag{1.19}$$

Taking the square root and keeping the lowest terms in $c$, we obtain the non-relativistic approximation

$$mc^2 + \frac{\hbar q}{4mc}\sigma^{\mu\nu}F_{\mu\nu} = \cdots + \frac{\hbar q}{4mc}\left[-\frac{2}{c}\sigma^{0i}E^i - 4s^k B^k\right]$$
$$\approx -\frac{\hbar q}{mc}s^k B^k = -\mathbf{d}_M \cdot \mathbf{B}. \tag{1.20}$$

---

[3] More detailed derivations can be found in several textbooks, see e.g. Ref. [4].

In the first passage we have singled out the factor of interest and used Eq. (1.18). In the second passage we have taken again the non-relativistic approximation by keeping the terms at the lowest order in $c$. In the third passage we have used the magnetic moment defined as

$$\mathbf{d}_M = \frac{\hbar q}{mc}\mathbf{s}. \tag{1.21}$$

We can say that in first approximation the particle behaves as a non-relativistic particle having a magnetic moment in addition to its charge $q$. Let us go back to natural units. The magnetic moment of an elementary particle of charge $q$, mass $m$, spin $\mathbf{s}$ and orbital angular momentum $\boldsymbol{\ell}$ is generally written as

$$\mathbf{d}_M = \frac{q}{2m}(\boldsymbol{\ell} + g\,\mathbf{s}) \tag{1.22}$$

where $g$ is the gyromagnetic ratio—the so-called $g$-factor. The Dirac equation predicts $g = 2$. The electron, having charge $q = -e$, has a magnetic moment $\mathbf{d}_M = -g_e\,\mu_B\,\mathbf{s}$, where $\mu_B \equiv e/(2m_e)$ is the Bohr magneton. The $g_e = 2$ result is a remarkable prediction of the Dirac equation. However, at high precision, one observes a small difference with the experimental value—the so-called anomalous magnetic moment. It was first measured accurately by Kusch and Foley in the 1940s [5]. It is due to relativistic effects (quantum loops) and requires quantum field theory to be calculated. In 1948 Schwinger showed that it can be calculated from quantum electrodynamics (QED), which was just being developed [6]. Since then there has been enormous progress both experimentally and theoretically.

## 1.2 The Idea of the Neutrino

Unlike the alpha spectra,[4] which consisted of one or several mono-energetic lines of nuclear origin, beta spectra showed up with quite complicated compositions, exhibiting a continuous distribution of electrons. In processes where radioactive nuclei decay with the emission of $\beta$ rays, called accordingly $\beta$ decays, the electrons should have a fixed energy, corresponding to the difference of the energies between the final nuclei and the initial one (neglecting the small electron mass). Instead, experiments showed that the electron has an energy spectrum (see Fig. 1.1). This peculiar result was known since 1914, thanks to an experiment performed by the English physicist James Chadwick [7]. Such experiment was repeated eight years

---

[4] Alpha radiation is the emission of alpha particles, namely helium nuclei. Gamma radiation is the term used for the emission of energetic photons; it represents a loss of energy by the nucleus, a desexcitation, much like an emission of light or X-rays by energetic atoms. It brings the nucleus down to a more stable energetic state.

## 1.2 The Idea of the Neutrino

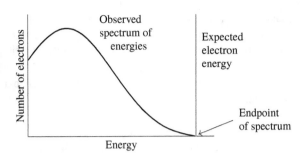

**Fig. 1.1** $\beta$ decay spectrum. The (qualitative) experimental plot of the energy distribution is compared with the one expected in a two body decay (not in scale)

later, in 1922, by Chadwick and his countryman Charles Ellis [8].[5] with a result supporting the view that "the continuous spectrum is emitted by the radioactive atoms themselves". For many years, the puzzling continuous $\beta$ spectrum was interpreted as the result of the loss of energy of electrons in the target, due, for instance, to undetected $\gamma$ rays. A step forward was made in 1927, when Ellis and William Wooster measured the total thermic energy released by a $\beta$-radioactive $^{210}$Bi source, which was put inside a calorimeter, and found that the average energy per one $\beta$ decay coincided with the average energy of the electrons [9]. It was demonstrated that the continuous $\beta$ spectra cannot be explained by an energy loss in the target. This experimental result revived speculation, advanced also by eminent scientists as Niels Bohr [10], about the non-conservation of energy. Another possible explanation was to assume that in $\beta$ decay the electron is produced together with a neutral particle so penetrating that its energy could not be transformed in thermic energy and revealed in experiments like the one by Ellis and Wooster. The total released energy is shared between the electron and the new particle. As a result, electrons produced in $\beta$ decay could have a continuous spectrum. There was a second reason in favour of an additional particle, namely angular momentum conservation. It was observed that if the mother atom carried an integer (fractional) spin then the same integer (fractional) spin was carried by the daughter, which cannot be explained by the emission of only one spin-1/2 electron.

In 1934, Enrico Fermi postulated that the $\beta$ decay was not a two-body, but a three body decay, adding in the final state a neutral particle, undetected and unknown, of spin 1/2. A suggestion for such a neutral particle had been made in 1930 by the Austrian physicist Wolfgang Ernst Pauli, in a famous letter to "Dear radioactive ladies and gentlemen" of a meeting in Tübingen.[6] Pauli assumed that these electrically neutral particles could exist *in the nucleus* and have mass of the same order of magnitude as the electron mass. According to this "desperate remedy", as Pauli defined it, the unseen particle, emitted along with the electron, could take away part of the energy—the continuous beta spectrum would then

---

[5] They were both confined in Ruhleben internment camp near Berlin during World War I.
[6] This letter, dated December 4th, 1930, was addressed to the participants of the meeting, and was written in German; an English translation can be found, for instance, in Ref. [11].

make sense and the principle of conservation of energy could be saved.[7] Pauli called this particle neutron, but in 1932 the name neutron was attributed to the electrically neutral particle, of mass comparable to the proton mass, discovered by Chadwick [12]. The origin of the name neutrino is attributed [13] to Enrico Fermi. It represents a word play on neutrone, the Italian name of the neutron, since in Italian the suffix -one and -ino stay, respectively, for big and small. Fermi's paper was first sent to the London journal Nature, that refused to publish it on the basis that it contained speculations too remote from reality to be of interest to the reader (sic!). Thus, Fermi published it on Nuovo Cimento [14] and soon afterwards another version on Zeitschrift fur Physik [15].

Since protons and neutrons have roughly the same mass, the $\beta$ decay could be interpreted as the conversion of a neutron into a proton,[8] accompanied by the emission of an electron to preserve the conservation of the electric charge—and by the undetected neutrino. Fermi postulated that the $\beta$ decay was due to a new kind of interaction, that he described through the Hamiltonian

$$H_F = G_F \, \bar{p}\gamma^\mu n \, \bar{e}\gamma_\mu \nu + h.c. \qquad (1.23)$$

where $h.c.$ stays for Hermitian conjugate, $G_F$ is the so-called Fermi constant, and $p, n\, e$ and $\nu$ are the proton, neutron, electron and neutrino Dirac spinor fields, respectively, all evaluated at the same point in space time. Let us observe that both currents in the interaction (1.23), $\bar{p}\gamma^\mu n$ and $\bar{e}\gamma_\mu \nu$ are charged, that is the electric charge of the initial state $(\nu, n)$ is not the same as that of the final state $(e, p)$. Since electromagnetic currents are always neutral, this Hamiltonian represents a different interaction. Because of the small value of $G_F \approx 10^{-5}/m_p^2$ ($m_p$ being the proton mass), it is called "weak" in contrast to the "strong" nuclear interaction. The horizon of the weak interactions was further extended by the discovery of the muon $\mu$ in 1937 [17, 18]. The observations of muon decay led Bruno Pontecorvo, an Italian physicist who had started his scientific work in 1932 in Rome as a student of Fermi, to propose in 1947 the universality of the Fermi interactions of electrons and muons [19].

Although one can safely say that the first observation of weak processes occurred when Henri Becquerel discovered the spontaneous radiation emitted by uranium in 1986, the understanding of weak interaction in its modern sense came only after Enrico Fermi postulated this physical mechanism for the $\beta$-decay. His theory,

---

[7] Right away after the 1964 discovery of $CP$ violation, Pauli's brilliant intuition was mimicked by suggesting that a so far unobserved very light neutral meson $U$ with $CP|U\rangle = -|U\rangle$ could restore $CP$ invariance: $K_L \to K_S + U \to (\pi\pi)_{K_S} + U$. Observing interference between $K_L \to \pi\pi$ and $K_S \to \pi\pi$ with $K_S$ being coherently regenerated from a $K_L$ beam ruled out this scenario. We could say, as ancient Roman people "Quod licet Jovi, non licet bovi" which means "What is permitted to Jove, is not permitted to a bull". It is not for everybody to get away with speculations like 'Jove = Jupiter = Pauli'.

[8] The existence of the proton had been proved by the New Zealander physicist Ernest Rutherford in 1919 [16].

and subsequent amendments,[9] has stood the ground very successfully for several decades, providing a useful guideline in the development of the gauge field theory which presently describes weak interactions in the Standard Model (SM) of particle physics.

In 1937, one year before his mysterious disappearance, the Italian physicist Ettore Majorana, a former student of Fermi, proposed a theory of particles with spin 1/2 which were truly neutral with respect to all charges, and thus identical to their antiparticles [20]. Today they are called Majorana particles. The same year Racah [21] suggested that the neutron could not be such a particle, because of its magnetic moment, while the neutrino could be—as of today, we still do not know.

## 1.3 The β Decay

The $\beta$ decay can occur in nuclei unstable with excess of neutrons, called radioactive isotopes or radionuclides. Nuclei decay in order to reach the lowest mass and highest binding energy. An example is the decay of the radioisotope of the cobalt into a stable isotope of the nickel, an electron $e^-$ and the anti-neutrino (the antiparticle of the neutrino) $\bar{\nu}$, namely the process

$$^{60}_{27}\text{Co} \rightarrow\, ^{60}_{28}\text{Ni} + e^- + \bar{\nu}. \tag{1.24}$$

The upper number is the atomic weight $A$, or mass number, that is the number of neutrons plus protons. The lower number is the atomic number $Z$, that is the number of protons. Schematically, we have

$$(A, Z) \rightarrow (A, Z+1) + e^- + \bar{\nu}. \tag{1.25}$$

Due to the $\beta$ decay, the atom moves one place forward in the periodic table. The final nickel nucleus is in an excited state and promptly decays to its ground state by emitting two gamma rays.

Given the nuclear $X$ state $(A, Z)$, one can define its atomic mass $m_X$ and its nuclear mass $m_X^N$. Their difference is the atomic binding energy $B_X$, that is the minimum energy required to disassemble an atom into free electrons and a nucleus, plus the rest masses of the electrons. It reads

$$m_X^N = m_X - Z m_e + B_X. \tag{1.26}$$

---

[9] The most notable, after the discovery of parity violation in weak interaction in the 1950s, was to replace the vector current of Fermi by an equal mixture of vector and axial vector currents, that is $\gamma^\mu \rightarrow \gamma^\mu(1-\gamma_5)$.

The $Q$-value of a nuclear reaction is the amount of energy released or absorbed during the nuclear reaction. For the decay to take place spontaneously, the energy must be released and the $Q$-value must be positive. The $Q$-value of the $\beta$ decay can be written as

$$Q = m_i^N - (m_f^N + m_e + m_{\bar{\nu}_e}) \tag{1.27}$$

$$= m_i - m_f + B_i - B_f - m_{\bar{\nu}_e} \tag{1.28}$$

where $i$ and $f$ refer to the initial and final nuclei. Neglecting the small atomic binding energies and the neutrino masses (less than a few eV), we can define the $Q$-value of $\beta$ decay as

$$Q \equiv m_i - m_f. \tag{1.29}$$

If the transition occurs into an excited energy level of the final nucleus, $m_f$ must be replaced with the appropriate energy. By setting (1.29), the mass and the binding energy of a nucleus are defined in terms of those for the corresponding neutral atom. In other terms, the mass difference between the parent and daughter nuclei in a $\beta$ decay includes the mass and the binding energy of an atomic electron as well.

The decay (1.24) is historically important since it was used in 1957 by the Chinese physicist Chien-Shiung Wu, known by many as Madame Wu, in the experiment where the violation of parity in weak decays was demonstrated for the first time [22].

Another example of $\beta$ decay is the decay of the heaviest isotope of hydrogen, tritium, transmuting into helium

$${}^{3}_{1}\text{H} \rightarrow {}^{3}_{2}\text{He} + e^- + \bar{\nu}. \tag{1.30}$$

The absolute value of the electron neutrino mass can be inferred from the investigation of $\beta$-spectra, since a finite neutrino mass would cause a truncation of the spectrum at its endpoint.[10] It turns out that the most sensitive direct searches for this value are based on the decay (1.30) [24, 25], as we will see in Sect. 10.1.1.

The conservation of angular momentum in the $\beta$ decay requires that $\mathbf{J}_p = \mathbf{J}_d + \mathbf{L} + \mathbf{S}$, with $\mathbf{J}_p$ and $\mathbf{J}_d$ being the spin of the parent and the daughter states, and $\mathbf{L}$ and $\mathbf{S}$ being the total orbital angular momentum and the total spin of the lepton pair, respectively. We distinguish two types of $\beta$ decay, the Fermi decay when $S = 0$ and the Gamow-Teller decay when $S = 1$.

The $\beta$ decay can also occur in nuclei unstable with excess of protons, and it is called positive $\beta$ decay ($\beta^+$), an example being magnesium-23 decaying into sodium-23

$${}^{23}_{12}\text{Mg} \rightarrow {}^{23}_{11}\text{Na} + e^+ + \nu \tag{1.31}$$

---

[10] This method of measurement of the neutrino mass was suggested by Francis Perrin [23] and Enrico Fermi [15].

## 1.3 The β Decay

that is

$$(A, Z) \to (A, Z-1) + e^+ + \nu. \tag{1.32}$$

The Fermi description provided by the Hamiltonian (1.23) is terms of particles, rather than nuclei, and β decays can be written as

$$\begin{aligned} n &\to p + e^- + \bar{\nu} & &\beta \text{ (or } \beta^-) \text{ decay} \\ p &\to n + e^+ + \nu & &\beta^+ \text{ decay}. \end{aligned} \tag{1.33}$$

The mass of the proton is slightly lower than the neutron mass ($m_p \sim 938$ MeV, $m_n \sim 939$ MeV), therefore the positive β decay can occur only in nuclei, where the proton can get the necessary additional energy to decay into the neutron. In the Standard Model, the free proton is stable due to the baryon number conservation, being the lightest baryon. Instead, the neutron can decay into a proton, by β decay, also outside the nucleus. A free neutron has a lifetime of about 14 min.

In the $\beta^+$ decay the atomic number decreases in the decay and the various charged lepton masses do not cancel out as in β decay, yielding a $Q$-value slightly different from the one in Eq. (1.29), namely $Q = m_i - m_f - 2m_e$.

In the Standard Model, the processes in Eq. (1.33) are viewed in terms of up and down quarks (see Fig. 1.2):

$$\begin{aligned} d &\to u + e^- + \bar{\nu} & &\beta \text{ (or } \beta^-) \text{ decay} \\ u &\to d + e^+ + \nu & &\beta^+ \text{ decay}. \end{aligned} \tag{1.34}$$

The β and $\beta^+$ decays are not the only processes allowed by the interaction (1.23). It is well-known that, given an allowed process, flipping a particle from one side of the equation to another and, in the process, switching it to its anti-particle, also

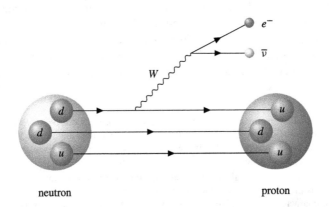

**Fig. 1.2** β decay in the Standard Model

results in an allowed process. By crossing, one obtains the charged-current mediated process from the neutron decay process[11]

$$\nu + n \to e^- + p \tag{1.35}$$

This means that the hadronic vertex can be treated with the same methods employed in the analysis of neutron decay.

Another possible process is the so-called electron capture (EC), that is:

$$e^- + p \to \nu + n \qquad \text{EC} \tag{1.36}$$

By crossing the neutrino in the positive $\beta$ decay, one obtains the interaction

$$\bar{\nu} + p \to e^+ + n \qquad \text{IBD} \tag{1.37}$$

known as inverse $\beta$ decay (IBD). Its cross section was calculated for the first time by Bethe and Peierls in 1934 [26], using the Fermi theory, and gave an extremely small value ($\sigma \approx 10^{-44}$ cm$^2$ at $E(\bar{\nu}) = 2$ MeV),[12] leading them to believe that the neutrino was an "undetectable particle".[13] In 1946 Bruno Pontecorvo challenged this opinion, suggesting a method which might make feasible direct observation of neutrinos through the inverse $\beta$ decay, by using nuclear reactors to produce a very intense neutrino source [27]. Pontecorvo's report was classified by the U. S. Atomic Energy Commission, because it was feared that the method could be used to measure the power output of reactors.

## 1.4 Neutrino Observations

In 1956, more than 20 years after the prediction of their existence, the inverse beta decay (Eq. (1.37)) was the one that allowed the first direct experimental evidence of neutrinos (actually, electron antineutrinos). It was given by two American physicists, Clyde L. Cowan and Frederick Reines [28, 29]. Their experiment exploited the fact that electron antineutrinos are produced in nuclear reactors when the radioactive products of nuclear fission undergo beta decay, and that their fluxes

---

[11] Crossing symmetry ensures that any of the interacting particles can be "crossed" over to the other side of the equation, provided it is turned into its antiparticle, and the resulting interaction will also be dynamically allowed. That is particles are indistinguishable from anti-particles with the opposite energy and momentum.

[12] A similar estimate can be obtained by dimensional considerations, setting $\sigma \sim \ell^3/(c\tau_n)$ where $\ell = \hbar/(\mu c)$, $c$ is the speed of light, $\hbar$ the reduced Planck constant, $\tau_n$ the mean life of the neutron and $\mu$ a mass characteristic of the IBD cross section, e.g. the mass of the electron.

[13] What they did not count on was the discovery of the nuclear fission, a few years later, which on a macroscopic scale produces a copious number of neutrinos (see Sects. 1.4 and 8.1).

## 1.4 Neutrino Observations

are far higher than any attainable neutrino flux from radioactive sources. It was the first reactor neutrino experiment. The search to detect the neutrino came to be known as "Project Poltergeist"—for the neutrino's ghostly and elusive nature [30]. For the detection of neutrino Reines (Cowan had passed away) was awarded one-half of the 1995 Nobel prize.

In 1957 Pontecorvo suggested the possibility of transition between a muonium, the bound state of a positive muon and an electron ($\mu^+ e^-$), and an antimuonium ($\mu^- e^+$) in analogy to the $K^0$ oscillations, even before the muonium atom had been formed for the first time [31].[14] In addition, he suggested the possibility of neutrino-antineutrino transition, that is the possibility that an oscillation could occur not only when the electrically neutral particle was a boson, like $K^0$ or the muonium, but when it was a fermion as well.

In 1958 the left handed helicity of neutrinos was confirmed by the Goldhaber, Grodzins and Sunyar experiment [32]. The same year Pontecorvo studied in detail for the first time the consequences of possible antineutrino-neutrino transitions, again guided by the analogy to the $K^0$ oscillations [33]. He considered oscillations of an active right-handed antineutrino into a right-handed neutrino which had no standard weak interaction, and that he named sterile [34, 35]. These were the only possible oscillations in the case of only one type of neutrino.

In 1962 the existence of two species of neutrinos, the electron neutrino ($\nu_e$), and the muon neutrino ($\nu_\mu$), was demonstrated at the Brookhaven National Laboratory (USA) particle accelerator AGS [36], where beams of neutrinos using a high energy accelerator were produced for the first time.[15] The same year Ziro Maki, Masami Nakagawa and Shoichi Sakata introduced the neutrino mixing, which defines the true neutrinos through orthogonal transformations [37].

Atmospheric neutrinos are generated when primary cosmic rays strike nuclei in the atmosphere and the hadron that results from these collision decays. They were first detected in 1965 by two independently working groups which placed detectors deep underground in the Kolar Gold Mines of South India [38] and a South African gold mine [39], respectively, to avoid cosmic-ray muons. They looked for upward-going muon events, generated through the interaction of atmospheric neutrino from the other side of the Earth with the surrounding rock.

After the discovery of the second neutrino, Pontecorvo applied his idea of neutrino oscillations to the case of two types of neutrinos, also in the context of solar neutrinos [34, 35]. In 1968 an experiment deep underground in the Homestake mine in South Dakota made the first observation of neutrinos from the Sun [40]. The radiochemical method of neutrino detection, proposed by Pontecorvo in 1946 [27],

---

[14] In 1950, Pontecorvo had moved to Russia and started to work at Dubna, where the largest accelerator in the world (at that time) was operating.

[15] In 1988, the leading scientists of the experiment, Leon M. Lederman, Melvin Schwartz and Jack Steinberger, shared the Nobel Prize in Physics "for the neutrino beam method and the demonstration of the doublet structure of the leptons through the discovery of the muon neutrino".

was used. Solar neutrinos were detected in this experiment via the observation of the reaction

$$\nu_e + {}^{37}\text{Cl} \rightarrow {}^{37}\text{Ar} + e^-. \tag{1.38}$$

The observed pattern of neutrino fluxes showed far fewer neutrinos than solar models had predicted (solar neutrino problem, see Sects. 7.1 and 7.2).

The first phenomenological scheme of neutrino mixing was proposed by Vladimir Gribov and Pontecorvo in 1969 [41]. The actual mixing matrix for neutrinos is known as the Pontecorvo-Maki-Nakagawa-Sakata (PMNS) o lepton mixing matrix in order to pay tribute to the pioneering contribution of these authors to the neutrino mixing and oscillations.

In the early 1970s, it was developed the Standard Model, which over time and through many experiments has become established as a well-tested physics theory and has successfully explained a wide variety of phenomena (see Chap. 4). In the Standard Model, the $\beta$ decay and the interactions introduced in Sect. 1.3 are mediated by the charged gauge particle of the weak force $W$, see Fig. 1.3.

In 1975 a new lepton, $\tau$ was discovered by a group led by physicist Martin Perl at the Stanford Linear Accelerator Center. In 1995, Perl won one half the Nobel Prize for this discovery.

In 1978, Lincoln Wolfenstein discovered that neutrinos propagating in matter are subject to a potential due to the coherent forward elastic scattering with the particles in the medium (electrons and nucleons), which modifies the mixing of neutrinos [42]. Other studies followed in the early 1980s, and in particular in 1985 Stanislav Mikheyev and Alexei Smirnov discovered that it is possible to have resonant flavour transitions when neutrinos propagate in a medium with varying density [43, 44], a phenomenon which is now known as the MSW mechanism.

In 1987 the first neutrinos from a supernova, the SN1987A in the Large Magellanic Cloud, about 160,000 light-years away from our solar system, were detected. Light from the explosion was discovered at the Las Campanas Observatory in northern Chile [45]. After hearing of the event, the experiment Kamiokande, lead by Masatoshi Koshiba,[16] and based on a large underground water Cherenkov detector in the Kamioka mine in Japan, immediately analysed the data in their possession, and found 16 neutrino events in a 30 s window [46]. Following the observation of Kamiokande, two experiments, IMB (Irvine-Michigan-Brookhaven), based on a water Cherenkov detector in the Morton salt mine in the U.S. [47], and BUST (Baksan Underground Scintillator Telescope) [48], an oil-based liquid scintillator in the Baksan Valley (Russia), detected 8 and 5 neutrino events, respectively, in the Kamiokande event time.

The Kamiokande experiment detected also clear solar neutrino signals, and, together with the Irvine-Michigan-Brookhaven detector (IMB) experiment,

---

[16] Koshiba shared the 2002 Nobel prize for pioneering contributions to astrophysics, in particular for the detection of cosmic neutrinos.

## 1.4 Neutrino Observations

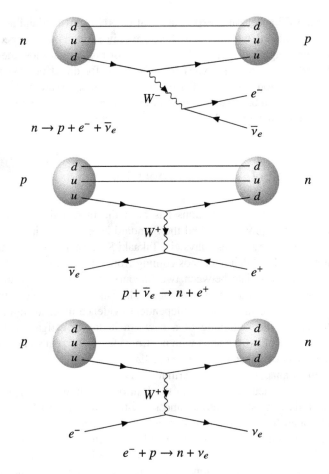

**Fig. 1.3** Tree level Standard Model diagrams for beta decay, inverse beta decay and electron capture

observed a deficit of atmospheric muon-neutrinos, the so called atmospheric neutrino anomaly [49, 50]. The discrepancy was attributed to neutrino oscillation, but the statistical evidence was insufficient to support this conclusion. It is interesting to remember that the original purpose of both these experiments was to search for nucleon decay predicted by Grand Unified Theories, and atmospheric neutrinos were one of the serious backgrounds of the nucleon decay search. However, nucleon decay was not discovered, thus neutrinos became the main target of research. The results of Kamiokande and IMB experiment mark the birth of neutrino astrophysics. As stated by Koshiba in his Nobel speech, by astrophysical observation one means that all the necessary information is available, e.g. the arrival direction, the arrival time and also the spectral information on the incoming neutrinos.

In 1989 the LEP accelerator experiments at CERN in Switzerland and the SLC ones at SLAC in USA determined that there are only 3 light neutrino species.[17]

In 2000, thirty eight years after the discovery of muon neutrinos, the scientists from the DONUT (the names comes from Detector for direct observation of tau neutrinos) collaboration working at Fermilab announced observation of tau particles produced by tau neutrinos, making the first direct observation of the third species of neutrino, the tau neutrino $\nu_\tau$ [52].

## 1.5 The Pursuit of Neutrino Properties

Discovery of the neutrino oscillations has been the first, and (still is) the only, evidence in favour of physics beyond the Standard Model in particle physics. The collaboration led by the Japanese physicist Takaaki Kajita at the Super-Kamiokande detector showed in 1998 that neutrinos coming from cosmic-ray interactions in the Earth's atmosphere oscillate between two flavours [53]. The change of identities for the neutrinos requires them to have mass. The Super-Kamiokande atmospheric neutrino result was the first model independent evidence of neutrino oscillations. This result marked the beginning of a new era in the investigation of neutrino oscillations—an era of experiments with neutrinos from different sources providing model independent evidence of neutrino oscillations.

In 2002 the Canadian physicist Arthur McDonald led the SNO collaboration to demonstrate, that neutrinos $\nu_e$ from the Sun do not disappear on their way to Earth. Instead, they transform into $\nu_\mu$ and $\nu_\tau$ by the time they reach the Earth [54]. This breakthrough resolved the longstanding solar neutrino problem, a significant disparity between the predicted and observed solar neutrino fluxes. Takaaki Kajita and Arthur McDonald were awarded the Nobel Prize in Physics in 2015 for their pioneering discoveries of oscillations in the atmospheric and solar neutrinos.

In 2003 the KamLAND reactor experiment provided the first compelling evidence of antineutrino oscillations [55]. In those years, neutrino oscillations were also detected in the accelerator experiments K2K [56] and MINOS [57], solidly corroborating the findings from the atmospheric Super-Kamiokande experiment.

Since the first neutrino detections at reactors and the first neutrino beams at the Brookhaven particle accelerator AGS (Alternating Gradient Synchrotron) [36], reactors and accelerators have been at the forefront, supplying intense beams of neutrinos. The increasingly precise measurements of mass-squared differences and mixing parameters, including the small parameters $\sin 2\theta_{13}$ from the accelerator experiment T2K [58], and reactor experiments Daya Bay [59], RENO [60], and Double Chooz [61], mark a significant advancement in neutrino physics, transitioning it into an era of high precision measurements. Present and future experiments

---

[17] For a review of these analyses see for example Ref. [51].

## 1.5 The Pursuit of Neutrino Properties

are focused on addressing the remaining open questions in neutrino physics, such as $CP$ violation in the lepton sector, determining the nature (Dirac or Majorana) of neutrinos, establishing the value of neutrino absolute masses, discerning the character (normal or inverted) of neutrino spectrum, and investigating the potential existence of as yet undiscovered sterile neutrinos.

Neutrinos originating from the Sun, supernovae, and nuclear reactors predominantly occupy the keV-MeV (1 keV $\equiv 10^3$ eV, 1 MeV $\equiv 10^6$ eV) energy range. Following on the energy scale are atmospheric neutrinos and neutrino from accelerators. Research on neutrinos is continually broadening its horizons to explore additional sources of neutrinos and encompass wider energy ranges. In the last two decades, neutrinos from radioactive decay processes in the interior of the Earth have been detected. These neutrinos, referred to as geo- or terrestrial neutrinos, typically fall within the keV-MeV energy range. The initial experimental indication of geo-neutrinos (approximately $2.5\sigma$ confidence level) was provided in 2005 by the KamLAND experiment [62]. Subsequently, five years later, the Borexino detector observed geo-neutrinos with a confidence level exceeding $3\sigma$ [63].

In the energy spectrum, the lowest energy range ($\mu$eV-meV) is occupied by cosmological (or relic) neutrinos. These neutrinos exist as a remnant from the early universe, analogous to the 2.7 Kelvin cosmic microwave background. At the opposite end of the energy spectrum, the highest energies are dominated by neutrinos originating from sources such as Active Galactic Nuclei (AGN).[18] or from interactions of ultra-energetic protons with the cosmic microwave background [64]. In 2013 the IceCube collaboration reported evidence for high-energy neutrinos, reaching orders of magnitude around the PeV scale (1 PeV equivalent to $10^{15}$ eV) [65, 66]. In 2017, IceCube registered a neutrino of likely astrophysical origin with a reconstructed energy of about 300 TeV. Within less than a minute from detection, IceCube's automatic alert system sent a notice to the astronomical community, triggering worldwide follow-up observations, in a multi-team effort which involved various observatories, instruments and cosmic messengers. The convergence of these observations and subsequent analyses by IceCube convincingly implicated the blazar[19] TXS 0506+056 as the most likely source [67]. The collaborative approach, where data from various cosmic messengers such as neutrinos, photons, cosmic rays, and gravitational waves are analyzed together to investigate astrophysical phenomena, is commonly referred to as multi-messenger.

The energy range spanning from keV to several GeV (1 GeV $\equiv 10^9$ eV) is primarily studied using underground detectors. As we move into the energy range from tens of GeV to about 100 PeV, characterized by much lower fluxes, the investigation shifts to employing massive Cherenkov light detectors deployed

---

[18] AGN are nuclei of active Galaxies with a central supermassive black hole, emitting non-thermal radiation at multiple wavelengths by accreting matter from the surrounding environment.

[19] A blazar is an AGN with its relativistic jet directed to Earth. They are considered potential neutrino emitters because their jets could provide suitable conditions to accelerate protons to very high-energies.

underwater and in ice, such as the IceCube detector. To explore even higher energies, beyond the PeV scale, requires the implementation of even more advanced techniques and experimental setups. Neutrino astronomy is an emerging field that involves analyzing neutrinos to gain insights into the universe's most energetic phenomena and explore regions of space that might be concealed from traditional electromagnetic observations.

# Chapter 2
# Global Symmetries

Neutrinos are studied in the formalism of quantum field theory, which assembles special relativity and quantum mechanics. Quantum field theory is the framework in which the theories of the electroweak and strong interactions, which together form the Standard Model of particle physics, are formulated.

In non-relativistic quantum mechanics, the solutions of the Schrödinger equation describe single-particle wave functions. The attempts to write equations of motion that are relativistic, with the same interpretation for their solutions, lead to inconsistencies. An important example is the Dirac equation for spin-1/2 systems. Contrary to the Schrödinger equation, the Dirac equation is invariant under Lorentz transformations, the coordinate transformations at the basis of the special theory of relativity. The solutions of the Dirac equation include states with negative energy. Negative-energy states make the system unstable, since each interaction, favouring a state of less energy, creates the condition for a cascade toward the state with infinite negative energy, emitting infinite radiative energy in the process. To overcome this problem, Dirac postulated that the negative-energy states were completely filled and that the Pauli exclusion principle prevented positive energy electrons from making transitions to them. On the contrary, it would have been possible for a negative energy electron to make a transition to a positive energy state, leaving a "hole" in the filled sea, or an antiparticle as it came to be known, with positive energy with respect to the sea. However, this assumption demands a many-particle picture in contradiction to the original single-particle interpretation. For example, if one negative-energy electron is excited to a positive-energy state, the final state is represented by both a positive-energy electron and a "hole", which corresponds physically to the electron-positron pair creation.

Once we walk out on the single-particle interpretation, quantum field theory comes to light as the framework of election to eliminate negative-energy states and similar difficulties. The variable $\psi(x)$ of the relativistic equations is interpreted as a quantum field $\psi(x, t)$. A field corresponds to a system with an infinite number of degrees of freedom, since at each continuously varying point $x$ an independent 'displacement' $\psi(x,t)$, which also varies with time, has to be determined. In quantum field theory, the description in terms of fields has to be combined with the application of quantum mechanics. Each field becomes an operator which, together with its conjugate field, obeys certain commutation relations. Radically different commutation relations distinguish particles with integral spins and those with half-integral spins. The commutation rules encode the quantum conditions of the fields. The concept of antiparticles arises naturally. Space and time are treated equally, as they are in relativity. In quantum mechanics, time is a parameter, while position is an observable. In quantum field theories, position is no longer an observable, but it is a parameter having the same status as time.

At the basis of quantum field theories and their development there is the concept of symmetry. The symmetry of a physical system is characterized by transformations of the system that leave invariant its dynamics. The dynamics of a quantum mechanical system is generally described in terms of the Hamiltonian, which determine the equation of motion. In both classical and quantum field theories, it is generally preferred the approach based on the principle of least action, with the action defined as $\int d^4x\, \mathscr{L}(x)$, where $\mathscr{L}(x)$ is the Lagrangian density. As a consequence, a symmetry transformation is a symmetry transformation of fields and coordinates that leaves the action invariant with respect to the transformation. This is enough in classical field theories, but not in quantum field theories, where in order to leave all physical observables unchanged it is necessary to check that the ground state, or "vacuum", and the quantization conditions remain invariant as well.

Symmetry transformations can be characterized by several adjectives: continuous or discrete, external or internal, global or local. Continuous symmetries are symmetries that leave the action invariant under some set of continuous transformations, namely described by a set of parameters that are allowed to vary continuously from one value to another, in some appropriate parameter space. An equivalent statement is that any continuous transformation can be connected to the identity transformation (no transformation at all) via the change of a continuous parameter. Discrete symmetries are related to non-continuous changes in a system, e.g. spatial reflections through the origin.

On the basis of the space where the transformations act, a symmetry can be external, internal, or both. External symmetry transformations change the space-time coordinates only, with notable examples given by the Lorentz and the Poincaré transformations, that we discuss in Sect. 2.4. Internal symmetry transformations transform states into each other, without reference to their dependence on space or time. They act on an "internal" space only. For example, proton and neutron represent two different states of the same particle, as already proposed by Heisenberg in 1932 [68]; they are distinguished by the third component of their 1/2 isospin and define a bi-dimensional space. The invariance of the strong interactions under transformations in the isospin space represents an internal symmetry.

The term symmetry without qualification generally refers to global symmetry. The parameters of global symmetry transformations are constants independent of spatio-temporal positions. For example, the Poincaré transformations in special relativity are global; their ten parameters are position independent. Another example of a global symmetry is the phase symmetry in non-relativistic mechanics, which corresponds to the freedom to shift the phase of the wave function, and refers to the exponential $e^{i\theta}$, with a constant phase $\theta$. A system is transformed as a unit under global transformations. When the previous conditions are not met, the symmetry is no more global, but local. The parameters of local symmetry transformations, equivalently named gauge symmetry, depend on spatio-temporal positions. Passing from a global to a local symmetry is known as "localise" or "gauging" the symmetry transformation. The Poincaré transformations become local in general relativity. The previous example of phase symmetry would now refer to $e^{i\theta(x)}$, where $\theta(x)$ depends on space-time.

In this chapter, after reminding a few notions of group theory in Sect. 2.1, we introduce the discrete transformations parity ($P$) and time reversal ($T$) in Sect. 2.2. The connection of $P$ and $T$ symmetries with magnetic and electric dipole moments is underlined in Sect. 2.3. Parity and time reversal belong to the set of time-spatial Lorentz transformations, discussed in Sect. 2.4, together with the Poincaré transformations (Sect. 2.5). They play a pivotal role in special relativity. Flavour quantum numbers are introduced in Sect. 2.6, while the discrete charge conjugation ($C$) transformation, as well as $G$-parity, are introduced in Sect. 2.7. In quantum field theory all non interacting theories are invariant under $P$, $T$ and $C$, but that is not always the case when interactions are switched on. Discrete symmetries are investigated more in depth from Sects. 2.8 to 2.13. We address helicity and chirality in Sect. 2.14 and we briefly introduce the $CPT$ theorem in Sect. 2.15.

## 2.1 Elements of Group Theory

A useful and common way to describe symmetry transformations makes use of group theory. A group $G$ is a set of elements $g_i$ with a single rule of composition, $g_i g_j = g_k$, which tells how each pair of elements $g_i$ and $g_j$ of the set is associated to get a third element $g_k$ in the same set.[1] Within this set, an identity element $I$ and an inverse for each group element, $g_i^{-1}$, need to be defined. The identity element is the unique group element such as for any $g_i \in G$ we have $g_i I = I g_i = g_i$. The composition of an element and its inverse gives the identity, that is $g_i g_i^{-1} = g_i^{-1} g_i = I$. The composition rule defines the group, independently on any particular way to write the group elements. It has to be associative, that is for any $g_i, g_j, g_k \in G$ we have $(g_i g_j) g_k = g_i (g_j g_k)$. An example is the set of three-dimensional rotations around a fixed axis, whose compositional rule is the succession of rotations. The identity element is the rotation by zero degrees. Another example is the general linear group $GL(V)$, where $V$ is a vector space on $\mathbb{R}$ or $\mathbb{C}$ of finite dimension, whose elements are all linear invertible maps $V \to V$, with the usual rule of composition.[2]

---

[1] The rule of composition is usually called multiplication, or product, and it is indicated with $\times$, $*$, $\cdot$ or nothing, just proximity, as we are doing here.

[2] The composition of two linear maps $T$ and $S \in GL(V)$ is defined as the linear map $TS \in GL(V)$ such that $(TS)v = T(Sv)$ for any vector $v \in V$.

A representation of finite dimension $n$ of the group $G$ is defined as an homomorphism D : $G \to GL(V)$, where the vector space $V$ has dimension $n$. If we pick a basis in the $n$-dimensional vector space $V$, we can identify $GL(V)$ with $GL_n$, the group of invertible $n \times n$ matrices.

While the structure of the abstract group is quite interesting, the representation of the group itself may be equally, if not more, important. The building blocks of representation theory are the irreducible representations, which act on vector spaces with no non-trivial invariant subspaces.

A relevant role in symmetry analyses is taken by the Lie groups and Lie algebras.[3] A finite-dimensional Lie algebra is a finite-dimensional vector space $\mathfrak{g}$ equipped with a binary operation, the Lie bracket $[,\ ]: \mathfrak{g} \times \mathfrak{g} \to \mathfrak{g}$. The Lie bracket needs to satisfy the following requirements:

1. bilinearity: $[aX + bY, Z] = a[X, Z] + b[Y, Z]$ and $[Z, aX + bY] = a[Z, X] + b[Z, Y]$ for arbitrary numbers $a, b$ and $\forall\, X, Y, Z \in \mathfrak{g}$
2. anticommutativity (skew symmetry): $[X, Y] = -[Y, X]$ $\quad \forall\, X, Y \in \mathfrak{g}$
3. the Jacobi Identity: $[X, [Y, Z]] + [Z, [X, Y]] + [Y, [Z, X]] = 0$ $\quad \forall\, X, Y, Z \in \mathfrak{g}$.

Note that the anticommutativity implies $[X, X] = 0$ for all $X \in \mathfrak{g}$. The bracket operation on a Lie algebra is not, in general associative; nevertheless, the Jacobi identity can be viewed as a substitute for associativity.

Let $\mathfrak{g}$ be a finite-dimensional Lie algebra, and let $X_1, X_2, \ldots, X_n$ be a basis for $\mathfrak{g}$ (as a vector space). Then the unique constants $c_{jkl}$ such that

$$[X_j, X_k] = \sum_{l=1}^{n} c_{jkl} X_l \tag{2.1}$$

are named the structure constants. This follows from the fact that the bracket is an element of $\mathfrak{g}$, and then it can be expressed in terms of a basis of elements of $\mathfrak{g}$. The structure constants satisfy the following two conditions:

1.

$$c_{jkl} + c_{kjl} = 0 \tag{2.2}$$

2.

$$\sum_{p} (c_{jkp} c_{plm} + c_{klp} c_{pjm} + c_{ljp} c_{pkm}) = 0 \tag{2.3}$$

for all $j, k, l, m$. The first of these conditions comes from the skew symmetry of the bracket, and the second comes from the Jacobi identity.

---

[3] They are named after the Norwegian mathematician of the nineteenth century Sophus Lie.

## 2.1 Elements of Group Theory

When restricting to matrix groups, one can give a simplified definition of a Lie group by identifying an associate Lie algebra as the collection of objects that give an element of the group when exponentiated.[4] In other terms, a Lie group $G$ whose elements are $n \times n$ matrices, with ordinary matrix multiplication as the composition rule, comes with a Lie algebra $\mathfrak{g}$ composed by all the $n \times n$ matrices X, such that $e^{rX} \in G$ for any real number $r$. For such Lie algebras the bracket is defined as $[X, Y] = XY - YX$, that is as the commutator of $X$ and $Y$. At a variance with the case of a group, the elements $XY$ and $YX$ need not to be part of the Lie algebra, but the difference $XY - YX$ always is. An example is the Lie algebra of the matrix group $SU(N)$, denoted as $\mathfrak{su}(N)$, which consists of all $N \times N$ complex, traceless, anti-hermitian matrices.

Given any matrix Lie group $G$, and the corresponding Lie algebra $\mathfrak{g}$, for any $g \in G$ one can define a linear map $\mathrm{Ad}_g : \mathfrak{g} \to \mathfrak{g}$ by

$$\mathrm{Ad}_g(X) = gXg^{-1} \qquad \forall X \in \mathfrak{g}. \tag{2.4}$$

It is called the adjoint map or adjoint representation of the group $G$. The group representation suggests the introduction of algebra representation in a similar way. In particular, for any $X \in \mathfrak{g}$, one can define a linear map $\mathrm{ad}_X : \mathfrak{g} \to \mathfrak{g}$ by

$$\mathrm{ad}_X(Y) = [X, Y] = (XY - YX) \qquad \forall Y \in \mathfrak{g}. \tag{2.5}$$

The map $X \to \mathrm{ad}_X$ is the adjoint map or adjoint representation of the algebra $\mathfrak{g}$.

These maps are connected by the following relation[5]

$$\exp(\mathrm{ad}_X) = \mathrm{Ad}_{\exp(X)} \tag{2.6}$$

By changing $e^{rX} \to e^{irX}$ we move to a slightly different, but widely used, notation.[6] More precisely, let us consider a matrix Lie group $G \subset GL_n$ with a related algebra $\mathfrak{g}$. Given a basis in the algebra, $T_1, T_2, \ldots, T_n$, we can write

$$g = e^{i \sum_{i=1}^{n} \theta_i T_i} \simeq 1 + ig \sum_{i=1}^{n} \theta_i T_i \qquad g \in G \tag{2.7}$$

where the quantities $\theta_i$ are $n$ real parameters. The number $n$, in its capacity as the number of parameters, therefore number of generators, is called the order (also

---

[4] Let us just mention that there is a formal and general definition of a Lie group as a differentiable manifold with specific properties and that is possible to give alternative definitions for a matrix Lie group.

[5] All notions recalled in this section are well known and can be found in standard group theory textbooks, for instance see Ref. [69].

[6] Let us observe that in this case the expressions for the Lie algebras of matrix Lie groups will differ by a factor of the imaginary unit $i$. For instance, the Lie algebra of the matrix group $SU(N)$ will consist of the set of $N \times N$ complex traceless Hermitean matrices.

the dimension) of the group. The matrices $T_i$ are usually called the infinitesimal generators (or generators, in short) of the group $G$, in some (in general reducible) $n$-dimensional matrix representation. The truncated expansion on the right holds only for $\theta_i$ infinitesimal. Lie groups represent continuous transformations, which can be thought of as repeated applications of infinitesimal steps, by the aid of generators. They have an infinite number of elements, but a finite number of generators. By definition, you can compose generators by adding and multiplying them, since they are in an algebra, while you can compose group elements only by multiplying them.

We expect the generators to satisfy the commutation relations

$$[T_i, T_j] = i \sum_{k=1}^{n} f_{ijk} T_k \tag{2.8}$$

where $f_{ijk}$ are the structure constants. The appearance of the imaginary unit i in (2.8), with respect to the definition (2.1), is correlated with the i in (2.7) which guarantees the hermiticity of the generators in unitary representations, a generally welcomed feature. Indeed, we have

$$g^\dagger = g^{-1} \Rightarrow T_i^\dagger = T_i. \tag{2.9}$$

By taking the hermitian conjugate of Eq. (2.8) and exploiting the hermiticity of the generators, it follows that $f_{ijk}$ are real numbers. When the generators are represented by matrices, we can take the trace of Eq. (2.8); by observing that the traces of $T_i T_j$ and $T_j T_i$ are equal, we conclude that the generators $T_k$ are traceless for any $k$. In view of the relation

$$\det(\exp M) = \exp(\mathrm{Tr} M) \tag{2.10}$$

where Tr$M$ stands for the trace of the matrix $M$, it follows that only unimodular (squared matrices with determinant 1) group elements can be expressed in the form (2.7).[7]

Because of the Jacobi identity

$$[[T_i, T_j], T_k] + [[T_j, T_k], T_i] + [[T_k, T_i], T_j] = 0 \tag{2.11}$$

which holds for any $T_i$, $T_j$, $T_k$ belonging to the algebra, one has generators $T_i^{adj}$ which are matrices whose elements are constructed out of the structure constants, according to

$$(T_i^{adj})_{jk} = -i f_{ijk} \qquad \forall i, j, k \in \{1, 2, \ldots, n\} \tag{2.12}$$

This is a representation of the Lie algebra by $n \times n$ matrices ($n$-dimensional) which corresponds to the adjoint representation.

---

[7] To make an explicit example, we observe that in this notation the Lie algebra of the matrix group $SU(N)$ becomes the set of $N \times N$ complex traceless Hermitean matrices.

## 2.2 Classical Physics: Parity and Time Reversal

In classical physics, a parity transformation ($P$) is defined as the transformation which changes the sign of the space coordinates $\mathbf{r} = (x, y, z)$

$$\mathbf{r} \xrightarrow{P} -\mathbf{r} \tag{2.13}$$

while the time $t$ remains unchanged. All the coordinates of *every* physical quantity are reversed. It is self-evident that parity transformations, also called space inversions, are space-time transformations. They belong to the set of Lorentz transformations, as we will see in Sect. 2.4. It is equally obvious that space inversion is a discrete symmetry, since it cannot be connected to the identity via the change of a continuous parameter, as it happens, for instance, with rotations. Let us underline that while there is some rotation that transforms one point of the space coordinates $(x, y, z)$ into $(-x, -y, -z)$ there is no one rotation, or one combination of rotations, that maps every $(x, y, z)$ into $(-x, -y, -z)$. For example, a rotation of $180^0$ along the $z$ axis transforms $(1,0,0)$ to $(-1,0,0)$, but does not transforms $(0,0,1)$ into $(0,0,-1)$. In parity transformations, *whatever* is at the point $(x, y, z)$ is moved to the point $(-x, -y, -z)$.

Parity symmetry consists in the invariance of the dynamics under parity transformations and it is an invariance of classical physics. It is sometimes called mirror symmetry, since a real mirror reflection reverses the space direction perpendicular to the plane of the mirror. To put it more precisely, in three dimensions, the parity transformation is equivalent to a mirror reflection with respect to an arbitrary plane, followed by a rotation with respect to an axis orthogonal to this plane (see Fig. 2.1). Since rotational symmetry, connected to the isotropy of space, is always assumed to hold in classical physics, symmetry under mirror reflections in a plane actually

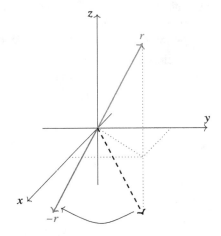

**Fig. 2.1** A reflection in the xy-plane followed by a rotation around the z-axis is equivalent to a parity transformation

corresponds to parity symmetry. As the mirror interchange left and right, parity symmetry is sometimes called left-right symmetry.

Since $\mathbf{r}$ change signs under parity, by the definition of momentum $\mathbf{p} = m\,d\mathbf{r}/dt$ it follows that $\mathbf{p} \xrightarrow{P} -\mathbf{p}$. There are also vectors that do not change sign under parity, for instance the angular momentum $\boldsymbol{\ell}$, given that $\boldsymbol{\ell} \equiv \mathbf{r} \times \mathbf{p} \xrightarrow{P} \boldsymbol{\ell}$. A vector is said polar if changes sign under parity, otherwise it is called axial. The term vector, with no adjectives, is a substitute for polar vector in common usage. The Newton's equation $\mathbf{F} = m\,d^2\mathbf{x}/dt^2$, that controls the motion of a point object of mass $m$, implies that the force $\mathbf{F}$ is a polar vector. The Lorenz relation $\mathbf{F} = q\mathbf{E} + \mathbf{v} \times \mathbf{B}$, which describes the motion of a particle of charge $q$ moving with velocity $\mathbf{v}$ in the electromagnetic field $\mathbf{E}$ and magnetic field $\mathbf{B}$, implies that $\mathbf{E}$ and $\mathbf{B}$ are polar and axial vectors, respectively.[8]

We also distinguish between scalars that do not change sign under parity, like $\mathbf{r}_1 \cdot \mathbf{r}_2$, and pseudoscalars that do, like $\mathbf{r} \cdot \boldsymbol{\ell}$. In electromagnetism, the well known relation $\mathbf{E} = -\nabla A^0 - \partial \mathbf{A}/\partial t$ implies that the classical $A^0$ is scalar and the vector potential $\mathbf{A}$ is a polar vector.

This characterisation carries over to the quantum domain as well. A typical quantum observable, the vector spin $\mathbf{s}$, can be considered an axial vector, since it can be interpreted as an angular momentum. In quantum mechanics, though, relevant variables are promoted to operators and classical transformations give way to operator transformations.

There is another discrete symmetry which is an invariant of classical physics, the time reversal symmetry, which is defined by the transformation that changes the time coordinate

$$t \xrightarrow{T} -t \qquad (2.14)$$

leaving the space coordinates unchanged. Thus, if $\mathbf{x}(t)$ is a solution of the classical dynamical equations, so is $\mathbf{x}(-t)$. Since $T$ leaves $\mathbf{r}$ unchanged, $\mathbf{p} \xrightarrow{T} -\mathbf{p}$ follows and time reversal can be interpreted as the reversal of motion. As before, we can use the expression of the Lorentz force to find the time reversal transformation rules for the electromagnetic field $\mathbf{E}$ and magnetic field $\mathbf{B}$, finding that $\mathbf{E} \xrightarrow{T} \mathbf{E}$ and $\mathbf{B} \xrightarrow{T} -\mathbf{B}$. Likewise, for the scalar and vector electromagnetic potential we can find that $A^0 \xrightarrow{T} A^0$ and $\mathbf{A} \xrightarrow{T} -\mathbf{A}$.

We summarise the transformation rules for commonly used physical quantities in Table 2.1. While passing from the quantities of the column on the left to the equivalent ones in the inner column (parity transformation) or in the right column (time reversal transformation), the total energy of a system does not change: both $P$

---

[8] Let us remark that, when not otherwise indicated, in the whole book we use natural units, where $c = 1$.

## 2.2 Classical Physics: Parity and Time Reversal

**Table 2.1** Transformation rules under parity $P$ and time reversal $T$ for commonly used physical quantities

|       | P      | T       |
|-------|--------|---------|
| **r** | −**r** | **r**   |
| $t$   | $t$    | $-t$    |
| **p** | −**p** | −**p**  |
| $\ell, s$ | $\ell, s$ | $-\ell, -s$ |
| **F** | −**F** | **F**   |
| **E** | −**E** | **E**   |
| **B** | **B**  | −**B**  |
| $A^0$ | $A^0$  | $A^0$   |
| **A** | −**A** | −**A**  |

and $T$ are symmetries of classical systems.[9] That also means that the transformed quantities remain solutions of the equations of motion.

Let us consider a trajectory $x(t)$ bringing the system from a state $A$ at a certain time $t_1$ to a state $B$ at $t_2$. If at the time $t_2$ one transforms $B$ into its temporarily reversed state, simultaneously temporarily reversing all the dynamical variables (for instance **p** → −**p**, but **E** → **E**)., the system flows back to the temporarily reversed state $A$. The laws of classical physics, which determine the properties of simple systems (e.g. Newton's law), appear to be invariant under the transformation (2.14). In order to have a genuine time reversed transformation, though, one has to interchange final and initial states as well. This seems in contradiction with what is normally observed, where time goes in only one directions. For example, a hole deflates a balloon full of gas, which expands in the external available volume, but we never observe the liberated gas to go back to the smaller volume through the same hole and spontaneously inflate the balloon again. In order to make it occur we should reverse simultaneously all the velocities of the molecules of the gas. This is difficult and therefore, for practical purposes, does not happen. In other words, if the $T$-invariant classical system consists of a large number of particles, although the time-reversed sequence is always possible, it is in general improbable. In our example, the probability of a molecule of gas to escape or to enter the balloon through a certain route A is the same; but starting the process in reverse, each molecule of gas has a very large choice of routes and it is very unlikely that will choose exactly A and go back to the primitive position. The macroscopic time irreversibility, that we observe linked to the increase of entropy of a thermodynamic system, does not originate in a microscopic time asymmetry of the basic microscopic laws, but from statistical considerations. Instead, in a system with a small number of particles, holding $T$ invariance, it is not possible in a statistical sense to differentiate between time-ordered and time-reversed sequences, both being similarly probable. Summarising, in classical physics time-reversal manifests itself very differently in macroscopic and a microscopic systems.

---

[9] This can be seen immediately in the simple case of a classical neutral particle of mass $m$, where $E = p^2/2m$.

## 2.3 Magnetic and Electric Dipole Moments

Classically, a charged particle with angular momentum $\ell$ possesses a magnetic dipole $\mathbf{d}_M$, and a term of the form $\mathbf{d}_M \cdot \mathbf{B}$ arises in the Lagrangian in presence of a uniform magnetic field of strength $\mathbf{B}$ (see e.g. Ref. [70]). For a point particle of charge $e$ and mass $m$, the magnetic dipole reads

$$\mathbf{d}_M = \frac{e}{2m}\boldsymbol{\ell}. \tag{2.15}$$

From the transformation rules of $\mathbf{B}$ and of the angular momentum under $P$ and $T$ (see Table 2.1), one finds the term $\mathbf{d}_M \cdot \mathbf{B}$ is invariant under both parity and time reversal, as expected for the classical Lagrangian.

These classical transformation rules extend to charged particles in quantum mechanics[10] which possess a non-zero orbital angular momentum and/or spin $\mathbf{s}$. In general, it is customary to write the magnetic moment of an elementary particle of charge $q$, mass $m$, spin $\mathbf{s}$ and orbital angular momentum $\boldsymbol{\ell}$ as

$$\mathbf{d}_M = \frac{q}{2m}(\boldsymbol{\ell} + g\,\mathbf{s}) \tag{2.16}$$

where $g$ is the gyromagnetic ratio. As seen in Sect. 1.1.1, the Dirac equation for a fermion in an external electromagnetic field predicts $g = 2$.

When there is a spatial separation of charges, an electric dipole moment $\mathbf{d}_E$ arises. In the classical Lagrangian the electric dipole interaction with the electric field is of the form $\mathbf{d}_E \cdot \mathbf{E}$. In order for this interaction to be invariant under parity and time reversal, $\mathbf{d}_E$ has to transform as the polar vector $\mathbf{E}$. Again, classical properties extends to quantum mechanics. The electric dipole interaction for an elementary particle is obtained from the magnetic one by exchanging $\mathbf{B}$ with $\mathbf{E}$, which in Eq. (1.18) corresponds to exchange $F^{\mu\nu}$ with its dual $\tilde{F}^{\mu\nu} = \epsilon^{\mu\nu\sigma\rho}F_{\sigma\rho}/2$, taken with the opposite sign. This gives a term $\propto i\bar{\psi}\gamma_5\sigma_{\mu\nu}\psi F^{\mu\nu}$ which in the non-relativistic limit can be written as $\propto d_E\,\boldsymbol{\sigma}\cdot\mathbf{E}$, namely it is proportional to the spin $\mathbf{s}$.

A permanent electric dipole moment (EDM) originates from a permanent charge separation inside the particle. In its center-of-mass frame, a subatomic particle's ground state has no available direction except its spin, which is an axial vector. Consequently, in this non-degenerate system, the EDM must align with its spin vector $\mathbf{s}$, as it represents the sole preferred direction in space. Since $\mathbf{s}$ and $\mathbf{E}$ transform in the opposite way under $P$ and $T$ (see Table 2.1), the interaction term is not invariant under these transformations. The existence of a permanent electric dipole moment for a non-degenerate system violates both $P$ and $T$, while magnetic dipole moments violate neither $P$ nor $T$. For the combined $CPT$ symmetry to hold, $T$ violation necessarily entails the breaking of combined $CP$ symmetry as well.

---

[10] A neutral particle as the neutron can still have a magnetic dipole moment thanks to the distributions of its charged fermion subcomponents (the quarks).

## 2.4 Lorentz Transformations

These conclusions may appear paradoxical, as there are well known examples of systems exhibiting an electric dipole moment in theories with both $P$ and $T$ symmetry. However, for the previous statement to remain true, the system must be 'non-degenerate': it can possess a (permanent) EDM without violating $P$ and $T$ invariance in the presence of degenerate states. An example of this are dumbbells, which have a positive charge at one end and a negative charge at the other, inherently constituting an electric dipole. This electric dipole doesn't imply $P$ (or $T$) violation since electrodynamics conserves both. The apparent paradox is resolved by the system's degeneracy. It's worth noting that the ground state of an atom is non-degenerate, thus it cannot possess an EDM when $P$ and $T$ symmetries are preserved. In this case, the observation of an EDM would be attributed to its constituents, signifying $P$ and $T$ violation in their dynamics.

The SM predicts the existence of EDMs, but their sizes (in the range of $10^{-31}$–$10^{-33}$ e·cm for nucleons) fall many orders of magnitude below current experimental limits.[11] An EDM observation at a much higher value would therefore be a clear and convincing sign of BSM physics. EDMs have been actively sought in neutral particles, particularly neutrons, for over 60 years, yet these experiments have thus far yielded only upper bounds.[12]

## 2.4 Lorentz Transformations

The space-time Lorentz and Poincaré transformations play a pivotal role in relativistic theories. Newtonian mechanics in Euclidean space is invariant under the Galileo transformations, time and space translation, rotations. In special relativity, the structure of the invariance is unified to Lorentz and Poincaré transformations.

Lorentz transformations are defined as the spatio-temporal transformations which preserve the proper distance $ds^2$ in the Minkowski space

$$x^\mu \to x'^\mu = \Lambda^\mu_\nu x^\nu$$
$$ds^2 \equiv g_{\mu\nu} dx^\mu dx^\nu = dt^2 - d\mathbf{x}^2 \to g_{\mu\nu} dx'^\mu dx'^\nu \qquad (2.17)$$

where $\Lambda^\mu_\nu$ is the Lorentz transformation matrix, $g_{\mu\nu}$ is the Minkowski metric tensor, and the standard convention of summing over repeated indexes holds.[13] The Minkowski coordinates $x^\mu$ can be written in terms of Cartesian coordinates plus time as $x^\mu = (t, \mathbf{x})$.

---

[11] In the SM the $CP$-violation is generated by the Kobayashi–Maskawa mechanism of weak interactions, as we will see in the next chapters.

[12] Further details can be found e.g. on Ref. [71].

[13] Unless explicitly mentioned, in the whole book we use natural units where $c = \hbar = 1$.

The Lorentz transformations can equivalently be described as the transformations leaving the metric tensor invariant from one inertial frame to another, that is

$$g_{\mu\nu} = g_{\alpha\beta} \Lambda^\alpha_\mu \Lambda^\beta_\nu \quad \Leftrightarrow \quad g = \Lambda^T g \Lambda \qquad (2.18)$$

The set of matrices which satisfy relations (2.18) define the Lorentz transformations and form a Lie group under matrix multiplication, known as the Lorentz group.

In Euclidean space, the metric tensor is merely a unit matrix, and Eq. (2.18) becomes $\Lambda^T \Lambda = 1$, which is the condition for orthogonal matrices. In that case, the group of matrices would be the orthogonal group $O(4)$, which includes the group of rotations in 4 dimensional space and reversal of orientation of spatial axis (parity transformations).

In Minkowski space, the metric tensor is defined as

$$g = \begin{pmatrix} 1 & 0 & 0 & 0 \\ 0 & -1 & 0 & 0 \\ 0 & 0 & -1 & 0 \\ 0 & 0 & 0 & -1 \end{pmatrix} \qquad (2.19)$$

that is as a diagonal matrix having $(1, -1, -1, -1)$ as elements on the diagonal.[14] Therefore the Lorentz group is often referred to as $O(1, 3)$, an orthogonal group (preserves a metric) corresponding to a metric with $(1, 3)$ signature. This metric is called Lorentzian, as opposite to the Euclidean one where all of the eigenvalues are positive.

By considering the determinant of Eq. (2.18) and the element of the metric tensor with $\mu = \nu = 0$, it is easy to demonstrate that

$$\det g = \det(\Lambda^T g \Lambda) = \det \Lambda^T \det g \det \Lambda \Rightarrow \det \Lambda = \pm 1$$

$$g^{00} = g^{\mu\nu} \Lambda^0_\mu \Lambda^0_\nu = \Lambda^0_0 \Lambda^0_0 - \sum_{k=1}^{3} \Lambda^0_k \Lambda^0_k = 1 \Rightarrow |\Lambda^0_0| \geq 1 \qquad (2.20)$$

If $\Lambda^0_0 \geq 1$, the Lorentz transformation preserves the direction of time and is called orthochronous. If $\Lambda^0_0 \leq -1$, the Lorentz transformation reverses the direction of time.

The Lorentz transformations which are orthochronous and have $\det \Lambda = 1$ describe rotations and boosts (and their combinations). The boosts corresponds to changing coordinates by moving to a frame that travels at a constant velocity. They constitute a subset of continuous transformations that can be formed by consecutive infinitesimal transformations starting from identity (they are 'continuously

---

[14] An equivalent definition of the metric, used by several references, reverts the signs.

## 2.4 Lorentz Transformations

connected' to the identity matrix) and has a group structure. This group is often referred to as the proper Lorentz group.

The transformations with $\det\Lambda = -1$ contain discrete spatio-temporal inversions. Indeed, we can single out the following matrix $\Lambda^\mu_\nu$ to define parity

$$P = \begin{pmatrix} 1 & 0 & 0 & 0 \\ 0 & -1 & 0 & 0 \\ 0 & 0 & -1 & 0 \\ 0 & 0 & 0 & -1 \end{pmatrix} \qquad (2.21)$$

and, analogously, the following matrix $\Lambda^\mu_\nu$ for time reversal

$$T = \begin{pmatrix} -1 & 0 & 0 & 0 \\ 0 & 1 & 0 & 0 \\ 0 & 0 & 1 & 0 \\ 0 & 0 & 0 & 1 \end{pmatrix} \qquad (2.22)$$

We can easily compare with Euclidean transformations in Eqs. (2.13) and (2.14).

On the whole, the Lorentz group consists of four sets, of those only one maintains a group structure, the set of proper Lorentz transformations:

$$L^\uparrow_+ = \{\Lambda \in O(1,3) | \det\Lambda = +1, \Lambda^0_0 \geq 1\} \qquad (2.23)$$

The transformations in the other sets can be written as the product of proper Lorentz transformations with parity ($L^\uparrow_-$: $\det\Lambda = -1$, $\Lambda^0_0 \geq 1$), with time reversal ($L^\downarrow_-$: $\det\Lambda = -1$, $\Lambda^0_0 \leq -1$), or with both $T$ and $P$ ($L^\downarrow_+$: $\det\Lambda = +1$, $\Lambda^0_0 \leq -1$). Since the identity transformation has $\det\Lambda = +1$ and $\Lambda^0_0 \geq 1$, it belongs only to the set of proper Lorentz transformations, and not to the other sets, preventing $L^\uparrow_-$, $L^\downarrow_-$ or $L^\downarrow_+$ to have a group structure. The proper (or restricted) Lorentz group $L^\uparrow_+$ in (2.23) is a Lie group and it is also called $SO(1, 3)$, where $S$ signifies special because of the requirement of a unit determinant.[15]

Formally, a scalar field $\phi(x)$ and a vector field are defined according to their transformation properties under the Lorentz group, that are

$$\phi'(x') = \phi(x) \qquad V'^\mu(x') = \Lambda^\mu_\nu V^\nu(x). \qquad (2.24)$$

---

[15] Let us underline that sometimes the adjective proper is used to indicate the Lie group formed of all Lorentz transformations with $\det\Lambda = 1$, therefore including the set $L^\downarrow_+$. In that case, this group takes the symbol $SO(1, 3)$, while the Lie group with $\det\Lambda = 1$ and $\Lambda^0_0 \geq 1$ is called the proper orthochronous Lorentz group, with symbol $SO^+(1, 3)$. Indeed, orthochronous refers to the additional property $\Lambda^0_0 \geq 1$ and means in physical terms that transformations with this property do not change the direction of time.

where $V^\mu(x)$ are the components of the vector.[16] The scalar does not change sign under a parity transformation, while the vector does for the spatial part. The same Lorentz transformation relations apply for pseudoscalar and axial vector fields, respectively, but they have an opposite behaviour under parity transformation. Vectors with upper indices are referred to as contravariant vectors, with lower indices as covariant ones. Quantities where contravariant and covariant index are all summed over (i.e. $V_\mu V^\mu$) are Lorentz invariant quantities.

### 2.4.1 The Proper Lorentz Group

The proper Lorentz group $SO(1, 3)$ consists of the Lorentz transformations that preserve the orientation of space and the direction of time, excluding parity, time-reversal transformations and their combination. It is a Lie group, with an infinite number of elements, but a finite number of generators.

The Minkowski space, the space-time of special relativity, is a four dimensional real vector space. A generic, real $4 \times 4$ matrix has 16 parameters. The relations (2.18) give 10 independent conditions,[17] which restrict the number of independent parameters of the Lorentz group to 6. Therefore, given Eq. (2.7), if we find 6 linearly independent generators, we have found a complete basis for the Lie algebra of the proper Lorentz group.[18]

In the 4-dimensional representation, also indicated as vectorial or 4-vector representation, the Lie algebra generators are represented by real $4 \times 4$ matrices acting on the space-time coordinates. We identify these matrices as the antisymmetric real matrices $V^{\mu\nu}$, with index running over the four dimensions, which satisfy the commutation relations

$$[V^{\mu\nu}, V^{\rho\sigma}] = i(g^{\nu\rho}V^{\mu\sigma} - g^{\mu\rho}V^{\nu\sigma} - g^{\nu\sigma}V^{\mu\rho} + g^{\mu\sigma}V^{\nu\rho}) \tag{2.25}$$

Because of anti-symmetry, there are only $n(n-1)/2 = 6$ such matrices. One can use the generators $J_i$ and $K_i$ instead of $V_{\mu\nu}$, by setting

$$V^{ij} = \epsilon^{ijk} J_k \qquad V^{0i} = K_i \tag{2.26}$$

---

[16] For simplicity, the vector (or more precisely 4-vector) of coordinates $V^\mu(x)$ is usually referred to as the vector $V^\mu(x)$.

[17] In the relations (2.18) the matrix $\Lambda$ appears together with its transpose. Therefore, the only independent conditions refer to the 4 conditions for the elements on the diagonal, plus $(16-4)/2 = 6$ conditions for the non diagonal elements.

[18] Let us underline that the Lorentz group $O(1, 3)$ and the proper Lorentz group $SO(1, 3)$ have the same Lie algebra, and therefore the same number of generators.

## 2.4 Lorentz Transformations

where $i, j, k$ run over the three space indices. It is equivalent to set

$$V^{\mu\nu} = \begin{pmatrix} 0 & K_1 & K_2 & K_3 \\ -K_1 & 0 & J_3 & -J_2 \\ -K_2 & -J_3 & 0 & J_1 \\ -K_3 & J_2 & -J_1 & 0 \end{pmatrix} \quad (2.27)$$

In this way, the six generators are identified with the three generators $J_i$ of the rotations and the three generators $K_j$ of the boosts. One can easily see that

$$[J_i, J_j] = i\epsilon_{ijk} J_k \quad [J_i, K_j] = i\epsilon_{ijk} K_k \quad [K_i, K_j] = -i\epsilon_{ijk} J_k \quad (2.28)$$

The first equation shows clearly that the spatial rotations $J_i$ form the subalgebra SO(3). The last equation shows that the Lorentz boosts $K_j$ do not form a subalgebra, i.e. they do not close in on themselves.

The irreducible representations of the Lorentz group can be constructed from irreducible representations of $SU(2)$. To see how it works, we can take the linear combinations of the generators $J_i$ and $K_i$

$$J_i^+ \equiv \frac{1}{2}(J_i + iK_i) \quad J_i^- \equiv \frac{1}{2}(J_i - iK_i) \quad (2.29)$$

and verify that they satisfy

$$[J_i^+, J_j^+] = i\epsilon_{ijk} J_k^+ \quad [J_i^-, J_j^-] = i\epsilon_{ijk} J_k^- \quad [J_i^+, J_j^-] = 0 \quad (2.30)$$

These commutation relations indicate that the Lie algebra for the Lorentz group has two commuting subalgebras, the three dimensional rotation algebras generated respectively by $J_i^+$ and $J_i^-$. This appears to be the same algebra as that of two independent $SU(2)$ algebras. This is why representations of algebra $SO(1,3)$ can be classified as representations of $SU(2) \oplus SU(2)$ algebra.[19]

This considerably simplifies the study of the irreducible representations. It can be used to discover further representations of the Lorentz group, besides the 4-vector representation described before. The new representations allow to describe physical systems that cannot be described by the vector representation. Each irreducible representation of $SU(2)$ is characterized by a Casimir operator $J^2$ (operator that commutes with all the generators[20]) with eigenvalue $j(j+1)$, where $j$ is a positive half-integer number. The representation acts on a vector space with $2j+1$ basis

---

[19] Notice that, since we are in the case of the proper Lorentz group, we have chosen to simplify the discussion not distinguishing among representations of the Lorentz group, of the Lorentz algebra or of the double cover of the Lorentz group. A more detailed discussion can be found for example in Ref. [72].

[20] It is a well-known result of group theory that Casimir elements give us linear operators with constant values for each representation, providing a natural way of labelling representations.

elements. In $SU(2) \oplus SU(2)$, we have two Casimir operators, $J_i^+ J_i^+$, and $J_i^- J_i^-$, characterized by eigenvalues $l(l+1)$ and $m(m+1)$, respectively, where $l$ and $m$ are positive half integer. It follows that each representations of the Lorentz group is characterized by the couple $(l,m)$ and has $(2l+1)(2m+1)$ degrees of freedom. Let us mention a few significant cases:

- The lowest order representation is $(0,0)$, the scalar representation. It acts on spin zero states with a well-defined parity, that can appear as scalars or pseudoscalars. In this representation the field transforms as $\phi(x) \to \phi(x'^\mu = \Lambda^\mu_\nu x^\nu)$.
- The spinor representations are the two non-equivalent $(1/2,0)$ and $(0,1/2)$ representations. They act on states that have two components (two degrees of freedom), called Weyl spinors, of spin 1/2.
- The linear combination $(1/2, 0) \oplus (0, 1/2)$ is a four-dimensional representation referred as the Dirac representation and acts on Dirac spinors, of spin 1/2. It describes spin 1/2 twice (a particle and its antiparticle) and is used in the Dirac equation. Lorentz transformations on Dirac spinors are represented as

$$S(\Lambda) = \exp(i\theta_{\mu\nu} S^{\mu\nu}) \qquad S^{\mu\nu} \equiv \frac{i}{4}[\gamma^\mu, \gamma^\nu] = \frac{1}{2}\sigma^{\mu\nu} \qquad (2.31)$$

where $\theta_{\mu\nu}$ are 6 real angles and $S^{\mu\nu}$ do not act on the space-time coordinates. One can check that $S^{\mu\nu}$ are generators, since they satisfy the Lorentz algebra (2.25), albeit in a different representation (Dirac) than the 4-vector representation.
- We can generate any other representation by multiplying spinor representations together. This procedure is equivalent to forming higher spin states by taking the product of many spin 1/2 states in the rotation group. For example, we recover the vectorial Lorentz representation as the $(1/2, 0) \otimes (0, 1/2) = (1/2, 1/2)$ representation (containing four degrees of freedom) which acts on 4-vectors, like photon fields, which have spin 1. In the vectorial representation the field index is a contravariant index and it transforms as in Eq. (2.24), that is $V^\mu(x) \to V'^\mu(\Lambda^\mu_\nu x^\nu) = \Lambda^\mu_\nu V^\nu(\Lambda^\mu_\nu x^\nu)$.
- The $(1, 0) \otimes (0, 1)$ representation is a six-dimensional representation used in the Maxwell equation. It describes $F_{\mu\nu}$, a second rank tensor in four dimensions containing spin 1 twice (the electric and magnetic fields).

## 2.5 Poincaré Transformations

The most general set of all transformations compatible with special relativity form the Poincaré group. If the Lagrangian is invariant under any transformation of the Poincaré group, the equations of motion take the same form in all frames of reference.

## 2.5 Poincaré Transformations

The Poincaré transformations can be indicated as

$$x'_\mu = \Lambda^\mu_\nu x^\nu + a^\mu \tag{2.32}$$

where $a^\mu$ is a constant 4-vector. Hence they are Lorentz transformations plus translations. The Poincaré group is defined as the group of Minkowski spacetime isometries (ISO(1,3)). It is a Lie group that preserves the proper distance $ds^2$ in Minkowski space. The Lorentz group is a subgroup of the Poincaré since Lorentz transformations are isometries that leave the origin fixed. In other terms, the 6 generators of the proper Lorentz group, plus the four generators of the space-time translations, with the corresponding appropriate commutation relations, define the Lie algebra of a 10 parameter group called group of Poincaré (sometimes also called the inhomogeneous Lorentz group).[21]

By indicating the space-time translations generators with $P_\mu$ and the Lorentz group generators with $V_{\mu\nu}$, the Poincaré algebra reads

$$[V_{\mu\nu}, V^{\rho\sigma}] = i(g_{\nu\rho}V_{\mu\sigma} - g_{\mu\rho}V_{\nu\sigma} - g_{\nu\sigma}V_{\mu\rho} + g_{\mu\sigma}V_{\nu\rho})$$
$$[V_{\mu\nu}, P_\rho] = -i\eta_{\rho\mu}P_\nu + i\eta_{\rho\nu}P_\mu$$
$$[P_\mu, P_\nu] = 0. \tag{2.33}$$

The Poincaré group has two independent Casimir operators. One is

$$P^\mu P_\mu = m^2 \tag{2.34}$$

where $m^2$ is the mass squared in the rest frame ($m^2 > 0$). Under Lorentz transformations, this operator transforms as a genuine scalar, and it is also invariant under translations. To construct the other Casimir operator, let us define the so-called Pauli-Lubanski axial vector

$$W^\mu = \frac{1}{2}\epsilon^{\mu\nu\rho\sigma}P_\nu V_{\rho\sigma} \tag{2.35}$$

where the tensor $\epsilon^{\mu\nu\rho\sigma}$ is antisymmetric on any pair of adjacent indexes and $\epsilon^{0123} = +1$. It is easily shown that the square of this vector is a Casimir operator

$$W^\mu W_\mu. \tag{2.36}$$

All physical states in quantum field theory can be labeled according to the eigenvalue of these two Casimir operators (since the Casimir commutes with all generators of the algebra). The physical significance of the Casimir (2.36) can be

---

[21] In reverse, what we call simply the Lorentz group is sometimes referred to as homogenous Lorentz group.

clarified by going to the rest frame of a massive particle, where the Pauli-Lubanski tensor coincides with the spin generator.[22] In other words, physical states in quantum field theory are described by irreducible representations of finite dimension of the Poincaré group, having certain values of spin and mass. The situation is quite different from the non-relativistic quantum mechanics, where the spin is introduced as an external degree of freedom, since it appears as the result of the space-time symmetry of the theory. It is remarkable that, by relativistic considerations only, mass and spin become indexes for the classification of free elementary particles and, at the same time, certain constraints on their values can be set. The masses of a physical particle can be zero or positive. If a particle has finite mass, then the value of the spin can be 0 or a positive number, integer or semi-integer. If the mass is zero, the spin is either 1 or 2.[23]

## 2.6 Flavour Quantum Numbers

Particle decays spontaneously into lighter particles, unless prevented from doing so by some conservation law. Conservation laws may be purely kinematic, like the conservation of energy and momentum and conservation of angular momentum. They apply to all interactions, since they derive from kinematics. The fact that a particle cannot spontaneously decay into particles heavier than itself is actually a consequence of conservation of energy. A notable consequence is that the photon, which is massless, is also stable, given that there is nothing lighter to decay into.

Other conservation laws have a dynamical character, since they proceeds from the gauge symmetry of a fundamental interaction. In that case we can use the Nöther's theorem [74], which leads to current and charge conservation. The Nöther's theorem states that for every continuous symmetry of a Lagrangian there is a corresponding conserved current $J_\mu^i$, where $\partial^\mu J_\mu = 0$ on any classical trajectory. The corresponding charge is $Q = \int J^0(x) d^3 x$. This theorem is very important since it tells us how to relate symmetries to conservation laws, as well as how to construct such currents and charges.[24] The conservation of the electric charge, that proceeds from QED gauge invariance, ensures that the electron is stable, since it is the lightest charged particle. Analogously, the conservation of the weak and strong charges proceeds from the gauge invariance of the weak and strong interactions, respectively.

---

[22] Let us underline that we are referring to massive particles, since for massless particles the physical interpretation of the Pauli-Lubanski vector is not so straightforward.

[23] For an extensive discussion see e.g. Ref. [73] Sect. 2.5.

[24] Several applications of the Nöther's theorem are well known, for instance, that translation invariance of the Lagrangian leads to linear momentum conservation and rotational invariance leads to angular momentum conservation.

## 2.6 Flavour Quantum Numbers

Some conservation laws, such as the conservation of the baryon and lepton number, do not seem to belong to the previous categories. They were stated as empirical rules coming from experiments. From the 1950s, it became possible to identify a set of physical quantities, which are conserved by at least one fundamental interaction and characterise particles together with the electric charge. They are referred to as (internal) quantum numbers (or charges as well). Discrete quantum numbers were assigned in decay processes to each particle and conservation of the total quantum numbers was assumed in the process, in analogy with the law of conservation of electric charge.

Quantum numbers are conserved or not depending on the details of the interactions that one considers. The isospin quantum number is conserved in strong interactions, but not in electroweak interactions. Weak interactions change a quark of one type (flavour) into another. Thus strong and electromagnetic (but not weak) interactions conserve the quantum numbers related to flavour, that is the strangeness $S_{qn}$ and the (flavour) quantum numbers related to heavier particles (charm $C_{qn}$, beauty (or bottomness) $B_{qn}$, and truth (or topness) $T_{qn}$).[25] There is some arbitrariness in the assignment of flavour quantum numbers, obviously. For historical reasons, the assigned strangeness to the $s$-quark and the assigned bottomness to the $b$-quark are $S_{qn} = B_{qn} = -1$, and to their antiparticles are $+1$. The opposite holds for the charm and top quark. The convention is that the flavour quantum number sign for the quark is the same as the sign of its electric charge. With this convention, any flavour carried by a charged meson has the same sign as its charge, e.g., the strangeness $S_{qn}$ of the $K^+$ is $+1$, the bottomness $B_{qn}$ of the $B^+$ is $+1$, and the charm $C_{qn}$ and strangeness of the $D_s^-$ are each $-1$. Antiquarks have the opposite flavour signs.

The total number of baryons, the heaviest hadrons, was observed to be always conserved, no matter what the interactions was, and the same for the total number of leptons. Analogous observations were made for anti-particles. The baryon number (or $B$-number) was defined as the number of baryons, such as protons, neutrons, and hyperons, minus the number of their antiparticles, in the initial or final state of the process. It is equivalent, but more convenient in practice, to say that each baryon and anti-baryon has assigned a baryon number $B = +1$ and $B = -1$, respectively, while a number $B = 0$ is assigned to mesons, leptons and bosons. In the SM, this is expressed defining $B = (n_q - n_{\bar{q}})/3$, where $n_{q/\bar{q}}$ is the total number of quarks/antiquarks. The baryon number of a state corresponds to the sum of the $B$-numbers of the particles which belong to the state; the B-number is additive by definition. Because of the total baryon number conservation in any physical process, to each newly produced particle could be assigned a $B$-number in a way that allowed matching the sum of baryon numbers on either side of the process. Baryon number conservation ensures that the proton is stable, since it is the lightest baryon.[26] Mesons are not constrained by conservation of baryon numbers

---

[25] In this section we use the unusual notation including the subscript $qn$ in order to avoid confusion.

[26] Our world is populated mainly by stable particles as protons, electrons, photons; the neutron is not stable, but it becomes stable in the protective environment of many atomic nuclei.

and a given collision or decay can produce as many mesons as it likes, consistent with conservation of energy.

The flavour quantum numbers of a quark are not completely independent, but are related to third component of its isospin, $I_3$, and its charge $Q$ by the Gell-Mann-Nishijima formula, which reads $Q = I_3 + (B + S_{qn} + C_{qn} + B_{qn} + T_{qn})/2$, in its generalisation to all quarks. Originally, this equation was based on a purely empirical observation, but in the context of the SM it follows simply from the isospin and charge assignments for quarks.

Since $Q$, $B$ and the flavour quantum numbers are all conserved in the electromagnetic interactions, it follows that $I_3$ is also conserved by the electromagnetic forces. This does not hold for the other two components, and hence for the isospin itself, which is not conserved in the electromagnetic interactions, see for instance the decay of a neutral pion in two photons, where the isospin goes from $I = 1$ to $I = 0$. This formula also tells us that $I_3$ is not conserved in weak interactions, that do not conserve the flavour quantum numbers.

In the lepton sector, similar selection rules were observed. To each (anti-)lepton $\ell$ was assigned a (anti-)lepton number $L_\ell = 1 \, (-1)$, and the same number was assigned to each (anti-)neutrino $\nu_\ell$. Equivalently, one defines $L_\ell = n_\ell - n_{\bar{\ell}}$, namely the difference between the number of leptons $n_\ell$ and antileptons $n_{\bar{\ell}}$. In the SM with massless neutrinos, the quantum number for each lepton family $L_{e/\mu/\tau}$ is separately conserved.

After the SM was established, the conservation of baryon and lepton numbers turned out to be connected to the global continuous symmetries of the SM Lagrangian, as we will see in Sect. 4.9.

## 2.7 Charge Conjugation

In 1932 the positron, the anti-particle of the electron, was detected in the cosmic rays [75]. In those years, the development of relativistic quantum physics, as described by the Dirac equation, led for the first time to the concept of anti-matter, a physical counterpart of ordinary matter. Conventionally, the negatively charged electrons and positively charged protons are labelled as particles, and their oppositely charged counterparts as anti-particles; compatible conventions are assigned to other particles. If a particle has no attributes beyond linear and angular momentum (which include energy and spin), then it is its own antiparticle, but if the particle has other attributes, such as charges or quantum numbers, then there is an antiparticle characterized by opposite attributes. The photon and the neutral pion are their own particles, while in the case of the proton the electric charge obviously distinguishes it from the anti-proton.

## 2.8 Parity Eigenstates

With the name charge conjugation $C$ one indicates a transformation that changes the signs of all charges and quantum numbers of a particle or set of particles, but only those. It does not affect other physical attributes as energy and mass or space-related quantities as position, momentum, spin, parity. Therefore charge conjugation really interchanges every distinct quantum number between particles and antiparticles. In this way, the charge conjugation transforms a particle into its anti-particle, and it is sometimes named particle-antiparticle conjugation. Thus, it arises naturally in the multi-particle framework of quantum field theory. It does not have analogue in classical physics or in non-relativistic quantum mechanics. Symmetry under charge conjugation for a physical state depends upon it being neutral, where in this context neutral does not only mean with zero electromagnetic charge, but with zero internal charges (quantum numbers). For instance, a photon can be considered truly neutral, but not a neutron, which has zero electric charge as well, but, additionally, carries a non-zero baryon quantum number.

### 2.7.1 G-Parity

Because so few particles are eigenstates of charge conjugation, its direct application is rather limited. Its power can be somewhat extended, if we confine our attention to the strong interactions, by combining it with an appropriate isospin transformation. Rotation by $\pi$ about the $y$-axis in isospin space will carry the isospin component orthogonal to the $y$-plane, that is $I_z$, into $-I_z$, converting, for instance, a proton into a neutron, or a $\pi^+$ into a $\pi^-$. If we then apply the charge conjugation operator, some particles, as the $\pi^-$, will go back to the original ones, the $\pi^+$ in this example. This combined operator is named $G$-parity. All mesons that carry no strangeness or charm, beauty, or truth quantum numbers, as the charged pions, are eigenstates of $G$-parity even though they are not eigenstates of charge conjugation alone. Instead, a strange particle as $K^+$, for example, is not a $G$-parity eigenstate, since by isospin transformation it goes to $K^0$, and by charge conjugation to $\bar{K}^0$. Likewise, a charged baryon as the proton cannot be a $G$-parity eigenstate, since charge conjugation changes sign to its baryon number, and so on.

## 2.8 Parity Eigenstates

In quantum mechanics, the parity transformation can be represented by a unitary operator $\hat{P}$, which performs a spatial inversion on the space coordinates of the wave function

$$\hat{P}\psi(\mathbf{r}) = \psi(-\mathbf{r}). \tag{2.37}$$

The unitarity of the operator ensures that probabilities are scalar, that is invariant under parity. Since $\hat{P}\psi(-\mathbf{r}) = \psi(\mathbf{r})$, applying twice $\hat{P}$ restores the initial state, that is

$$\hat{P}^2 = 1. \tag{2.38}$$

Let us indicate with $|\pi\rangle$ the eigenstates of the parity operators. By definition, they satisfy the equation

$$\hat{P}|\pi\rangle = P|\pi\rangle. \tag{2.39}$$

According to Eq. (2.38), $P^2 = 1$ holds. Hence the parity of the eigensystem is either odd or even, with eigenvalues $P = -1$ and $P = +1$, respectively.[27]

Under parity, an operator $\hat{O}$ transforms as

$$\hat{O} \xrightarrow{P} \hat{P}\hat{O}\hat{P}^\dagger. \tag{2.40}$$

In a system where the energy levels are non-degenerate, the invariance of the Hamiltonian operator under parity implies that the parity operator commutes with the Hamiltonian, and viceversa:

$$\hat{P}H\hat{P}^\dagger = H \quad \Longleftrightarrow \quad [H, \hat{P}] = 0. \tag{2.41}$$

It follows immediately that the energy eigenstates have assigned definite parity. But even when an energy level is degenerate, we can always pick a basis of energy eigenstates which have definite parity.

In nature, the only interactions which do not display parity symmetry are the weak interactions. If weak effects are neglected, the Hamiltonian of all the other interactions commutes with the parity operator, giving to the parity eigenvalue the status of a conserved quantum number. The existence of a conservation law for parity does not have an analogue in classical physics. Classically, parity symmetry implies that the dynamical laws of physics, as for example, Newton's law, remain invariant under parity transformation, but this does not results in any constant of motion.

In agreement with classical mechanics the momentum changes its sign under parity inversion. Parity and momentum operators do not commute and, therefore, cannot have simultaneous eigenstates unless the eigenvalue of momentum vanishes. Thus a single particle can be in an eigenstate of $\hat{P}$ only if it is at rest. The value of the eigenvalue of $\hat{P}$ for a particle in its rest frame is called intrinsic parity.

Analogously to quantum mechanics, in quantum field theory the parity transformation is represented by a unitary operator $\hat{P}$ in Hilbert space. The concept of intrinsic parity plays a significant role only in quantum field theory, where particles

---

[27] Let us observe that we reach the same conclusion if we include an arbitrary phase $\alpha$ in the definition of the parity transformation, that is if we define $\hat{P}\psi(r) = e^{i\alpha}\psi(-r)$. In this case, the condition $\hat{P}^2 = 1$ implies $\alpha \in \{0, \pm\pi, \pm 2\pi, \ldots\}$.

## 2.8 Parity Eigenstates

can be created and destroyed: in single-particle quantum mechanics the intrinsic parities are identical in the initial and final states for any physical process and therefore do not add to our knowledge of the process.

It is conventional to assign positive intrinsic parity (+) to spin 1/2 fermions. Thus lepton and quarks have positive parity. The sign here is simply due to convention, since $\hat{P}$ and $-\hat{P}$ are defined by the same equations and therefore any of them could be chosen as the parity operator. However, while the intrinsic parity of a particle is a matter of convention, the relative parity of a particle and its anti-particle is not. According to quantum field theory, the parity of fermions and anti-fermions are opposite, while bosons and their anti-bosons have the same parity.[28] Consequently anti-leptons and anti-quarks have negative intrinsic parity. The different behaviour of fermions and bosons also explains why mesons and anti-mesons can be grouped in a single spin-parity multiplet, while baryons and anti-baryons cannot.

A quantum scalar (pseudoscalar) field $\phi(x)$ has intrinsic parity $+1$ ($-1$), and transforms according to Eq. (2.24)

$$\phi(t, \mathbf{x}) \xrightarrow{P} \phi(t, -\mathbf{x}) \qquad \text{scalar}$$

$$\phi(t, \mathbf{x}) \xrightarrow{P} -\phi(t, -\mathbf{x}) \qquad \text{pseudoscalar.} \qquad (2.42)$$

The same Eq. (2.24) defines the transformation rule of a quantum vector or axial field $V^\nu(x)$. Since parity here refers to the spatial components $V^i(x)$, where $i = 1, 2, 3$, that reverse sign in a polar vector and do not in an axial vector, the convention is slightly different than in the case of scalars. By taking as the $\Lambda^\mu_\nu$ matrix the one given in Eq. (2.21) we have

$$V^\nu(t, \mathbf{x}) \xrightarrow{P} \Lambda^\mu_\nu V^\nu(t, -\mathbf{x}) = V_\nu(t, -\mathbf{x}) \qquad \text{vector}$$

$$V^\nu(t, \mathbf{x}) \xrightarrow{P} -\Lambda^\mu_\nu V^\nu(t, -\mathbf{x}) = -V_\nu(t, -\mathbf{x}) \qquad \text{axial vector} \qquad (2.43)$$

The notation is that if we have $V^i(x) \xrightarrow{P} -V^i(x)$ then $V^\nu$ has intrinsic parity $-1$ and it is a vector, if $V^i(x) \xrightarrow{P} V^i(x)$ then $V^\nu$ has intrinsic parity 1 and we call it a pseudovector, or axial vector.

Strictly speaking, we could not assign an intrinsic parity to the photon, not existing a reference frame where the photon can be considered at rest. However, the photon represent a quantic equivalent of the classical vector potential **A**, that has negative parity, as seen in Sect. 2.2. Thus a negative intrinsic parity is conventionally assigned to the photon. In field theoretical terms, this is equivalent to postulate the following transformation properties for the quantized electromagnetic field

$$\hat{P} A_\mu(t, \mathbf{x}) \hat{P}^\dagger = A^\mu(t, -\mathbf{x}) \qquad (2.44)$$

---

[28] For a demonstration see for example Sect. 5.5. of Ref. [73].

In an analogous way, one could assign an intrinsic parity to the gluon, but that is useless since the gluon is not observable as a free particle. Equally useless is to assign an intrinsic parity to the weak gauge bosons $W^\pm$ and $Z$, since weak interactions do not respect parity symmetry. In quantum field theory, it is also postulated that the vacuum state is invariant under parity, i.e. that it is an eigenstate of the parity operator with eigenvalue +1.

Parity is a multiplicative quantum number, i.e. a quantum number of a composite system is given by the product of the quantum numbers of all the constituents. The parity transformation acts simultaneously in all states that is applied on. A multiplicative composition law is typical of quantum numbers associated with discrete symmetry transformations.[29] The parity of a composite system in its ground state is the product of the parities of its constituents. Thus, in the ground state, baryons have positive parity, $P = (+1)^3 = 1$, while mesons have negative parity, $P = (-1)(+1) = -1$. The parity of a composite excited state depends on the product of the intrinsic parities of the particles of the state, times any collective parity of the composite particle state. Also a single particle, when in an orbital angular momentum state, has an orbital parity associated with it, since parity commutes with angular momentum (both position and momentum change sign). If the overall wave function of a particle (or system of particles) contains spherical harmonics, the parity associated with orbital angular momentum is due to the properties of transformation of the spherical harmonics under parity. It is well known that for a wave function containing a radial part $R(r)$ and spherical harmonics $Y_m^L(\theta, \phi)$ we have, under a parity transformation

$$\psi(r,\theta,\phi) = R(r)Y_m^L(\theta,\phi) \xrightarrow{P} R(r)Y_m^L(\pi-\theta, \phi+\pi) = \\ = (-1)^L R(r)Y_m^L(\theta,\phi) = (-1)^L \psi(r,\theta,\phi). \quad (2.45)$$

Thus, for a system of one or more particles, with intrinsic parities $p_i$, in a state of defined angular momentum $L$, the parity of the system is[30]

$$P = p_1 p_2 \ldots (-1)^L. \quad (2.46)$$

---

[29] Continuous symmetries are normally associated to quantum numbers that are additive, i.e. the quantum number associated with a given symmetry of a composite system is obtained by adding together (algebraically or vectorially) the corresponding quantum numbers of all the components of the system. The difference has its root in the fact that additive quantum numbers are the eigenvalues of the generators of continuous transformations, which act sequentially, while multiplicative quantum numbers are the eigenvalues of discrete transformations themselves.

[30] This relation can be modified in obvious ways, for instance for a system of three particles in a state in which $L$ is the orbital angular momentum of two of the particles in their centre of mass, and $l$ that of the third particle with respect to the centre of mass of the first two, one has $P = p_1 p_2 p_2 (-1)^L (-1)^l$.

## 2.8 Parity Eigenstates

We can determine the parities of states by observation of physical process mediated by strong or electromagnetic interactions, in which parity is a conserved quantum number. For instance, let us consider the strong process

$$\pi^- + {}^2\text{H} \to n + n \tag{2.47}$$

where a low energy pion is absorbed by the spin 1 deuteron ${}^2\text{H}$ producing a pair of neutrons.[31] Parity conservation gives

$$p_{\pi^-}\, p_{{}^2\text{H}}\,(-1)^{L_i} = p_n\, p_n\,(-1)^{L_f} = (-1)^{L_f} \tag{2.48}$$

where $p_x$ is the intrinsic parity of the $x$ particle, and $L_{i,f}$ are the orbital angular momenta of the initial and final state, respectively. The pion, being a meson in its ground state, has negative intrinsic parity $p_{\pi^-} = -1$, as seen before. Since we are at low energies in the initial state, only the $S$ partial wave will be excited, and $L_i = 0$. The deuteron is a bound state of two nucleons, a neutron and a proton, which are in an S wave, namely their relative angular momentum is zero. Hence the intrinsic parity of the deuteron is positive, $p_{{}^2\text{H}} = 1$.

We conclude that the parity of the initial state is negative, and, by conservation of parity by strong interactions, this is also the parity of the final state. The confirmation comes by observing that the final state consists of two identical neutron and thus must be antisymmetric under the interchange of the neutrons. We also observe that the meson is a pseudoscalar, i.e. $J^P = 0^-$, while for a deuteron $J^P = 1^+$.[32] Then the total angular momentum $J_{tot}$, which is conserved by strong interactions, has the value $J_{tot} = 1$ in the initial state. The pair of nucleons are into a final state of spin $s$, which can be an antisymmetric state with $s = 0$ or one of the symmetric states with $s = 1$. In the former case, the value of $L_f$ should be even, but this is not allowed by the conservation of $J_{tot}$, which requires $L = 1$. Then this possibility is ruled out. In the case of $s = 1$, the minimal orbital antisymmetric state with $L = 1$ is allowed, since summing $L = 1$ and $s = 1$ can result into a state with $J_{tot} = 1$. Hence we have $L_f = 1$ and negative parity of the final state.

Another interesting example is given by the $\eta$ meson, which is heavy enough to decay in two or three pions. Let us show that the former decay cannot occur via the strong interaction, which respects the parity. From Eq. (2.46), the parity of the final state with two pions is $P = (-1)^L$ and of three pions is $P = -(-1)^L$ for any orbital angular momentum $L$, since pions have, as all mesons included the $\eta$ one, intrinsic negative parity. The meson $\eta$ is in the ground state, with zero total angular momentum. The $\eta$ meson and the pions have no spin. Conservation of the total angular momentum implies $L = 0$, so $P = 1$ for the two pions state. Adding

---

[31] The deuteron is the nucleus of a deuterium atom, a stable isotopes of hydrogen.

[32] The notation $J^P$ is commonly used in the classification of hadrons in their rest frame according to their total spin $J$ and their intrinsic parity $P$. $J$ is expressed in unit of $\hbar$, that in natural units correspond to 1.

a third pion but keeping $L = 0$, gives $P = -1$. Hence, the two pion decay violates parity conservation, whereas the three pion decay not.

### 2.8.1 Observations of Parity Violation

Electromagnetic and strong interaction (and gravity) conserve parity, and until the 1950s most physicists simply carried the assumption that the same would be true in weak interactions. It took an experimental anomaly to shake that assumption, the so-called $\theta/\tau$ puzzle. In 1953, cosmic ray analyses and the first experiments at accelerators had shown the existence of two apparently identical particles of spin zero, named $\theta^+$ and $\tau^+$, which differed only in their decay modes, $\theta^+ \to \pi^+\pi^0$ and $\tau^+ \to \pi^+\pi^+\pi^-$. These two final states have defined parity, that is they are parity eigenstates, and the eigenvalues are opposite, positive for the two pion state and negative for the three pion state.[33] The apparent solution, which respected the rule of conservation of parity, as observed until then, was to interpret $\theta^+$ and $\tau^+$ as two different particles, with different intrinsic parities, and having by accident the same mass, lifetime and spin. But in 1956, the Chinese theoretical physicists Tsung-Dao (T.D.) Lee and Chen-Ning Yang performed a thorough examination of the available experimental evidence and put forward a new interpretation, arguing that, unlike strong and electromagnetic interactions, no experimental proof existed of parity conservation in weak interactions [76].[34] To fill this gap, they proposed experimental tests of parity conservation in $\beta$ decays induced by a polarised nucleus.

Putting aside the doubts raised by such a novel hypothesis, experimentalists seized on Lee and Yang's idea. In 1957, as mentioned in Chap. 1, the Chinese physicist Chien-Shiung Wu[35] and her collaborators at the National Bureau of Standards in Washington, DC (USA) demonstrated for the first time parity violation in weak interactions [22]. They used a sample of nuclei of cobalt-60, which has spin-parity $J^P = 5^+$, placed inside a solenoid to be polarised. The magnetic field of the solenoid oriented the nuclear magnetic moments, which are parallel to the spins. The sample was cooled to a temperature of about $10^{-3}$ K. At such temperatures,

---

[33] To simplify we limit the discussion to the case in which the total orbital angular momentum for the three pion state is $L = 0$, assuming higher values disfavoured by the quantum mechanical angular barrier.

[34] T.D. Lee and Chen-Ning Yang came to know each other in 1946 when both, on Chinese fellowships, became graduate students of Fermi in Chicago. Fermi had moved from Italy to USA in 1938, after having received, the same year, the Nobel prize. The year in which Fermi moved to USA was the year in which racial laws were first promulgated in Italy by the Fascist government. T.D. Lee and Chen-Ning Yang were awarded the Nobel Award in physics in 1957, when they were in their 30s and working in US. T.D. Lee, at barely 31 years of age, became the second youngest scientist ever to receive this distinction (the youngest was Sir Lawrence Bragg, who shared the Physics Prize with his father in 1915, at the age of 25).

[35] C.S. Wu, often respectfully called Madame Wu, had received her Ph.D. in 1940 at Berkeley, and joined Columbia University in 1944. She never received the Nobel prize, but was honoured with several awards, including the inaugural Wolf prize in Physics in 1978.

## 2.8 Parity Eigenstates

**Fig. 2.2** Pictorial representation of the directions of the spin of the cobalt-60 nucleus and of the emitted electron momentum, before and after a parity transformation

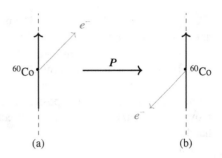

the interaction of the magnetic moments of the nuclei with the magnetic field overcomes the tendency to thermal disorder, and the nuclear spins align parallel to the field direction. Cobalt-60 is an unstable isotope of cobalt which undergoes $\beta$ decay. Indeed, the $^{60}$Co nucleus decays to an excited state of nickel-60 of spin-parity $J'^P = 4^+$ by the process

$$^{60}_{27}\text{Co} \to ^{60}_{28}\text{Ni} + e^- + \bar{\nu}_e \tag{2.49}$$

The daughter nucleus keeps the polarisation of the parent nucleus. We assume that the $^{60}$Co nucleus is fully polarised, i.e. $|J_z| = J = 5$. The $\beta$ decay is followed by decay of the excited nickel, maintaining polarisation, i.e. $|J'_z| = 4$, in a lower energy state with the emission of two photons. Since the $\beta$ decay products carry no orbital angular momentum, by the conservation of the third component of the angular momentum the 1/2 electron spin and 1/2 neutrino spin have to be aligned. The experiment recorded the number of emitted electrons with respect to the direction of the spin of the decaying nuclei. A very simplified representation of the electron decays is given in Fig. 2.2. On the left the polarisation of the $^{60}$Co nucleus is indicated with a black arrow; when the nucleus undergoes $\beta$ decay, it emits an electron which flies out of the nucleus in some direction, indicated in the figure with the red arrow. To see why this is relevant to parity, we look on the right at the mirror image of the same decay. Parity transformation reverses all particle momenta while leaving spin and angular momenta unchanged. If the angle between the electron trajectory and the direction of the nucleus spin is $\alpha$ then in the mirror the angle between the two is $\pi + \alpha$. Symmetry under parity would imply that the probability of one interaction is the same as the probability of the same interaction in the mirror image, and so we should see the same numbers of electrons in both directions. The experiment shows that this is not true and that one of these angles is favoured. It turns out that electrons leave the nucleus preferentially away from the direction of the magnetic moment, with an angle of $\pi/2 < \alpha < \pi$.

In other terms, the result of the experiment indicated that the electrons are emitted in directions (almost) opposite to the nuclear spin much more frequently than (almost) along it.[36] A forward-backward asymmetry in the decay could be observed.

---

[36] In the actual experiment, the field direction was moved from upward to downward. In the former configuration the detectors counted the electrons emitted at about 0°, in the latter at about 180°.

Were parity conserved, there would be no preferred direction; the forward-backward asymmetry would be zero. Since the spin of the electron has to be parallel to the $^{60}$Co spin, the electron preferential direction of emission is also antiparallel to its spin. Thus the electron preferentially has negative helicity—it is preferentially left-handed. As the nuclei decay at rest, the decay is practically a two body decay, and the antineutrino moves in the opposite direction to the electron. Thus the antineutrino moves preferentially in the same direction of its spin; it is preferentially right-handed.[37]

Thus, parity violation in weak interactions was proved for the first time. It was not small—almost all of the electrons were emitted in only one direction (almost "maximal" violation). To put it in the words of the Nobel prize Richard Feynman [77]: "The magnet has grown hairs! The south pole of a magnet is of such a kind that the electrons in a $\beta$-disintegration tend to go away from it; that distinguishes, in a physical way, the north pole from the south pole".

It became obvious to solve the $\theta/\tau$ puzzle assuming that $\theta^+$ and $\tau^+$ were the same particle ($K^+$), with parity-conserving and parity violating decay modes. This epochal discovery was almost instantly confirmed by other groups [78, 79].

## 2.9 Charge Conjugation Eigenstates

In quantum theory, the charge conjugation transformation introduced in Sect. 2.7 is represented by the unitary operator of charge conjugation $\hat{C}$. Its effect on a state with definite momentum $p$, spin projection $s$ and internal quantum numbers (or charges) denoted collectively with $Q$ is

$$\hat{C}|p, s, Q> = e^{i\alpha}|p, s, -Q> \qquad (2.50)$$

where $\alpha$ is an arbitrary real phase. Charge conjugation $\hat{C}$ acts on internal properties (the quantum numbers) and not on spatial or temporal coordinates, at a variance with parity and time reversal. It is obvious that while only particles with zero *electric* charge can be $\hat{C}$ eigenstates, not all of them are. They have to be *completely* neutral, which is not equivalent to electrically neutral, but means having all (internal) quantum numbers equal to zero. In other terms, only particles that are their own antiparticles, called self-conjugate, can be eigenstates of $\hat{C}$. That is not the case, for example, of the neutral $B^0$ meson, which does not coincide with its anti-particle $\bar{B}^0$ since they have opposite quantum number beauty. Instead, the photon and neutral bosons as $\pi^0$, $\eta$, $\eta'$ $\rho^0$, $J/\psi$, $\Upsilon$ are self-conjugate and eigenstates of $\hat{C}$.

---

[37] In the SM this is an obvious consequence of the V-A structure of the week charged currents mediating the $\beta$ decay.

## 2.9 Charge Conjugation Eigenstates

The charge conjugation eigenvalue is called (intrinsic) charge conjugation $C$ or $C$-parity. Since applying twice the charge conjugation operator we return to the initial state, the only possible eigenvalues are $C = \pm 1$, as in the case of parity. In classical electromagnetism, changing the sign of all electrical charges results in sign reversal of charge density ($\rho = \int d^3x\, j^0(t, \mathbf{x}) \to -\rho$), current ($\mathbf{j} \to -\mathbf{j}$), electric field ($\mathbf{E} \to -\mathbf{E}$) and magnetic field ($\mathbf{B} \to -\mathbf{B}$). We see that charge conjugation leads to $j^\mu \to -j^\mu$ and $A^\mu \to -A^\mu$. These transformations do not affect the charge symmetric description, as it is manifest by checking the invariance of the Maxwell equations. It is also evident the charge conjugation invariance of the interaction Lagrangian density $j_\mu A^\mu$. In the quantum theory, we can generalise the classical result by setting[38]

$$\hat{C}^\dagger A^\mu(t, \mathbf{x})\hat{C} = -A^\mu(t, \mathbf{x}). \tag{2.51}$$

Thus we say that the $C$-parity of the photon as well as the current is negative, and we can write

$$\hat{C}|\gamma> = -|\gamma>. \tag{2.52}$$

Charge conjugation parity, as parity, is a multiplicative quantity. For a system composed of several free particles which are charge conjugation eigenstates, the $C$-parity is the product of $C$-parities for each particle. Therefore, for a system of $n$ photons

$$\hat{C}|n\gamma> = (-1)^n|n\gamma>. \tag{2.53}$$

A system composed of a charged particle and its antiparticle, for instance $e^+e^-$, $q\bar{q}$ or $n\bar{n}$, can be truly neutral and therefore a $\hat{C}$ eigenstate, even if the two particles are not $\hat{C}$ eigenstates. Even if the two particles are distinct by their internal charge, they transform into each other by charge conjugation, and the system as a whole stays the same. In the centre of mass rest frame, such transformation is equivalent to exchange the position (by a mirror reflection), the spin and the charge labels. In a state of defined angular momentum $L$, a mirror reflection of the two particles brings an additional factor to the angular part of the spatial wave function, as seen in Eq. (2.45), that gives a factor $(-1)^L$. As for the spin exchange, it introduces a sign depending on the symmetry or anti-symmetry of the total spin $S$ state. We distinguish two possibilities

- if the system is made by a fermion and an antifermion, the exchange of spin labels introduces a sign $(-1)^{S+1}$.[39] Moreover, since we are switching fermions

---

[38] This transformation law can be extended to vector fields which represent neutral states either than a photon.

[39] One can easily see that by adding two spin 1/2; the final state S=0 is in an antisymmetric configuration $(1/\sqrt{2}(|\uparrow\downarrow\rangle - |\downarrow\uparrow\rangle))$, while the final states S=1 are in a symmetric one (for example $|\uparrow\uparrow\rangle$).

which anti-commute, there is an additional − sign, and the spin factor becomes $(-1)^S$.
- For a system boson-antiboson, the factor coming from the spin exchange gives directly $(-1)^S$.[40]

In conclusion, charge conjugation eigenvalues for a system particle-antiparticle are given by the relation

$$C = (-1)^{L+S} \tag{2.54}$$

in a configuration with orbital angular momentum $L$ and total spin $S$.[41]

Applying relation (2.54) to a pseudoscalar ($S = 0$) meson (quark-antiquark) in the ground state ($L = 0$), as $\pi^0$ (or $\eta$ or $\eta'$), we immediately observe that its charge conjugation eigenvalue is positive. On the contrary, for vector mesons like $\rho^0$, $\phi$ and $\omega$, with angular momentum 0 and spin 1, the value of the charge conjugation is negative. In case of two identical neutral bosons, as two pions $\pi^0 \pi^0$, the total angular momentum and spin must be even, so the $C$-parity is always +1. In general, since $C(\pi^0) = +1$, a state composed by any number of neutral pions is always even under charge conjugation, independently of its spatial configuration.

As parity, charge conjugation is conserved in strong and electromagnetic interactions, but not in weak interactions. The difficulty to prepare antiparticle states limits the possibility to perform direct tests of charge conjugation invariance, and most tests are done indirectly, by verifying the constraints imposed by charge conjugation symmetry. Due to the connection between parity and charge conjugation symmetries in the weak interactions theory, several early experiments on parity violation also had implications on $C$ violation. Phenomenological consequences of charge conjugation invariance are common selection rules, for instance we observe the electromagnetic decay $\pi^0 \to \gamma\gamma$, but do not observe the decay $\pi^0 \to 3\gamma$.

### 2.9.1 Charge Symmetry and G-Parity

In Sect. 2.7.1 we have defined the $G$-parity as the product of the charge conjugation and the rotation in the isospin space ($I$-space) by $180^0$ around y-axis.

---

[40] Let us consider the example of two spin 1 particles. The total spin can have the values $S = 0, 1$ or 2. It is easy to check that the states of total spin $S = 0$ or 2 are symmetric, while the state with $S = 1$ is antisymmetric.

[41] Let us underline that the sum $L + S$ is the sum of two numbers, not the composition of the corresponding angular momenta, i.e. it is not in general the total angular momentum of the system. Let us also remark that sometimes $C$ is listed as though it were a valid quantum number for an entire multiplet, while it pertains only to the neutral members.

## 2.9 Charge Conjugation Eigenstates

In order to be more precise, let us consider an $SU(2)$ isospin doublet for quarks

$$q = \begin{pmatrix} u \\ d \end{pmatrix}. \tag{2.55}$$

In the lowest-dimension nontrivial representation isospin rotation group ($I = 1/2$) we have

$$\hat{I}_i = \frac{\hat{\sigma}_i}{2} \qquad i = \{1, 2, 3\} \tag{2.56}$$

where the $\sigma_i$ are the Pauli matrices

$$\sigma_1 = \begin{pmatrix} 0 & 1 \\ 1 & 0 \end{pmatrix} \qquad \sigma_2 = \begin{pmatrix} 0 & -i \\ i & 0 \end{pmatrix} \qquad \sigma_3 = \begin{pmatrix} 1 & 0 \\ 0 & -1 \end{pmatrix} \tag{2.57}$$

already introduced in Sect. 1.1.

The isospin rotation of the quark isospin doublet of an angle $\pi$ around the y-axis (the second one: $I_2 = I_y$) gives

$$e^{i\pi \hat{I}_2} q = e^{i\pi \hat{\sigma}_2/2} q = (\cos\frac{\pi}{2} + i\hat{\sigma}_2 \sin\frac{\pi}{2})q = i\hat{\sigma}_2 q =$$

$$= i \begin{pmatrix} 0 & -i \\ i & 0 \end{pmatrix} \begin{pmatrix} u \\ d \end{pmatrix} = \begin{pmatrix} d \\ -u \end{pmatrix}. \tag{2.58}$$

The relation between the exponential function and the trigonometric functions in the first line of Eq. (2.58) can be found by expanding the exponential function and using the properties of the $\sigma_2$ matrix, namely $(\sigma_2)^2 = 1$, $(\sigma_2)^3 = \sigma_2$ and so on. It holds for any exponent $\alpha$. In general we have

$$e^{-i\alpha\sigma_i/2} = \cos\frac{\alpha}{2} - i\sigma_i \sin\frac{\alpha}{2} \tag{2.59}$$

that can be easily verified

$$e^{-i\alpha\sigma_i/2} = \sum_{n=0}^{\infty} \frac{1}{n!} \left(\frac{-i\alpha\sigma_i}{2}\right)^n =$$

$$= \sum_{n=0}^{\infty} \frac{1}{2n!} \left(\frac{-i\alpha\sigma_i}{2}\right)^{2n} + \sum_{n=0}^{\infty} \frac{1}{(2n+1)!} \left(\frac{-i\alpha\sigma_i}{2}\right)^{2n+1} =$$

$$= \sum_{n=0}^{\infty} \frac{(-1)^n}{2n!} \left(\frac{\alpha}{2}\right)^{2n} - i\sigma_i \sum_{n=0}^{\infty} \frac{(-1)^n}{(2n+1)!} \left(\frac{\alpha}{2}\right)^{2n+1} =$$

$$= \cos\frac{\alpha}{2} - i\sigma_i \sin\frac{\alpha}{2}. \tag{2.60}$$

We have represented the exponential, the sin and cos functions as series, and used the properties of the Pauli matrices in Eq. (1.9).

Under this isospin rotation $u \to d$ and $d \to -u$, thus it changes the proton into the neutron ($p = uud \to ddu = n$) and vice versa ($n = udd \to -duu = -p$), or $\Sigma^+$ into $\Sigma^-$ and so on. Invariance under such rotation is called charge symmetry.[42] It is respected by strong interaction, but broken by the effects of light quark mass difference and electromagnetism.

Since strong interactions are invariant under $C$ and isospin, neglecting mass differences among $u$ and $d$ quarks, they are also invariant under G-parity. In terms of the charge operator $\hat{C}$ and the operator $\hat{I}_y$, we write the G-parity operator as

$$\hat{G} = \hat{C} e^{i\pi \hat{I}_y} \tag{2.61}$$

The multiplicative quantum number G-parity was first used to classify the multipion states in $pp$ and $\pi p$ collisions. Indeed, the entire isomultiplet of pions is an eigenstate of G-parity. The neutral pion is an eigenstate of both charge conjugation, it goes into itself by $\hat{I}_y$ rotation, and is therefore an eigenstate of $G$ as well. The positive pion is changed to the negative one, which has $I_z = -1$, by the conjugation operation $C$, but the specific isospin rotation inverts the value of $I_z$, returning it back to an $I_z = +1$ state, i.e., to a positive pion back again. Thus $\pi^+$, though not an eigenstate of either $C$ or $\hat{I}_y$ rotation, is an eigenstate of the $G$ operation. So it is $\pi^-$. Similar conclusions can be reached for other isomultiplets of mesons. The G-parity of a multiplet of isospin $I$ is easily found by the formula

$$G = (-1)^I C_0 \tag{2.62}$$

where $C_0$ is the charge conjugation parity of the truly neutral member of the multiplet.

For a single pion, $G = -1$, since the C parity of $\pi^0$ is $+1$. For a state with $n$ pions we have $G = (-1)^n$. This tells you how many pions can be emitted in a particular strong decay. For example, the $\rho$ meson, which belongs to a multiplet of isospin 1, with $C_0 = -1$, has $G = 1$ and can decay to two pions, but not to three.

Conversely from observation, for instance, of the $\omega$ particle decaying into three pions, we can infer that $\omega$ is negative under G-parity. Since $\omega$ is an isosinglet which is unaffected by the isospin rotation $e^{i\pi \hat{I}_y}$, it follows that $\omega$ is odd under charge conjugation. The negative G-parity of the $\omega$ particle tells us that it cannot decay into two pions, but of course this is true only if G-parity is conserved in the interactions. Indeed we observe the decay of the $\omega$ particle into two pions through electromagnetic interactions. Incidentally, that explains why the amplitude for the two-pion decay mode is much smaller than the one for the decay into a three pion state, even if the latter is expected to be suppressed from phase space

---

[42] Charge symmetry should not be confused with symmetry under charge conjugation $C$, which refers to replacing a particle with its anti-particle.

considerations. We have seen that it is because the two-pion decay can occur only through electromagnetic interactions, while the three-pion decay can occur through strong interactions. Another usage for $G$-parity is to classify weak hadronic currents, as we will see in Sect. 9.2.

## 2.10  $CP$ Invariance

Putting together the parity and the charge-conjugation transformations, we obtain the $CP$ transformation, effected by the operator $CP$. When it was realised that, upsetting as it was at the time, parity and charge conjugation could be violated, a solace came from the hypothesis that the combined transformation $CP$ was still a symmetry. Violation of $C$ implies asymmetry between particles and antiparticles, but if $CP$ is conserved, the difference between particles and antiparticles can be resolved by looking at them through a mirror. So $C$ violation alone, in this respect, is not an essential symmetry and its role in creating the asymmetry between matter and antimatter in the universe is no more primary.

The consequences of unbroken $CP$ symmetry can be clarified further by considering how we could communicate, in words, our definition of right and left to someone living on a distant galaxy [77]. If $P$ is unbroken, the difference between left and right is a matter of convention, but since weak interaction does not conserve parity we can find a way by using a weak interaction decay as the $\beta$ decays of cobalt-60, described in Sect. 2.8.1. In order to define left or right we could say to our hypothetical correspondent to perform the experiment, check in which directions the electrons are preferably emitted, and then observe the verse of rotation associated with the nuclear spin. This way we can unambiguously define what can be called clock-wise or anticlockwise, which in turn defines right and left. Now let us also assume that our alien correspondent is made of antimatter. If charge conjugation was an unbroken symmetry, "right" matter and "right" antimatter would behave the same way, which is not the case since charge conjugation is violated in weak interactions. However, if $CP$ symmetry is conserved, the "right" matter works the same way as antimatter to the "left". Therefore, returning to our Martian, this time made of anti-matter, if we give him instructions to find what is "right", he will find the opposite direction. At long last, right and left symmetry is still maintained in this sense and $CP$ may be considered to be more fundamental than the individual $P$ and $C$ symmetries.

The neutral kaons provide a perfect experimental system for testing $CP$ invariance. For a non self-conjugated particle as a neutral kaon $K^0$, after a combined transformation of $C$ and $P$, we can set, with an appropriate choice of phases[43]

$$CP|K^0\rangle = |\bar{K}^0\rangle \quad CP|\bar{K}^0\rangle = |K^0\rangle \tag{2.63}$$

---

[43] The sign can be reversed adopting a different phase, $e^{i\alpha} = 1$, in Eq. (2.50).

As in the case of $C$ and $P$ transformations we have $(CP)^2 = 1$. The kaon $K^0(\equiv d\bar{s})$ and its antiparticle $\bar{K}^0(\equiv \bar{d}s)$ have opposite strangeness, respectively 1 and $-1$. For strong and electromagnetic interactions, $CP$ is a valid symmetry and the physical states, written as

$$|K_1\rangle = \frac{1}{\sqrt{2}}(|K^0\rangle + |\bar{K}^0\rangle) \quad |K_2\rangle = \frac{1}{\sqrt{2}}(|K^0\rangle - |\bar{K}^0\rangle) \tag{2.64}$$

are the even and odd eigenstates of the $CP$ transformation

$$CP|K_1\rangle = |K_1\rangle \quad CP|K_2\rangle = -|K_2\rangle. \tag{2.65}$$

Since the two $CP$ eigenstates are not a particle-antiparticle system, they can exhibit different lifetimes and masses. It was predicted [80] that both $|K_1\rangle$ and $|K_2\rangle$ existed and could be observed through weak decays to different final states. If all interactions respect $CP$ symmetry, the $|K_1\rangle$ only decays to states with $CP$ even, and the $|K_1\rangle$ to states with $CP$ odd. The $CP$ eigenvalues of the two neutral kaons were determined by observing the $CP$-parities of the final states in the decays and finding an assignment compatible with $CP$ conservation. Typically, neutral kaons decay into neutral states of two or three pions, which can be $CP$ eigenstates. Kaons and pions are pseudoscalar particles ($S=0$). Due to angular momentum conservation, the two pions in a $K \to 2\pi$ decay must be in a state of orbital angular momentum $l = 0$. The parity of the final state is $P = (-1)^2(-1)^l = +1$. Both the systems $\pi^0\pi^0$ and $\pi^+\pi^-$ are particle-antiparticle systems and applying relation (2.54) we find that any neutral $2\pi$ state is $CP$-even ($CP = +1$). For the $K \to 3\pi$ decay, the neutral states $\pi^+\pi^-\pi^0$ and $\pi^0\pi^0\pi^0$ can be considered as the combined state of a particle-antiparticle system plus an extra $\pi^0$. Let us call $L$ the angular momentum of the two pions in their center mass system and $\bar{L}$ the angular momentum of the third pion with respect to the $2\pi$ sub-system in the overall centre of mass frame. The total angular momentum of the three pion system is the sum of the two and must be zero, for the conservation of the total angular momentum, implying $\bar{L} = L$. The parity is thus $P = (-1)^3(-1)^{L+\bar{L}} = -(-1)^{2L} = -1$. As for the charge conjugation, we have $C(\pi^+\pi^-) = (-1)^L$ and $C(\pi^0) = 1$, then since the charge conjugation is a multiplicative quantum number we have that the $C$-parity is $C = (-1)^L$ for the $\pi^+\pi^-\pi^0$ state. The value $L = 0$ can always be assumed, since, in any case, this value is expected to be favoured. Indeed, the small difference between the $K$ mass and the mass of three pions implies a small phase space volume in the decay, which strongly favours the S wave. The same holds for the $\pi^0\pi^0\pi^0$ state.[44] We conclude that the $CP$-parity of the $3\pi$ final state is $-1$.

Two different neutral kaon states, undergoing weak interactions decays, were observed; their large lifetime difference (a factor $\sim 600$) led to indicate the two eigenstates as short-lived ($K_S$), with lifetime $\tau(K_S) \simeq 10^{-10}$ s, and long-lived, ($K_L$)

---

[44] Besides, as seen in Sect. 2.9, the $C$-parity is +1 for any system composed only of neutral pions.

## 2.10  CP Invariance

with $\tau(K_L) \simeq 10^{-8}$ s.[45] Their mass difference was found to be quite small, $m_{K_L} - m_{K_S} \sim 10^{-6}$ eV. The short-lived neutral kaon was seen to decay into two pions, while the long-lived one into three-body final states, hence assuming $CP$ invariance the $K_S$ was identified with the CP=+1 eigenstate, and the $K_L$ with the CP=-1 eigenstate. The lifetime difference between the two $CP$ eigenstates originates from phase space volume. The pion mass is about 136 MeV and the phase space for $3\pi$ is restricted to about $3m_\pi \sim 420$ MeV versus a kaon mass of about $\sim 500$ MeV. Thus it is natural to expect the lifetime for the CP odd state to be much longer than for the CP even one.

In 1964, about a decade after the discovery of parity violation, $CP$ violation was discovered in kaon decays [82]. If $CP$ is conserved, by starting with a beam of $K^0$

$$|K^0\rangle = \frac{1}{2}(|K_S\rangle + |K_L\rangle) \tag{2.66}$$

the $K_S$ component will quickly decay away, and down the line we shall have a beam of pure $K_L$. That means that near the source (a few cm) we observe a lot of two pions events, but farther along (at distance of meters) we observe only three pions decays. By using a long enough beam, we can produce an arbitrarily pure sample of $K_L$; if at this point we observe a two pion decay, we shall know that $CP$ has been violated. Such was observed by the first experiment that reported evidence of $CP$ violation [82]. We are forced to accept that the $K_L$ is not a perfect eigenstate of CP, but contains a small admixture of $K_S$.

The violation of $CP$ is called indirect when it proceeds not via explicit breaking of the $CP$ symmetry in the decay itself but via the admixture of the $CP$ state with opposite $CP$ parity to the dominant one. Instead, the so-called direct $CP$ violation is realised via a direct transition of a $CP$ odd to a $CP$ even state.

The consistency of the experimental results with the general scheme of charged weak interactions and $CP$ violation ($\mathcal{CP}$) in the SM is non-trivial. Several theoretical analyses were advanced on the origin of $\mathcal{CP}$, as the possibility of a superweak interaction, until in the 70s the development of a renormalizable gauge theory of the weak interactions and experimental discoveries of new flavours gave a special position to the six-quark model, where $\mathcal{CP}$ arises from flavour-mixing.

$CP$ violation has been experimentally established in the $B$-meson decays in 2001 by the Belle experiment [83] at KEK and the Babar experiment [84] at SLAC. In 2019 $\mathcal{CP}$ has been observed in the $D$-meson decays by the LHCb collaboration [85] at CERN. $CP$ violation has not been firmly established in the leptonic sector yet, where is actively searched in current and upcoming neutrino experiments.

---

[45] The long-lived neutral, strange meson was discovered by Lederman and his collaborators in 1956 [81].

## 2.11 Time Reversal

As seen in Sect. 2.2, time reversal $T$ is a discrete, spacetime operation, which corresponds to the inversion of the time coordinate

$$t \xrightarrow{T} -t \tag{2.67}$$

leaving the space coordinates unchanged. In analogy with classical mechanics, as seen in Table 2.1, we expect in quantum mechanics that the following connections of the time reversal operator $\hat{T}$ with the position and momentum operators $\hat{\mathbf{X}}$ and $\hat{\mathbf{P}}$ hold

$$\hat{T}^{-1}\hat{\mathbf{X}}\hat{T} = +\hat{\mathbf{X}} \quad \text{or} \quad [\hat{\mathbf{X}}, \hat{T}] = 0$$
$$\hat{T}^{-1}\hat{\mathbf{P}}\hat{T} = -\hat{\mathbf{P}} \quad \text{or} \quad \{\hat{\mathbf{P}}, \hat{T}\} = 0 \tag{2.68}$$

together with the following transformation for angular momentum $\hat{\mathbf{J}}$

$$\hat{T}^{-1}\hat{\mathbf{J}}\hat{T} = -\hat{\mathbf{J}} \quad \text{or} \quad \{\hat{\mathbf{J}}, \hat{T}\} = 0 \tag{2.69}$$

where $[\hat{J}_a, \hat{J}_b] = i\,\varepsilon_{abc}\,\hat{J}_c$ with $a, b \in \{1, 2, 3\}$. By analogy, one assumes the same transformation properties for spin.[46]

Let us observe that the commutation relations (2.68) imply the following transformation

$$\hat{T}^{-1}[X_a, P_b]\hat{T} = -[\hat{X}_a, \hat{P}_b] \qquad a \neq b. \tag{2.70}$$

It seems not possible to hold the correspondence with the classical transformations and at the same time leave the quantum mechanic commutation relations $[\hat{X}_a, \hat{P}_b] = i\,\delta_{ab}$ unchanged. The apparent contradiction is solved by assuming that $\hat{T}$ are *anti-unitary* operators, leading to

$$\hat{T}^{-1}\,i\,\hat{T} = -i. \tag{2.71}$$

Indeed, by definition, an anti-unitary operator $\hat{T}$ in the Hilbert space is antilinear, that is for any complex number $\alpha$ and any state $|a\rangle$

$$\hat{T}(\alpha|a\rangle) = \alpha^*\hat{T}|a\rangle \tag{2.72}$$

---

[46] Sometimes time reversal symmetry is referred to as motion-reversal symmetry since it implies what is also known as the reciprocity relation, namely given a configuration of energy-momenta and spin, the probability of an initial state being transformed into a final state with the same probability of the process with initial and final states inverted and with momenta and spins reversed. However, there are subtleties affecting this general definition, and we will not dwell on them. For instance one could also define $T$-odd transformations, which refer only to changing the sign of all odd variables under $t \to -t$ in the Hamiltonian, without exchanging initial and final states.

## 2.11 Time Reversal

holds. Moreover, anti-unitary operators conserve probabilities, since by definition we have

$$\langle (a|\hat{T})|(\hat{T}|b)\rangle = \langle a|\hat{T}^\dagger\hat{T}|b\rangle = \langle a|b\rangle^* = \langle b|a\rangle. \tag{2.73}$$

An anti-unitary operator $\hat{T}$ can be defined by splitting the two actions of $T$, namely the mapping of the states into their motion-reversed counterparts and the complex conjugation of complex numbers. Hence we can write the product

$$\hat{T} = \hat{U}\hat{K} \tag{2.74}$$

where $\hat{U}$ is unitary and $\hat{K}$ is the (anti-unitary) complex conjugation operator defined by the antilinearity relation

$$\hat{K}(\alpha|a\rangle + \beta|b\rangle) = \alpha^*\hat{K}|a\rangle + \beta^*\hat{K}|b\rangle. \tag{2.75}$$

$T$ invariance in classical physics as in quantum mechanics means that the rates for processes $a \to b$ and $b \to a$ are the same if $b$ (the final state in the first process) is arranged identically in all aspects as the initial state for the second one. Such an arrangement requires fine-tuning the initial conditions, that is typically very difficult to achieve. A more feasible method is to resort to oscillations of neutral particle and anti-particle into each other due to quantum mechanics, as we will see in the next section.

### 2.11.1 Observations of Time Reversal Violation

Evidence of $T$ violation has been claimed for the first time in 1998 by the CPLEAR collaboration in the system of neutral kaons $K^0 - \bar{K}^0$ [86]. This observation was made by comparing the probabilities of a $K^0$ state transforming into a $\bar{K}^0$ and vice-versa. Time reversal symmetry requires that the probability that a $K^0$ (at an initial time) observed at a later time as a $\bar{K}^0$ state should be equal to the probability that an initial $\bar{K}^0$ is observed as a $K^0$ after the same time. Any difference between these two probabilities is a signal for $T$ violation.

Under the LEAR programme, four machines—the Proton Synchrotron (PS), the Antiproton Collector (AC), the Antiproton Accumulator (AA), and the Low Energy Antiproton Ring (LEAR)—all located at CERN—worked together. Protons from the PS created antiprotons in collisions with a fixed target. Antiprotons can be produced in any collision that is sufficiently energetic to produce a proton-antiproton pair. The AC and the AA worked to collect antiprotons in sufficient numbers, and LEAR, the final element in the chain, passed the antiprotons on to experiments.[47] The experiment CPLEAR observed initial neutral kaons with defined strangeness produced from the annihilation at rest of the low-energy $\bar{p}$ beam on a hydrogen target, via the reactions $p\bar{p} \to K^-\pi^+K^0$ and $p\bar{p} \to K^+\pi^-\bar{K}^0$. The quantum number strangeness is conserved in the strong interactions, hence a pair strange-antistrange has to be created in the final state since the initial state has zero strangeness. The neutral kaon strangeness at the production time ($t = 0$) was determined (tagged) by the charge of the accompanying charged kaon. Since weak interactions do not conserve strangeness, the $K^0$ and $\bar{K}^0$ may subsequently transform into each other. The strangeness of the neutral kaon at decay time $\tau$ was determined through the semileptonic decays $K^0 \to e^+\pi^-\nu_e$ and $\bar{K}^0 \to e^-\pi^+\bar{\nu}_e$, respectively, that tags the $K^0$ ($\bar{K}^0$) state with the positive (negative) associated lepton charge. In this way, the following asymmetry was measured

$$A = \frac{R[\bar{K}^0(t=0) \to e^+\pi^-\nu_e(t=\tau)] - R[K^0(t=0) \to e^-\pi^+\bar{\nu}_e(t=\tau)]}{R[\bar{K}^0(t=0) \to e^+\pi^-\nu_e(t=\tau)] + R[K^0(t=0) \to e^-\pi^+\bar{\nu}_e(t=\tau)]}$$
(2.76)

where $R[...]$ is the decay rate of the process. This ratio compares the probability that a flavour eigenstate $\bar{K}^0$ in the initial state oscillates to a final state $K^0$, with the probability that an initial state $K^0$ oscillates to a final state $\bar{K}^0$. Since the states $K^0$ and $\bar{K}^0$ are particle and antiparticle, the two transitions are connected by both $T$ and $CP$ violation.

The asymmetry in Eq. (2.76) was found different than zero by $4\sigma$ and this has been interpreted by the collaboration CPLEAR as the first direct measurement of time-reversal non-invariance [86]. However, some authors doubt whether the experiment does provide such a direct evidence for $T$ violation. The basic argument is that decay processes enter in the observables, making $CP$-violation manifest.[48] The observed effect is then attributed to these irreversible processes, rather than $T$-violation.[49]

The bypass to this argument came from the strategy of not including explicitly the decay products in the time reversal violating asymmetry and using the decay as a filtering measurement of the projected meson state only. This strategy exploits the quantum entanglement of pairs of neutral mesons produced in the facilities built at the SLAC National Accelerator Laboratory in California, USA, and at the KEK Laboratory at Tsukuba in Japan, called B Factories (because they produce lots of B

---

[47] The antiproton machines were closed down in 1996 to free resources for the LHC. In order to continue experiments with slow antiprotons, a new machine—the Antiproton Decelerator (AD)—started operation in 2000.

[48] In general each example of difference in the probability of oscillation $i \to f$ from $f \to i$ immediately implies a test of $CP$ violation (conceptual if not practical) by going to the antiparticles. In contrast the simplest tests of $CP$ violation have no direct relation to $T$ violation; for example, $\Gamma(A \to f) = \Gamma(\bar{A} \to \bar{f})$ involves a rate which has nothing to do with a time reversal violable observables.

[49] For opposite viewpoints see for instance Refs. [87] and [88].

## 2.11 Time Reversal

mesons). These factories collide electrons and positrons at just the right energy to produce the $\Upsilon(4S)$ meson, a $b\bar{b}$ state which has mass about 10.6 GeV, spin $J = 1$, $P = -1$, charge conjugation $C = -1$ and, most importantly, beauty quantum number zero. The final state from the $\Upsilon(4S)$ decay by strong interaction maintain the same beauty quantum number zero, hence an hadron in the final state containing a $b$ (or $\bar{b}$) quark has to be produced only in pair with its antiparticle. Since the mass of the $\Upsilon(4S)$ is only slightly higher than twice the mass of the $B$ meson, the $\Upsilon(4S)$ can at the most decay into a pair of $B\bar{B}$ mesons by strong interactions. The $B\bar{B}$ pairs are approximately 49% $B^0(\equiv \bar{b}d) \bar{B}^0(\equiv b\bar{d})$ pairs, when the hadronization process picks up a $d\bar{d}$ quark pair, and 51% $B^+(\equiv \bar{b}u) B^-(\equiv b\bar{u})$ pair, when it picks up a $u\bar{u}$ pair. Only about 4% of the $\Upsilon(4S)$ decays are non $B\bar{B}$ decays. The state composed by the $B\bar{B}$ pair conserves both parity and $C$-parity of the decaying meson. The two $B$ mesons have low momenta (about 330 MeV) and are produced almost at rest in the $\Upsilon(4S)$ reference frame with no additional particles besides those associated to the $B$ decays.

The system $B^0$ and $\bar{B}^0$ represents an example of quantum entanglement: the two pseudoscalar states lose their individuality and can be treated as a unique quantum system. Their properties are related because together they are in agreement with the properties of the $\Upsilon(4S)$. Since the parity is odd, the (entangled) quantum state of the pair can be expressed as an anti-symmetric combination of the flavour eigenstates $B^0$ and $\bar{B}^0$[50]

$$|\psi_{pair}\rangle = \frac{1}{\sqrt{2}}(|B^0\rangle_1|\bar{B}^0\rangle_2 - |\bar{B}^0\rangle_1|B^0\rangle_2) \quad (2.77)$$

The initial state of the neutral meson system is a singlet state in the flavour space. The antisymmetric entanglement is essential: there is no trace of combinations $B^0 B^0$ or $\bar{B}^0 \bar{B}^0$, and only $B^0 \bar{B}^0$ states appear at any time. The observation of the decay of one meson through a weak decay at time $t$, e.g. $B^0$, signals that the (still alive) partner meson state at this time is the $B$-state not decaying, namely $\bar{B}^0$. The decay has filtered the $B^0$-state, and the orthogonal state is then tagged at time $t$. For subsequent times, we have a single state time evolution from $\bar{B}^0$. Quantum entanglement of the neutral meson system ensures the transfer of the information from the decaying meson to its (still alive) orthogonal partner. The $T$ asymmetry can be searched in the time evolution of this last partner.

The $T$ violation found this way has been established in the $B^0 - \bar{B}^0$ system [89]. Let us make an example, pictured in Fig. 2.3. As you can see in the top of this figure, the collision electron-positron produce a $\Upsilon(4S)$ resonance, that decays to an entangled pair of orthogonal states $B^0$ and its anti-particle $\bar{B}^0$.

Among $B^0$ meson weak decay modes, there are decays into leptons accompanied by hadrons with no beauty quantum number (weak interactions do not conserve

---

[50] They are in a $p$-wave state, as can be deduced also because of Bose statistics and angular momentum conservation.

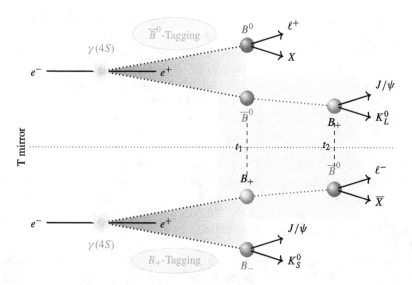

**Fig. 2.3** $T$ violation experiment with $B$ mesons

it), that can be searched for. At quark level this decay is driven by a weak decay of a positive anti-quark $\bar{b}$ into a negative anti-quark $\bar{u}$ or $\bar{c}$, accompanied by a positive lepton (because of charge conservation) and the corresponding anti-lepton. The weak decay conserves baryon and lepton quantum number. The final state can be indicated with $\ell^+ X$; we are interested only to identify the positive lepton $\ell^+$ and indicate with a generic $X$ all other particles in the final state. That is because the presence of a positive lepton tags the $B^0$ meson; the positive lepton charge cannot originate by the $\bar{B}^0$ meson decays, driven at quark level by a weak decay of a negative quark $\bar{b}$. If this decay occurs at time $t_1$, because of the entanglement we know that the other meson produced at the $\Upsilon(4S)$ resonance is a $\bar{B}^0$ meson at $t = t_1$. We use the decay of one meson to tag at $t = t_1$ the surviving one as $\bar{B}^0$, namely as a flavour eigenstate with beauty quantum number $-1$. It would also be a mass eigenstate, in the absence of weak interactions; that not being the case, it can evolve from this initial condition and decay into a $CP$ eigenstate final state at a later time $t = t_2$.[51] In Fig. 2.3 we have considered the decay into the $CP$-even eigenstate $J/\psi K_L^0$, but one could equivalently search for $CP$-odd final states as $c\bar{c}K_S$, $\psi(2S)K_S$ or $\xi_{c_1}K_S$. This second decay at $t = t_2$ filters the neutral $B$-states in

---

[51] One can build orthogonal $CP$ eigenstates $B_+$ and $B_-$ in the neutral meson system $B^0\bar{B}^0$, in analogy to what done for the kaon system in Sect. 2.10. The entangled, antisymmetric system can be expressed in terms of their linear combinations

$$|\psi_{pair}>= \frac{1}{\sqrt{2}}(|B_+>_1 |B_->_2 - |B_->_1 |B_+>_2) \qquad (2.78)$$

We neglect for the purpose of discussion $CP$ violation in both the kaon and $B$ systems.

a $CP$ eigenstate $B_+$. The net result is a transition $\bar{B}^0 \to B_+$. An event reconstructed in the time-ordered final states $(\ell^+ X, J/\psi K_L^0)$ identifies this transition. To study time reversal, we have to compare the rate at which these transitions occurs with the rate of the reversed transitions $B_+ \to \bar{B}^0$, pictured at the bottom of Fig. 2.3. Here the first decay selects the $CP$-odd eigenstate $B_-$, so the surviving state is the other eigenvector $B_+$ because of the quantum entanglement. A weak decay of the surviving $B_+$ into leptonic states with negative leptons identifies (tags) a $\bar{B}^0$ meson. The full transition corresponds to an event reconstructed in the time-ordered final states $(J/\psi K_S^0, \ell^- \bar{X})$. Any difference in these two rates is evidence for $T$-symmetry violation.

The main difference between the CPLEAR and $B$-factory experiments is that the initial and final states of the processes under study in the CPLEAR experiment are the $CP$-conjugate of each other, namely $K^0$ and $\bar{K}^0$, while initial and final states $\bar{B}^0$ and $B_+$, are not, belonging to different sets of orthogonal states. This experiment could provide direct evidence of $T$ non-invariance, without using an observation which also violates $CP$. It has been performed for the first time by the SLAC $B$-factory BABAR [90], providing the first direct observation of $T$-violation through the exchange of initial and final states in transitions that can only be connected by a $T$-symmetry transformation.

One could in principle consider neutrino $\nu_e$ to $\nu_\mu$ mixing, but in practice this would require currently not available very high-luminosity.

## 2.12 Lorentz Transformations for Spin 1/2 Fields

The Dirac equation is relativistic and by construction respects Lorentz invariance. Let us indicate with $\Lambda^\mu_\nu$ the Lorentz transformation and with $S(\Lambda)$ the operator that acts on the spinor in the same transformation

$$x^\mu \to x'^\mu = \Lambda^\mu_\nu x^\nu \qquad \psi(x) \to \psi'(x') = S(\Lambda)\psi(x). \qquad (2.79)$$

This rule of transformation for $\psi(x)$ under Lorentz transformations can also be taken as the definition of a generic spinor $\psi(x)$.

The left side of the Dirac equation for the spinor $\psi(x)$ (Eq. (1.11)) transforms as

$$\left(i\gamma^\mu \frac{\partial}{\partial x^\mu} - m\right)\psi(x) = \left(i\gamma^\mu \frac{\partial x'_\alpha}{\partial x^\mu}\frac{\partial}{\partial x'_\alpha} - m\right) S^{-1}(\Lambda)\psi'(x') =$$

$$= \left(i\Lambda_{\alpha\mu}\gamma^\mu \frac{\partial}{\partial x'_\alpha} - m\right) S^{-1}(\Lambda)\psi'(x'). \qquad (2.80)$$

Hence the Dirac equation holds for $\psi'(x')$ as well if we assume that

$$S(\Lambda)\Lambda_{\alpha\mu}\gamma^\mu S^{-1}(\Lambda) = \gamma_\alpha \qquad (2.81)$$

that is

$$S^{-1}(\Lambda)\gamma^\mu S(\Lambda) = \Lambda^\mu_\nu \gamma^\nu. \tag{2.82}$$

This rule can be considered a consequence of the vector nature of the $\gamma$ matrices, see Eq. (2.24).

The Dirac adjoint, also called Dirac conjugate

$$\bar{\psi}(x) = \psi(x)^\dagger \gamma^0 \tag{2.83}$$

transforms as

$$\bar{\psi}(x) \to \bar{\psi}'(x') = \bar{\psi}(x) S(\Lambda)^{-1} \tag{2.84}$$

as expected, since $\bar{\psi}(x)\psi(x)$ is Lorentz invariant. Indeed we have

$$\bar{\psi}'(x') = (\psi'(x'))^\dagger \gamma^0 = (S(\Lambda)\psi(x))^\dagger \gamma^0 = \psi(x)^\dagger S(\Lambda)^\dagger \gamma^0 = \\
= \psi(x)^\dagger \gamma^0 \gamma^0 S(\Lambda)^\dagger \gamma^0 = \psi(x)^\dagger \gamma^0 S(\Lambda)^{-1} = \bar{\psi}(x) S(\Lambda)^{-1}. \tag{2.85}$$

Here we have used the equalities

$$S(\Lambda)^\dagger = \gamma^0 S(\Lambda)^{-1} \gamma^0 \tag{2.86}$$

which can be easily checked by using the explicit form of the Lorentz group elements in the Dirac representation, given in Eq. (2.31). Although the finite dimensional matrices $S(\Lambda)$ provide a representation of the Lorentz group, the representation is not unitary ($S(\Lambda)^\dagger S(\Lambda) \neq 1$) in conformity with the fact that the Lorentz group is non-compact.

By exploiting the transformation properties of the spinor and its conjugate we can attribute a (pseudo-)scalar, (axial) vector or tensor nature to the following bilinear forms in the spinor fields, as defined by their transformation properties under the Lorentz group

$$\bar{\psi}'_2(x')\psi'_1(x') = \bar{\psi}_2(x) S(\Lambda)^{-1} S(\Lambda) \psi_1(x) = \bar{\psi}_2(x) \psi_1(x) \quad \text{(scalar)}$$
$$\bar{\psi}'_2(x')\gamma^\mu \psi'_1(x') = \bar{\psi}_2(x) S(\Lambda)^{-1} \gamma^\mu S(\Lambda) \psi_1(x) = \Lambda^\mu_\nu \bar{\psi}_2(x) \gamma^\nu \psi_1(x) \quad \text{(vector)}$$
$$\bar{\psi}'_2(x')\sigma^{\mu\rho} \psi'_1(x') = \bar{\psi}_2(x) S(\Lambda)^{-1} \sigma^{\mu\rho} S(\Lambda) \psi_1(x) \\
= \Lambda^\mu_\nu \Lambda^\rho_\sigma \bar{\psi}_2(x) \sigma^{\nu\sigma} \psi_1(x) \quad \text{(tensor)}. \tag{2.87}$$

In order to continue, we define

$$\gamma_5 = i\gamma^0 \gamma^1 \gamma^2 \gamma^3 = -\frac{i}{4!} \epsilon_{\mu\nu\lambda\rho} \gamma^\mu \gamma^\nu \gamma^\lambda \gamma^\rho \tag{2.88}$$

where $\epsilon_{\mu\nu\lambda\rho}$ is the completely anti-symmetric Levi-Civita tensor with four indexes. We have

$$\varepsilon^{0123} = +1 \qquad \varepsilon^{ijkl} = (-1)^{n_p} \qquad (2.89)$$

with $n_p$ being the number of permutations that bring $\varepsilon^{ijkl}$ to $\varepsilon^{0123}$. The Levi-Civita tensor defines the determinant of a square matrix $A$, $\det A$, through the relation

$$(\det A)\varepsilon^{ijkl} = A^i_m A^j_n A^k_p A^l_q \varepsilon^{mnpq}. \qquad (2.90)$$

Then one can show that

$$\bar{\psi}'_2(x')\gamma_5\psi'_1(x') = -\frac{i}{4!}\epsilon_{\mu\nu\lambda\rho}\bar{\psi}_2(x)S(\Lambda)^{-1}\gamma^\mu\gamma^\nu\gamma^\lambda\gamma^\rho S(\Lambda)\psi_1(x) =$$

$$= -\frac{i}{4!}\epsilon_{\mu\nu\lambda\rho}\Lambda^\mu_\alpha\Lambda^\nu_\beta\Lambda^\lambda_\sigma\Lambda^\rho_\omega\bar{\psi}_2(x)\gamma^\alpha\gamma^\beta\gamma^\sigma\gamma^\omega\psi_1(x)$$

$$= (\det\Lambda)\bar{\psi}_2(x)\gamma_5\psi_1(x). \qquad (2.91)$$

A similar derivation can be followed in the case of an axial vector. Summarising, we add to the list in equation (2.87) the following bilinear forms

$$\bar{\psi}'_2(x')\gamma_5\psi'_1(x') = (\det\Lambda)\bar{\psi}_2(x)\gamma_5\psi_1(x) \qquad \text{(pseudo-scalar)}$$
$$\bar{\psi}'_2(x')\gamma_5\gamma^\mu\psi'_1(x') = (\det\Lambda)\Lambda^\mu_\nu\bar{\psi}_2(x)\gamma_5\gamma^\nu\psi_1(x) \quad \text{(axial vector)} \qquad (2.92)$$

with $\det\Lambda$ assuming the value $\pm 1$ according to the specific transformation $\Lambda$ (for instance, in case of a parity transformation the pseudo-scalar bilinear form changes sign).

Since the spinor field has 4 components, there are sixteen independent $4 \times 4$ matrices $\Gamma$ which can be used to combine spinors, $\bar{\psi}_2(x)\Gamma\psi_1(x)$. Since the previous combinations give $1 + 4 + 6 + 4 + 1 = 16$ independent contractions, the above are exhaustive; any other matrix sandwiched between $\bar{\psi}_2(x)$ and $\psi_1(x)$ must be a linear combination of $1$, $\gamma^\mu$, $\sigma^{\mu\rho}$, $\gamma_5$ and $\gamma_5\gamma^\mu$.

## 2.13 Field Transformations Under C, P and T

In this section we discuss how $C$, $P$ and $T$ transformations affect scalar, vector and spinor fields.

Let us start from parity. In Sect. 2.8 we have already discussed parity transformations of scalar and vector fields, so here we limit to fields in the Dirac spinor representation. Let us consider the matrix representing parity transformations in

Lorentz transformations

$$\Lambda^\mu_\nu = \begin{pmatrix} 1 & 0 & 0 & 0 \\ 0 & -1 & 0 & 0 \\ 0 & 0 & -1 & 0 \\ 0 & 0 & 0 & -1 \end{pmatrix}. \tag{2.93}$$

In the Dirac spinor representation, we can use relation (2.82), which implies

$$S(\Lambda)^{-1}\gamma^0 S(\Lambda) = \Lambda^0_\mu \gamma^\mu = \gamma^0$$
$$S(\Lambda)^{-1}\gamma^i S(\Lambda) = \Lambda^i_\mu \gamma^\mu = -\gamma^i \qquad i = 1, 2, 3 \tag{2.94}$$

which is satisfied if we choose

$$S(\Lambda) = \gamma^0 \eta_P \tag{2.95}$$

where $\eta_P$ is an arbitrary phase not observable.

Let us pass to time reversal $\hat{T}$.[52] In non relativistic quantum mechanics the Schrödinger equation reads

$$i\hbar \frac{d}{dt}\psi(t, \mathbf{x}) = H\psi(t, \mathbf{x}) \tag{2.96}$$

It is evident that what would have been a näive choice for the time reversed state, namely $\psi(-t, \mathbf{x})$, is not a solution of the Schrödinger equation. Instead, assuming a real Hamiltonian, we have invariance under time reversal when

$$\psi(t, \mathbf{x}) \xrightarrow{\hat{T}} \psi^*(-t, \mathbf{x}) \tag{2.97}$$

with a possible arbitrary phase. This mapping sends particles to particles (not antiparticles). As an example, time reversal transforms a plane wave $e^{-i(Et - p \cdot x)}$ of energy $E$, travelling with momenta $\mathbf{p}$, into $e^{i(-Et - p \cdot x)}$, which describes a plane wave of the same (positive) energy travelling in the opposite direction.

In field theory one defines the time-reversed counterpart of a scalar field $\phi(t, \mathbf{x})$ equal to itself, if not for its space-time dependence.[53] We have

$$\phi(t, \mathbf{x}) \xrightarrow{\hat{T}} \eta_T \phi(-t, \mathbf{x}) \qquad \phi^\dagger(t, \mathbf{x}) \xrightarrow{\hat{T}} \eta_T^* \phi^\dagger(-t, \mathbf{x}) \tag{2.98}$$

---

[52] In this chapter we indicate time reversal with $\hat{T}$ to distinguish it from the transpose symbol, that we indicate with a simple $T$.

[53] In analogy to the case of the wave-function in quantum mechanics, we are tempted to describe the effect of time reversal as $\phi(t, \mathbf{x}) \to \phi^\dagger(-t, \mathbf{x})$, but this is not admissible, since the Hermitian conjugate would transform, for instance, a particle at rest into an antiparticle.

## 2.13 Field Transformations Under C, P and T

where $\eta_T$ is an arbitrary phase. The change of the phase factor is due to the anti-linear implementation of time reversal. A vector field $A^\mu$ transforms as[54]

$$A^\mu(t,\mathbf{x}) \xrightarrow{\hat{T}} A_\mu(-t,\mathbf{x}) \qquad (2.99)$$

For a spinor state $\psi(x)$, one can start from the Dirac equation

$$(i\gamma^\mu \partial_\mu - m)\psi(t,\mathbf{x}) = 0. \qquad (2.100)$$

By complex conjugation of Eq. (2.100) we find

$$(-i\gamma^{\mu*}\partial_\mu - m)\psi^*(t,\mathbf{x}) = \left(-i\gamma^{0*}\frac{\partial}{\partial t} - i\gamma^{i*}\frac{\partial}{\partial x^i} - m\right)\psi^*(t,\mathbf{x}) = 0$$

$$\left(-i\gamma^{0*}\frac{\partial}{\partial(-t)} - i\gamma^{i*}\frac{\partial}{\partial x^i} - m\right)\psi^*(-t,\mathbf{x}) =$$

$$= \left(i\gamma^{0*}\frac{\partial}{\partial t} - i\gamma^{i*}\frac{\partial}{\partial x^i} - m\right)\psi^*(-t,\mathbf{x}) = 0. \qquad (2.101)$$

Let us name $\psi^{\hat{T}}(t,\mathbf{x})$ the solution of the time reversed Dirac equation, and define, apart from an arbitrary phase

$$\psi^{\hat{T}}(t,\mathbf{x}) = \hat{T}\psi^*(-t,\mathbf{x}). \qquad (2.102)$$

By definition, we should have

$$(i\gamma^\mu \partial_\mu - m)\psi^{\hat{T}}(t,\mathbf{x}) = 0$$

$$\left(i\gamma^0 \frac{\partial}{\partial t} + i\gamma^i \frac{\partial}{\partial x^i} - m\right)\hat{T}\psi^*(-t,\mathbf{x}) = 0. \qquad (2.103)$$

Therefore, comparing the two equations (2.103) and (2.101), we see that the operator $\hat{T}$ has to satisfy

$$\hat{T}\gamma^{0*}\hat{T}^{-1} = \gamma^0 \qquad \hat{T}\gamma^{i*}\hat{T}^{-1} = -\gamma^i \qquad (2.104)$$

---

[54] This transformation rule, which on the basis of the derivation made for parity transformations in Sect. 2.8 we were led to expect with an overall minus sign, is forced by the antiunitarity propriety of the time reversal operator (see for instance Ref. [91]). An intuitive way to understand it is by analogy with the transformations for the electromagnetic vector field $A^\mu$ in the classic case, see Table 2.1.

where $i = 1, 2, 3$. Since $\gamma^0$ is hermitian, and $\gamma^i$ are anti-hermitian, we can invert these relations obtaining

$$\hat{T}^{-1}\gamma^0\hat{T} = \gamma^{0*} = (\gamma^{0\dagger})^T = (\gamma^0)^T$$
$$\hat{T}^{-1}\gamma^i\hat{T} = -\gamma^{i*} = -(\gamma^{i\dagger})^T = (\gamma^i)^T \qquad (2.105)$$

where $T$ stays for transpose. These equalities can be written in a more compact form as

$$\hat{T}^{-1}\gamma^\mu\hat{T} = \gamma^{\mu T}. \qquad (2.106)$$

One possibility in the Dirac representation is to set

$$\hat{T} = i\gamma^1\gamma^3 \qquad (2.107)$$

to within a phase. With this choice, the $\hat{T}$ matrix satisfies

$$\hat{T} = \hat{T}^\dagger = \hat{T}^{-1}. \qquad (2.108)$$

That follows easily from the properties

$$(\gamma^0)^2 = I \qquad (\gamma^0)^{-1} = \gamma^0 \qquad (\gamma^i)^2 = -I \qquad (\gamma^i)^{-1} = -\gamma^i \qquad (2.109)$$

where $I$ is the identity matrix. In order to check Eq. (2.108) one can consider that in the Dirac representation

$$\gamma^0 = \gamma^{0T} \qquad \gamma^1 = -\gamma^{1T} \qquad \gamma^2 = \gamma^{2T} \qquad \gamma^3 = -\gamma^{3T}. \qquad (2.110)$$

In the case of quantum field theory one finds similar results. One complex conjugates numbers, but does not transform a spinor field $\psi(t, \mathbf{x})$ into its conjugate, in analogy to the scalar case. We have[55]

$$\psi(t, \mathbf{x}) \xrightarrow{\hat{T}} \eta_T \hat{T}\psi(-t, \mathbf{x}) \qquad \bar{\psi}(t, \mathbf{x}) \xrightarrow{\hat{T}} \eta_T^* \bar{\psi}(-t, \mathbf{x})\hat{T}^\dagger \qquad (2.111)$$

where $\eta_T$ is an arbitrary phase and $\hat{T}$ is defined as in Eq. (2.104).

Finally, let us consider charge conjugation. We have discussed photon and vector fields in Sect. 2.9. In the scalar case, a free field, corresponding to what can be regarded as a neutral particle, does not change under charge conjugation

$$\phi \xrightarrow{C} \phi. \qquad (2.112)$$

---

[55] For an explicit demonstration see for example Ref. [92].

## 2.13 Field Transformations Under C, P and T

For a charged scalar field, which can be regarded as a combination of two neutral scalar fields $1/\sqrt{2}(\phi_1 + i\phi_2)$, one has

$$\phi \xrightarrow{C} \phi^\dagger \tag{2.113}$$

or equivalently

$$\phi_1 \xrightarrow{C} \phi_2 \qquad \phi_2 \xrightarrow{C} -\phi_2 \tag{2.114}$$

setting to zero arbitrary phase factors. This is the case of a field satisfying the Klein-Gordon equation.

Let us next analyse fermions of charge $e$ in an electromagnetic field. If the fermions are minimally coupled to the photons, the Dirac equation takes the form

$$\left[\gamma^\mu(i\partial_\mu - eA_\mu) - m\right]\psi(x) = 0. \tag{2.115}$$

We look for a solution $\psi^C(x)$ of the Dirac equation that describes a particle of opposite electric charge and same mass

$$\left[\gamma^\mu(i\partial_\mu + eA_\mu) - m\right]\psi^C(x) = 0. \tag{2.116}$$

By complex conjugating Eq. (2.115) and multiplying for $\gamma_0$ we find

$$\bar\psi(x)\left[\gamma^\mu(-i\partial_\mu - eA_\mu) - m\right] = 0. \tag{2.117}$$

Taking the transpose of the adjoint equation, we obtain

$$\left[\gamma^{\mu T}(-i\partial_\mu - eA_\mu) - m\right]\bar\psi^T(x) = 0 \tag{2.118}$$

where $T$ denotes the transpose of a matrix. Given a matrix $C$ such that

$$C\gamma^{\mu T}C^{-1} = -\gamma^\mu \tag{2.119}$$

we have

$$\left[\gamma^\mu(i\partial_\mu + eA_\mu) - m\right]C\bar\psi^T(x) = 0. \tag{2.120}$$

It is easy to see that Eq. (2.116) is satisfied if

$$\psi \to \eta_C \psi^C \tag{2.121}$$

where $\eta_C$ is an arbitrary phase, and the charge conjugate field $\psi^C$ is expressed as

$$\psi^C(t, \mathbf{x}) = C\bar\psi^T = C\gamma^0\psi^*. \tag{2.122}$$

In the Dirac-Pauli representation of Eq. (1.6), a possible choice of $C$ satisfying Eq. (2.119) is

$$C = i\gamma^2\gamma^0 \tag{2.123}$$

up to an arbitrary phase not fixed by Eq. (2.119). It is now easy to show that the matrix $C$ satisfies the relations

$$C^\dagger = C^{-1} = C^T = -C$$
$$(C^{-1})^\dagger = C$$
$$C^{-1}\gamma_5 C = \gamma_5^T \quad C^{-1}\gamma_\mu C = -\gamma_\mu^T \quad C^{-1}\gamma_\mu\gamma_5 C = (\gamma_\mu\gamma_5)^T$$
$$\sigma_{\mu\nu} C = -C\sigma_{\mu\nu}^T \quad \gamma^{0*}C^*\gamma^0 = C^{-1} \tag{2.124}$$

where we have the usual definition for the $\gamma_5$ matrix

$$\gamma_5 \equiv i\gamma^0\gamma^1\gamma^2\gamma^3. \tag{2.125}$$

The explicit form of the spinor obtained by using Eq. (2.123) is

$$\psi^C(t, \mathbf{x}) = C\bar{\psi}^T = C\gamma^0\psi^* = i\gamma^2\psi^*. \tag{2.126}$$

One can write the adjoint of the charge conjugate spinor as

$$\overline{\psi^C} = -\psi^T C^{-1}. \tag{2.127}$$

In the Dirac-Pauli representation, given Eqs. (2.107) and (2.123), and remembering that

$$(\gamma^0)^2 = -(\gamma^i)^2 = I \quad (i = 1, 2, 3) \tag{2.128}$$

where $I$ is the identity matrix, we find a relation between the time reversal and charge conjugation operator

$$\hat{T} = -i\gamma_5 C. \tag{2.129}$$

Let us apply the charge conjugation to a particular Dirac spinor $\psi(x) = u(p, s)e^{-ip\cdot x}$. In the representation (2.123), we immediately see that

$$\psi^C = i\gamma^2[u(p)e^{-ip\cdot x}]^* = u(-p)e^{ip\cdot x} = v(p)e^{ip\cdot x} \tag{2.130}$$

## 2.13 Field Transformations Under C, P and T

where we have also exploited Eqs. (1.14).[56] It becomes evident the passage to the antiparticle state induced by the charge conjugation operator.

Even if we have used the Dirac equation in the one particle formalism, previous results can be extended to field theory;[57] summarising, we have

$$\psi^C = \eta_C \, C \bar{\psi}^T \qquad \overline{\psi^C} = -\eta_C^* \, \psi^T C^{-1} \qquad (2.131)$$

where $\eta_C$ is a phase factor ($|\eta_C|^2 = 1$).

In order to investigate the invariance of a Lagrangian under discrete symmetries, it is useful to observe that a Lagrangian can be written in terms of bilinear forms, and consider their parity, charge conjugation and time reversal transformations.[58] For two different fields $\psi_a$ and $\psi_b$ and for a generic combination of $\gamma$ matrices, $\Gamma$, let us make the following identifications[59]

$$\begin{aligned}
&\bar{\psi}_a^P(t,\mathbf{x}) \, \Gamma \, \psi_b^P(t,\mathbf{x}) \equiv \bar{\psi}_a(t,-\mathbf{x}) \, \Gamma^P \, \psi_b(t,-\mathbf{x}) \\
&\bar{\psi}_a^C(t,\mathbf{x}) \, \Gamma \, \psi_b^C(t,\mathbf{x}) \equiv \bar{\psi}_b(t,\mathbf{x}) \, \Gamma^C \, \psi_a(t,\mathbf{x}) \\
&\bar{\psi}_a^{\hat{T}}(t,\mathbf{x}) \, \Gamma \, \psi_b^{\hat{T}}(t,\mathbf{x}) \equiv \bar{\psi}_a(-t,\mathbf{x}) \, \Gamma^{\hat{T}} \, \psi_b(-t,\mathbf{x}) \\
&\bar{\psi}_a^{CP}(t,\mathbf{x}) \, \Gamma \, \psi_b^{CP}(t,\mathbf{x}) \equiv \bar{\psi}_b(t,-\mathbf{x}) \, \Gamma^{CP} \, \psi_a(t,-\mathbf{x}) \\
&\bar{\psi}_a^{CPT}(t,\mathbf{x}) \, \Gamma \, \psi_b^{CPT}(t,\mathbf{x}) \equiv \bar{\psi}_b(-t,-\mathbf{x}) \, \Gamma^{CPT} \, \psi_a(-t,-\mathbf{x}). \quad (2.132)
\end{aligned}$$

In this way, we transfer the effect of the transformations $S$ on the states to the matrices

$$\psi \xrightarrow{S} \psi^S \Rightarrow \Gamma \xrightarrow{S} \Gamma^S. \qquad (2.133)$$

The transformations of basic matrices are summarized in Table 2.2. In the Dirac representation they can be explicitly written as

$$\Gamma \xrightarrow{P} \gamma^0 \Gamma \gamma^0 \qquad \Gamma \xrightarrow{C} (\gamma^2 \gamma^0 \Gamma \gamma^2 \gamma^0)^T \qquad (T : \text{transpose})$$

$$\Gamma \xrightarrow{\hat{T}} -\gamma^1 \gamma^3 \Gamma^* \gamma^1 \gamma^3 \qquad \qquad \Gamma \xrightarrow{CP} -(\gamma^2 \Gamma \gamma^2)^T$$

$$\Gamma \xrightarrow{CP\hat{T}} \gamma_5 \gamma^0 \Gamma^\dagger \gamma^0 \gamma_5. \qquad (2.134)$$

---

[56] Let us observe that the index $s$ change, and the net results is $u(p,1(2))e^{-ip\cdot x} \xrightarrow{C} v(p,2(1))e^{ip\cdot x}$.
[57] An explicit demonstration is provided, example, by Ref. [92].
[58] We always assume that operators that are products of field operators are normally ordered.
[59] We use the symbol $T$ for time reversal, but when there is the possibility of confusion with the transpose symbol we switch to $\hat{T}$.

**Table 2.2** Transformation rules for bilinear forms under parity $P$, charge conjugation $C$, time reversal $T$ and their combinations

| $\Gamma$ | $\Gamma^P$ | $\Gamma^C$ | $\Gamma^T$ | $\Gamma^{CP}$ | $\Gamma^{CPT}$ |
|---|---|---|---|---|---|
| $1$ | $1$ | $1$ | $1$ | $1$ | $1$ |
| $\gamma_\mu$ | $\gamma^\mu$ | $-\gamma_\mu$ | $\gamma^\mu$ | $-\gamma^\mu$ | $-\gamma_\mu$ |
| $\gamma_5$ | $-\gamma_5$ | $\gamma_5$ | $\gamma_5$ | $-\gamma_5$ | $-\gamma_5$ |
| $\gamma_\mu \gamma_5$ | $-\gamma^\mu \gamma_5$ | $\gamma_\mu \gamma_5$ | $\gamma^\mu \gamma_5$ | $-\gamma^\mu \gamma_5$ | $-\gamma_\mu \gamma_5$ |
| $\sigma_{\mu\nu}$ | $\sigma^{\mu\nu}$ | $-\sigma_{\mu\nu}$ | $-\sigma^{\mu\nu}$ | $-\sigma^{\mu\nu}$ | $\sigma_{\mu\nu}$ |

Let us observe that $\bar{\psi}\psi$, $\bar{\psi}\gamma_\mu\psi$, $\bar{\psi}\gamma_5\psi$, $\bar{\psi}\gamma_\mu\gamma_5\psi$, $\bar{\psi}\sigma_{\mu\nu}\psi$ have the same parity, charge conjugation, time reversal transformation properties of scalar, vector, pseudoscalar, axial vector and tensor fields, respectively. The transformation properties of a Lagrangian written in terms of bosons (including the postulated graviton, that has a tensor structure and spin 2) is thus precisely the same as that for a Lagrangian involving fermion bilinears. For instance, for a charged $W$ boson vector

$$(CP)W^{+\mu}(t,\mathbf{r})(CP)^\dagger = -W_\mu^-(t,-\mathbf{r}) \tag{2.135}$$

apart from an arbitrary CP conjugation phase.

Let us consider the QED Lagrangian for a spinor field $\psi$ of charge $Q$ in units of $e$, the charge of the proton

$$\mathscr{L}_{QED} = -\frac{1}{4}F^{\mu\nu}F_{\mu\nu} + eA_\mu J^\mu_{em}. \tag{2.136}$$

$A^\mu$ is the vector field of the photon and the current $J^\mu_{em}$ is a vector bilinear form

$$J^\mu_{em} = Q\bar{\psi}\gamma^\mu\psi. \tag{2.137}$$

The Lagrangian is invariant for $C$, $P$, and $T$ transformations.[60] We can easily check that

$$P\mathscr{L}_{QED}(t,\mathbf{x})P^\dagger = \mathscr{L}_{QED}(t,-\mathbf{x}) \tag{2.138}$$

and so on. The same can be said for the classical QCD Lagrangian. Moreover, the vacuum of QED is invariant for $C$, $P$, and $T$ transformations, and the quantization conditions as well.

Let us make another example, by considering the Fermi Lagrangian of weak interactions

$$\mathscr{L}_F = -\frac{G_F}{\sqrt{2}}J^\mu J^\dagger_\mu \tag{2.139}$$

---

[60] As it is well known, also the partial derivative $\partial^\mu$ behaves as a vector, for instance $P\partial_\mu P^\dagger = \partial^\mu$, where $P$ is the parity operator.

where we consider for simplicity only one fermion generation ($\psi_1, \psi_2$) and define the vector current $V_\mu$ and the axial current $A_\mu$ as

$$J_\mu = \bar{\psi}_1 \gamma_\mu (1 - \gamma_5) \psi_2 \equiv V_\mu - A_\mu. \tag{2.140}$$

Parity violation comes from the fact that the behaviour of the vector and axial vector currents under a parity transformation are different. As you can see from Table 2.2, the vector current flips sign under parity whereas the axial vector doesn't. The interference between these two terms creates the parity violation. One can see this by writing schematically

$$\mathcal{L}_F \sim VV^\dagger - AV^\dagger - VA^\dagger + AA^\dagger \xrightarrow{P} VV^\dagger + AV^\dagger + VA^\dagger + AA^\dagger. \tag{2.141}$$

$\mathcal{L}_F$ violates maximally parity. "Maximal" parity violation means that the vector and axial vector currents occur with equal coupling constants. The same goes for charge conjugation $C$. From Table 2.2, one can also see that $\mathcal{L}_F$ is $CP$ and $CPT$ invariant.

Let us consider more in detail similar interference terms with generic complex coefficients. One can write

$$k V_\mu^+(t, \mathbf{x}) A^{\mu-}(t, \mathbf{x}) + h.c. = k V_\mu^+(t, \mathbf{x}) A^{\mu-}(t, \mathbf{x}) + k^* A^{\mu+}(t, \mathbf{x}) V_\mu^-(t, \mathbf{x}) \tag{2.142}$$

where $k$ is a constant, $V_\mu^{+(-)}$ and $A_\mu^{+(-)}$ are positive (negative) vector and axial vector fields. We immediately see that both terms on the right side behave as a pseudoscalar. Their sum transforms under $CP$ into

$$k V_\mu^-(t, -\mathbf{x}) A^{\mu+}(t, -\mathbf{x}) + k^* A^{\mu-}(t, -\mathbf{x}) V_\mu^+(t, -\mathbf{x}) \tag{2.143}$$

hence it is $CP$ invariant if the coupling constant $k$ is real. $CPT$ is always conserved, no matter what the coupling $k$ is.

## 2.14 Helicity and Chirality

In this section we define and distinguish two important physical quantities that, in a relativistic theory, are strictly associated with spin: helicity and chirality. In two dimensions, the spin operator is $\mathbf{S} = \boldsymbol{\sigma}/2$, where $\sigma_i$ ($i = 1, 2, 3$) is a Pauli matrix (see Eq. (2.57)). In four dimension, we can define the spin operator $\mathbf{S}$ acting on each fermion as

$$\mathbf{S} \equiv \frac{1}{2} \boldsymbol{\Sigma} \qquad \text{where} \qquad \boldsymbol{\Sigma} \equiv \begin{pmatrix} \sigma & 0 \\ 0 & \sigma \end{pmatrix} = \gamma_5 \gamma^0 \boldsymbol{\gamma}. \tag{2.144}$$

The operator **S** has all the properties of spin in quantum mechanics: it has the same commutation relationships of the bi-dimensional $1/2\,\sigma$ and both $S^2$ and $S_z$ are diagonal.

The helicity $h$ is defined as

$$h = \frac{1}{2}\Sigma \cdot \hat{p} \qquad (2.145)$$

where $\hat{p} = \mathbf{p}/|\mathbf{p}|$ is the unit vector pointing in the direction of the momentum. It represents the projection of the particle spin on the direction of motion. The helicity is a good quantum number for a free fermion, since it commutes with the Dirac Hamiltonian[61]

$$[h, H] = 0. \qquad (2.146)$$

This is evident in the Dirac-Pauli representation, where

$$\gamma^0 = \begin{pmatrix} I_2 & 0 \\ 0 & -I_2 \end{pmatrix} \qquad \gamma^i = \begin{pmatrix} 0 & \sigma^i \\ -\sigma^i & 0 \end{pmatrix} \qquad (i = 1, 2, 3). \qquad (2.147)$$

The matrix $I_2$ is the two-by-two unit matrix, and the Dirac Hamiltonian for a state of momentum **p** reads (see Eq. (1.15))

$$H = \begin{pmatrix} m & \sigma \cdot \hat{p} \\ \sigma \cdot \hat{p} & -m \end{pmatrix} \qquad (2.148)$$

where Eq. (2.146), a set of eigenfunctions diagonalising $H$ and $h$ simultaneously exists. The standard solutions of the Dirac equation are Dirac spinors $\psi$ with four components, that are also eigenvectors of the helicity operator. The helicity eigenstates are the eigenstates of a freely propagating fermion.

By using the same Dirac-Pauli representation and relations (1.8) one can show that the spin does not commute with the Hamiltonian and precisely $[H, \mathbf{S}] = i\alpha \times \mathbf{p}$, where $\alpha$ is defined in Eq. (1.6). The spin is not conserved thus it is not a good quantum number. By exploiting also the commutation relations between position and momenta in quantum mechanics, we can show that the orbital angular momentum $\mathbf{L} = \mathbf{r} \times \mathbf{p}$ is not conserved as well, since $[H, \mathbf{L}] = -i\alpha \times \mathbf{p}$. While **L** and **S** are not conserved, it is easy to see that their combination, the total angular momentum $\mathbf{J} = \mathbf{L} + \mathbf{S}$, is conserved.

Since it is a product of two vectors (not quadrivectors), the helicity is obviously not Lorentz invariant. Observers in different reference frames can measure different

---

[61] Helicity is not necessarily preserved by interactions, e.g. in QED, if a left-handed spinor has its direction reversed by an electric field, its helicity changes.

## 2.14 Helicity and Chirality

values for the helicity. In particular, when the particle is overtaken, $\mathbf{p} \to -\mathbf{p}$ and the helicity is reversed.

Particles with spin greater than 1/2 can also be eigenstates of helicity. For example, for a photon (spin 1), the projection of the rotation generator $J_3$ has eigenstates $\pm 1$ corresponding to the states of circularly polarised light in the z direction, which are helicity eigenstates. Massless states have two helicity states, regardless of the value of their spin. Also the polarisations of massless gravitons (with spin 2) can be taken to be helicity eigenstates.[62]

The chirality (or handedness[63]) operator is represented by the Dirac matrix $\gamma_5$. Since $\gamma_5^2 = I$, where $I$ is the identity matrix in four dimension, the eigenvalues of $\gamma_5$ can assume only the values $\pm 1$. Let us define the so-called left-handed and right-handed projection operators as

$$P_L \equiv \frac{1-\gamma_5}{2} \qquad P_R \equiv \frac{1+\gamma_5}{2} \psi. \qquad (2.149)$$

By definition these projection operators satisfy

$$P_L^2 = P_L \qquad P_R^2 = P_R$$
$$P_L P_R = P_R P_L = 0$$
$$P_L + P_R = 1 \qquad (2.150)$$

which implies that any four component spinor can be uniquely decomposed into a right-handed and a left-handed component

$$\psi_L \equiv P_L \psi \qquad \psi_R \equiv P_R \psi. \qquad (2.151)$$

It is immediate to check that they are eigenstates of $\gamma_5$ (chiral eigenstates)

$$\gamma_5 \psi_{R(L)} = \pm \psi_{R(L)}. \qquad (2.152)$$

The chirality of a spinor (in an eigenstate) is defined as the eigenvalue of $\gamma_5$. The eigenstate with chirality $-1(+1)$ is called left(right)-handed and vice versa.[64] The left handed and right handed states have each two degrees of freedom. The Dirac spinor has four degrees of freedom and can be written as

$$\psi = \psi_L + \psi_R. \qquad (2.153)$$

---

[62] Let us remind that it is impossible to have interacting theories with massless fields of spin greater than 2, see for instance Ref. [91]).

[63] The term chirality is derived from the Greek word for hand. In chemistry and in mathematics, roughly speaking, chiral objects are identified as the ones that are not identical to their mirror image, as for instance clockwise is different from counterclockwise. Sometimes, in physical literature, the term handedness is associated to helicity, rather than chirality.

[64] It is self-evident that, unlike helicity, the chirality is defined for spinors only. We generally use the same symbol for both spinor fields and spinors.

In the Weyl (or chiral) representation, the matrices $\gamma^i$ are the same than in Eq. (2.147), but $\gamma^0$ changes

$$\gamma^0 = \begin{pmatrix} 0 & I_2 \\ I_2 & 0 \end{pmatrix}. \tag{2.154}$$

Therefore $\gamma_5$ is also different, and diagonal

$$\gamma_5 = i\gamma^0\gamma^1\gamma^2\gamma^3 = \begin{pmatrix} -I_2 & 0 \\ 0 & I_2 \end{pmatrix}. \tag{2.155}$$

In this representation it is easy to see that the chiral eigenstates $\psi_{L(R)}$ have two degrees of freedom. We can write the Dirac field as[65]

$$\psi = \begin{pmatrix} \psi_L \\ \psi_R \end{pmatrix}. \tag{2.156}$$

$P_{L(R)}$ project on the chiral components

$$P_L \psi = \begin{pmatrix} \psi_L \\ 0 \end{pmatrix} \qquad P_R \psi = \begin{pmatrix} 0 \\ \psi_R \end{pmatrix}. \tag{2.157}$$

In Sect. 2.4, we have introduced two inequivalent spin 1/2 representation of the Lorentz group. The $(0, 1/2)$ representation acts on the right-handed spinors and the $(1/2, 0)$ on the left-handed spinors. The chiral eigenstates are called Weyl fields.

In contrast with helicity, the chirality is invariant by proper Lorentz transformations, but it is not a good quantum number for massive particles. In fact, going back to the Dirac-Pauli representation where

$$\gamma_5 = i\gamma^0\gamma^1\gamma^2\gamma^3 = \begin{pmatrix} 0 & I_2 \\ I_2 & 0 \end{pmatrix} \tag{2.158}$$

we can immediately check that

$$[\gamma_5, H] = 2m \begin{pmatrix} 0 & -I_2 \\ I_2 & 0 \end{pmatrix} \tag{2.159}$$

that is, the chirality operator does not commute with the Hamiltonian unless the mass is zero. This is the case independently of the representation. Chirality is conserved and corresponds to a good quantum number only if the mass is

---

[65] We might as well define the Dirac spinor taking different combinations of $\psi_{L(R)}$; the present choice defines the chiral representation.

## 2.14 Helicity and Chirality

zero. In this case, the chirality and the Dirac Hamiltonian have a common set of eigenfunctions.

When the particle mass is zero, the chirality operator is the same as the helicity operator, which is observable and gives it a physical meaning. In fact, in momentum space and in the massless limit the Dirac equation Eq. (1.11) becomes

$$\gamma^\mu p_\mu \psi = (\gamma^0 p_0 - \boldsymbol{\gamma} \cdot \mathbf{p})\psi = 0. \tag{2.160}$$

Hence

$$\gamma_5 p_0 \psi = \gamma_5 \gamma^0 \boldsymbol{\gamma} \cdot \mathbf{p}\, \psi = \boldsymbol{\Sigma} \cdot \mathbf{p}\, \psi$$
$$\gamma_5 \psi = \boldsymbol{\Sigma} \cdot \hat{p}\, \psi. \tag{2.161}$$

For massless particles chirality and helicity coincide; they are both Lorentz invariant, conserved and have a common set of eigenvectors with the Dirac Hamiltonian, the Weyl fields. A right-(left-)handed massless Weyl fermion has helicity $+(-)1/2$.

One can obtain the two component Weyl fields also by writing the two bi-dimensional representation of a Dirac field as in Eq. (2.156) and observing that the Dirac equation decouples in two independent equations in the massless limit. That is easy to demonstrate in the Weyl representation, where the Dirac equation in Eq. (1.11) becomes

$$\begin{pmatrix} -m & i\partial_0 + i\bar{\sigma} \cdot \bar{\nabla} \\ i\partial_0 - i\bar{\sigma} \cdot \bar{\nabla} & -m \end{pmatrix} \begin{pmatrix} \psi_L \\ \psi_R \end{pmatrix} = 0 \tag{2.162}$$

In the massless limit this turn into

$$\begin{pmatrix} 0 & i\partial_0 + i\bar{\sigma} \cdot \bar{\nabla} \\ i\partial_0 - i\bar{\sigma} \cdot \bar{\nabla} & 0 \end{pmatrix} \begin{pmatrix} \psi_L \\ \psi_R \end{pmatrix} = 0 \tag{2.163}$$

that is equivalent to the two independent equations for the Weyl fields

$$i(\partial_0 + \bar{\sigma} \cdot \bar{\nabla})\psi_R = 0 \qquad i(\partial_0 - \bar{\sigma} \cdot \bar{\nabla})\psi_L = 0. \tag{2.164}$$

A Weyl spinor can describe only free massless particles, because mass terms mix chiralities, as seen in Eq. (2.16). Thus, if a fermion is massive, the helicity eigenstates do not coincide with the chirality eigenstates.

A Weyl field, say $\psi_L$, annihilates a left-handed particle or create a right-handed antiparticle. Let us observe that the Dirac adjoint of the Weyl field $\psi_{L(R)}$ changes its chirality

$$\overline{\psi_{L(R)}} = \overline{\psi} P_{R(L)} \tag{2.165}$$

since $\gamma_5^\dagger = \gamma_5$ and $\{\gamma_5, \gamma^\mu\} = 0$ for $\mu = \{0, 1, 2, 3\}$.

Under parity transformation, left-handed Weyl spinors transform into right-handed ones and vice versa. In fact, by following Eqs. (2.79) and (2.95), where the parity transformation operator is $S(\Lambda) = \gamma^0 \eta_P$, with $\eta_P$ being an unobservable phase factor, and $x' = (x^0, -\mathbf{x})$, we have

$$\psi_L \xrightarrow{P} \psi_L^P(x') = S(\Lambda) \left( \frac{1 - \gamma_5}{2} \right) \psi(x) =$$

$$= \eta_P \gamma^0 \left( \frac{1 - \gamma_5}{2} \right) S(\Lambda)^{-1} S(\Lambda) \psi(x) =$$

$$= \left( \frac{1 + \gamma_5}{2} \right) S(\Lambda) \psi(x) = P_R \psi^P(x') \qquad (2.166)$$

and vice versa. We have also used the relations $\gamma^0 \gamma_5 \gamma^0 = -\gamma_5$ and $(\gamma^0)^2 = 1$. Massless spin 1/2 particles were first considered by Weyl in the 1930s, but were ignored because by parity transformation the chirality of the two component states changed; this objection turned to dust with the discovery in the 1950s that parity is violated in the weak interactions and the observation of only left-handed neutrinos and right-handed antineutrinos.

Also the charge conjugation switches the chirality of the Weyl spinor

$$\psi_{L(R)} \xrightarrow{C} \psi_{L(R)}^C = P_{R(L)} \psi^C. \qquad (2.167)$$

As seen in Eq. (2.122), antiparticle spinors are obtained from particle spinors $\psi$ through charge conjugation

$$\psi^C = C \bar{\psi}^T = C \gamma^0 \psi^*. \qquad (2.168)$$

Then we have

$$\psi_{L(R)}^C = C(\psi^\dagger P_{L(R)}^\dagger \gamma^0)^T = C(\psi^\dagger \gamma^0 P_{R(L)})^T = C(\bar{\psi} P_{R(L)})^T$$

$$= C P_{R(L)}^T C^{-1} C \bar{\psi}^T = P_{R(L)} \psi^C \qquad (2.169)$$

where we have used one of the properties of $C$ operator listed in Eqs. (2.124) and the fact that $\gamma_5$ is hermitian. Therefore, an interaction containing only left-handed fields violates charge conjugation maximally. In other terms

$$\psi_{L(R)}^C = (\psi_{L(R)})^C \neq (\psi^C)_{L(R)} \qquad (2.170)$$

and the operations of charge-conjugation and projection onto chirality components do not commute. Let us also observe that, according to Eq. (2.122), we have

$$\bar{\psi} = -(\psi^C)^T (C^{-1})^T \qquad (2.171)$$

We also have

$$\overline{(\psi_{L(R)})^C} = \overline{\psi^C} P_{L(R)}. \tag{2.172}$$

At the level of the classical Lagrangian, we can easily build theories in which the left and right-handed components of a fermion field couple differently to gauge bosons. A theory that distinguishes between states of different chirality is sometimes said chiral. An example is the Standard Model for electroweak interactions, described in detail in Chap. 4. Since helicity is the projection of the spin along the direction of motion, it can be measured directly, but its value depends on the frame from which it is viewed. In contrast, handedness is a relativistically invariant quantity, but it is not a constant of the motion for a free particle and cannot be measured directly. Nevertheless, handedness is the quantity that describes the properties of the weak interactions and of the particle states that interact through the weak force and have definite weak charges.

## 2.15 The CPT Theorem

Although any relativistic field theory must be invariant under the proper Lorentz group, it does not need to be invariant under $P$ or $T$. Also charge conjugation $C$ is not necessarily a symmetry of physical systems. The free Dirac Lagrangian is invariant under $C$, $P$ and $T$ separately. In QED and in QCD, $C$, $P$ and $T$ are also conserved, and the Lagrangian can be built by assuming Lorentz invariance. The weak interactions violate $P$ and $C$ separately; the Lagrangian contain terms that violate $P$ and $C$, but are invariant with respect to other transformations of the Lorentz group.

In usual physical quantum field theories, as the SM, an important theorem, known as the $CPT$ theorem, holds. The $CPT$ theorem states that the combined transformation of charge conjugation $C$, parity $P$ and time reversal $T$, taken in any order, can always be defined, as an antiunitary operator, in such a way that it represents an exact symmetry of any quantum field theory based on a reasonable Lagrangian.

Let's see what happens, roughly speaking. Any generic local Lorentz-invariant Lagrangian density is a Lorentz scalar, so all tensor indexes are contracted and no minus sign comes from the discrete transformation, as one can see from Table 2.2. $PT$ transformation, also called strong reflection, inverts all the space-time coordinates and space-time derivatives, while charge conjugation $C$ exchanges the fields with their Hermitian conjugates. Numerical coefficients are transformed into their complex conjugates (because of $T$). When the Lagrangian density transforms into its Hermitian conjugate, it does not change since it is Hermitian. The Hamiltonian density shares the same property of the Lagrangian one. In the Hamiltonian, which is the integral over space-time of the Hamiltonian density, the sign difference in the coordinates wipes away, leaving the Hamiltonian unchanged under $CPT$.

Several proofs of the $CPT$ theorem exist [93–97], based on slightly different hypotheses. In general, the assumptions are that the Lagrangian is Hermitian, local and Lorentz invariant, the existence of an appropriate (unique) vacuum state[66] and the usual field commutation and anti-commutation rules. The requirement of locality in field theory means that no action at a distance is allowed, and that the Lagrangian is composed by terms containing the fields evaluated at the same space-point or derivates at finite order. Non local products of fields as $\phi(x)\phi(y)$ are not allowed.[67]

There is a closed link between Lorentz invariance and the $CPT$ theorem and indeed $CPT$ violation in realistic field theories implies Lorentz violation [98]. In general, theories disobeying any of the assumptions that enter this theorem could violate $CPT$ invariance. For example, since the $CPT$ theorem holds within quantum field theory, it has been suggested it may be theoretically feasible to violate $CPT$ invariance when conventional quantum mechanics fails in the context of quantum gravity [99].

By applying the anti-unitary operator $CPT$, a single particle state is transformed into an antiparticle state with reversed spin

$$CPT|\psi> = \eta_{CPT}|\bar\psi> \quad (2.173)$$

where $\eta_{CPT}$ is a phase. It is worth to underline that, since the $CPT$ theorem holds even when charge-conjugation is violated, several statements which can be made about the relation between particles and antiparticles do not really depend on exact $C$-symmetry, but rather on the more fundamental $CPT$-symmetry.

The $CPT$ theorem has important consequences, one of which is the equality of masses of particles and antiparticles [100]. It can be grasped intuitively by looking at Eq. (2.173) and considering that, since $CPT$ is a symmetry, the corresponding operator commutes with the Hamiltonian of the theory. Another relevant consequence is the equality of total widths or lifetimes for particles and antiparticles [100]— which is also not startling, since decay widths can be interpreted as imaginary part of masses.

When the $CPT$ theorem holds, violation of one of the discrete symmetries implies violation of the complementary one. For instance, if $CP$ is violated, then $T$ must be violated too. We should remark, however, that the $CPT$ theorem applies to quantum field theories, not to specific observables, which may violate a symmetry without violating the complementary one. For instance, a particle with an electric dipole moment may represent a violation of $T$ (and of $P$), without representing a violation of $CP$. However, the Lagrangian responsible for the electric dipole moment of the electron must violate both $T$ and $CP$.

---

[66] A not appropriate vacuum may be, for instance, in string theory, the case of vector fields requiring a nonzero field value in the lowest-energy state, implying a preferred direction in the vacuum, and violating rotation invariance.

[67] The term $\phi(x)\phi(y)$ indeed contains an infinite number of derivatives, as can be verified by Tailor expanding $\phi(y)$ around $x$.

## 2.15 The CPT Theorem

The $CPT$ theorem has withstood numerous high-precision experimental tests, since its violation represent an excellent candidate signature for unconventional physics. $CPT$ remains to date the only combination of $C$, $P$ and $T$ that is observed as an exact symmetry in nature.

# Chapter 3
# Gauge Theories

The roots of gauge invariance go back to the nineteenth century when electromagnetism was discovered and the first electrodynamic theory was proposed [101]. In classical electromagnetism, the gauge transformations are particular transformations which relate scalar and vector potentials describing the same electric and magnetic fields. The gauge transformations are local, that is they depend on the particular point in space and time.

In quantum theories, gauge transformations are related to local phase transformations of the fields. This realisation started in 1926 with Vladimir Fock, who extended the freedom of choosing the electromagnetic potentials related by gauge symmetry in classical electrodynamics to quantum systems of charged particles interacting with electromagnetic fields [102]. It was understood for the first time that the form of the quantum equations remains unchanged by gauge transformations if the wave function is multiplied by a space and time dependent phase factor.[1] What was not anticipated was how that procedure would evolve into a general principle that defines what it is now called a quantum gauge field.[2]

As a matter of fact, each of the fundamental quantum interactions of nature, the electromagnetic, weak, and strong interactions, can be related to a different gauge transformation which is, at the very least, a symmetry of the corresponding Lagrangian. Equivalently, one can say that a major requirement to build the Lagrangian for a fundamental interaction is the invariance under an internal gauge symmetry.[3] The electromagnetic and weak interactions are invariant under the gauge

---

[1] See below Sect. 3.1.

[2] The name gauge invariance (or Eichinvarianz in German) was coined in 1929 by Hermann Weyl [103], although it was only with the advent of modern quantum field theory, and the construction of the electroweak theory and quantum chromodynamics, that its deep significance emerged. For an historical review of early times see e.g. [104].

[3] Gauge theories also occupy center stage in the analyses of condensed matter and solid state systems, in statistical and mathematical physics.

$SU(2) \times U(1)$ symmetry. The strong interactions are symmetric under the gauge group of color $SU(3)$. The description of the whole set of strong, electromagnetic, and weak interactions. in terms of gauge theories, is at the basis of the so-called "Standard Model" of particle interactions (SM).

Gauge transformations arise naturally if one tries to make local a global phase symmetry. This procedure is often referred to as "gauging a symmetry" or "localise". It means to make the transformation parameters vary spatio-temporally so that the system does not behave as a unit under the transformations. A global transformation seems to be somewhat in contrast with the spirit of special relativity, where locality plays a major role, for instance in distinguishing time-like and space-like intervals (where information can be respectively exchanged or not), and in the recourse to field theory. A local symmetry is more complicated than a global one because it demands the global invariance of the entire system under transformations which are carried out differently at different space-time points. Theories that are globally invariant do not necessarily remain invariant if the same symmetry becomes local. However, by introducing new force fields and endowing them with specific transformation rules under the local transformations, local invariance may be restored.

Local symmetries are ubiquitous in quantum field theories. The theory of gravitational interactions, general relativity, can be arrived at by generalising the global (space–time independent) coordinate transformation of special relativity to local ones [105]. Gauge invariance requirements entail the existence of a new field—the gravitational one—and constrain its form of interaction. It is in order to remark that, despite this 'gauge' property, no consistent quantum field theory of general relativity is known yet.

Quantum field theory is the result of combining quantum mechanics with special relativity, the latter bringing along Lorentz symmetry. In the case of gauge theories, the procedure for the quantization is complicated from the fact that not all degrees of freedom of gauge fields are physical, because fields that are related by gauge transformation are not considered different, but physically equivalent. In other terms, a gauge invariant theory is kind of redundant, in the sense that certain local degrees of freedom do not enter the dynamics. In order to quantize a gauge field theory, we have the choice to preserve manifest Lorentz invariance, adopting a so called covariant gauge, like the Lorenz gauge, or to give up formal covariance and work in a non covariant gauge, like the Coulomb or the axial gauges. In the former case, one introduces covariant supplementary conditions, called gauge fixing conditions, which reduce the number of degrees of freedom. In other terms, a gauge fixing procedure is used to render dynamical all degrees of freedom. In the latter case, one directly quantize the physical degrees of freedom.

When it was realised that the weak interactions could be unified with electromagnetism in a non-abelian $SU(2) \times U(1)$ gauge theory [106–108] and that the resulting theory was renormalizable [109, 110], the importance of gauge theories grew enormously. A non abelian gauge theory for Lie groups as $SU(N)$ is generally denoted as Yang-Mills theory.

Let us remark that the quantization of the non-abelian case presents additional complications with respect to the abelian case. In an exact Yang-Mills theory, the assumption that the same Feynman rules hold for tree and loop diagrams leads to a violation of unitarity in loop diagrams [111]. Unitarity can be restored by introducing, for each massless gauge boson loop, a loop of anticommuting scalar (spin zero) fields, also massless. These additional contributions cancel the contribution of the spurious polarisations introduced by the gauge. The introduction of anticommuting spin zero particles is in contradiction to the spin statistics theorem, which states that fermionic particles (odd-half-integer spin) are antisymmetric under exchange (the fields anticommute), whereas bosonic particles (integer spin) are symmetric under exchange (the fields commute). The spin–statistics theorem can be violated by allowing the state to have a negative norm. A state of this kind is not considered "real", but fictitious, and dubbed a ghost state, or simply a ghost. They only propagate in loops and only couple to the vector boson mediators. The need for ghosts arises independently of the previous arguments as a necessary step in the quantization of non-abelian theories in terms of the path integral (Faddeev–Popov ghosts) [112]. In the Feynman gauge, the Faddeev–Popov ghosts are in fact the scalar fields needed to restore the unitarity of the amplitudes.

We start in Sect. 3.1 by discussing electromagnetism, and pass to the general formalism of gauge theories in Sect. 3.2. In Sect. 3.3 we introduce the spontaneously symmetry breaking mechanism of gauge theories.

## 3.1 Electromagnetism

The local invariance connected to classical electromagnetism is the well-known gauge invariance of Maxwell's equations. In its quantum version, the electromagnetism becomes 'quantum electrodynamics' or QED.

In classical electromagnetism, the gauge transformation acts on the potential $A^\mu = (A^0, \mathbf{A})$,

$$A^\mu \to A'^\mu = A^\mu - \partial^\mu \alpha(x) \tag{3.1}$$

where $\alpha(x)$ is a generic function depending on the space-time variable $x$. This transformation leaves unchanged the field strength tensor

$$F^{\mu\nu} \equiv \partial^\mu A^\nu - \partial^\nu A^\mu. \tag{3.2}$$

As a consequence it leaves unchanged the electric and magnetic fields, and the Maxwell equations, which are written, in a manifestly Lorentz covariant form, as

$$\partial_\mu F^{\mu\nu} = J^\nu \tag{3.3}$$

where $J^\nu = (\rho, \mathbf{J})$ is the electromagnetic current, which does not depend on $A^\mu$.

In non-relativistic quantum mechanics, the Schrödinger equation for a spinless particle of charge $q$ moving in an electromagnetic field is

$$\left(\frac{1}{2m}(-i\nabla - q\mathbf{A})^2 + qA^0\right)\psi(x) = i\frac{\partial \psi(x)}{\partial t}. \tag{3.4}$$

It may be obtained from the classical Hamiltonian by the usual prescription $\mathbf{p} \to -i\nabla$ for Schrödinger's wave mechanics. If we carry on the gauge transformation (3.1) on the potential in Eq. (3.4), we obtain an equation that cannot be satisfied by the same wavefunction $\psi(x)$. Unlike the Maxwell equations, the Schrödinger equation is not gauge invariant. However, wave functions are not directly observable quantities as the electromagnetic fields, and it is possible to describe the same physics by allowing them to change as well. The required change is

$$\psi(x) \to \psi'(x) = e^{iq\alpha(x)}\psi(x). \tag{3.5}$$

This can be easily seen by introducing the so-called covariant derivative for electromagnetic interactions, namely

$$D^\mu = \partial^\mu + iqA^\mu \tag{3.6}$$

and observing that

$$D'^\mu \psi'(x) = e^{iq\alpha(x)} D^\mu \psi(x) \tag{3.7}$$

where $D'^\mu = \partial^\mu + iqA'^\mu = \partial^\mu + iqA^\mu - iq\partial^\mu \alpha(x)$. The Schrödinger equation (3.4) can be written as

$$\frac{1}{2m}(-i\mathbf{D})^2 \psi(x) = iD^0 \psi(x) \tag{3.8}$$

It is easy to show that it is is gauge covariant, that is it maintains the same form under a gauge transformation

$$\frac{1}{2m}(-i\mathbf{D}')^2 \psi'(x) = iD'^0 \psi'(x) \tag{3.9}$$

The wave functions $\psi$ and $\psi'$ describe the same physics, since the probability densities are not affected by a phase factor.

Gauge invariance implies that the transformation (3.1) cannot affect the motion of particles. This can lead to undervalue the physical meaning of the potential with respect to fields. One should remark that there are gauge-invariant properties of the potentials (apart from the fields) that specify the motion of particles. A significative example in quantum mechanics is the so-called Aharonov-Bohm effect [113]. It refers to the motion of quantum mechanical charged particles, which are affected in the form of a phase shift by the existence of electromagnetic potential in the region where they travel, even if in that region the magnetic field is zero. This effect can be observed by shifts in interference patterns.

We have started taking it as known the form of the Schrödinger equation and checking its gauge invariance under the combined transformations of the potential and the wavefunction. It is possible and intriguing to reverse the argument, and to start by demanding that a theory is invariant under local phase transformation Eq. (3.5). It follows that such a phase invariance is not possible for a free theory, but rather requires an interacting theory involving a (4-vector) field $A^\mu$ (the photon) whose interactions with the charged particle are precisely determined, and which undergoes the transformation (3.1). This corresponds to substitute the derivative with the covariant derivative (3.6) in the Schrödinger equation, and write directly Eq. (3.9)—or, equivalently, perform the substitution in the Lagrangian formalism. Requesting gauge invariance has therefore dictated the form of the Lagrangian— this is the essence of the so-called gauge principle. The gauge principle implies the presence of the vector field $A^\mu$, which is called gauge field and interacts in the same way with any particle of charge $q$. The gauge field is massless as a consequence of gauge invariance. Indeed, a mass term would be proportional to $A_\mu A^\mu$, which is excluded since it is not gauge invariant. The gauge principle has represented a guiding principle to devise the form of other interactions, and to explain the universality of their interactions.

## 3.2 Gauge Theories and the Covariant Derivative

In this section we generalise the rules for building a gauge theory that we have outlined in Sect. 3.1. Let us consider a generic field $\psi(x)$ whose dynamics is defined by a Lagrangian which is invariant under a $n$ dimensional Lie group (see Sect. 2.1).[4] Under such group, the field $\psi(x)$ transforms as

$$\psi(x) \rightarrow \psi'(x) = U(\theta^i)\psi(x) \qquad (i = 1, 2, ..., n) \tag{3.10}$$

where (see Eq. (2.7))

$$U(\theta^i) = e^{ig \sum_{i=1}^{n} \theta^i T^i} \simeq 1 + ig \sum_{i=1}^{n} \theta^i T^i. \tag{3.11}$$

The quantities $\theta^i$ are continuous numerical parameters, $g$ is the coupling constant and $T^i$ are matrices representing the infinitesimal generators (or generators, in short) of the group of transformations (3.10), in some (in general reducible)

---

[4] An example could be the Lie group $U(1)$ ($n = 1$) of quantum electrodynamics.

representation of the fields $\psi(x)$. The truncated expansion on the right holds only for $\theta^i$ infinitesimal. We restrict ourselves to the case of internal symmetries, so the $T^i$ are independent of the space-time coordinates, and the arguments of the fields $\psi(x)$ and $\psi'(x)$ are the same. We expect the representation matrices (and the generators as well) to satisfy the commutation relations

$$[T^i, T^j] = iC^{ijk}T^k \tag{3.12}$$

where $C^{ijk}$ are the structure constants, which are completely antisymmetric in their indices. We adopt the standard notation with sum over repeated indexes. We gauge the theory by making the parameters $\theta^i$ depend on the spacetime coordinates, that is $\theta^i \to \theta^i(x_\mu)$. If all generators in Eq. (3.12) commute, the gauge theory is said to be abelian (in this case all the structure constants vanish), otherwise it is a non-abelian gauge theory. The original Lagrangian which is invariant under a global transformation (3.10), it is generally no longer invariant under the gauge transformations $U(\theta^i(x_\mu))$, because of the derivative terms. We recover gauge invariance if we follow a procedure similar to the one used in Sect. 3.1, that is if we build a covariant derivative and replace each ordinary derivative by the covariant derivative. The covariant derivative, as the name says, transforms in a covariant way, that is at the same way than the fields on which it acts. As a consequence, by definition, the covariant derivative transforms as

$$D^\mu \psi(x) \to D'^\mu \psi'(x) = U(\theta^i(x_\mu))D^\mu \psi(x) \tag{3.13}$$

We introduce the vector gauge fields $A^i_\mu$ with $i = 1, 2, \ldots n$ (one for each group generator) and form the gauge covariant derivative through the minimal coupling

$$D_\mu = \partial_\mu - ig A^i_\mu T^i \equiv \partial_\mu - ig \mathbf{A}_\mu \cdot \mathbf{T} \tag{3.14}$$

The $A^i_\mu$ represent the gauge bosons associated to each generator of the group $T^i$. It is remarkable that only a constant $g$ is needed (coupling universality).

The transformation property (3.13) implies that

$$\left(\partial_\mu - ig A'^i_\mu T^i\right)[U(\theta^i(x_\mu))\psi] = U(\theta^i(x_\mu))\left(\partial_\mu - ig A^i_\mu T^i\right)\psi \tag{3.15}$$

that is

$$\mathbf{A}'_\mu \cdot \mathbf{T} = U(\theta^i)\,\mathbf{A}_\mu \cdot \mathbf{T}\,U^{-1}(\theta^i) - \frac{i}{g}(\partial_\mu U(\theta^i))\,U^{-1}(\theta^i) \tag{3.16}$$

## 3.2 Gauge Theories and the Covariant Derivative

which defines the transformation law for the gauge fields. For infinitesimal $\theta^i$, we can use the last term of Eq. (3.11) and the transformation property becomes

$$\mathbf{A}'_\mu \cdot \mathbf{T} = \mathbf{A}_\mu \cdot \mathbf{T} + ig\,\theta^j A^k_\mu [T^j, T^k] + T^i\,\partial_\mu \theta^i \qquad (3.17)$$

plus terms higher order in $\theta^i$. By using the relation (3.12) we have

$$A'^i_\mu = A^i_\mu + \partial_\mu \theta^i - g\,C^{ijk}\theta^j A^k_\mu \qquad (3.18)$$

for infinitesimal parameters. At this stage, the gauge fields $A^i_\mu$ appear as external fields that do not propagate. In order to construct a gauge invariant kinetic energy term for the gauge fields $A^i_\mu$, we define

$$[D_\mu, D_\nu] = -ig\left(\partial_\mu(\mathbf{A}_\nu \cdot \mathbf{T}) - \partial_\nu(\mathbf{A}_\mu \cdot \mathbf{T}) - g^2[\mathbf{A}_\mu \cdot \mathbf{T}, \mathbf{A}_\nu \cdot \mathbf{T}]\right)$$

$$\equiv -igT^i F^i_{\mu\nu} \equiv -ig\mathbf{T} \cdot \mathbf{F}_{\mu\nu} \qquad (3.19)$$

for infinitesimal parameters, where

$$F^i_{\mu\nu} \equiv \partial_\mu A^i_\nu - \partial_\nu A^i_\mu + gC^{ijk}A^j_\mu A^k_\nu. \qquad (3.20)$$

This relation defines $n$ second-rank tensors of the gauge fields $F^i_{\mu\nu}$ (strength tensors), which are antisymmetric in the Lorentz indices. From the fact that $D_\mu \psi$ has the same gauge transformation property as $\psi$, we see that

$$[D'_\mu, D'_\nu]\psi' = U(\theta^i)[D_\mu, D_\nu]\psi. \qquad (3.21)$$

Substituting the definition (3.19) of the tensors $F^i_{\mu\nu}$ on both sides, we find

$$\mathbf{T} \cdot \mathbf{F}'_{\mu\nu} U(\theta^i)\psi = U(\theta^i)\mathbf{T} \cdot \mathbf{F}_{\mu\nu}\psi \Rightarrow \mathbf{T} \cdot \mathbf{F}'_{\mu\nu} = U(\theta^i)\mathbf{T} \cdot \mathbf{F}_{\mu\nu} U(\theta^i)^{-1}. \qquad (3.22)$$

Unlike the abelian case, $F^i_{\mu\nu}$ transform nontrivially. The transformation properties are those of a tensor of the adjoint representation. For the infinitesimal transformation, this translates into

$$F'^i_{\mu\nu} = F^i_{\mu\nu} - g\,C^{ijk}\theta^j F^k_{\mu\nu} \qquad (3.23)$$

showing again that the tensor is not gauge invariant. The right side of relation (3.22) indicates that instead the product

$$\text{Tr}[(\mathbf{T} \cdot \mathbf{F}_{\mu\nu})(\mathbf{T} \cdot \mathbf{F}_{\mu\nu})] \qquad (3.24)$$

is gauge invariant. This term, indicated as the pure Yang-Mills term, include the kinetic terms for the gauge bosons, it is gauge invariant and four-dimensional and therefore it has to be added to the gauge invariant, four-dimensional, Lagrangian. The Yang-Mills term also includes factors that are trilinear and quadrilinear in $A^\mu$, such as

$$gC^{ijk}\partial_\mu A_\nu^i A^{\mu j} A^{\nu k} \quad g^2 C^{ijk} C^{ilm} A_\mu^j A_\nu^k A^{\mu l} A^{\nu m} \tag{3.25}$$

that correspond to self-couplings of non-abelian gauge fields. This is completely different than the abelian case, where the gauge bosons have no self-interactions. The properties of universality are enhanced, since any rescaling in the coupling to the field $\psi$ would affect the strength of the boson gauge self-interactions. On the other side, in analogy with the abelian case, a mass term, which would be proportional to

$$A_\mu A^\mu \tag{3.26}$$

is forbidden by the gauge invariance of the Lagrangian. There are no gauge invariant mass terms for the gauge bosons, who remain massless also in non-abelian gauge theories.

A relevant example of a non-abelian Lie group is the unitary and special group in $N$ dimensions, $SU(N)$. The transformation under $SU(N)$ of a field $\psi$ in the general $d$-dimensional representation of $SU(N)$ reflects the one in Eq. (3.11) and it is

$$\psi(x) \to \psi'(x) = e^{ig\,\theta^i(x)T^i}\psi(x)$$

where $T^i$ represent the non commuting generators of the group, acting on the $d$-dimensional representation, $g$ is the associated charge, and a sum over $i$ is assumed as usual. One can define the covariant derivative as in Eq. (3.14)

$$D_\mu = \partial_\mu - ig A_\mu^i T^i \tag{3.27}$$

which transforms, as seen above, as

$$D^\mu \psi(x) \to D'^\mu \psi'(x) = e^{ig\,\theta^i(x)T^i} D^\mu \psi(x).$$

The generators are normalized such that

$$\text{Tr}_d(T^I T^J) = T(d)\delta^{ij} \tag{3.28}$$

where $T(d)$ is the Dynkin index of the representation. $T(d)$ is 1/2 for the fundamental representation and N for the adjoint representation of $SU(N)$. By

## 3.3 Spontaneous Symmetry Breaking

assuming that the matter fields are in the fundamental representation of $SU(N)$, as they normally are in the SM, we have

$$\text{Tr}(T^i T^J) = \frac{1}{2}\delta^{ij}. \tag{3.29}$$

By using the previous relation we can transform the Yang-Mill term as

$$\text{Tr}[(\mathbf{T} \cdot \mathbf{F}_{\mu\nu})(\mathbf{T} \cdot \mathbf{F}^{\mu\nu})] = \text{Tr}(T^i T^j) F^i_{\mu\nu} F^{\mu\nu\, j} = \frac{1}{2} F^i_{\mu\nu} F^{\mu\nu\, i} = \frac{1}{2} \mathbf{F}_{\mu\nu} \cdot \mathbf{F}^{\mu\nu}. \tag{3.30}$$

The addition of the term $-1/4\, \mathbf{F}_{\mu\nu} \cdot \mathbf{F}^{\mu\nu}$ in the Lagrangian represents the gauge invariant way to add kinetic terms of the form $\partial_\mu A_\nu \partial^\mu A^\nu$ to the corresponding Hamiltonian.[5]

Let us now consider a specific case, the gauge group $SU(3)$. We can identify the generators of the fundamental representation of $SU(3)$ with $T^i = \frac{\lambda^i}{2}$, where $\lambda^i$'s represent the eight Gell-Mann matrices. For the $\bar{3}$ representation, that is the complex conjugate representation of the (3-dimensional) fundamental represntation, indicated by 3, the generators are $\bar{T}^i = -(T^i)^T$. Indeed, since the eight gauge bosons $A^a_\mu$ of $SU(3)$ are real and the generators are hermitian, the transformation law of $\psi^*$ reads

$$\psi \to e^{ig\theta^i(x) T^i} \psi \qquad \psi^* \to e^{-ig\theta^i(x)(T^i)^T} \psi*$$

so the generators of the $\bar{3}$ representations are $-(T^i)^T$. Now that we have specified the generators in some representations, we can write explicitly the covariant derivative in these cases

- triplet: $D_\mu = \partial_\mu - ig A^i_\mu T^i$;
- antitriplet: $D_\mu = \partial_\mu + ig A^i_\mu (T^i)^T$;
- singlet: $D_\mu = \partial_\mu$;

where $g$ denotes the (universal) coupling constants for $SU(3)$. The triplet and antitriplet refer to the 3-dimensional vector space of the fundamental representation and its complex conjugate. A singlet state is the one which is not affected by the group transformations.

## 3.3 Spontaneous Symmetry Breaking

In quantum field theories there are different ways to realise (and break) symmetries. The symmetry is fully realised when it leaves invariant (1) the Lagrangian (more generally, the action), and as a consequence the dynamics of the system (2) the

---

[5] The additional factor $-1/2$ is conventionally added in analogy with QED, since the Maxwell equation in the standard form (3.3) correspond to a Lagrangian where the kinetic term is $-1/4\, F_{\mu\nu} F^{\mu\nu}$.

ground state (generally called vacuum), that is the symmetry of the fundamental (equilibrium) state of the system (3) the quantization conditions, which affect the statistics of the system.

When condition (1) is not fulfilled, and the Lagrangian is not invariant under a particular symmetry, such symmetry is said explicitly broken. A different, and distinct, possibility is spontaneous symmetry breaking (SSB), which affects condition (2). A symmetry is spontaneously broken when the corresponding transformation is a symmetry of a Lagrangian but not of the ground state of the system. The attribute spontaneous follows from the fact that the symmetry is broken without any need to introduce a term explicitly breaking the symmetry in the Lagrangian.

In general, there is no reason why a symmetry of the Lagrangian (or equivalently of the Hamiltonian) of a quantum-mechanical system should also be a symmetry for the ground state of the system. An example is provided by the nuclear forces and the nucleus: the latter are rotationally invariant, but the former not necessarily so, as it happens when the spin is different from zero. This is a triviality for nuclei, but it has highly non-trivial consequences if we consider systems which, unlike nuclei, are of infinite spatial extent. The typical example is the Heisenberg ferromagnet, an infinite crystalline array of spin 1/2 magnetic dipoles $s_i$, with spin-spin interactions between nearest neighbours such that neighbouring dipoles tend to align. The phenomenological Hamiltonian can be written as

$$H = -J \sum_{i,j} \mathbf{s_i} \cdot \mathbf{s_j} \qquad J > 0 \tag{3.31}$$

where the summation is over the nearest neighbours. This Hamiltonian is rotationally symmetric: if we rotate all spins by the same angle, the Hamiltonian does not change. The ground state is the state in which all the dipoles are parallel to each other, and the energy is minimised. The original rotational invariance is lost in the ground state, where a single unique direction has been chosen along which all the spins have lined up. We can observe infinitely many ground states, since all the dipoles can be aligned in some arbitrary direction. However, in a sense, the system preserves the rotational symmetry because the direction of alignment is random, and all the degenerate ground states may be reached from a given one by rotation. Nevertheless, someone living inside the ferromagnet would have a hard time to detect the rotational invariance, since experiments would be affected and biased by the ground state, which is not invariant. The symmetry is not manifest and it is said to be spontaneously broken. Moreover, direct detection of which direction have the ground state dipoles, namely what is the specific ground state, would be prevented by the infinite spatial extent of the ferromagnet. Being of finite extent, one could only change the direction of a finite number of dipoles at a time, while, to go from one ground state of the ferromagnet to another, one should change the directions of an infinite number of dipoles. The ferromagnet is not the only system that exhibits spontaneous symmetry breaking, there are other systems as superfluids, superconductors, Bose-Einstein condensates, and so on.

The previous picture can be easily generalised to relativistic quantum mechanics, passing from the Hamiltonian of a ferromagnet to the Hamiltonian of a quantum

## 3.3 Spontaneous Symmetry Breaking

field theory, from rotational invariance to internal symmetries and from the ground state of the ferromagnet to the vacuum state. In a quantum field theory, let us consider a symmetry described by a Lie group, and denote a generic element of the group by $U = e^{i\theta^i T^i}$, where $\theta^i$ are the continuous parameters and $T^i$ the generators of the group. If the vacuum state $\phi_0$ respects the symmetry, it is invariant under any group transformation and we have

$$U\phi_0 = \phi_0 \quad \forall \theta^i \Rightarrow T^i \phi_0 = 0 \quad \forall i. \tag{3.32}$$

These relations are no more satisfied in the case of SSB, when the vacuum state is not invariant; in that case, at least one generator acting on the vacuum does not give zero

$$\exists j : T^j \phi_0 \neq 0. \tag{3.33}$$

In most cases, the breaking of a certain symmetry leaves invariant a different, lower symmetry. The vacuum which is no more invariant under the original symmetry group may still be invariant under one of its subgroups. The choice of the vacuum decides the way how the spontaneous symmetry is broken, selecting the subgroup where the vacuum is still invariant.

Let us restate the previous example of a rotational symmetry which is spontaneously broken in group theoretic terms. The rotational symmetry corresponds to the $SO(3)$ Lie group, whose generators are the three angular momentum operators $J_i$ with $i \in (x, y, x)$. In the fundamental representation, they can be expressed by $3 \times 3$ matrices and a suitable basis, which satisfies the commutation relations of (2.28), is

$$J_x = \begin{pmatrix} 0 & 0 & 0 \\ 0 & 0 & -1 \\ 0 & 1 & 0 \end{pmatrix} \quad J_y = \begin{pmatrix} 0 & 0 & 1 \\ 0 & 0 & 0 \\ -1 & 0 & 0 \end{pmatrix} \quad J_z = \begin{pmatrix} 0 & -1 & 0 \\ 1 & 0 & 0 \\ 0 & 0 & 0 \end{pmatrix}. \tag{3.34}$$

They act on a scalar multiplet $\phi$ of three components (a triplet). Since $J_z$ is the only matrix to have all zeros on the third line and column, a vacuum expectation value of the triplet $\phi_0$ chosen as

$$\phi = \begin{pmatrix} \phi_1 \\ \phi_2 \\ \phi_3 \end{pmatrix} \Longrightarrow \phi_0 = \begin{pmatrix} 0 \\ 0 \\ v \end{pmatrix} \quad (v \neq 0) \tag{3.35}$$

breaks spontaneously the generators $J_{x,y}$, but not $J_z$. Indeed, $J_x\phi_0 \neq 0$ and $J_y\phi_0 \neq 0$, but $J_z\phi_0 = 0$, since rotations around the $z$ axis still leave the vacuum state invariant. Thus the $SO(3)$ group is broken to the $SO(2)$ group of rotations around the $z$ axis.

In the following we will illustrate the SSB in the case of a discrete symmetry and continue by examining the consequences of SSB in continuous symmetries, global and local. The SSB mechanism acts in different ways on global and local continuous symmetries in quantum field theory. If a global symmetry is spontaneously broken, a massless boson, said Goldstone boson, appears for each generator of the spontaneously broken symmetry. If the broken symmetry is local, the SSB may occur without implying the existence of massless particles (Higgs mechanism). The Higgs mechanism is responsible for mass terms of gauge bosons and fundamental fermions in the SM Lagrangian.

### 3.3.1 Internal Discrete Symmetry

Let us consider a Lagrangian density for the neutral scalar field $\phi$ in a potential $V(\phi)$

$$\mathscr{L} = \frac{1}{2}\partial^\mu \phi \, \partial_\mu \phi + V(\phi). \tag{3.36}$$

The kinetic terms are invariant under the global, internal and discrete transformation

$$\phi \to -\phi. \tag{3.37}$$

If we choose an equally invariant potential, this transformation represents a symmetry of the system. In order to describe SSB, we pass to the Hamiltonian formalism of field theory, where the ground state is identified with the eigenvalue of minimum energy. The conjugate momentum is defined as

$$\pi(x) \equiv \frac{\partial \mathscr{L}}{\partial \dot{\phi}} = \dot{\phi}(x) \tag{3.38}$$

where $\dot{\phi}(x) = \partial^0 \phi(x)$. The Hamiltonian reads

$$H = \int d^3x \left( \pi(x)\dot{\phi}(x) - \mathscr{L} \right) = \int d^3x \left( \frac{1}{2}\pi^2 + \frac{1}{2}(\nabla\phi)^2 - V(\phi) \right). \tag{3.39}$$

A potential built out of terms $\phi^{2n}$ is invariant under the transformation system (3.37). We limit to the $n = 1, 2$ terms, in order to preserve renormalization, and write the potential energy $V_E(\phi)$ as

$$V_E(\phi) \equiv -V(\phi) = \pm \frac{\mu^2}{2}\phi^2 + \frac{\lambda}{4}\phi^4 \tag{3.40}$$

## 3.3 Spontaneous Symmetry Breaking

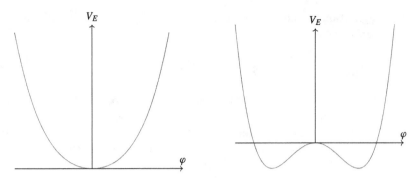

**Fig. 3.1** Two possible behaviours of the potential energy. Left: $V_E(\phi) = +\frac{\mu^2}{2}\phi^2 + \frac{\lambda}{4}\phi^4$. Right: $V_E(\phi) = -\frac{\mu^2}{2}\phi^2 + \frac{\lambda}{4}\phi^4$

where the arbitrary parameters $\mu^2$ and $\lambda$ are assumed to be positive.[6] The Hamiltonian density reads

$$\mathscr{H} = \frac{1}{2}\pi^2 + \frac{1}{2}(\nabla\phi)^2 \pm \frac{\mu^2}{2}\phi^2 + \frac{\lambda}{4}\phi^4. \qquad (3.41)$$

The classical configuration of minimum energy is a uniform field $\phi(x) = \phi_0$ which minimises the Hamiltonian. In quantum theory, this value is the classical limit ($\hbar \to 0$) of $\langle 0|\phi|0\rangle$, where $|0\rangle$ is the vacuum state. Thus, we can identify this classical constant value with a quantum vacuum expectation value (VEV) $\langle 0|\phi|0\rangle = \phi_0$, evaluated at tree-level.[7] The VEV of $\phi$ is generally indicated with $\langle \phi \rangle$. Since $\langle \phi \rangle = \phi_0$ is constant, the kinetic terms give zero, and maxima and minima of the Hamiltonian are found by imposing the following condition over the potential $V(\phi)$

$$\left.\frac{\partial V}{\partial \phi}\right|_{\phi=\langle\phi\rangle} = 0. \qquad (3.42)$$

According to the sign of the $\mu^2$ coefficient in Eq. (3.40), we have two different solutions. If the sign is positive, the potential energy include the usual mass term $+\mu^2\phi^2/2$ of the scalar meson and it has the form depicted in Fig. 3.1 (left). The only real solution of Eq. (3.42) is a constant field which is zero everywhere, that is $\langle\phi\rangle = \phi_0 = 0$. The ground state is unique and fully symmetric, and the symmetry is manifest.

---

[6] In the free field case the potential for a massive scalar field is given by the mass term $V(\phi) = -\mu^2\phi^2/2$. Both the factor 1/2 (also in the kinetic term) and the negative sign of the mass in the Lagrangian are required to return the Klein-Gordon equation of motion from the Euler-Lagrange equation.

[7] We do not discuss quantum corrections to the VEV.

If the sign of the $\mu^2$ coefficient in Eq. (3.40) is negative, the potential energy takes the form

$$V_E(\phi) = -\frac{\mu^2}{2}\phi^2 + \frac{\lambda}{4}\phi^4. \tag{3.43}$$

By considering the first terms as a mass term, that would be negative. A negative mass squared implies that the momentum is space-like. Space-like momenta can be used to communicate faster than the speed of light, and therefore negative mass-squared particles are called tachyons. The extremal condition becomes

$$\left.\frac{\partial V}{\partial \phi}\right|_{\phi=\langle\phi\rangle} = -\mu^2\phi + \lambda\phi^3\Big|_{\phi=\langle\phi\rangle} = 0 \tag{3.44}$$

which is satisfied not only by the solution $\langle\phi\rangle = \phi_0 = 0$, but by other two uniform solutions

$$\langle\phi\rangle = \pm\frac{\mu}{\sqrt{\lambda}} \equiv \pm v. \tag{3.45}$$

The second derivative of the potential

$$\frac{\partial^2 V}{\partial \phi^2} = -\mu^2 + 3\lambda\phi^2 \tag{3.46}$$

is negative when $\phi = 0$, which is therefore a local maximum, and no more the ground state. Instead, it is positive for $\phi = \pm v$, and both these values are minima of the potential. The form of the potential energy is depicted in Fig. 3.1 (right). Since the ground state is degenerate, which minimum we choose as the VEV is irrelevant to the physics of the system, because of the symmetry (3.37). However, this symmetry transforms each minimum into the other one—the symmetry is spontaneously broken. Choosing one or the other of the minima breaks the symmetry spontaneously and has noticeable consequences in the proximity of the minimum itself.

Given the definition of the real parameter $v$ in Eq. (3.45), the potential (3.43) can be written in a more convenient form, that differs from (3.43) only for an (irrelevant) constant

$$V_E(\phi) = \frac{\lambda}{4}(\phi^2 - v^2)^2. \tag{3.47}$$

Let us choose $\langle\phi\rangle = v$ as the vacuum. The physical fields, which are excitations above the vacuum, are then attained by performing perturbations about $\langle\phi\rangle = v$, not about $\langle\phi\rangle = 0$. To investigate physics near the asymmetric vacuum, let us define a new field

$$\phi' = \phi - v. \tag{3.48}$$

## 3.3 Spontaneous Symmetry Breaking

In terms of the new defined field, the potential (3.47) becomes

$$V_E(\phi) = \frac{\lambda}{4}(\phi'^2 + 2v\phi')^2 = \frac{\lambda}{4}\phi'^4 + \mu^2\phi'^2 + \sqrt{\lambda}\,\mu\phi'^3 \qquad (3.49)$$

which describes a scalar field of positive mass $\mu = \sqrt{\lambda}v$. The theory is now free of tachyons, at the price of the introduction of a cubic self-coupling in the shifted field $\phi'$. The $\phi'^3$ term is obviously not invariant under the symmetry (3.37). In the Lagrangian expressed in terms of the field $\phi'$, the symmetry (3.37) is explicitly broken. The net result of SSB is that, near the asymmetric vacuum, the global symmetry is no more manifest or easy to detect.[8]

### 3.3.2 Continuous Global Symmetry

Near the asymmetric vacuum, the SSB of a continuous global symmetry brings a new general phenomena. Let us consider for example the following Lagrangian density of two fields $\phi_1$ and $\phi_2$

$$\mathcal{L} = \frac{1}{2}\partial^\mu\phi_1\,\partial_\mu\phi_1 + \frac{1}{2}\partial^\mu\phi_2\,\partial_\mu\phi_2 - \frac{\lambda}{4}(\phi_1^2 + \phi_2^2 - v^2)^2. \qquad (3.50)$$

It is invariant under $SO(2)$, the group of rotations in two dimension, that is under the continuous transformation

$$\phi_1 \to \cos\theta\,\phi_1 + \sin\theta\,\phi_2$$
$$\phi_2 \to \cos\theta\,\phi_2 - \sin\theta\,\phi_1. \qquad (3.51)$$

The minima of the potential are all the fields that lie on a continuous line, the circle described by

$$\phi_1^2 + \phi_2^2 = v^2. \qquad (3.52)$$

They are all equivalent (degeneracy) and it is irrelevant which one we choose as vacuum. Whatever the choice, the rotational symmetry is spontaneously broken. Let us choose

$$\langle\phi_1\rangle = v \qquad \langle\phi_2\rangle = 0. \qquad (3.53)$$

We define new fields $\phi'_{1,2}$ by shifting the old ones

$$\phi'_1 = \phi_1 - v \qquad \phi'_2 = \phi_2. \qquad (3.54)$$

---

[8] To express this concept, a symmetry spontaneously broken is sometimes called hidden.

In terms of the new fields the Lagrangian becomes

$$\mathcal{L} = \frac{1}{2}\partial^\mu \phi'_1 \, \partial_\mu \phi'_1 + \frac{1}{2}\partial^\mu \phi'_2 \, \partial_\mu \phi'_2 - \frac{\lambda}{4}(\phi'^2_1 + \phi'^2_2 + 2v\,\phi'_1)^2. \qquad (3.55)$$

The term $v^2 \lambda\, \phi'^2_1$ represents a mass term for $\phi'_1$. It has the same mass of the field $\phi'$ in Sect. 3.3.1. Instead, there are no terms proportional to $\phi'^2_2$. After SSB, two spinless particles have been generated, and one of them is massless. This phenomena is general, and it is not a consequence of the particular form of the potential, but only of the spontaneous breaking of the continuous global symmetry group, in this case $SO(2)$.

The same result follows if we change variables in Eq. (3.50) and pass to different fields $\rho$ and $\theta$

$$\phi_1 = \rho \cos\theta \qquad \phi_2 = \rho \sin\theta. \qquad (3.56)$$

The Lagrangian becomes

$$\mathcal{L} = \frac{1}{2}\partial^\mu \rho\, \partial_\mu \rho + \frac{1}{2}\rho^2\, \partial^\mu \theta\, \partial_\mu \theta - \frac{\lambda}{4}(\rho^2 - v^2)^2. \qquad (3.57)$$

The symmetry now corresponds to the invariance under the change

$$\rho \to \rho \qquad \theta \to \theta + \alpha \qquad (3.58)$$

where $\alpha$ is a generic angle. It is immediately evident its rotational invariance, since the potential does not depend on $\theta$. The change of coordinate is ill defined at the origin, and this is reflected in the singular form of the kinetic terms. This is of no interest to us, since the origin is not a minimum and we are interested in the field behaviour around the VEV. By choosing for instance

$$\langle \rho \rangle = v \qquad \theta = 0 \qquad (3.59)$$

and introducing the shifted field as before

$$\rho' = \rho - v \qquad \theta' = \theta \qquad (3.60)$$

we have

$$\mathcal{L} = \frac{1}{2}\partial^\mu \rho'\, \partial_\mu \rho' + \frac{1}{2}(\rho' + v)^2 \partial^\mu \theta'\, \partial_\mu \theta' - \frac{\lambda}{4}(\rho'^2 + 2v\rho')^2. \qquad (3.61)$$

It is not surprising that the particle $\theta$ does not have a mass term, since the potential does not depend on $\theta$. From a geometric viewpoint, it is even more explicit. If the point we have chosen as vacuum is not invariant under $SO(2)$ transformations, then we can move from that point by $SO(2)$ rotations, and we obtain a curve. On this

## 3.3 Spontaneous Symmetry Breaking

curve, the (rotationally invariant) potential is constant. In terms of angular variables, this is the curve of constant $\rho$. By expanding the potential around the vacuum, no terms can appear involving the variable $\theta$ that measures displacement along this curve, where the potential is constant. Hence the corresponding field will always represent a massless particle.

Let us now generalise the previous reasoning to a Lie symmetry group, with $n$ generators $T_i$. An element of the group acts on a set of $m$ real fields $\phi_j$ according to

$$\phi_j \to e^{i\alpha_i T_i} \phi_j \tag{3.62}$$

where the $\alpha_i$ are $n$ arbitrary continuous real parameters, independent of space-time coordinates. Given the potential $V(\phi_j)$, the invariance of the Lagrangian implies

$$V(\phi_j) = V(e^{i\alpha_i T_i} \phi_j) \tag{3.63}$$

for every field $\phi_j$. The invariance of the vacuum is expressed by the condition (3.32). Some generators may form a subgroup which leaves the vacuum $\langle \phi_j \rangle$ invariant. Let us call $p$ their number—we can always choose them so they represent the first $p$ generators, that is

$$T_i \langle \phi_j \rangle = 0 \quad i \leq p. \tag{3.64}$$

By definition, the remaining $m - p$ generators do not leave the vacuum invariant, and break the symmetry spontaneously. By the same argumentations as before, we have, passing through the vacuum, a $(m - p)$-dimensional surface of constant potential. Therefore, the theory contains $m - p$ massless spinless bosons, one for each spontaneously broken infinitesimal symmetry. According to the parity-transformation properties of the spontaneously broken generator, the corresponding boson may be either scalar or pseudoscalar. It may even have no well-defined parity at all, if the original Lagrangian is not parity conserving or if the parity is itself spontaneously broken.

Summarising, the spontaneous breaking of a continuous global symmetry entails the existence of a massless particle. This is a general feature, which follows from the so-called Goldstone's theorem. The Goldstone's theorem states that for every spontaneously broken continuous global symmetry, the theory must contain a massless spinless particle, said Goldstone (or Nambu-Goldstone) boson [114, 115]. The Goldstone's theorem holds in classical and in quantum field theory, and Goldstone bosons cannot acquire mass from any order of quantum corrections. SSB was introduced in relativistic quantum field theory in analogy to the BCS (Bardeen, Cooper, Schrieffer) theory of superconductivity [116–118]. The problem studied was the spontaneous breaking of chiral symmetry induced by a fermion condensate and the pions were interpreted as Goldstone bosons.

**Fig. 3.2** The behaviour of the potential energy
$V_E(\phi) = -\mu^2\phi^*\phi + \frac{\lambda}{4}(\phi^*\phi)^2$

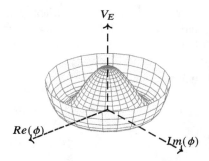

### 3.3.3 Linear Sigma Model

Another relativistic theory where the Goldstone's theorem finds application describes a complex field with Lagrangian

$$\mathscr{L} = \partial_\mu \phi^* \partial^\mu \phi + \mu^2 \phi^* \phi - \frac{\lambda}{4}(\phi^*\phi)^2 \tag{3.65}$$

where both $\mu^2$ and $\lambda$ are positive. This Lagrangian is comparable with the Lagrangian of two real fields in Eq. (3.50). The potential energy is the same given in (3.43), but for a complex field, and it is sometimes called Mexican hat potential, from its form (see Fig. 3.2).

Given a constant $\alpha$, $\mathscr{L}$ is invariant under the global $U(1)$ symmetry

$$\phi \to e^{i\alpha}\phi \tag{3.66}$$

which is spontaneous broken.[9] An infinite set of degenerate minima, that we indicate generically with $\phi_0$, occurs, which satisfy the relation

$$|\phi_0|^2 = \phi_0^*\phi_0 = 2\mu^2/\lambda. \tag{3.67}$$

This condition fixes only the norm of the minimum $\phi_0$, not its phase. All the minima are equivalent (by symmetry), and they can be written in a general form as

$$\phi_0 = \sqrt{\frac{2\mu^2}{\lambda}} e^{i\theta} \tag{3.68}$$

---

[9] Let us observe that since $e^{i\alpha} = \cos\alpha + i\sin\alpha$, one can define an isomorphism among elements of $U(1)$ and 2-by-2 rotation matrices, i.e. elements of $SO(2)$. This way we resume the SSB described in Sect. 3.3.2.

## 3.3 Spontaneous Symmetry Breaking

for any constant $\theta$. The choice of a VEV among these equivalent minima spontaneously breaks the symmetry. It is conventional to choose a VEV which is real, setting

$$\langle\phi\rangle \equiv v \equiv \mu \frac{\sqrt{2}\mu}{\sqrt{\lambda}}. \tag{3.69}$$

Since the vacuum is real, the usual shift of the field around it, that is $\phi' = \phi - v$, where $\phi'$ is a complex field (see Eq. (3.48)), corresponds to a shift in the real part only. Thus, it is often convenient to parameterise the field around the minimum $v$ as

$$\phi(x) = \left(v + \frac{1}{\sqrt{2}}\sigma(x)\right) e^{\frac{i\pi(x)}{F_\pi}} \tag{3.70}$$

where we have introduced two real fields $\sigma(x)$ and $\pi(x)$, as well as a real number $F_\pi$. In this way, the potential depends only on $\sigma(x)$ and not on $\pi(x)$. Using Eq. (3.70), the Lagrangian (3.65) yields

$$\mathcal{L} = \frac{1}{2}(\partial_\mu \sigma)^2 + \frac{1}{F_\pi^2}\left(v + \frac{1}{\sqrt{2}}\sigma(x)\right)^2 (\partial^\mu \pi)^2$$
$$+ \frac{\mu^2}{\lambda} - \mu^2 \sigma^2 - \frac{\mu\sqrt{\lambda}}{2}\sigma^3 - \frac{\lambda}{16}\sigma^4. \tag{3.71}$$

Choosing $F_\pi \equiv \sqrt{2}v$ implies $v^2/F_\pi^2 (\partial^\mu \pi)^2 \to 1/2(\partial^\mu \pi)^2$, that is the canonically normalized form for the $\pi$ kinetic term. The Lagrangian (3.71) describes a massless particle $\pi$ and a massive particle $\sigma$ with mass $\sqrt{2}\mu$. The $\pi$ field is the massless Goldstone boson predicted by the Goldstone's theorem.

By definition, a phase rotation of the field $\phi(x)$ amounts to a shift in the field $\pi(x)$. A global shift in the $\pi$ field

$$\pi(x) \to \pi(x) + F_\pi \alpha \tag{3.72}$$

with the $\sigma$ field invariant, is a symmetry of the spontaneously broken Lagrangian (3.71). This symmetry maintains the $\pi$ field in a massless mode, forbidding the presence of a mass term for $\pi(x)$. From a geometric viewpoint, the massless mode, the $\pi$ field, corresponds to excitations along the symmetry direction, where the potential is flat. Instead, the mode with positive mass-squared, the $\sigma$ field, corresponds to excitations along the radial direction, where we feel an approximately quadratic rise of the potential energy.

The Lagrangian (3.71) describes the so-called linear sigma model. The massless $\pi$ shares the same symbol with the pion because this model has been used to describe pions, interpreted as Goldstone bosons.

## 3.3.4 The Higgs Mechanism

In the case of a local (non global) continuous symmetry, the presence of SSB does *not* imply the existence of physical massless states. We will illustrate this fact by considering the complex scalar field $\phi$ and the vector field $A_\mu$, and gauging the global $U(1)$ gauge transformations (3.66) and (3.1)

$$\phi(x) \to e^{ie\alpha(x)}\phi(x) \qquad A_\mu \to A_\mu - \partial_\mu \alpha(x) \tag{3.73}$$

where $e$ is the adimensional $U(1)$ constant charge. By substituting the partial derivative with the $U(1)$ covariant one $D_\mu$

$$D_\mu = \partial_\mu + ieA_\mu \tag{3.74}$$

we obtain a gauge invariant Lagrangian for the free fields

$$\mathscr{L} = \frac{1}{2}(D^\mu \phi)^* D_\mu \phi - \frac{1}{4}F_{\mu\nu}F^{\mu\nu} \tag{3.75}$$

where

$$F_{\mu\nu} = \partial_\mu A_\nu - \partial_\nu A_\mu. \tag{3.76}$$

In order to break spontaneously the symmetry, we add to this Lagrangian the same potential of the linear sigma model in Eq. (3.65), which happens to respect not only the global $U(1)$ symmetry, but the $U(1)$ gauge symmetry as well. It reads

$$V(\phi) = +\mu^2 \phi^* \phi - \frac{\lambda}{2}(\phi^* \phi)^2 \tag{3.77}$$

with $\mu^2, \lambda > 0$. The full Lagrangian describes a complex, scalar field $\phi$ with a negative mass term in the potential energy and a massless vector field $A_\mu$. We note at once that in this system there are four degrees of freedom, two from the complex $\phi$ and two from the massless $A_\mu$ (the two degrees of freedom correspond to the two independent transverse modes).

The potential $V$ has degenerate minima in the scalar field $\phi$ and the symmetry can be spontaneously broken by selecting a particular vacuum. In the context of a local symmetry, the scalar field $\phi$, with the wrong mass term in the Lagrangian and a non-vanishing vacuum expectation value, is called a Higgs (or BEH, Brout-Englert-Higgs) field (*before* the SSB). Let us now break spontaneously the symmetry by choosing the following VEV

$$\langle \phi \rangle = \frac{\mu}{\sqrt{\lambda}} \equiv \frac{v}{\sqrt{2}}. \tag{3.78}$$

## 3.3 Spontaneous Symmetry Breaking

By definition, $v$ has the dimension of a mass. We want to study the excitations around the chosen VEV, that we expect in correspondency to the physical fields in the observable spectrum after SSB. The complex field $\phi$ can be decomposed into two real fields $\phi_1$ and $\phi_2$

$$\phi = \frac{1}{\sqrt{2}}(\phi_1 - i\phi_2). \tag{3.79}$$

The shifted fields $\phi'_{1,2}$ can be defined as (see Sect. 3.3.2)

$$\phi_1 = v + \phi'_1 \qquad \phi_2 = \phi'_2 \tag{3.80}$$

or, alternatively, separating out the phase and the modulus of $\phi(x)$ via

$$\phi(x) = \frac{1}{\sqrt{2}}(v + \rho(x))e^{-i\theta(x)/v} \tag{3.81}$$

as in the linear sigma model described in Sect. 3.3.3. We choose the latter description, where the Goldstone modes are precisely the quanta associated with the phase field $\theta(x)$.

The gauge transformation (3.73) modifies the $\phi(x)$ field as

$$\phi'(x) = \frac{1}{\sqrt{2}}(v + \rho'(x))e^{-i\theta'(x)/v} = \frac{1}{\sqrt{2}}e^{ie\alpha(x)}(v + \rho(x))e^{-i\theta(x)/v} \tag{3.82}$$

and the fields $\rho(x)$ and $\theta(x)$ as

$$\rho'(x) = \rho(x) \qquad \theta'(x) = \theta(x) - ev\alpha(x). \tag{3.83}$$

While the $\rho$ fields does not change, the Goldstone field $\theta$ is translated of a local quantity at all space-time points. It is possible to set $\theta'(x) = 0$ by choosing the particular gauge function:

$$\alpha(x) = \frac{1}{ev}\theta(x). \tag{3.84}$$

Thus the massless Goldstone excitation is not present in the physical particle spectrum. In other terms, we have used the gauge freedom to remove one degree of freedom from the complex field $\phi$. In the global $U(1)$ case, $\alpha$ was a constant and the Goldstone field could not be cancelled away at all space-time points.

In this gauge, called the unitary gauge, the vector and scalar fields become

$$A'_\mu(x) = A_\mu(x) - \frac{1}{ev}\partial_\mu\theta(x) \qquad \phi'(x) = \frac{1}{\sqrt{2}}(v + \rho(x)) \tag{3.85}$$

while the Lagrangian becomes

$$\begin{aligned}
\mathscr{L}_U &= \frac{1}{2}(\partial_\mu - ieA'_\mu)\phi'^*(\partial^\mu + ieA'^\mu)\phi' - \frac{1}{4}F'_{\mu\nu}F'^{\mu\nu} + \mu^2\phi'^*\phi' - \frac{\lambda}{2}(\phi'^*\phi')^2 \\
&= \frac{1}{2}(\partial_\mu\rho\partial^\mu\rho) + \frac{e^2}{2}A'_\mu A'^\mu(v+\rho)^2 - \frac{1}{4}F'_{\mu\nu}F'^{\mu\nu} + \left(\frac{\mu^2}{2}\rho^2 + \mu^2 v\rho\right) \\
&\quad - \frac{\lambda}{8}(\rho^4 + 6v^2\rho^2 + 4v^3\rho + 4v\rho^3) + \text{cost.} \\
&= \frac{1}{2}(\partial_\mu\rho\partial^\mu\rho) - \frac{1}{4}F'_{\mu\nu}F'^{\mu\nu} + \frac{e^2}{2}v^2 A'_\mu A'^\mu - \mu^2\rho^2 + \frac{e^2}{2}A'_\mu A'^\mu(\rho^2 + 2v\rho) + \\
&\quad - \frac{\lambda}{8}\rho^4 - \frac{\lambda}{2}v\rho^3 + \text{cost.} \quad (3.86)
\end{aligned}$$

In analogy with the Lagrangian after the SSB of a global symmetry in Eq. (3.71), $\mathscr{L}_U$ has lost the original $U(1)$ (gauge) symmetry. Another similarity stays in the fact that the real $\rho$ field has acquired the status of a massive physical particle, due to the presence of the massive term with the right sign $-\mu^2\rho^2$. This physical field is again called the Higgs (or BEH) boson (*after* SSB). Noteworthy differences with the case of global symmetry are that the Goldstone boson $\theta(x)$ has now vanished, and the vector field $A'_\mu$ has acquired a mass term $\frac{e^2}{2}v^2 A'_\mu A'^\mu$. The mass term has the right sign, since its spatial part is $-\frac{e^2}{2}v^2 A'^2_i$. Let us remark that by choosing the unitary gauge we have ensured a Lagrangian that returns immediately the physical spectrum, since it does not include terms of difficult interpretation, as e.g. terms of kinetic mixing between $\theta$ and $A_\mu$, or bilinear terms mixing $\theta$, $\rho$ and $A_\mu$.

Summarising, the SSB of the local symmetry does not give rise to an observable massless scalar particle; instead, the gauge field acquires a mass proportional to the value of the chosen VEV and, becoming massive, acquires a longitudinal component as well. The four degrees of freedom with which we started have been distributed into three degrees of freedom for the massive spin-1 $A_\mu$ and one for the massive neutral scalar field $\rho$. In a graphic way, one says that, in the unitary gauge, the gauge boson $A_\mu$ has eaten the Goldstone boson in order to acquire mass.

This is an example of a general mechanism occurring in case of SSB of gauge theories: Goldstone bosons disappear from the spectrum and gauge bosons become massive. It is named the Higgs mechanism.[10] One could wonder what there is to gain by SSB, when one could give up gauge symmetry from the beginning and just add mass terms for the gauge vector fields. The answer is that the system after SSB preserves a number of good properties of the unbroken theory, the ones connected to the gauge invariance of the Lagrangian, and, in particular, remains renormalizable, as proven in 1971 [123].

---

[10] It was first discussed by Peter Higgs in the $U(1)$ case [119, 120], by François Englert and Robert Brout for non-abelian symmetries [121], and by Gerald Guralnik et al. [122].

# Chapter 4
# The Standard Model

The elementary particles interact through electromagnetic, weak and strong interactions.[1] The theory of the electroweak (EW) interactions, also known as Glashow-Weinberg-Salam (GWS) theory, is based on the gauge symmetry $SU(2) \otimes U(1)$, spontaneously broken to the electromagnetic gauge group $U(1)_{em}$. The strong interactions satisfy exactly the color gauge symmetry $SU(3)_c$. As a whole, these fundamental interactions are described by a gauge field theory, called the Standard Model (SM). The SM Lagrangian is invariant under $SU(2) \otimes U(1) \otimes SU(3)_c$ gauge transformations, and the group representations correspond to the elementary particles. Being renormalizable, the SM is expected to describe, for any given quantum scale, all the appropriate fundamental physics.

Until now, the SM has supported calculations of physical quantities with unparalleled precision. Quarks and leptons are classified according irreducible representations of this gauge group. In this chapter, we discuss the structure and the main characteristics of the SM as far as the description of EW interactions is concerned. Since we are mostly interested in neutrinos, that are not affected by strong interactions, we will not discuss them.

The SM has been formulated after a great deal of experimental data and theoretical conjectures. The earlier studies analyzed processes mediated by charged currents, e.g. the $\beta$-decay $n \to pe\bar{\nu}_e$, discussed in Chap. 1. Fermi postulated that the charged current matrix elements were of vector type; more than twenty years after, it was established that they were both of vector and axial vector type. The recognition of the existence of processes mediated by neutral currents, in the 1970s, was a major step towards the building of the GWS theory of the EW interactions. Other ones were the experimental evidences of the universality of the coupling, signalling an underling gauge theory, and the finite range of the interactions. Lower ranges of the interaction correspond to higher bounds on the masses of particles mediating

---

[1] Elementary particles also feel the effect of gravitational interactions, but this effect, being proportional to the tiny masses of the particles, can be safely neglected.

the interaction. The phenomenon of spontaneous symmetry breaking provided a solution to the problem of how the weak vector bosons could be massive, while still being (like the photon) gauge fields.[2]

## 4.1 The Electroweak Gauge Symmetry

As already said, the SM Lagrangian is invariant under the gauge group $SU(3)_c \otimes SU(2) \otimes U(1)$ and particles can be classified according to its irreducible representations. Here we focus on the EW Lagrangian, which is invariant under the $SU(2) \otimes U(1)$ group.

The detailed analyses of the energy and angular distributions in weak flavour-changing processes, such as $\beta$ decays, has shown that only the left-handed (right-handed) fermion (antifermion) chiralities participate in those transitions. The different behaviour of left and right-handed components of a fermion field implies that they cannot be assigned to the same representation. In the SM, left-handed fermions, that is fermions with chirality $-1$, are assigned to the fundamental representation of $SU(2)$, the doublet. Thus the group $SU(2)$, also known as weak isospin, is generally indicated with $SU(2)_L$.[3] The $U(1)$ group is called (weak) hypercharge group (in analogy with the charge of the $U(1)$ electromagnetic group) and it is often indicated with $U(1)_Y$.

For the $SU(2)_L \otimes U(1)_Y$ symmetry, the covariant derivative, introduced in Sect. 3.2, reads

$$D_\mu = \partial_\mu - ig \sum_{i=1}^{3} W_\mu^i T_i - ig' B_\mu Y \tag{4.1}$$

where $g$ and $g'$ denote the $SU(2)_L$ and $U(1)_Y$ coupling constants, or charges, respectively. They are universal, that is they do not depend on the representation. The fields $W_\mu^i$ and $B_\mu$ are the carriers of the gauge interaction, four real vector boson fields of spin one. The operators $T_i$ are the three generators of the non-abelian $SU(2)_L$ group and $Y$ is the neutral generator of the abelian $U(1)_Y$ group. The $SU(2)_L$ generators satisfy the commutation relations

$$[T_i, T_j] = i\varepsilon_{ijk} T_k \tag{4.2}$$

---

[2] In this chapter we never discuss the effects of higher orders in the quantum mechanics perturbation theory, but let us just mention a problem that arises with the Higgs mechanism of SSB, the so-called SM fine-tuning (or hierarchy) problem (see for example Chapter 3 of [124]).

[3] The fact that the gauge symmetry $SU(2)$ is imposed only on left-handed states leads to a weak interaction that is completely left-right asymmetric.

## 4.1 The Electroweak Gauge Symmetry

where $\varepsilon_{ijk}$ is the Levi-Civita symbol and $i, j, k \in \{1, 2, 3\}$. In the gauge group $SU(2)_L \otimes U(1)_Y$ the groups $SU(2)_L$ and $U(1)_Y$ are factorised, hence $Y$ commutes with each $T_i$. The hypercharge $Y$ is a real multiple of the unit matrix for each irreducible representation of $SU(2)_L \otimes U(1)_Y$.

There is another $U(1)$ subgroup of $SU(2)_L \times U(1)_Y$, distinct from $U(1)_Y$, which can be identified with the electromagnetic group $U(1)_{em}$, and it is generated by a linear combination of $T_3$ and $Y$. By denoting with $Q$ the generator of the electromagnetic group $U(1)_{em}$, we assume the following relation among the charges[4]

$$Q = T_3 + Y. \tag{4.3}$$

The covariant derivative can be written in a different form, by replacing the two real fields with two complex fields $W_\mu^+$, $W_\mu^-$ [5]

$$W_\mu^\pm \equiv \frac{W_\mu^1 \mp i W_\mu^2}{\sqrt{2}} \tag{4.4}$$

We can also act on the neutral sector, by defining the fields $Z_\mu$ and $A_\mu$ through a bi-dimensional rotation of an angle $\theta_W$, called weak mixing (or Weinberg) angle

$$\begin{pmatrix} B_\mu \\ W_\mu^3 \end{pmatrix} = \begin{pmatrix} \cos\theta_W & -\sin\theta_W \\ \sin\theta_W & \cos\theta_W \end{pmatrix} \begin{pmatrix} A_\mu \\ Z_\mu \end{pmatrix}. \tag{4.5}$$

We aim at singling out the electromagnetic charge $Q$ and identifying the field $A_\mu$ with the electromagnetic field.[6] We introduce the electric charge unit $e$ through the relations

$$g = \frac{e}{\sin\theta_W} \qquad g' = \frac{e}{\cos\theta_W} : \tag{4.6}$$

We have traded the two bare gauge couplings $g$ and $g'$ with two different bare parameters, a bare charge $e$ and the weak mixing angle. None of these parameters is determined within the SM, that only predicts some relations among them. Their values have to be inferred phenomenologically. After the substitutions (4.4), (4.5)

---

[4] Notice that sometimes the covariant derivative is defined as $D_\mu = \partial_\mu - ig \sum_{i=1}^{3} W_\mu^i T^i - ig' B_\mu Y/2$, therefore the relation among the operators (and respective charges) becomes $Q = T_3 + Y/2$.

[5] There are several reason to prefer complex fields in field theory. A complex field leads naturally to an associated charge and current conservation (see e.g. Sect. 13.6 in Ref. [125]), and it represents a charged particle whose complex conjugates is a particle of opposite charge (see e.g. Sect. 6 in Ref. [126]).

[6] More details on this identification will be given in Sect. 4.4.

and (4.6), the covariant derivative (4.1) is expressed in terms of the new fields $W^\pm$, $Z_\mu$, $A_\mu$ and reads

$$D_\mu = \partial_\mu - i\frac{g}{\sqrt{2}}(W_\mu^+ T^+ + W_\mu^- T^-) - ieQA_\mu - i\frac{g}{\cos\theta_W}(T_3 - \sin^2\theta_W\, Q)Z_\mu \tag{4.7}$$

where $T^\pm = T_1 \pm iT_2$.

The doublet is the fundamental representation of $SU(2)_L$, and it accommodates left-handed fermion pairs of weak isospin 1/2. For simplicity, let us consider for now only one pair, the $u$ and $d$ quarks. Both their left-handed projections $u_L$ and $d_L$ belong to a weak doublet $q_L$, whose component $u_L$ has $T_3 = 1/2$ and $d_L$ has $T_3 = -1/2$. By using a matricial notation we can write

$$q_L = \begin{pmatrix} u_L \\ d_L \end{pmatrix} \qquad Y = \frac{1}{6}. \tag{4.8}$$

Since the electromagnetic charges of $u_L$ and $d_L$, in units of $e$, are 2/3 and $-1/3$, respectively, the value of the hypercharge, $Y = 1/6$, follows from the relation (4.3). In the SM the right-handed quarks are accommodated into SU(2) singlets

$$\begin{aligned} u_R &\qquad Y = Q = 2/3 \\ d_R &\qquad Y = Q = -1/3. \end{aligned} \tag{4.9}$$

Let us underline that the fermion fields have been decomposed into chirality eigenstates, as in Eq. (2.153), that is

$$u = u_L + u_R \qquad d = d_L + d_R \tag{4.10}$$

where

$$u_{L(R)} = P_{L(R)} u \qquad d_{L(R)} = P_{L(R)} d. \tag{4.11}$$

The electric charge value $Q$ is the same for both fermion chiralities. On the contrary, the hypercharge value, being the same for each multiplet (the doublet, and the two singlets), is forced to be different for both fermion chiralities. Thus the $B_\mu$ field cannot be identified with the electromagnetic field, since the photon has the same interaction with both the left and right-handed components of the fermion field.

The fundamental doublet $q_L$ transforms under the gauge $SU(2)_L \times U(1)_Y$ group as

$$q_L \to e^{i\sum_{i=1}^3 \alpha_i(x) T_i} e^{i\beta(x) Y} q_L \tag{4.12}$$

## 4.1 The Electroweak Gauge Symmetry

where $\alpha_i(x)$ and $\beta(x)$ are real, arbitrary parameters. The generators acting on the fundamental representation are

$$T_i = \frac{\tau_i}{2} \qquad (4.13)$$

where $\tau_i$ are the Pauli matrices (see Eq. (2.57)).[7] The $SU(2)_L$ transformations only affects the doublet field $q_L$, therefore the gauge $SU(2)_L \times U(1)_Y$ group acts on the right-handed fields as

$$u_R(d_R) \rightarrow e^{i\beta(x)Y} u_R(d_R). \qquad (4.14)$$

The pair $(u, d)$ is labelled as a quark family or generation. Two other quark families have been experimentally identified, the pair $(c, s)$ and the pair $(t, b)$.[8] For each of these families, that differ for the relative masses of the quarks, the previous reasoning holds at the same way. The additional quark doublets and singlets are

$$\begin{pmatrix} c_L \\ s_L \end{pmatrix} \quad \begin{pmatrix} t_L \\ b_L \end{pmatrix} \quad c_R \quad s_R \quad t_R \quad b_R. \qquad (4.15)$$

In analogy with Eq. (4.8), we can write the weak doublet of chirality $-1$ for the lepton family composed by the negatively charged electron $e$ and the neutral electron neutrino $\nu_e$

$$\ell_L = \begin{pmatrix} \nu_{eL} \\ e_L \end{pmatrix} \qquad Y = -\frac{1}{2}. \qquad (4.16)$$

The charged lepton of chirality $+1$ is a SU(2) singlet

$$e_R \qquad Y = Q = -1. \qquad (4.17)$$

In the SM the neutrinos are assumed to be massless and only left-handed—there are no right-handed neutrinos.[9] An hypothetical right-handed neutrino would have both electric charge and weak hypercharge equal to zero. Since it would not couple to the weak bosons, such a particle would not have any kind of SM interaction.

In analogy to the quark sector, the SM includes other lepton doublets (families or generations) of different masses, all affected in the same way by the gauge

---

[7] The Pauli matrices, generally indicated with $\sigma_i$, are often denoted by $\tau_i$ when used in connection with isospin symmetries.

[8] As it is well known, in the SM the quarks are named up, down, strange, charm, bottom and top $(u, d, s, c, b, t)$.

[9] Now we know that neutrinos have indeed a tiny mass; we will discuss neutrino mass and its consequences starting from Chap. 5.

transformations. The previous reasoning and quantum numbers apply to any lepton family. The additional lepton doublets and singlets are.[10]

$$\begin{pmatrix} \nu_{\mu L} \\ \mu_L \end{pmatrix} \quad \begin{pmatrix} \nu_{\tau L} \\ \tau_L \end{pmatrix} \quad \mu_R \quad \tau_R. \tag{4.18}$$

Charge conjugation reverses the sign of the eigenvalues of all generators of symmetry transformations. This is due to the operation of complex conjugation which is part of charge conjugation. The charge conjugation changes the chirality, as seen in Eqs. (2.167) and (2.172), but it does not change the representation of the SM gauge group. Therefore, for anti-quarks, the $SU(2)$ doublets will be right-handed with hypercharge $Y = -1/6$. The left-handed anti-quarks are singlets.

Let us consider the quark doublet $q_L$. The right-handed doublet of anti-fermions belong to the conjugate of the fundamental representation—under the gauge group $SU(2)_L$, it transforms as

$$q_R^C \rightarrow e^{-i \sum_{i=1}^{3} \alpha_i(x) \tau_i^*/2} q_R^C. \tag{4.19}$$

In SU(2), the fundamental representation 2 and its conjugate $\bar{2}$ are equivalent, in that we can find a unitary matrix $U_C$ such that

$$U_C \, e^{-i \sum_{i=1}^{3} \alpha_i(x) \tau_i^*/2} \, U_C^{-1} = e^{i \sum_{i=1}^{3} \alpha_i \tau_i /2}. \tag{4.20}$$

One matrix $U_C$ that satisfies the previous equality corresponds to a particular $SU(2)$ transformation, a rotation through $\pi$ about the 2-axis

$$U_C = e^{-i\pi \tau_2/2} = -i\tau_2. \tag{4.21}$$

This relation follows from the equality (see Eq. (2.59))

$$e^{-i\alpha \tau_i/2} = \cos\frac{\alpha}{2} - i\tau_i \sin\frac{\alpha}{2}. \tag{4.22}$$

By using the properties of the Pauli matrices in Eq. (1.9), one easily verifies Eq. (4.20). This implies that the anti-fermion doublet

$$U_C \begin{pmatrix} u_R^C \\ d_R^C \end{pmatrix} = -i\tau_2 \begin{pmatrix} u_R^C \\ d_R^C \end{pmatrix} = \begin{pmatrix} 0 & -1 \\ 1 & 0 \end{pmatrix} \begin{pmatrix} u_R^C \\ d_R^C \end{pmatrix} = \begin{pmatrix} -d_R^C \\ u_R^C \end{pmatrix}. \tag{4.23}$$

---

[10] Until now, six leptons have been observed, the three negative charged leptons, the electron, the muon, the tau ($e, \mu, \tau$), and the three neutral neutrinos, named the electron, muon and tau neutrinos ($\nu_e, \nu_\mu, \nu_\tau$).

transforms in exactly the same way as $q_L$ in Eq. (4.8). Thus, the charge-conjugated isospin doublet is

$$q_R^C \equiv \begin{pmatrix} -d_R^C \\ u_R^C \end{pmatrix} \qquad Y = -\frac{1}{6} \tag{4.24}$$

with $T_3 = 1/2$ and $T_3 = -1/2$ for the upper and lower components, respectively. It follows the same transformation rule, Eq. (4.12), under $SU(2)_L$, as the fundamental doublet. The most positively charged particle in the doublet has positive $T_3$, as in the fundamental doublet. The previous discussion can be repeated for the lepton pair $\ell_L$ in Eq. (4.16), and it gives the anti-fermion lepton doublet composed by the positron and the anti-neutrino, both right-handed, as

$$\ell_R^C \equiv \begin{pmatrix} -e_R^C \\ v_{eR}^C \end{pmatrix} \qquad Y = \frac{1}{2}. \tag{4.25}$$

## 4.2 The EW Lagrangian for Massless Fermions

By using the covariant derivative $D^\mu$ defined in Eq. (4.7), one can immediately write the $SU(2)_L \otimes U(1)_Y$ gauge invariant EW Lagrangian for massless fermions. By restricting for simplicity to the first family, the kinetic term for the quark doublet $q_L$ in Eq. (4.8) and the quark singlets $u_R$ and $d_R$ reads

$$\mathcal{L}_q = i\bar{q}_L \slashed{D} q_L + i\bar{u}_R \slashed{D} u_R + i\bar{d}_R \slashed{D} d_R. \tag{4.26}$$

The fermion fields are decomposed into chirality eigenstates $u_{L(R)} = P_{L(R)} u$, $d_{L(R)} = P_{L(R)} d$, as in Eq. (2.153), and $\bar{u}_{L(R)} \equiv \overline{u_{L(R)}} = \bar{u} P_{R(L)}$, $\bar{d}_{L(R)} \equiv \overline{d_{L(R)}} = \bar{d} P_{R(L)}$, $\bar{q}_L = (\bar{u}_L, \bar{d}_L)$.

Equation (4.26) can be rewritten as

$$\mathcal{L}_q = i\bar{q}_L \slashed{\partial} q_L + i\bar{u}_R \slashed{\partial} u_R + i\bar{d}_R \slashed{\partial} d_R +$$
$$+ g(W_\mu^+ J_W^{\mu+} + W_\mu^- J_W^{\mu-} + Z_\mu^0 J_Z^\mu) + eA_\mu J_{em}^\mu \tag{4.27}$$

where the electromagnetic and weak currents are

$$J_W^{\mu+} = \frac{1}{\sqrt{2}} \bar{u}_L \gamma^\mu d_L$$

$$J_W^{\mu-} = \frac{1}{\sqrt{2}} \bar{d}_L \gamma^\mu u_L$$

$$J_Z^\mu = \frac{1}{\cos\theta_W}\Big[\sum_{f=u_L,d_L} \bar{f}\gamma^\mu \left(T_3 - Q_f \sin^2\theta_W\right) f$$

$$+ \sum_{f=u_R,d_R} \bar{f}\gamma^\mu \left(-Q_f \sin^2\theta_W\right) f\Big]$$

$$= \frac{1}{\cos\theta_W}\Big[\overline{u_L}\gamma^\mu\left(\frac{1}{2} - \frac{2}{3}\sin^2\theta_W\right) u_L + \overline{d_L}\gamma^\mu\left(-\frac{1}{2} + \frac{1}{3}\sin^2\theta_W\right) d_L$$

$$+ \overline{u_R}\gamma^\mu\left(-\frac{2}{3}\sin^2\theta_W\right) u_R + \overline{d_R}\gamma^\mu\left(+\frac{1}{3}\sin^2\theta_W\right) d_R\Big]$$

$$= \frac{1}{\cos\theta_W}\Big[J_3^\mu - \sin^2\theta_W\, J_{em}^\mu\Big]$$

$$J_{em}^\mu = \sum_{f=u,d} \bar{f}\, Q\, \gamma^\mu f = \frac{2}{3}\bar{u}\gamma^\mu u - \frac{1}{3}\bar{d}\gamma^\mu d$$

$$J_3^\mu = \sum_{f=u_L,d_L} \bar{f}\, T_3\, \gamma^\mu f = \frac{1}{2}(\overline{u_L}\gamma^\mu u_L - \overline{d_L}\gamma^\mu d_L). \tag{4.28}$$

The EW Lagrangian analogous to Eq. (4.26) for leptons in the first family (see Eq. (4.16)) is

$$\mathcal{L}_\ell = i\overline{\ell_L}\,\slashed{D}\,\ell_L + i\overline{e_R}\,\slashed{D}\,e_R =$$

$$= i\overline{\ell_L}\,\slashed{\partial}\,\ell_L + i\overline{e_R}\,\slashed{\partial}\,e_R +$$

$$+ g\left(W_\mu^+ J_W^{\ell+\mu} + W_\mu^- J_W^{\ell-\mu} + Z_\mu^0 J_Z^{\ell\mu}\right) + eA_\mu J_{em}^{\ell\mu} \tag{4.29}$$

where

$$J_W^{\ell+\mu} = \frac{1}{\sqrt{2}}\overline{\nu_L}\gamma^\mu e_L$$

$$J_W^{\ell-\mu} = \frac{1}{\sqrt{2}}\overline{e_L}\gamma^\mu \nu_L$$

$$J_Z^{\ell\mu} = \frac{1}{\cos\theta_W}\Big[\sum_{f=e_L,\nu_L} \bar{f}\gamma^\mu\left(T_3 - Q_f \sin^2\theta_W\right) f$$

$$+ \sum_{f=e_R} \bar{f}\gamma^\mu\left(-Q_f \sin^2\theta_W\right) f\Big]$$

$$= \frac{1}{\cos\theta_W}\Big[\frac{1}{2}\overline{\nu_L}\gamma^\mu \nu_L + \left(\sin^2\theta_W - \frac{1}{2}\right)\overline{e_L}\gamma^\mu e_L + \sin^2\theta_W\, \overline{e_R}\gamma^\mu e_R\Big] =$$

$$= \frac{1}{\cos\theta_W}\Big[J_3^{\ell\mu} - \sin^2\theta_W\, J_{em}^{\ell\mu}\Big]$$

$$J_{em}^{\ell\mu} = -\bar{e}\gamma^\mu e = Q\bar{e}\gamma^\mu e$$

## 4.2 The EW Lagrangian for Massless Fermions

$$J_3^{\ell\mu} = \frac{1}{2}(\overline{\nu_L}\gamma^\mu \nu_L - \overline{e_L}\gamma^\mu e_L) = \sum_{f=\nu_L, e_L} \bar{f}\, T_3\, \gamma^\mu f. \tag{4.30}$$

It is immediate to observe that the weak interactions behave differently towards left-handed and right-handed states. For instance, the $W$ bosons couple only to left-handed fermion fields, according to the form of the charged fermion current

$$J_W^{\mu+} = \frac{1}{\sqrt{2}}\overline{f'_L}\gamma^\mu f_L = \frac{1}{\sqrt{2}}\overline{f'}\gamma^\mu P_L f = \frac{1}{2\sqrt{2}}\overline{f'}\gamma^\mu(1-\gamma_5)f \tag{4.31}$$

and of its Hermitian conjugate $J_W^{\mu-}$, for the $(f', f)$ quark or lepton family. Due to the presence of $\gamma^\mu$ and $\gamma^\mu \gamma_5$, this coupling is said $V - A$ (vector-axial vector). The coupling $V - A$, and the absence of compensating charged right-handed currents $V + A$, implies that the the weak interactions violate the parity symmetry and that the violation is maximal (see Sect. 2.14).

The vector nature of the coupling demands that the fermion handedness is conserved at each charged vertex, since

$$\overline{u_L}\gamma^\mu d_R = \overline{u} P_R \gamma^\mu P_R d = \overline{u}\gamma^\mu P_L P_R d = 0$$
$$\overline{u_R}\gamma^\mu d_L = \overline{u} P_L \gamma^\mu P_L d = \overline{u}\gamma^\mu P_R P_L d = 0. \tag{4.32}$$

However, we know from Sect. 2.14 that chirality is conserved and corresponds to a good quantum number only if the fermion mass is zero. Therefore, it is only when interactions involving massless fermions, or in the ultra-relativistic limit, where masses can be approximately considered zero, that we can state that only left-handed fermions and right-handed antifermions participate in charged current weak interactions.

The EW Lagrangian built by summing the terms $\mathscr{L}_q$ and $\mathscr{L}_\ell$ is not complete. Because of the presence of the gauge bosons mediating the gauge symmetries, an additional piece $\mathscr{L}_{kin}^G$, the gauge invariant kinetic terms for gauge bosons, must be added, obtaining

$$\mathscr{L}_{EW}^{m=0} = \mathscr{L}_q + \mathscr{L}_\ell + \mathscr{L}_{kin}^G. \tag{4.33}$$

The gauge invariant kinetic terms for the gauge bosons have been discussed in Sect. 3.2, and read

$$\mathscr{L}_{kin}^G = -\frac{1}{4} W_{\mu\nu}^i W^{\mu\nu i} - \frac{1}{4} B_{\mu\nu} B^{\mu\nu} \tag{4.34}$$

where the field strength tensors are

$$W^i_{\mu\nu} = \partial_\mu W^i_\nu - \partial_\nu W^i_\mu + g\varepsilon^{ijk} W^j_\mu W^k_\mu \qquad B_{\mu\nu} = \partial_\mu B_\nu - \partial_\nu B_\mu \qquad (4.35)$$

and $\varepsilon^{ijk}$ are the structure constant of the non-abelian $SU(2)$ gauge group (Levi-Civita symbol). The kinetic terms are necessary to give a dynamics to the gauge fields. With the substitutions (4.4), (4.5) and (4.6), we could write also $\mathscr{L}^G_{kin}$ in terms of the fields $A_\mu$, $Z_\mu$ and $W^\pm_\mu$.

So far we have built a Lagrangian that is not distinguishable by a classical gauge field Lagrangian. A particular care is required if we want to quantize a gauge field theory: we have to make sure we count separately fields which differ only by a gauge transformation, since those are considered physically equivalent. This procedure is particularly complex in the case of non-Abelian gauge theories as the SM. One way to proceed is to request that a certain gauge condition holds, that is to fix the gauge, another one is to make use of Faddeev–Popov (FP) ghosts (covariant quantization). In the latter case, the Lagrangian must include two additional terms, the so-called gauge-fixing term and the FP ghost term.

In the quantization procedure, a symmetry of the classical Lagrangian may not be explicitly preserved in the intermediate steps, and at the end of the quantization procedure it may happen that such a symmetry is not recovered. In that case, the symmetry is called anomalous, and the theory is said to have an anomaly.[11] If a symmetry is anomalous, the associated current and charges will not be conserved. Quantum corrections violate the classical conservation law, and the Nöther current $J^{Ni}_\mu$ for continuous symmetries (see Sect. 2.7) has no more zero divergence. It receives a contribution arising from quantum corrections, $\partial^\mu J^{Ni}_\mu = \mathscr{A}^i$, where the quantity $\mathscr{A}^i$ is called the anomaly. The current $J^i_\mu$ (or its divergence) is said to be anomalous or to contain an anomaly.

In a unitary gauge theory, like the SM, the anomalies of the gauge symmetries must cancel, since they can induce violations of the unitarity. A theory without gauge anomalies is said anomaly free.

One example in the SM is the so-called triangle anomaly, which may occur since fermions couple to both vector and axial vector currents. As seen in Sect. 4.1, the fermion content of the SM consists of six quarks and six leptons, organised into three families of quarks $(u, d)$, $(c, s)$, $(t, b)$, and leptons $(\nu_e, e)$, $(\nu_\mu, \mu)$, $(\nu_\tau, \tau)$. The number of quark and lepton families is not predicted by the SM and it has been experimentally determined—however, the cancellation of the triangle anomaly requires an equal number of families of leptons and quarks.

---

[11] We observe that the commonly used term "anomalous symmetry" may appear misleading, since in the quantum world the symmetry is simply never there.

## 4.2 The EW Lagrangian for Massless Fermions

### 4.2.1 Isospin and CVC Hypothesis

Let us consider the charged weak currents and the electromagnetic one listed in Eq. (4.28)

$$J_W^{\mu+} = \frac{1}{2\sqrt{2}} \bar{u}\gamma^\mu(1-\gamma_5)d$$

$$J_W^{\mu-} = \frac{1}{2\sqrt{2}} \bar{d}\gamma^\mu(1-\gamma_5)u$$

$$J_3^\mu = \frac{1}{4}(\bar{u}\gamma^\mu(1-\gamma_5)u - \bar{d}\gamma^\mu(1-\gamma_5)d)$$

$$J_{em}^\mu = = \frac{2}{3}\bar{u}\gamma^\mu u - \frac{1}{3}\bar{d}\gamma^\mu d. \qquad (4.36)$$

They can be re-expressed in terms of the isospin doublet $q$ in (2.55)

$$J_W^{\mu\pm} = \frac{1}{2\sqrt{2}} \bar{q}\gamma^\mu(1-\gamma_5)\tau_\pm q$$

$$J_3^\mu = \frac{1}{4}\bar{q}\gamma^\mu(1-\gamma_5)\tau_3 q$$

$$J_{em}^\mu = \frac{1}{2}\bar{q}\gamma^\mu \tau_3 q + \frac{1}{6}\bar{q}\gamma^\mu q \qquad (4.37)$$

where the $\tau_i = \sigma_i$ are the Pauli matrices in Eq. (2.57) and $\tau_\pm \equiv 1/2(\tau_1 \pm i\tau_2)$. The electromagnetic current contains an isovector portion $\bar{q}\gamma^\mu \tau_3 q$ and an isoscalar one $\bar{q}\gamma^\mu q$.

If we can single out the vector part of the weak currents

$$V_W^{\mu\pm} = \bar{q}\gamma^\mu \tau_\pm q$$

$$V_3^\mu = \bar{q}\gamma^\mu \tau_3 q \qquad (4.38)$$

we observe that the isovector portion of the electromagnetic current belongs to this triplet, related by isospin rotation. If isospin invariance holds, the weak currents are conserved as the electromagnetic one. Thus we have the conserved vector current (CVC) hypothesis, originally proposed at the end of the fifties, that states that the vector hadronic weak current, its Hermitian conjugate, and the isovector portion of the EM current form a single isospin triplet of conserved currents. CVC requires that the vector coupling constant remains unmodified by hadronic complications at zero momentum transfer.

## 4.3 Charged and Neutral Currents

In the general case of three families, the EW massless Lagrangian (4.33) becomes[12]

$$\mathcal{L}_{EW}^{m=0} = \mathcal{L}_{kin}^{G} + \mathcal{L}_{kin}^{f} + \mathcal{L}_{CC} + \mathcal{L}_{NC} + \mathcal{L}_{em} \tag{4.39}$$

where $\mathcal{L}_{kin}^{G}$ is the kinetic Lagrangian for the gauge fields in Eq. (4.34), $\mathcal{L}_{kin}^{f}$ is the kinetic Lagrangian for the fermions and the other terms describe fermion-gauge bosons interactions. We have

$$\mathcal{L}_{kin}^{f} = i \sum_{\substack{f=u,d,s,\\c,b,t,\\e,\mu,\tau}} \left(\bar{f}_L \not{\partial} f_L + \bar{f}_R \not{\partial} f_R\right) + \sum_{f=e,\mu,\tau} \bar{\nu}_{fL} \not{\partial} \nu_{fL}$$

$$\mathcal{L}_{CC} = g(W_\mu^+ J_W^{\mu+} + W_\mu^- J_W^{\mu-})$$

$$\mathcal{L}_{NC} = g Z_\mu^0 J_Z^\mu$$

$$\mathcal{L}_{em} = e A_\mu J_{em}^\mu. \tag{4.40}$$

We have distinguished among charged current interactions, mediated by the exchange of a charged $W$ boson, neutral currents, mediated by the neutral boson $Z^0$, and electromagnetic currents, mediated by photons.

Let us generalise to case of three families of quark and leptons the EW currents found in Sect. 4.2 for the one generation. By indicating the up-type quarks with $u_i = (u, c, t)$ and the down-type quark as $d_i = (d, s, b)$, the charged weak current, $J_W^{\mu+}$, can be written as

$$J_W^{\mu+} = \frac{1}{\sqrt{2}} \left( \sum_{i=1,2,3} \bar{u}_{iL} \gamma^\mu d_{iL} + \sum_{f=e,\mu,\tau} \bar{\nu}_{fL} \gamma^\mu f_L \right). \tag{4.41}$$

The two neutral EW currents differs in their Lorentz structure and in the respective couplings. The neutral electromagnetic current $J_{em}^\mu$ couples to the photon with the proton charge $e$, and the neutral current $J_Z^\mu$ couples to the neutral boson $Z^0$ with the electroweak charge $g$, related to the proton charge by the relation (4.6) $g = e/\sin\theta_W$. The $J_{em}^\mu$ is a vector current—it reads

$$J_{em}^\mu = \sum_{\substack{f=u,d,s,\\c,b,t,\\e,\mu,\tau}} Q_f \bar{f} \gamma^\mu f \tag{4.42}$$

---

[12] We omit possible gauge-fixing and FP terms, mentioned in Sect. 4.2.

## 4.3 Charged and Neutral Currents

where $Q_f$ is the charge of the fermion $f$ in units of positive $e$. Since neutrinos have no charge, photons cannot mediate their interactions. The current $J_Z^\mu$ has also a vector-axial component. The comparison with the electromagnetic case can be more easily appreciated by writing $J_Z^\mu$ in the form

$$J_Z^\mu = \sum_{\substack{f=u,d,s,\\c,b,t,\\e,\mu,\tau}} \left( g_L^f \bar{f}_L \gamma^\mu f_L + g_R^f \bar{f}_R \gamma^\mu f_R \right) + \sum_{f=e,\mu,\tau} g_L^\nu \bar{\nu}_{fL} \gamma^\mu \nu_{fL} =$$

$$= \sum_{\substack{f=u,d,s,\\c,b,t,\\e,\mu,\tau\\\nu_e,\nu_\mu,\nu_\tau}} \left( g_V^f \bar{f} \gamma^\mu f - g_A^f \bar{f} \gamma^\mu \gamma_5 f \right) \tag{4.43}$$

In the last passage the vector and axial couplings have been defined as

$$g_V^f = \frac{g_L^f + g_R^f}{2} \qquad g_A^f = \frac{g_L^f - g_R^f}{2}. \tag{4.44}$$

Let us observe that, for quarks and leptons

$$g_L^{f,\nu} = \frac{1}{\cos\theta_W} \left( T_3 - Q_f \sin^2\theta_W \right)$$

$$g_R^f = \frac{1}{\cos\theta_W} \left( -Q_f \sin^2\theta_W \right) \tag{4.45}$$

where $T_3$ is the third component of the weak isospin generator. The explicit couplings for each fermion field are reported in Table 4.1.

**Table 4.1** Values for the $g_L^f$ (left), $g_R^f$ (right) $g_V^f$ (vector), $g_A^f$ (axial) coupling constants for the fundamental fermion fields

| $f$ | $g_L^f$ | $g_R^f$ | $g_V^f$ | $g_A^f$ |
|---|---|---|---|---|
| $\nu_e, \nu_\mu, \nu_\tau$ | $+\frac{1}{2} \frac{1}{\cos\theta_W}$ | $0$ | $+\frac{1}{2} \frac{1}{\cos\theta_W}$ | $+\frac{1}{2} \frac{1}{\cos\theta_W}$ |
| $e, \mu, \tau$ | $\frac{1}{\cos\theta_W}\left(-\frac{1}{2} + \sin^2\theta_W\right)$ | $\frac{\sin^2\theta_W}{\cos\theta_W}$ | $\frac{1}{\cos\theta_W}\left(-\frac{1}{2} + 2\sin^2\theta_W\right)$ | $-\frac{1}{2}\frac{1}{\cos\theta_W}$ |
| $u, c, t$ | $\frac{1}{\cos\theta_W}\left(+\frac{1}{2} - \frac{2}{3}\sin^2\theta_W\right)$ | $-\frac{2}{3}\frac{\sin^2\theta_W}{\cos\theta_W}$ | $\frac{1}{\cos\theta_W}\left(+\frac{1}{2} - \frac{4}{3}\sin^2\theta_W\right)$ | $+\frac{1}{2}\frac{1}{\cos\theta_W}$ |
| $d, s, b$ | $\frac{1}{\cos\theta_W}\left(-\frac{1}{2} + \frac{1}{3}\sin^2\theta_W\right)$ | $+\frac{1}{3}\frac{\sin^2\theta_W}{\cos\theta_W}$ | $\frac{1}{\cos\theta_W}\left(-\frac{1}{2} + \frac{2}{3}\sin^2\theta_W\right)$ | $-\frac{1}{2}\frac{1}{\cos\theta_W}$ |

## 4.4 The Higgs Mechanism in the SM

Gauge invariance in the SM Lagrangian leads one to expect massless vector bosons, in contrast with the experimental evidence that the weak interactions are short ranged and therefore require massive mediators. Such impasse, mostly debated in the sixties, is surmounted by the Higgs mechanism, discussed in Sect. 3.3.4. The EW gauge group $SU(2)_L \times U(1)_Y$ has four generators. Not all the four gauge bosons must become massive, one must remain massless, and be identified with the photon. Hence we require that the EW gauge symmetry is spontaneously broken to $U(1)_{em}$, meaning that after the SSB the VEV is still symmetric under $U(1)_{em}$ gauge transformations.

To break the symmetry spontaneously, we add to the EW Lagrangian the gauge invariant term

$$\mathcal{L}_\phi = (D_\mu \phi)^\dagger D^\mu \phi + V(\phi) \tag{4.46}$$

where $V$ is the Higgs potential

$$V = +\mu^2 \phi^\dagger \phi - \frac{\lambda}{4}(\phi^\dagger \phi)^2 \tag{4.47}$$

and $\mu^2$ and $\lambda$ are positive real arbitrary coefficients. The Higgs (or Brout-Englert-Higgs) field is a complex doublet and belongs to the fundamental (spinor) representation of $SU(2)$

$$\phi = \begin{pmatrix} \phi^+ \\ \phi^0 \end{pmatrix} \qquad Y = \frac{1}{2}. \tag{4.48}$$

One member of the doublet must be neutral in order to have the possibility of a $U(1)_{em}$-invariant VEV.

The kinetic term in $\mathcal{L}_\phi$ connects the Higgs field to the EW gauge bosons through the covariant derivative (4.7)

$$D_\mu = \partial_\mu - i\frac{g}{\sqrt{2}}(W_\mu^+ T^+ + W_\mu^- T^-) - ieQA_\mu - i\frac{g}{\cos\theta_W}(T_3 - \sin^2\theta_W \, Q)Z_\mu. \tag{4.49}$$

The minimum of the potential energy is at the non-zero values

$$|\phi| = \sqrt{\frac{2}{\lambda}}\mu \equiv \frac{v}{\sqrt{2}}. \tag{4.50}$$

Let us choose a VEV of the form

$$\langle \phi \rangle = \frac{1}{\sqrt{2}} \begin{pmatrix} 0 \\ v \end{pmatrix} \tag{4.51}$$

## 4.4 The Higgs Mechanism in the SM

and check that it is invariant under $U(1)_{em}$. It is easier to look at infinitesimal $U(1)_{em}$ transformations

$$\delta \phi = i\epsilon(x) Q \phi. \tag{4.52}$$

By translating the relation among charges $Q = T_3 + Y$ given in (4.3)[13] to generators in the fundamental representation, it is clear that the transformation

$$\delta \phi = i\epsilon(x) \frac{1}{2}(\mathbb{1} + \tau_3) \phi \tag{4.53}$$

is a symmetry of the VEV since

$$\delta \langle \phi \rangle \propto (\mathbb{1} + \tau_3) \langle \phi \rangle = \begin{pmatrix} 0 \\ 0 \end{pmatrix}. \tag{4.54}$$

Oscillations about the VEV are conveniently parameterised by

$$\phi(x) = \frac{1}{\sqrt{2}} \begin{pmatrix} 0 \\ v + h(x) \end{pmatrix}. \tag{4.55}$$

We have set the phase to zero by an appropriate gauge transformation, moving into the unitary gauge, as done in Sect. 3.3.4. In other terms, we can always make a gauge transformation to move to the unitary gauge, where $\phi$ reduces to the field (4.55) with one physical degree of freedom. If we substitute the Higgs field in proximity of the VEV value into the Higgs Lagrangian $\mathcal{L}_\phi$, we obtain

$$D_\mu \phi = \frac{1}{\sqrt{2}} \left[ \begin{pmatrix} \partial_\mu & 0 \\ 0 & \partial_\mu \end{pmatrix} - i \frac{g}{\sqrt{2}} \begin{pmatrix} 0 & W_\mu^+ \\ W_\mu^- & 0 \end{pmatrix} \right.$$
$$\left. - i \frac{g}{\cos \theta_W} \begin{pmatrix} \frac{1}{2} Z_\mu & 0 \\ 0 & -\frac{1}{2} Z_\mu \end{pmatrix} \right] \begin{pmatrix} 0 \\ v + h \end{pmatrix} \tag{4.56}$$

where we have expressed the generators in the fundamental representation as Pauli matrix and remembered the zero charge of the Higgs doublet. Then we have

$$(D_\mu \phi)^\dagger D^\mu \phi = \frac{1}{2} \partial_\mu h \, \partial^\mu h + \frac{g^2}{4}(v+h)^2 \, W_\mu^+ W^{\mu -} + \frac{g^2}{8 \cos^2 \theta_W}(v+h)^2 Z_\mu Z^\mu. \tag{4.57}$$

---

[13] It is even more immediate when $Q = T_3 + Y/2$ is taken as the definition of the relation among the operators (and respective charges).

By the substitution of the field (4.55) in the potential (4.47), we have

$$V(\phi) = +\mu^2 \phi^\dagger \phi - \frac{\lambda}{4}(\phi^\dagger \phi)^2 = \frac{\mu^2 v^2}{4} - \mu^2 h^2 - \frac{\mu^2}{v} h^3 - \frac{\mu^2}{4v^2} h^4 \qquad (4.58)$$

where we have expressed $\lambda$ in terms of $\mu$ by using the relation (4.50). In the kinetic factor $(D_\mu \phi)^\dagger D^\mu \phi$ one can identify the mass terms for the gauge bosons $W^\pm$ and $Z^0$, and the interactions of the gauge bosons with the Higgs field $h(x)$. From the potential $V(\phi)$ one obtains a mass term for the Higgs field and its self-interactions. These identifications will be done explicitly in the next section.

Finally a few consideration of the vacuum. In classical electrodynamics the ground state (or vacuum), being identified with the lowest-energy configuration of a system, is the field-free configuration. The Maxwell vacuum is empty. In the case of SM interactions, the Higgs potential forces a non-vanishing VEV of the Higgs in the ground state. The vacuum for a system containing a Higgs-type field is not empty, but filled with a scalar condensate. The Higgs field is a scalar and does not select a preferred direction in spacetime: $CPT$ and Lorentz invariance still hold.

### 4.4.1 Massive Gauge Fields and Higgs Interactions

By adding to the Higgs Lagrangian $\mathscr{L}_\phi$ the gauge boson kinetic terms $\mathscr{L}_{kin}^G$ in Eq. (4.34), we obtain the EW Lagrangian without fermion fields $\mathscr{L}_{EW}^{f=0}$.

Let us substitute the field (4.55) into $\mathscr{L}_{EW}^{f=0}$, express the kinetic parts in terms of the gauge bosons fields $W_\mu^\pm$, $Z_\mu^0$ and $A_\mu$, and retain only terms which are second order in the fields (i.e. kinetic energies or mass terms). We find that $\mathscr{L}_{EW}^{f=0}$ can be decomposed into three parts

$$\mathscr{L}_{EW}^{f=0} = \mathscr{L}_\phi + \mathscr{L}_{kin}^G \simeq \mathscr{L}_H + \mathscr{L}_W + \mathscr{L}_{Z/A}. \qquad (4.59)$$

The first term

$$\mathscr{L}_H = \frac{1}{2}\partial_\mu h(x) \partial^\mu h(x) - \mu^2 h(x)^2 \qquad (4.60)$$

tells us that we have a scalar field, the Higgs boson $h(x)$, of mass

$$M_H = \sqrt{2}\mu. \qquad (4.61)$$

## 4.4 The Higgs Mechanism in the SM

The second term represents a kinetic and mass term for a complex massive vector field, that is[14]

$$\mathcal{L}_W = -\frac{1}{2}F^{W+}_{\mu\nu}F^{W-\,\mu\nu} + \frac{g^2v^2}{4}W^+_\mu W^{-\mu} \tag{4.62}$$

where

$$F^{W^\pm}_{\mu\nu} = \partial_\mu W^\pm_\nu - \partial_\nu W^\pm_\mu. \tag{4.63}$$

The field $W^+$ describes an incoming positive or an outgoing negative W boson, and its Hermitian conjugate describes an incoming negative or an outgoing positive W boson. The mass of the charged $W^\pm$ bosons is

$$M_W = \frac{gv}{2}. \tag{4.64}$$

In the remaining term we have

$$\mathcal{L}_{Z/A} = -\frac{1}{4}F_{\mu\nu}F^{\mu\nu} - \frac{1}{4}F^Z_{\mu\nu}F^{Z\mu\nu} + \frac{1}{8}v^2(g^2+g'^2)Z_\mu Z^\mu \tag{4.65}$$

where

$$F_{\mu\nu} = \partial_\mu A_\nu - \partial_\nu A_\mu \qquad F^Z_{\mu\nu} = \partial_\mu Z_\nu - \partial_\nu Z_\mu. \tag{4.66}$$

$\mathcal{L}_{Z/A}$ describes one massless field $A_\mu$, the photon, and one massive Z neutral gauge boson, of mass

$$M_Z = \frac{1}{2}\frac{gv}{\cos\theta_W} = \frac{1}{2}v\sqrt{g^2+g'^2} = \frac{M_W}{\cos\theta_W} \tag{4.67}$$

where we have used the relations (4.6). Now we see explicitly as the mixing of the fields $W^3_\mu$ and $B_\mu$ described in Eqs. (4.5) allows to identify the two physical neutral bosons, the photon and the $Z^0$.

From Eq. (4.67) we obtain the relation

$$\rho \equiv \frac{M_W^2}{M_Z^2\cos^2\theta_W} = 1 \tag{4.68}$$

---

[14] This term reproduces the equation of motion for a free complex vector massive field, in analogy to what happens for a free complex scalar massive field, whose Lagrangian is $\mathcal{L} = \partial_\mu\phi^\dagger\partial^\mu\phi - m^2\phi^\dagger\phi$.

**Fig. 4.1** Tree level Feynman diagram for muon decay into an electron

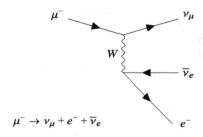

$\mu^- \to \nu_\mu + e^- + \bar{\nu}_e$

which represents an accurate constraint on $\cos\theta_W$ and a firm prediction of $M_Z > M_W$. In the absence of a mixing angle $\theta_W$, the masses of the three EW bosons would be the same, in contrast with experimental observations.

Summarising, after SSB the SM Lagrangian describes massive gauge bosons $W^\pm$ and $Z^0$, a massless photon and a massive neutral scalar field, the Higgs. The measured values of the massive gauge boson are approximately $M_Z \simeq 91$ GeV and $M_W \simeq 80$ GeV [127], which from the relation in Eq. (4.68) implies $\sin^2\theta_W = 1 - M_W^2/M_Z^2 \simeq 0.22$. The Higgs boson was discovered using the ATLAS and CMS detectors at the LHC accelerator at CERN, and officially announced on July 4th, 2012. It has a mass of about 125 GeV.[15]

The vacuum expectation value $v$ is directly connected to the Fermi constant $G_F$. We can find this connection by analysing the decay where a muon $\mu^-$ decays into an electron $e^-$, a muon neutrino $\nu_\mu$ and an electron antineutrino $\bar{\nu}_e$, that is the $\mu^- \to e^-\bar{\nu}_e\nu_\mu$ decay. The momentum transfer $q^2 = (p_\mu - p_{\nu_\mu})^2 = (p_e + p_{\nu_e})^2$ cannot be larger than $m_\mu^2$, with the muon mass being $m_\mu \simeq 105$ MeV, about 207 times the mass of the electron. Hence $q^2 \ll m_W^2$ and the W propagator in Fig. 4.1 can be well approximated through a local four-fermion interaction, that is

$$\frac{g^2}{M_W^2 - q^2} \xrightarrow{approx} \frac{g^2}{M_W^2} \equiv 4\sqrt{2} G_F. \qquad (4.69)$$

where the last passage follows by the definition of the Fermi constant $G_F$. $G_F$ has been very accurately determined from muon lifetime measurements [128] and its value is approximately $G_F \simeq 10^{-5}$ GeV$^{-2}$. By exploiting the equality (4.69) and relation between the vacuum expectation value $v$ and the $M_W$ mass in Eq. (4.64),

---

[15] In 2013, the Nobel Prize in Physics was awarded jointly to François Englert and Peter W. Higgs (Brout had passed away two years before) for their work of about 50 years earlier [119–121], where they described what we now call the Higgs mechanism. The recognition of their research came after the Higgs discovery and indeed the Nobel motivation was "for the theoretical discovery of a mechanism that contributes to our understanding of the origin of mass of subatomic particles, and which recently was confirmed through the discovery of the predicted fundamental particle, by the ATLAS and CMS experiments at CERN's Large Hadron Collider".

## 4.4 The Higgs Mechanism in the SM

the Fermi coupling provides the value of the vacuum expectation value $v$

$$v = \frac{1}{\sqrt{\sqrt{2}G_F}}. \tag{4.70}$$

The value of $v \simeq 246$ GeV [127] sets the scale of electroweak symmetry breaking. Since by definition $\lambda = 4\mu^2/v^2$, knowing the value of $v$ allows the replacement of the self-coupling $\lambda$ in the potential $V(\phi)$, leaving only the parameter $\mu$, and consequently the Higgs mass $M_H = \sqrt{2}\mu$, undetermined.

Let observe that Eq. (4.69) can be recast in the form

$$\frac{e^2}{\sin^2 \theta_W M_W^2} = 4\sqrt{2}G_F. \tag{4.71}$$

The measured values of the electron charge, $M_W$ and $G_F$ provide another estimate of $\sin^2 \theta_W$, which is independent of the one based on Eq. (4.68).

Until now, we have analyzed only terms that are second order in the fields; at higher orders, the EW Lagrangian contains also third and fourth-order interactions of vector bosons, as well as terms of Higgs-boson interactions. The Higgs self-interacting terms originate by Eq. (4.58) and read

$$-\frac{1}{16}\lambda h^4(x) - \sqrt{\frac{\lambda}{8}} M_H h^3(x) \tag{4.72}$$

remembering that the Higgs mass is $M_H = \sqrt{2}\mu$. The expansion of the kinetic energy term in the unitarity gauge yields additional terms coupling the Higgs boson field to $W^\pm$ and $Z^0$. Indeed, from Eq. (4.57) we have

$$+ g^2 \frac{1}{\sqrt{2\lambda}} M_H h W_\mu^- W^{\mu -} + \frac{g^2}{\cos^2 \theta_W} \frac{1}{\sqrt{8\lambda}} M_H h Z_\mu Z^\mu +$$

$$+ \frac{g^2}{4} h^2 W_\mu^+ W^{\mu -} + \frac{g^2}{8 \cos^2 \theta_W} h^2 Z_\mu Z^\mu. \tag{4.73}$$

Let us observe that since $g = 2M_W/v$ and $g/\cos \theta_W = 2M_Z/v$, all Higgs couplings to the charged gauge bosons are proportional to $m_W^2$ and all Higgs couplings to the $Z^0$ are proportional to $m_Z^2$, scaled by $1/v$ or $1/v^2$ if it is a three-bosons or a four-bosons interaction, respectively.

The particular form of the Lagrangian we have discussed corresponds to a choice of gauge, namely the unitary one. Since the SM is a gauge invariant theory, any other gauge would lead to the same observable quantities, but the theory would be more difficult to interpret. The Higgs field apparently would have more degrees of freedom which, however, turn out to be nonphysical, since they can be removed by a gauge transformation. We can use other gauges, when we find it advantageous, for instance for the purpose of the renormalization of the theory.

## 4.5 The Yukawa Lagrangian

In the EW Lagrangian built so far there are no terms that could represent fermion mass terms. We could add them, by including in the EW Lagrangian a four dimensional sector $\mathscr{L}_m$, whose terms are bilinear in the fermion fields. Restricting to the first quark generation, we would have

$$\mathscr{L}_m = -m_d \bar{d} d - m_u \bar{u} u = -m_d \bar{d}_L d_R - m_u \bar{u}_L u_R \qquad (4.74)$$

where the couplings $m_{d(u)}$ represent the masses. In the last passage we have decomposed as usual the fermion fields into chirality eigenstates $u_{L(R)} = P_{L(R)} u$, $d_{L(R)} = P_{L(R)} d$ and $\bar{u}_{L(R)} = \bar{u} P_{R(L)}$, $\bar{d}_{L(R)} = \bar{d} P_{R(L)}$. We have seen in Sect. 4.1 that left-handed doublets and right-handed singlets transform differently under the gauge group $SU(2)_L$ of the SM. In particular, the singlets $u_R$ and $d_R$ do not change, while for the doublet $q_L$ defined in Eq. (4.8) the following transformation rule holds

$$\begin{pmatrix} u_L \\ d_L \end{pmatrix} \rightarrow e^{i\alpha_i(x)\tau_i/2} \begin{pmatrix} u_L \\ d_L \end{pmatrix} \qquad i \in \{1, 2, 3\} \qquad (4.75)$$

where by repeated indexes we imply, as usual, the sum over the three indexes. It is self-evident that $\mathscr{L}_m$ is not invariant under the gauge group $SU(2)_L$ of the SM, and therefore it cannot belong to the SM Lagrangian.

The simplest way being precluded, we take another route, that exploits the SSB of the EW interactions. We start by building a new term in the EW Lagrangian, that includes the complex (Higgs) doublet $\phi$ introduced in Sect. 4.4 and it is bilinear in the fermion fields. We require that this term is four-dimensional, because of renormalizability, and a singlet under $SU(2)_L \otimes U(1)_Y$, in order to remain invariant and not break the explicit gauge invariance of the Lagrangian. The most general $SU(2)_L \otimes U(1)_Y$, Lorentz invariant term of dimension four involving the Higgs doublet $\phi$ and quarks is given, for the first generation, by

$$\mathscr{L}_{Y_d} = -Y^{(d)} \bar{q}_L \phi \, d_R + h.c. \qquad (4.76)$$

where h.c. stands for Hermitian conjugate and $\bar{q}_L = (\bar{u}_L \; \bar{d}_L)$. The boson-fermion $Y^{(d)}$ coupling, referred as Yukawa coupling, is a complex number completely arbitrary. The fundamental scalar doublet $\phi$ and the antifermion doublet $\bar{q}_L$ transform under $SU(2)_L$ as

$$\begin{pmatrix} \phi^+ \\ \phi^0 \end{pmatrix} \rightarrow e^{i\alpha_i(x)\tau_i/2} \begin{pmatrix} \phi^+ \\ \phi^0 \end{pmatrix}$$

$$(\bar{u}_L \; \bar{d}_L) \rightarrow e^{(i\alpha_i(x)\tau_i/2)^\dagger} (\bar{u}_L \; \bar{d}_L) \rightarrow e^{-i\alpha_i(x)\tau_i/2} (\bar{u}_L \; \bar{d}_L) \qquad (4.77)$$

## 4.5 The Yukawa Lagrangian

where in the last passage we have exploited the hermiticity of the $\tau_i$ matrices $i\tau_i \to -i\tau_i^\dagger = -i\tau_i$. The hypercharges of $\bar{q}_L$, $\phi$ and $d_R$, as defined in Eq. (4.3), are $Y = -1/6, 1/2, -1/3$, respectively, thus their sum, the total hypercharge of $\mathscr{L}_{Y_d}$, is zero.

As seen in Sect. 4.4, the Higgs field $\phi$ is the agent of spontaneous breaking the $SU(2)_L \otimes U(1)_Y$ symmetry to $U(1)_{em}$ in the SM. In the unitary gauge, choosing the VEV as in Eq. (4.51)

$$\langle \phi \rangle = \langle 0|\phi|0 \rangle = \frac{1}{\sqrt{2}} \begin{pmatrix} 0 \\ v \end{pmatrix} \tag{4.78}$$

implies that oscillations about the VEV are parametrised by

$$\phi(x) = \frac{1}{\sqrt{2}} \begin{pmatrix} 0 \\ v + h(x) \end{pmatrix}. \tag{4.79}$$

The $d$-quark acquires a finite mass when the $\phi$-doublet acquires a non-zero VEV. In fact, after the insertion of the field $\phi(x)$ in Eq. (4.79), the fermion mass terms for the $d$ quark follow naturally, since

$$\begin{aligned}\mathscr{L}_{Y_d} &= -Y^{(d)} \bar{q}_L \phi d_R - Y^{(d)*} \bar{d}_R \phi^\dagger q_L \\ &\to -\frac{v}{\sqrt{2}} \left( Y^{(d)} \bar{d}_L d_R + Y^{(d)*} \bar{d}_R d_L \right) + \cdots\end{aligned} \tag{4.80}$$

Additional terms describe an interaction between the Higgs field $h(x)$ and fermions.

In order to give mass to the $u$-quark as well, we define $\tilde{\phi}$ as

$$\tilde{\phi} \equiv i\tau_2 \phi^* = i\tau_2 \begin{pmatrix} \phi^- \\ \phi^{0*} \end{pmatrix} = \begin{pmatrix} \phi^{0*} \\ -\phi^- \end{pmatrix}. \tag{4.81}$$

The fields $\phi$ and $\tilde{\phi}$ transform in the same way under $SU(2)$. Indeed, under $SU(2)$ transformations, we have

$$i\tau_2 \phi^* \to i\tau_2 e^{-i\alpha_i \tau_i^*/2} \phi^* \to e^{i\alpha_i \tau_i/2} i\tau_2 \phi^* \tag{4.82}$$

where the last passage exploits the properties of the Pauli matrices (see Sect. 2.9.1). We have just shown that $\tilde{\phi}$ is a doublet of $SU(2)$, with $Y = -1/2$. The vacuum value $v$ in $\tilde{\phi}$ appears in the upper component

$$\langle \tilde{\phi} \rangle = \langle 0|\tilde{\phi}|0 \rangle = \frac{1}{\sqrt{2}} \begin{pmatrix} v \\ 0 \end{pmatrix}. \tag{4.83}$$

In analogy to Eq. (4.76) we can write

$$\mathscr{L}_{Y_u} = -Y^{(u)} \bar{q}_L \tilde{\phi} u_R + h.c. \tag{4.84}$$

which after SSB describes mass terms for the $u$-quarks and fermion-Higgs interaction terms.

Summarising, the most general Yukawa Lagrangian involving scalars and quarks is given, for the first generation, by

$$\mathscr{L}_{Y_q} = -Y^{(d)} \bar{q}_L \phi d_R - Y^{(u)} \bar{q}_L \tilde{\phi} u_R + h.c. \tag{4.85}$$

where $\phi$ and $\tilde{\phi}$ are Higgs doublets of hypercharge $Y = 1/2$ and $Y = -1/2$, respectively. The boson-fermion $Y^{(d)}$ and $Y^{(u)}$ couplings, referred as Yukawa couplings, are complex numbers completely arbitrary. They are free parameters in the SM. The arbitrariness of the Yukawa couplings is not shared the gauge couplings of the interactions terms, which are real, determined by the gauge group and enter through the covariant derivative (4.1).

In the leptonic sector, we can repeat the above reasonings for the lepton doublet and singlet defined in Eqs. (4.16) and (4.17). Neutrinos are massless, due to the fact that no right-handed neutrino is introduced, preventing a mass term of the form of Eq. (4.84). Thus in the SM lepton mass terms concern only charged leptons: we have, for the first lepton generation

$$\mathscr{L}_{Y_\ell} = -Y^{(\ell)} \bar{\ell}_L \phi e_R + h.c.. \tag{4.86}$$

Also in the lepton case, the Yukawa coupling is arbitrary and has to be determined phenomenologically.

In the case of $n_g$ fermion generations, the most general Yukawa Lagrangian involving scalars and fermions is

$$\mathscr{L}_Y = \mathscr{L}_{Y_q} + \mathscr{L}_{Y_\ell} =$$
$$= - \sum_{i,j=1}^{n_g} \left( Y_{ij}^{(d)} \bar{q}_L^i \phi d_R^j + Y_{ij}^{(u)} \bar{q}_L^i \tilde{\phi} u_R^j + Y_{ij}^{(\ell)} \bar{\ell}_L^i \phi e_R^j + h.c. \right). \tag{4.87}$$

The Yukawa couplings $Y^{(d)}$ and $Y^{(u)}$ are generalised to matrices $n_g \times n_g$.

## 4.6 Fermion Mass Eigenstates

After spontaneous symmetry breaking, the mass terms of the Yukawa Lagrangian for the first quark generation are (see Eq. (4.80))

$$\mathscr{L}_Y = -M^{(d)} \bar{d}_L d_R - M^{(u)} \bar{u}_L u_R + h.c. \tag{4.88}$$

where

$$M^{(d(u))} \equiv Y^{(d(u))} \frac{v}{\sqrt{2}}. \tag{4.89}$$

## 4.6 Fermion Mass Eigenstates

Let us generalise to the case of $n_g$ families $(u^i, d^i)$, $i \in \{1, .., n_g\}$. We have

$$\mathscr{L}_Y = -\sum_{i,j=1}^{n_g} \left( M_{ij}^{(d)} \overline{d}_L^i d_R^j + M_{ij}^{(u)} \overline{u}_L^i u_R^j \right) + h.c.. \tag{4.90}$$

The complex coefficients $M^{(d(u))}$ are generalised to matrices $n_g \times n_g$. By adopting a matrix notation, we can write

$$\mathscr{L}_Y = -\overline{\hat{d}}_L M^{(d)} \hat{d}_R - \overline{\hat{u}}_L M^{(u)} \hat{u}_R + h.c. \tag{4.91}$$

where $\hat{u}$ and $\hat{d}$ refers collectively to the $n_g$ up and down family states. In the SM, $n_g = 3$, and $\hat{u}^T \equiv (u^1, u^2, u^3) \equiv (u, c, t)$ and $\hat{d}^T \equiv (d^1, d^2, d^3) \equiv (d, s, b)$, where $T$ stays for transpose. These are flavour eigenstates, that is states participating in gauge interactions, but not yet mass eigenstates. Indeed, the $M^{(u)}$ and $M^{(d)}$ matrices are not necessarily Hermitian, nor there is an a priori theoretical reason that they should be diagonal in the generation index. It can be shown [129, 130] that they can be both made hermitian and diagonal by a bi-unitary transformation

$$U_L^{u\dagger} M^{(u)} U_R^u = M_D^u \qquad U_L^{d\dagger} M^{(d)} U_R^d = M_D^d \tag{4.92}$$

where $U_{L(R)}^{u(d)}$ are unitary matrices and $M_D^u$ and $M_D^d$ are diagonal with positive eigenvalues $M_D^u = \text{diag}(m_1^u, m_2^u, m_3^u)$ and $M_D^d = \text{diag}(m_1^d, m_2^d, m_3^d)$. Basically the proof follows from the fact that each product $M^{(u(d))} M^{(u(d))\dagger}$, being Hermitian, can be diagonalized by one unitary matrix, and that the phase degrees of freedom can be used to ensure that the eigenvalues are positive.[16] The transformation (4.92) is equivalent to transform the quark states as

$$\begin{aligned} \hat{u}_L &\to U_L^u \hat{u}_L & \hat{u}_R &\to U_R^u \hat{u}_R \\ \hat{d}_L &\to U_L^d \hat{d}_L & \hat{d}_R &\to U_R^d \hat{d}_R. \end{aligned} \tag{4.93}$$

The new states are the mass eigenstates, since the mass terms are diagonal in that basis. We have

$$\begin{aligned} \mathscr{L}_Y &= -\sum_{i=1}^{3} (m_i^d \overline{d}_L^i d_R^i + m_i^u \overline{u}_L^i u_R^i) + h.c. \\ &= -\sum_{i=1}^{3} (m_i^d \overline{d}^i d^i + m_i^u \overline{u}^i u^i). \end{aligned} \tag{4.94}$$

---

[16] See e.g. Ref. [131] or [132].

This change in quark eigenstates should inevitably be registered by the other sectors of the EW Lagrangian containing quarks. Let us check the fermion currents in Eq. (4.27). We observe that the neutral current and electromagnetic terms remain invariant after the transformations (4.93). For example, for the electromagnetic current of the $up$-type quarks (see Eq. (4.42)), omitting the common charge factor $Q = 2/3\, e$ by simplicity, we have

$$\sum_{i=1}^{3} \bar{u}_L^i \gamma^\mu u_L^i = \bar{\hat{u}}_L \gamma^\mu \hat{u}_L \to \bar{\hat{u}}_L U_L^{u\dagger} \gamma^\mu U_L^u \hat{u}_L \to \bar{\hat{u}}_L \gamma^\mu \hat{u}_L \qquad (4.95)$$

since $U_L^{u\dagger} U_L^u = 1$ by unitarity. Neutral currents are the same weather expressed in terms of flavour or mass eigenstates. There are no flavour changing neutral currents in the SM at tree level. On the contrary, the charged current interactions are affected by the change of basis (4.93). As an example, let us express the EW current $J_W^{\mu+}$ for the quarks in Eq. (4.41) (factor omitted)

$$\sum_{i=1}^{3} \bar{u}_L^i \gamma^\mu d_L^i = \bar{\hat{u}}_L \gamma^\mu \hat{d}_L \to \bar{\hat{u}}_L U_L^{u\dagger} \gamma^\mu U_L^d \hat{d}_L = \bar{\hat{u}}_L \gamma^\mu V \hat{d}_L \qquad (4.96)$$

in terms of the quark mass eigenstates. The change can be parameterised by a new matrix $V$ defined as

$$V \equiv U_L^{u\dagger} U_L^d. \qquad (4.97)$$

It is a unitary matrix, being the product of unitary matrices. In conclusion, the hadronic charged current terms in the Lagrangian become

$$\mathscr{L}_{CC} = \frac{g}{\sqrt{2}} (W_\mu^+ \bar{\hat{u}}_L \gamma^\mu V \hat{d}_L + W_\mu^- \bar{\hat{d}}_L V^\dagger \gamma^\mu \hat{u}_L) \qquad (4.98)$$

in terms of the quark mass eigenstates. In this new Lagrangian the fields are physical, mass eigenstates, and in the SM they can reassume, without ambiguity, the previous names, that is $\hat{u}^T \equiv (u, c, t)$ and $\hat{d}^T \equiv (d, s, b)$.

In the SM, the unitary matrix $V$ is a three by three matrix called Cabibbo-Kobayashi-Maskawa (CKM) matrix, whose entries represent couplings of the charged quark currents to the charged bosons

$$V = \begin{pmatrix} V_{ud} & V_{us} & V_{ub} \\ V_{cd} & V_{cs} & V_{cb} \\ V_{td} & V_{ts} & V_{tb} \end{pmatrix}. \qquad (4.99)$$

The values $V_{ij}$ are completely arbitrary in the SM, and they have to be inferred by the experiments. Non diagonal terms allow weak interaction transitions between

## 4.6 Fermion Mass Eigenstates

quark generations. At tree level, the SM allows flavour changing currents, as long as they are also charge changing ones, i.e. not-neutral.

The leptonic sector mimics the quark one, if not for the fact that the mass terms concern only charged leptons. We can write

$$\mathcal{L}_{Y_\ell} = -\bar{\hat{e}}_L M^\ell \hat{e}_R + h.c. \tag{4.100}$$

where

$$M^\ell \equiv Y^{(\ell)} \frac{v}{\sqrt{2}} \tag{4.101}$$

and $\hat{e}$ refers collectively to the $n_\ell$ lepton family states. In the SM, $\hat{e} \equiv (e, \mu, \tau)$ and $M^\ell$ is the $3 \times 3$ lepton mass matrix. In order to diagonalize the lepton mass matrix, we choose the unitary matrices $U^\ell_{L(R)}$ such that

$$U_L^{\ell\dagger} M^\ell U_R^l = M_D^\ell = \mathrm{diag}(m_e, m_\mu, m_\tau) \tag{4.102}$$

where we have indicated the eigenvalues in the diagonal $M_D^\ell$ matrix. The equivalent transformation on the states is

$$\hat{e}_L \to U_L^\ell \hat{e}_L \qquad \hat{e}_R \to U_R^\ell \hat{e}_R. \tag{4.103}$$

At a variance with the quark case, we can take advantage from the freedom given by the absence of the neutrino mass matrix and transform the left-handed neutrino states with the same matrix of the left-handed charged states

$$\hat{v}_L \to U_L^\ell \hat{v}_L \tag{4.104}$$

where $\hat{v}^T \equiv (v_e, v_\mu, v_\tau)$. By imposing this choice, we ensure that not only neutral but also charged currents preserve their form with the weak eigenstates substituted by the mass eigenstates. Indeed, in terms of mass eigenstates, for $n_\ell$ generations, we have

$$\sum_{i=1}^{n_l} \bar{e}^i_L \gamma^\mu v^i_L = \bar{\hat{e}}_L \gamma^\mu \hat{v}_L \to \bar{\hat{e}}_L U_L^{\ell\dagger} \gamma^\mu U_L^\ell \hat{v}_L \to \bar{\hat{e}}_L \gamma^\mu \hat{v}_L \tag{4.105}$$

and the charged current in the lepton sector remains (see Eqs. (4.29) and (4.30))

$$\mathcal{L}_{CC\nu} = \frac{g}{\sqrt{2}}(W^+_\mu \bar{\hat{v}}_L \gamma^\mu \hat{e}_L + W^-_\mu \bar{\hat{e}}_L \gamma^\mu \hat{v}_L). \tag{4.106}$$

In the SM the zero neutrino mass prevents the appearance of a mixing matrix in the leptonic sector.

Finally, let us remark that we have derived the fermion mass terms in the Lagrangian, after SSB, diagonalizing the Yukawa matrix, substituting the Higgs field (4.79) into the Yukawa Lagrangian (4.87) and selecting terms at the lowest order in fields. If we now consider also higher orders, we obtain the terms in the Lagrangian which couple the physical Higgs $h(x)$ to the mass eigenstate fermions in the unitary gauge. Also in this case we can diagonalize the Yukawa matrix and observe that, for any quark or lepton flavour $f$, the Higgs boson couples according to

$$-m_f \frac{1}{v} \bar{f} f h \qquad (4.107)$$

where the fermions participating to the interaction are mass eigenstates. The couplings of the Higgs boson to fermions are proportional to their masses.

## 4.7 The CKM Matrix

In the Standard Model, the entries of the CKM are, by definition, not predicted, but have to be inferred from the experiments. We only know that the CKM is a complex, unitary matrix, and we exploit this property in order to compute the number of its independent parameters.

Real and complex $n \times n$ matrices have respectively $n^2$ and $2n^2$ independent, real parameters. Unitarity implies $\sum_l V_{il} V_{lm}^\dagger = \delta_{im}$, that is $n^2$ constraints for the free parameters. Thus a (complex) unitary $n \times n$ matrix is left with only $2n^2 - n^2 = n^2$ real independent parameters. We would like to characterise them as phase or angle parameters. Let us first count the number of independent parameters for a general, orthogonal, $n \times n$ matrix $O$. The number of free parameters is constrained by the condition of orthogonality, $O O^T = I$, where $I$ is the unit matrix. There are $n$ equations which come from the diagonal terms $\sum_l O_{il} O_{li}^T = 1$ and give $n$ constraints. The equations for off-diagonal terms $\sum_l O_{il} O_{lj}^T = 0$ ($i \neq j$) in the upper right corner of the matrix are identical to those off-diagonal terms in the lower left corner of the matrix. Then the number of the off diagonal constraints is $(n^2 - n)/2$ (half the total number of matrix elements minus the diagonal terms). We conclude that the total number of independent parameters of a general (real) orthogonal matrix is $n^2 - (n + (n^2 - n)/2) = n^2 - (n^2 + n)/2 = n(n-1)/2$. By definition, an orthogonal matrix preserves the dot product of vectors, and the free parameters can be identified with independent rotation angles. Hence, we can interpret $n_\theta = n(n-1)/2$ of the $n^2$ real independent parameters of the unitarity matrix as rotation angles, that is

$$n_\theta = \frac{n(n-1)}{2}. \qquad (4.108)$$

## 4.7 The CKM Matrix

The remaining $n^2 - n(n-1)/2 = n(n+1)/2$ parameters can be interpreted as phases, although not all of them have a physical meaning. In quantum mechanics one has the freedom to re-phase the $n$ quark fields, that is to set

$$u_i \rightarrow e^{i\alpha_i^u} u_i \qquad u_i \in \hat{u}$$
$$d_i \rightarrow e^{i\alpha_i^d} d_i \qquad d_i \in \hat{d}$$

where $\alpha_i^{u(d)}$ are global, free parameters. The fact that these phases can be transformed away by field redefinitions means that they can never be determined in any physical process. On the CKM matrix elements $V_{ij}$ that implies

$$V_{ij} \rightarrow e^{i(\alpha_i^u - \alpha_j^d)} V_{ij}. \qquad (4.109)$$

In this way, one may change, and in particular eliminate, $2n-1$ relative phases. One cannot eliminate $2n$ phases since the overall phase is redundant.[17] Summarising, the CKM matrix $V$ contains $n_\alpha = n(n+1)/2 - (2n-1)$ physical phases, that is

$$n_\alpha = \frac{(n-1)(n-2)}{2}. \qquad (4.110)$$

### 4.7.1 Two Fermion Generations

In the two generation case, there is no physical phase ($n_\alpha = 0$), and only one physical parameter ($n_\theta = 1$). The CKM matrix is a $2 \times 2$ unitary matrix, which is in general a complex matrix and can be parameterised as

$$V_C = \begin{pmatrix} e^{i(\alpha_1^u - \alpha_1^d)} \cos\theta_C & e^{i(\alpha_1^u - \alpha_2^d)} \sin\theta_C \\ -e^{i(\alpha_2^u - \alpha_1^d)} \sin\theta_C & e^{i(\alpha_2^u - \alpha_2^d)} \cos\theta_C \end{pmatrix} \qquad (4.111)$$

where four un-physical phases are present (and three of them are independent, since the overall phase is redundant). We have expressed the phases as differences, in order to make explicit their cancellation by the re-phasing of the quark fields, as previously explained. In other terms, we can use the freedom to change the phase of the quark fields and express the $V_C$ matrix as a real matrix

$$V_C = \begin{pmatrix} \cos\theta_C & \sin\theta_C \\ -\sin\theta_C & \cos\theta_C \end{pmatrix}. \qquad (4.112)$$

---

[17] The invariance of the Lagrangian under an overall phase is related to the conservation of the baryon number.

This is the well-known Cabibbo matrix, involving just one angle, the so-called Cabibbo angle $\theta_C$. In this approximation $\sin\theta_C \simeq 0.22$, corresponding to a value $\theta_C \simeq 13°$.

The Cabibbo matrix has an history of its own. In the sixties, observations of the $\beta$ decay and muon decay cross sections showed that the weak interaction couplings could be considered universal. However, it was also observed that the strangeness-changing coupling was weaker than the strangeness-preserving coupling, in contrast with the presumed universality. In 1963, Nicola Cabibbo [133] suggested that universality could be restored assuming that the down-type quark was not the mass-eigenstates $d$, but a linear combination including a strange component

$$d' = \cos\theta_C d + \sin\theta_C s \tag{4.113}$$

where $\theta_C$ was to be experimentally determined. The assumption is that the quarks entering the charged weak interactions are not in states belonging to the orthonormal basis of mass-eigenstates $(d, s)$, but to another orthonormal basis $(d', s')$, where

$$s' = -\sin\theta_C d + \cos\theta_C s. \tag{4.114}$$

In other terms, the states are connected by a bidimensional rotation

$$\begin{pmatrix} d' \\ s' \end{pmatrix} = \begin{pmatrix} \cos\theta_C & \sin\theta_C \\ -\sin\theta_C & \cos\theta_C \end{pmatrix} \begin{pmatrix} d \\ s \end{pmatrix}. \tag{4.115}$$

The coefficients of $d$ and $s$ states satisfy the normalization condition, namely the sum of their square is one. Provided that the quark mixing phenomenon is taken into account, the charged weak interactions maintain the universality feature in the quark sector. With a small angle $\theta_C$, the transition probabilities changing strangeness, which are proportional to $\sin^2\theta_C$, are smaller than the ones conserving the strangeness, that have the factor $\cos^2\theta_C$, in agreement with the observations.

In 1970, prior to the discovery of the charm quark, Sheldon Glashow, John Iliopoulos and Luciano Maiani [134] hypothesised its existence. The possibility of accommodating two doublets $(u, d')$, $(c, s')$ in the weak interactions, rather than one, with the flavour and mass eigenstates connected by the Cabibbo matrix (4.115), is the essence of the so-called GIM mechanism, which motivated several experimental observations. At that time, only three quarks, $u$, $d$ and $s$, were known, and the latter therefore was not considered part of an $SU(2)$ doublet. It followed that the neutral gauge $Z^0$-boson could couple not only to currents whose quark content was $(\bar{u}, u)$ or $(\bar{d}, d)$, but also to neutral currents $(\bar{d}, s)$ and $(\bar{s}, d)$, of different initial and final flavour, just like the charged $W$-boson could couple to charged currents joining the $u$-quark to either a $d$ or an $s$ quark. However, the so called flavour changing neutral currents (FCNC) were not observed experimentally.

## 4.7 The CKM Matrix

By introducing doublets, and the rotation of the negatively charged quarks as in Eq. (4.115), the $Z^0$ couples to the pairs

$$\bar{u}u + \bar{c}c + \bar{d}'d' + \bar{s}'s' = \bar{u}u + \bar{c}c + \bar{d}d + \bar{s}s \tag{4.116}$$

and the FCNC currents are automatically cancelled at this order. An experimental consequence is, for example, the strong suppression (by a factor of $10^{-8}$) of the $K_L^0 \to \mu^+\mu^-$ decay rate with respect to the $K^+ \to \mu^+\nu_\mu$. The former process would be induced at tree level by the $Z^0$ coupling

$$\cos\theta_C \sin\theta_C Z^0 \bar{d}_L s_L \tag{4.117}$$

at a rate comparable to the $K^+ \to \mu^+\nu_\mu$ decay rate. Instead, this strangeness-changing neutral coupling is precisely cancelled in tree approximation by the $Z^0$ coupling induced by $s'$

$$-\cos\theta_C \sin\theta_C Z^0 \bar{d}_L s_L. \tag{4.118}$$

Another experimental observation is the suppression of the $K^+ \to \pi^+\bar{\nu}\nu$ decay rate with respect to the $K^+ \to \pi^0 e^+\nu$ one, which is likewise compatible with the absence of neutral currents mediating strangeness-changing processes at tree level, induced by the GIM mechanism.

At the time of the proposal, the importance of the GIM mechanism was not immediately recognised. It seemed, a little far fetched to introduce a new quark just to fix some aspects of a theory still in the making. But the skeptics were silenced by the discovery of the $J/\Psi$ in 1974, the charm-anticharm bound state.[18] The absence of FCNCs at tree level is maintained in the three generations case, since it is based on the unitarity of the mixing matrix. It is nowadays one of the most important test of the SM.

### 4.7.2 Three Generations

In the SM, the CKM matrix is a key element in describing the flavour dynamics. As seen above, it is unitary, but this is its only theoretical constraint. The parameters of the CKM have to be determined experimentally, and there is no a priori theoretical way to determine their values within the SM framework. The CKM matrix induces

---

[18] On November 10, 1974, SLAC's Burton Richter and colleagues found evidence for a particle they called the $\Psi$ [135]. Meanwhile, at the Brookhaven National Laboratory, Samuel Ting of MIT and his colleagues found comparable evidence for a particle they called the $J$ [136]. Both were the same particle and papers from both groups were published on the same issue of Physical Review Letters on 2 December, 1974, as the first evidence for what is now known as the $J/\Psi$. Richter and Ting were awarded the 1976 Nobel Prize in Physics.

flavour-changing transitions inside and between generations in the charged sector at tree level. By contrast, there are no flavour-changing transitions in the neutral sector at tree level.

In the SM, there are three generations of fermions. The bottom quark was discovered in 1977 at the Fermi National Accelerator Laboratory (Fermilab or FNAL, Illinois, USA), in 400 GeV proton nucleus collisions that produced a bound $b\bar{b}$ state (bottomonium or $\Upsilon$) at about 9.5 GeV, by a research group led by Leon Lederman [137]. The other member of the isospin doublet, the top quark, much higher in mass ($\sim$173 GeV), was observed almost twenty years later, in 1995, at the Tevatron (Fermilab) [138, 139].

The CKM matrix of three generations has 4 independent physical parameters, three angles and a phase. This brings about an important difference with the two family case. We know that the operation of $CP$ interchanges the operators of the Lagrangian with their Hermitian conjugates, taking the complex conjugate of couplings and coefficients. Therefore, the charged current Lagrangian $\mathscr{L}_{CC}$ in Eq. (4.98) is $CP$ invariant only in case of real couplings, including the entries of the CKM matrix. A complex CKM matrix implies a $CP$ violating Lagrangian. With three quark families the CKM matrix cannot be made real by any re-phasing of the quark fields. This non-cancellable, physical phase generates $CP$ violation, as it was first pointed out by Kobayashi and Maskawa [140].

The CKM matrix can be parameterised in several equivalent ways in terms of three rotation angles and one phase. The position in which complex terms appear in the CKM matrix is not physically significant, since the complex phase can appear in different matrix elements in different parameterisations. Clearly, the physics does not depend on the choice, and physical quantities are independent on the position of the phase. A common parametrisation is

$$V = \begin{pmatrix} V_{ud} & V_{us} & V_{ub} \\ V_{cd} & V_{cs} & V_{cb} \\ V_{td} & V_{ts} & V_{tb} \end{pmatrix}$$

$$= \begin{pmatrix} c_{12}c_{13} & s_{12}c_{13} & s_{13}e^{-i\delta} \\ -s_{12}c_{23} - c_{12}s_{23}s_{13}e^{i\delta} & c_{12}c_{23} - s_{12}s_{23}s_{13}e^{i\delta} & s_{23}c_{13} \\ s_{12}s_{23} - c_{12}c_{23}s_{13}e^{i\delta} & -c_{12}s_{23} - s_{12}c_{23}s_{13}e^{i\delta} & c_{23}c_{13} \end{pmatrix}. \quad (4.119)$$

Here $c_{ij} = \cos\theta_{ij}$ and $s_{ij} = \sin\theta_{ij}$, with $i$ and $j$ labelling families that are coupled through that angle ($i, j = 1, 2, 3$). The rotation angles may be restricted to lie in the first quadrant, provided one allows the phase $\delta$ to be free. As a consequence, $c_{ij}$ and $s_{ij}$ can all be chosen to be positive. The angle $\theta_{12}$ is generally called the Cabibbo angle ($\theta_C$), and $\sin\theta_C \simeq 0.22$, corresponding to a value $\theta_C \simeq 13°$. The angle of mixing between the second and the third family is $\theta_{23} \simeq 2°$, and between the first and the third is $\theta_{13} \simeq 0.2°$. The phase $\delta$ is constrained by measurements of the CP violation in $K$ decays to be in the range $0 < \delta < \pi$. Its value is approximately $\delta \simeq 1.2$.

## 4.7 The CKM Matrix

This parametrisation can be seen as the product of three rotations, with the phase put on the smallest element

$$V = \begin{pmatrix} 1 & 0 & 0 \\ 0 & c_{23} & s_{23} \\ 0 & -s_{23} & c_{23} \end{pmatrix} \begin{pmatrix} c_{13} & 0 & s_{13}e^{-i\delta} \\ 0 & 1 & 0 \\ -s_{13}e^{i\delta} & 0 & c_{13} \end{pmatrix} \begin{pmatrix} c_{12} & s_{12} & 0 \\ -s_{12} & c_{12} & 0 \\ 0 & 0 & 1 \end{pmatrix}. \quad (4.120)$$

In this parametrisation, the $s_{ij}$ are simply related to directly measurable quantities

$$s_{13} = |V_{ub}|$$

$$s_{12} = |V_{us}|/\sqrt{1 - |V_{ub}|^2} \sim |V_{us}|$$

$$s_{23} = |V_{cb}|/\sqrt{1 - |V_{ub}|^2} \sim |V_{cb}| \quad (4.121)$$

where we have set $|V_{ub}| \ll 1$, as indicated by data.

Let us remark that $\beta$ decays place one of the most stringent tests of SM through precision measurements of the first-row CKM matrix elements. In the $\beta$ decay the parity of the initial nuclear state $P_i$ connects to the parity of the nuclear final state $P_f$ as $P_i = P_f(-1)^L$, where $L$ is the magnitude of the angular momentum of the lepton pair. Hence one has $\Delta P = P_i P_f = (-1)^L$.

Transitions wherein the outgoing lepton pair carries off zero orbital angular momentum, $L = 0$, so that the parity of the initial and final nuclear states is the same, are named allowed. The term superallowed is used when the initial and final nuclei are members of the same isomultiplet. When the orbital momentum transfer is different than zero, the transitions are termed forbidden and they occur with less probability compared to the allowed ones. Superallowed beta decays $J^P = 0^+ \to J^P = 0^+$ occur only via the vector current of the weak interaction.[19] They are particularly well suited to extract the value of $V_{ud}$.

According to experimental evidence, the CKM matrix has a hierarchical structure. Transitions within the same generation are characterized by matrix elements of order $O(1)$. Transitions between the first and second generations are suppressed by a factor of $O(10^{-1})$, between the second and third generations by a factor of $O(10^{-2})$ and between the first and third generations by a factor of $O(10^{-3})$. This hierarchy prompted Wolfenstein to suggest, after 10 years from the original KM proposal, an approximate parametrisation [141] based on a series expansion in the small parameter $\lambda = |V_{us}|$. At the lowest order

$$V = \begin{pmatrix} 1 - \lambda^2/2 & \lambda & A\lambda^3(\rho - i\eta) \\ -\lambda & 1 - \lambda^2/2 & A\lambda^2 \\ A\lambda^3(1 - \rho - i\eta) & -A\lambda^2 & 1 \end{pmatrix} + O(\lambda^4). \quad (4.122)$$

---

[19] The demonstration follows the procedure described in Sect. 4.8.1.

This parametrisation corresponds to a particular choice of phase convention which eliminates as many phases as possible and puts the one remaining complex phase in the matrix elements $V_{ub}$ and $V_{td}$. In this parametrisation the unitarity of the matrix is explicit, up to $\lambda^3$ corrections. The real independent magnitude parameters $A$, $\rho$ and $\eta$ are known to be roughly of order unity, while $\lambda$, that is essentially the sine of the Cabibbo angle, is a small number, of order 0.2. Relative sizes of amplitudes depending on CKM parameters can be roughly estimated by counting powers of $\lambda$ in the Wolfenstein parametrisation. For instance, it is evident from (4.122) that all $B$-decay amplitudes are suppressed by at least two powers of $\lambda$.

It is convenient to express the Wolfenstein parameters through phase convention-independent quantities

$$s_{12}^2 = \lambda^2 = \frac{|V_{us}|^2}{|V_{ud}|^2 + |V_{us}|^2}$$

$$s_{23}^2 = A^2\lambda^4 = \frac{|V_{cb}|^2}{|V_{ud}|^2 + |V_{us}|^2}$$

$$\bar{\rho} + i\bar{\eta} = -\frac{V_{ud}V_{ub}^*}{V_{cd}V_{cb}^*}. \tag{4.123}$$

where $\bar{\rho}$ and $\bar{\eta}$ are two new parameters that substitute $\rho$ and $\eta$. These relations ensure that the CKM matrix written in terms of $\lambda$, $A$, $\bar{\rho}$, and $\bar{\eta}$ is unitary to all orders in $\lambda$ [142]. When terms of the $\mathcal{O}(\lambda^6)$ are neglected, we have

$$\mathbf{V}_{\text{CKM}} \simeq \begin{pmatrix} 1 - \frac{1}{2}\lambda^2 - \frac{1}{8}\lambda^4 & \lambda & A\lambda^3(\bar{\rho} - i\bar{\eta}) \\ -\lambda + \frac{1}{2}A^2\lambda^5[1 - 2(\bar{\rho} + i\bar{\eta})] & 1 - \frac{1}{2}\lambda^2 - \frac{1}{8}\lambda^4(1 + 4A^2) & A\lambda^2 \\ A\lambda^3[1 - (\bar{\rho} + i\bar{\eta})] & -A\lambda^2 + \frac{1}{2}A\lambda^4[1 - 2(\bar{\rho} + i\bar{\eta})] & 1 - \frac{1}{2}A^2\lambda^4 \end{pmatrix}. \tag{4.124}$$

Since we have defined

$$s_{13}e^{i\delta} = V_{ub}^* = A\lambda^3(\rho - i\eta) \tag{4.125}$$

the following relation holds

$$\rho + i\eta = \left(1 + \frac{\lambda^2}{2}\right)\bar{\rho} + i\bar{\eta} + O(\lambda^4). \tag{4.126}$$

Thus one can reproduce the CKM matrix (4.124) at the same order in $\rho$ and $\eta$ by the substitutions $\bar{\rho} \to \rho$ and $\bar{\eta} \to \eta$ in all entries, except $V_{td}$ where the substitution is $(\bar{\rho} + i\bar{\eta}) \to (1 - \frac{1}{2}\lambda^2)(\rho + i\eta)$.

## 4.8 C, P and CP Symmetries in the EW Lagrangian

Now that we have built the gauge invariant EW Lagrangian, we can check its invariance under the global, discrete $C$, $P$ and $CP$ transformations introduced in Chap. 2. The operative method is to transform each term in the EW Lagrangian under $C$, $P$ and $CP$, with the aid of the Eqs. (2.132), and check its invariance.

In Sect. 2.13, we have already seen that the electromagnetic interactions, which are of vector nature only, are $P$ and $C$-invariant.[20] The tensor structure of neutral and charged currents in EW interactions is of vector or vector axial nature. For $V \pm A$ structures, Table 2.2 gives

$$\overline{\psi}_a(t, \mathbf{x})\,(\gamma_\mu \pm \gamma_\mu \gamma_5)\,\psi_b(t, \mathbf{x}) \xrightarrow{C} -\overline{\psi}_a(t, -\mathbf{x})\,(\gamma_\mu \mp \gamma_\mu \gamma_5)\,\psi_b(t, -\mathbf{x})$$

$$\overline{\psi}_a(t, \mathbf{x})\,(\gamma_\mu \pm \gamma_\mu \gamma_5)\,\psi_b(t, \mathbf{x}) \xrightarrow{P} \overline{\psi}_b(t, \mathbf{x})\,(\gamma^\mu \mp \gamma^\mu \gamma_5)\,\psi_a(t, \mathbf{x}).$$

(4.127)

In the Lorentz invariant Lagrangian, the currents are accompanied by vector gauge fields, which transform as $V^\mu \xrightarrow{C} -V^\mu$ and $V^\mu \xrightarrow{P} V_\mu$. The V-A structure of EW currents breaks explicitly the $P$ and $C$ symmetries, and induces a chirality flip for $C$ and $P$ transitions on the fermion states.

In the SM, the $W$ bosons couple only to the left-handed fermion fields. Since $P$ and $C$ transform fields of a given chirality into each other, it is immediate to see that the chirally asymmetric weak interactions violate parity (and charge-conjugation) The violation, originated by $\alpha V + \beta A$ currents, where $|\alpha| = |\beta| = 1$, is said maximal. In 1958, Maurice Goldhaber, Lee Grodzins and Andrew Sunyar at BNL (Brookhaven National Laboratory) in USA demonstrated that a massless neutrino has left-handed helicity [32]. At the time, the experiment helped to distinguish among different forms of weak interactions.

Under the combined $CP$, the EW currents transform as

$$\overline{\psi}_a(t, \mathbf{x})\,(\gamma_\mu \pm \gamma_\mu \gamma_5)\,\psi_b(t, \mathbf{x}) \xrightarrow{CP} -\overline{\psi}_b(t, -\mathbf{x})\,(\gamma^\mu \pm \gamma^\mu \gamma_5)\,\psi_a(t, -\mathbf{x}). \quad (4.128)$$

There is no chirality flip for $CP$ transitions. In contrast to $C$ and $P$ taken singularly, $CP$-transformations affect vector and pseudovector currents in the same way. The operation of $CP$ interchanges these operators with their Hermitian conjugates, that still belong to the Hermitian Lagrangian. However, this is not enough to ensure $CP$ invariance, but there is an additional subtlety. In the Lagrangian, Hermitian conjugate operators have complex conjugate coefficients. The $CP$ transformation interchanges the operators with their Hermitian conjugates but leaves invariant the

---

[20] One can easily check that fermion (e.g. $\bar{\psi}\partial\!\!\!/\psi$) and EW gauge bosons kinetic terms are $P$ and $C$-invariant as well.

coefficients, which do not transform under $C$ or $P$. We conclude that the EW Lagrangian is invariant under $CP$ unless there are coefficients of the operators in the Lagrangian which are complex. In the SM, the only complex coefficients may come from the Yukawa sector. After SSB, the charged current terms in the Lagrangian are (see Eq. (4.98))

$$\mathscr{L}_{CC} = \frac{g}{\sqrt{2}} (W_\mu^+ \overline{\hat{u}_L} \gamma^\mu V \hat{d}_L + W_\mu^- \overline{\hat{d}_L} V^\dagger \gamma^\mu \hat{u}_L) \qquad (4.129)$$

where $V$ is the CKM matrix. Under $CP$, $W^+$ and $W^-$ switch place since they are each other's antiparticle and we have

$$\mathscr{L}_{CC} \xrightarrow{CP} \frac{g}{\sqrt{2}} (W_\mu^- \overline{\hat{d}_L} V^T \gamma^\mu \hat{u}_L + W_\mu^+ \overline{\hat{u}_L} \gamma^\mu (V^\dagger)^T \hat{d}_L). \qquad (4.130)$$

Hence $\mathscr{L}_{CC}$ is invariant under $CP$ if $V = V^\dagger$, that is if $V$ is real.

The CKM matrix is the only place where a physical, non eliminable, complex phase can occur in the EW Lagrangian. The EW Lagrangian would be CP invariant only in case of real coefficients of the CKM matrix.[21] As we have seen, for one and two families this matrix can always be made real by suitable fermion rotations, but not for three families or more. Hence the fact that there are three families in nature gives us a natural mechanism for $CP$-violation in the SM.

### 4.8.1 Matrix Elements of Hadronic Currents

As seen in Chap. 2, under parity, a vector current $V^\mu$ transforms as $V^0 \to V^0$ for its scalar part and $\mathbf{V} \to -\mathbf{V}$ for its vector part. Hence, we can summarise the spin and parity properties of a polar vector current by writing

$$J^P(V^0) = 0^+ \quad J^P(\mathbf{V}) = 1^- \qquad (4.131)$$

For axial vector currents $A^\mu$, the rotation properties are the same but parity properties are opposite, i.e.,

$$J^P(A^0) = 0^- \quad J^P(\mathbf{A}) = 1^+ \qquad (4.132)$$

One can use these properties to find the general form for the matrix elements of quark currents between different kinds of states.

The simplest example refers to the matrix elements of a vector current $V^\mu$ between the vacuum and any meson $M$ in the $0^-$ octet, namely $<0|V^\mu|M>$.

---

[21] It is obvious that mass terms after SSB, for instance $m\bar{\psi}\psi$, are invariant under $P$ and $C$.

The vacuum is assumed to be always in a $0^+$ state. Hence the time component of this matrix element should be in a combination of $0^+$ (for the vacuum), $0^+$ (for the current) and $0^-$ (for the meson), i.e. should have $J^P = 0^-$. Analogously, the spatial components of the matrix element should be in a combination of $0^+$ (for the vacuum), $1^-$ (for the current) and $0^-$ (for the meson), i.e. should have $J^P = 1^+$. In other terms, the matrix element should be some quantity whose time and space components transform like $0^-$ and $1^+$ respectively. However, this matrix element can depend only on the meson 4-momentum $p^\mu$, and neither $p^\mu$ nor any quantity that depends only on it can transforms this way. It follows that this matrix element must be zero. By substituting an axial vector current $A^\mu$ to the vector current, the same arguments show that the time and space components of the matrix element should have $J^P = 0^+$ and $J^P = 1^-$. These are the same $J^P$ values of the components of $p^\mu$ itself, and therefore the matrix element mediated by the axial vector current can be non-zero and proportional to $p^\mu$. Summarising, the analysis of parity properties of the currents has provided us with the information that only the axial vector current can contribute to the matrix element between the vacuum and a pseudoscalar meson, the polar vector current cannot.

We can proceed in an analogous way for other hadronic matrix elements. Another example is a matrix element of the form $< M_1 | J^\mu | M_2 >$, where we take both $M_1$ and $M_2$ to be $0^-$ mesons, and $J^\mu$ stands for either $V^\mu$ or $A^\mu$. In this case, the time and space components of the matrix element of the polar vector current should transform like $0^+$ and $1^-$ respectively. That is the way 4-momentum vectors transform, and we conclude that $< M_1(p_1) | V^\mu | M_2(p_2) > = f_1 p_1^\mu + f_2 p_2^\mu$, where $f_1$ and $f_2$ are form factors, while $< M_1(p_1) | A^\mu | M_2(p_2) > = 0$.

## 4.9 Accidental Symmetries

Let us consider the gauge invariant, massless fermion Lagrangian of the SM in Eqs. (4.26) and (4.29) in the three generation case, before SSB

$$\mathscr{L}_f = \mathscr{L}_q + \mathscr{L}_\ell = i \sum_{i=1}^{3} \left( \overline{q}_L^i \slashed{D} q_L^i + \overline{u}_R^i \slashed{D} u_R^i + \overline{d}_R^i \slashed{D} d_R^i + \overline{\ell}_L^i \slashed{D} \ell_L^i + \overline{e}_R^i \slashed{D} e_R^i \right). \tag{4.133}$$

Here $i$ is the generation (or flavour) index for the fermions doublets $q_L^i$ and $\ell_L^i$, and the fermion singlets $u_R^i$, $d_R^i$ and $e_R^i$. These covariant derivatives do not depend on the flavour index, i.e., the gauge interactions are independent of the flavour of the fermion fields. One can make global, linear, unitary transformations among the matter fields in a given SM representation without altering the Lagrangian. Since there are 3 copies of each SM representation, these transformations can be represented by unitary $3 \times 3$ matrices, that is by elements of the global Lie group

$U(3)$. In other terms, the Lagrangian $\mathscr{L}_f$, is invariant under the global Lie group

$$U(3)^5 = U(3)_{q_L} \times U(3)_{d_R} \times U(3)_{u_R} \times U(3)_{\ell_L} \times U(3)_{e_R} \qquad (4.134)$$

where the subscript of the unitary $U(3)$ groups indicate the matter fields in each SM representation. This is the largest group of unitary field transformations that commutes with the EW gauge group [143], and it can be decomposed into $SU(3)^5 \times U(1)^5$. This invariance takes the name of flavour symmetry.

Flavour symmetry is not exact in the SM, since it is explicitly broken by the Yukawa interactions, described by Eq. (4.87)

$$\mathscr{L}_{Y_f} = - \sum_{i,j=1}^{3} \left( Y_{ij}^{(d)} \overline{q^i}_L \phi d_R^j + Y_{ij}^{(u)} \overline{q^i}_L \tilde{\phi} u_R^j + Y_{ij}^{(\ell)} \overline{\ell^i}_L \phi e_R^j + h.c. \right). \qquad (4.135)$$

The flavour violation is not complete, though, since the symmetries $U(1)$ and $U(1)^3$ survive in the quark and lepton sector, respectively. It is easy to verify these symmetries in the EW Lagrangian after SSB. The mass terms in the quark sector of the Lagrangian, in term of the mass eigenstates, are given by Eq. (4.94)

$$\mathscr{L}_Y = - \sum_{i=1}^{3} \mathscr{L}_Y^i = - \sum_{i=1}^{3} (m_i^d \overline{d^i} d^i + m_i^u \overline{u^i} u^i). \qquad (4.136)$$

Each $\mathscr{L}_Y^i$ is invariant only under two global $U(1)$ transformations (or phase transformations). However, after SSB, the Lagrangian possesses also charged currents terms (see Eqs. (4.98))

$$\mathscr{L}_{CC} = \frac{g}{\sqrt{2}} (W_\mu^+ \overline{\hat{u}_L} \gamma^\mu V \hat{d}_L + W_\mu^- \overline{\hat{d}_L} V^\dagger \gamma^\mu \hat{u}_L) \qquad (4.137)$$

where $\hat{u}$ and $\hat{d}$ refer collectively to the up and down family states. Due to the presence of the CKM matrix, only one overall $U(1)$ symmetry is respected, and we identify the corresponding charge with the baryon number. Each phase transformations must correspond, by Nöther's theorem, to a globally conserved charge. To each quark is assigned a baryon number $B = 1/3$, and $B = -1/3$ to each anti-quark, since a baryon, which conventionally has $B = 1$, is composed by three constituent quarks.

We can repeat the previous reasoning for leptons. Due to the masslessness of neutrino in the SM, and the absence of the mixing matrix in the lepton sector, the charged lepton term in the SM Lagrangian is (see Eq. (4.106))

$$\mathscr{L}_{CC\nu} = \frac{g}{\sqrt{2}} \left[ W_\mu^+ \sum_{i=1}^{3} \overline{\nu}_L^i \gamma^\mu e_L^i + W_\mu^- \sum_{i=1}^{3} \overline{e}_L^i \gamma^\mu \nu_L^i \right] \qquad (4.138)$$

## 4.9 Accidental Symmetries

where $(e^1, e^2, e^3) = (e, \mu, \tau)$ and $(\nu^1, \nu^2, \nu^3) = (\nu_e, \nu_\mu, \nu_\tau)$ are the mass eigenstates. The neutral lepton term in the EW Lagrangian can be read from Eqs. (4.42) and (4.43)

$$\mathcal{L}_{NC} + \mathcal{L}_{em} = gZ^0_\mu \left[ \sum_{i=1}^{3} \left( g^\ell_L \bar{e}^i_L \gamma^\mu e^i_L + g^\ell_R \bar{e}^i_R \gamma^\mu e^i_R + g^\nu_L \bar{\nu}^i_L \gamma^\mu \nu^i_L \right) \right] +$$

$$+ eA_\mu \sum_{i=1}^{3} \bar{e}^i \gamma^\mu e^i \qquad (4.139)$$

where $g^{\ell,\nu}_L$ and $g^\ell_R$ are given in Table 4.1. There are global symmetries $U(3)_L$ and $U(3)_R$ in the lepton flavour. The mass matrix terms for charged leptons in the SM are

$$\mathcal{L}_{Y_l} = -\sum_{i=1}^{3} m^\ell_i \, \overline{e_L}^i e_R^i + h.c. = -\sum_{i=1}^{3} m^\ell_i \, \bar{e}^i e^i \qquad (4.140)$$

where $m^\ell = (m^\ell_1, m^\ell_2, m^\ell_3) = (m_e, m_\mu, m_\tau)$. By the addition of this term, only three overall $U(1)$ symmetries remains. Each global $U(1)$ symmetry refers to an individual lepton family, and the corresponding charges are identified with $L_e$, $L_\mu$ and $L_\tau$. Each charged (anti-)lepton $\ell$ is assigned a lepton number $L_\ell = +(-)1$, and the same lepton number is assigned to the (anti-)neutrino in the same family. Summarising, in the SM the flavour symmetries are not entirely broken by the Yukawa couplings, but leave the following global symmetries

$$U(1)_B \times U(1)_{L_e} \times U(1)_{L_\mu} \times U(1)_{L_\tau} \qquad (4.141)$$

where $B$ is the baryon number and $L_e$, $L_\mu$, $L_\tau$ are the three lepton family numbers. If the Lagrangian includes mass terms for neutrinos (see Sect. 5.2), the lepton family numbers $L_\ell$ are not conserved. The total lepton number $L$ is the sum of the individual lepton numbers, that is $L \equiv L_e + L_\mu + L_\tau$. In case of Dirac mass terms, the $U(1)^3$ symmetry is lost, but the lepton flavour number $L$ is conserved. In analogy with the baryon number in the quarks sector, it corresponds to a $U(1)$ symmetry.

The global symmetries we have just exposed are not as fundamental as the gauge symmetries, and are often denominated accidental symmetries.[22] One might suspect that the corresponding quantum numbers are not absolutely conserved, as it

---

[22] In particular, let us emphasise that the $U(1)$ global symmetries in Eq. (4.141) are not on the same footing of the weak hypercharge $U(1)$ symmetry of the SM, which is required by gauge invariance. Accidental symmetries are not symmetries of the theory, complete in the ultraviolet regime, but they are approximate symmetries of a partial theory.

is indeed the case. At the end of Sect. 4.2 we have mentioned that a symmetry can be anomalous, and that the anomalies which break local gauge symmetries cancel in a consistent gauge quantum field theory. In the case of global symmetries, the breaking of symmetries by quantum effects does not lead to inconsistencies, but it gives real physical effects. The electroweak forces of the SM by themselves violate baryon number through a quantum anomaly. Also the total lepton number $L$ is violated at a quantum level. The $B$ and $L$ symmetries are anomalous, but in the same way, thus $B - L$ is not anomalous. For baryon and lepton currents one has $\partial^\mu J_\mu^B = \partial^\mu J_\mu^L = \mathscr{A} \neq 0$. The anomaly term $\mathscr{A}$ can be written as a total derivative, so it cannot contribute at any order in perturbation theory (any Feynman diagram with this vertex would have a factor of $\sum p_\mu = 0$). Fermion number violating processes are inherently non-perturbative and occur with extremely suppressed probabilities, absolutely negligible at our age of the Universe.[23]

That is not always the case if we consider models which extend the SM; some models, for instance Grand Unified Theories (GUTs),[24] predict observable baryon number violation.

Even if the lepton family numbers $L_\ell$ are not conserved in the neutrino sector, there is still no evidence of transitions involving charged leptons and violating $L_\ell$, as it would be for instance the $\mu \to e\gamma$ decay. Since charged lepton flavour violating processes are forbidden in the SM, their existence would be considered a clear signal of physics extending the SM. The search for such a violation has been pursued to date in a host of channels both at dedicated and general purpose experiments.

As we will see in Sect. 5.4, depending on their nature, massive neutrinos could also induce another kind of new interactions, violating altogether the total lepton number $L$. One possible evidence, which is actively hunted experimentally, would be the observation of the so-called neutrino-less double $\beta$ decay, that we will discuss in Sect. 10.3.

## 4.10 Lepton Universality

We have already mentioned in Chap. 3 that universality is a fundamental property of gauge interactions. The gauge symmetry of the EW interactions $SU(2)_L \times U(1)_Y$ is characterized by only two couplings, $g$ and $g'$ (or, equivalently, by the more commonly used $g$ and $e$). These couplings are the same, no matter what fermion field is coupled to the gauge bosons (universality). The Higgs mechanism for the breakdown of the EW gauge symmetry does not affect the universality of the

---

[23] For more details see, for instance, Sect. 21.4 of [144].

[24] GUTs try to describe the fundamental interactions by means of a unique gauge group $G$, which contains the SM gauge group. The basic idea is that at energies higher than a certain energy threshold the group symmetry is $G$ and that, at lower energies, the symmetry is broken down to the SM gauge symmetry.

## 4.10 Lepton Universality

gauge couplings. The only difference between the three fermion families comes from the Yukawa interactions. The diagonalization of the Yukawa matrices yields mass terms that are different for each particle and break flavour universality in the SM. Moreover, in the quark sector, the mixing matrices between weak and mass eigenstates motivate the appearance of the CKM matrix, which modifies the couplings of quarks to the weak gauge bosons $W^\pm$ and originates a differences between the three generations. In the case of the charged leptons, the universality pattern is known as lepton universality (LU): other than effects related to their different masses, all the SM interactions treat the three charged leptons identically. In the SM, the EW gauge interactions respect LU, when mass terms can be neglected.[25]

LU can be tested by analysing what happens when leptons of different generations couple to gauge bosons in otherwise identical or similar decays. Tests of LU have been performed on neutral currents at the $e^+e^-$ colliders LEP [145] and SLC [146], which were designed to operate at centre-of-mass energies of approximately 91 GeV, close to the mass of the $Z^0$ boson. The ratios of the leptonic partial widths in the processes $Z^0 \to e^+e^-$, $Z^0 \to \mu^+\mu^-$ and $Z^0 \to \tau^+\tau^-$ have been measured at the $Z^0$ pole, giving [147]

$$\frac{\Gamma[Z^0 \to \mu^+\mu^-]}{\Gamma[Z^0 \to e^+e^-]} = 1.0009 \pm 0.0028$$

$$\frac{\Gamma[Z^0 \to \tau^+\tau^-]}{\Gamma[Z^0 \to e^+e^-]} = 1.0019 \pm 0.0032. \tag{4.142}$$

Under the LU assumption and in case of massless leptons, the partial widths for all decays $Z^0 \to \ell^+\ell^-$ are expected to be equal in the SM.[26] The results are in agreement with the SM. There are also measurements performed at the hadron $pp$ collider LHC that support LU in $Z^0$ decays, see e.g. [148].

The LEP $e^+e^-$ collider operated from 1989 to 2000, and until 1995 the running was dedicated to the $Z^0$ boson region. From 1996 to 2000, the centre-of-mass energy was increased to 161 GeV and ultimately to 209 GeV, allowing the production of pairs of W bosons by the process $e^+e^- \to W^+W^-$. The first period is referred to as LEP-I, and the period beginning in 1996 as LEP-II. LU can be tested in charged currents by measurements of ratios of partial widths in the decays $W^- \to e^-\bar{\nu}_e$, $W^- \to \mu^-\bar{\nu}_\mu$ and $W^- \to \tau^-\bar{\nu}_\tau$, which are expected to be the same assuming massless leptons, as in the case of $Z^0$ decays. These ratios have been measured at LEP-II [149], at LHC [148, 150] and at the hadronic $p\bar{p}$ collider Tevatron [151]. All experimental results are in good agreement with LU, but the precision is generally

---

[25] A mixing matrix, analogous to the CKM, is introduced by extending the SM to massive neutrinos. However, its effect is not relevant on the present discussion of LU, as we will see in Sect. 6.1.

[26] Due to the large mass of the $\tau$ lepton, we expect the prediction for the $Z \to \tau^+\tau^-$ partial width to differ the most from the value computed in the massless limit.

**Fig. 4.2** Tree level diagram of $\tau$ decays into a $\nu_\tau$ and one of the pairs $(e^-, \nu_e)$, $(\mu^-, \nu_\mu), (\bar{u}, d), (\bar{u}, s)$

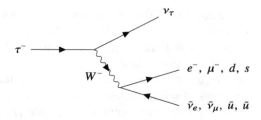

worse than that obtained with $Z^0$ boson decays. Tests involving ratios of the first two lepton families are more precise with respect to the ones comparing with the third one, due to the more challenging reconstruction of final states with $\tau$ leptons.

LU in charged currents can also be tested by employing pure leptonic $\tau$ decays, where several final states are kinematically allowed. The decays of the $\tau$ lepton proceed at tree level through a $W$-exchange mechanism into kinematically allowed states (see Fig. 4.2). At each vertex both charge and lepton number are conserved in the SM. Due to the universality of the $W$-couplings, all these decay modes have equal amplitudes, if QCD interactions and final fermion masses in the phase space factors are neglected, except for an additional $3 V_{ui}$ ($i \in \{d, s\}$) factor in the semileptonic channels, where $V$ is the CKM matrix of Eq. (4.99) and 3 is the number of quark colours. In order to check LU, one can distinguish the couplings according to the generation, re-naming the weak coupling $g_\ell$, where $\ell \in \{e, \mu, \tau\}$. If LU holds $g_e = g_\mu = g_\tau = g$. Constraints on the couplings for electron and muon are obtained by comparing measured branching ratios in

$$\frac{\Gamma[\tau^- \to \nu_\tau e^- \bar{\nu}_e]}{\Gamma[\tau^- \to \nu_\tau \mu^- \bar{\nu}_\mu]} \propto \frac{|g_\tau|^2 |g_e|^2}{|g_\tau|^2 |g_\mu|^2} = \frac{|g_e|^2}{|g_\mu|^2}. \quad (4.143)$$

The last term is 1 if LU holds. One obtains analogous constraints on the ratio $|g_\tau|/|g_\mu|$ by comparing tau and muon electronic decay in the ratio $\Gamma[\tau^- \to \nu_\tau e^- \bar{\nu}_e]/\Gamma[\mu^- \to \nu_\mu e^- \bar{\nu}_e]$. Similarly, the ratio of the couplings to the third and first family can be obtained from the combination of the measurements of the $\Gamma[\tau^- \to \nu_\tau \mu^- \bar{\nu}_\mu]$ and of the $\tau$ and $\mu$ lifetimes. All the above results are in agreement with LU. The LU tests employing $\tau$ decays are among the most stringent experimental tests available today [152].

Other processes where LU in charged EW currents can be precisely tested are leptonic decays of kaons and pions. Constraints on the coupling of the $W$ boson to the first two families of leptons have been obtained from $\pi$ and $K$ leptonic decays, using the ratios $\Gamma[\pi^- \to e^- \bar{\nu}_e]/\Gamma[\pi^- \to \mu^- \bar{\nu}_\mu]$ and $\Gamma[K^- \to e^- \bar{\nu}_e]/\Gamma[K^- \to \mu^- \bar{\nu}_\mu]$, measured at the experiments PIENU [153] and NA62 [154], respectively. One can perform analogous tests for $D$-mesons, or probe LU by analysing different combinations of partial widths for pseudoscalar mesons and $\tau$ leptonic decays, for example, $\Gamma(\tau^- \to \nu_\tau \pi^-)/\Gamma(\pi^- \to \mu^- \bar{\nu}_\mu)$ or $\Gamma(\tau^- \to \nu_\tau K^-)/\Gamma(K^- \to \mu^- \bar{\nu}_\mu)$. All the experimental results so far are in good agreement with LU, albeit with different degrees of precision. Other LU tests employ quarkonia resonances.

## 4.10 Lepton Universality

The most precise tests compare $J/\psi \to e^+e^-$ and $J/\psi \to \mu^+\mu^-$ partial widths measured by the BES-III [155, 156], CLEO [157] and KEDR [158] collaborations. In summary, there are no indications of LU violation in all the processes presented above.

In the past decade, measurements of semileptonic decays of beauty hadrons at the LHCb experiment (CERN, Switzerland), Belle at KEK (Japan) and BaBar at PEP-II (USA), have reached an unparalleled degree of precision. In such decays, for the first time, some experimental results have hinted at potential deviations from the SM. Such deviations are collectively referred to as "flavour anomalies", and they typically feature tensions at the level of 2–3 standard deviations between experimental results and SM predictions. An interesting aspect of these anomalies lies in the fact that they all seem to point towards the presence of LU violation in the interactions mediating the processes. Mostly studied observables are ratios of branching fractions of semileptonic $B$ decays. As of today, there are experimental results favouring values of the ratios $R_{K^{(*)}}$ slightly below the SM prediction, and values of the ratios $R_{D^{(*)}}$ slightly above, as well as values of other ratios or angular observables, related to semileptonic decays, in tension with SM predictions. The accuracy of the predictions for these branching fractions is generally higher than the one for hadronic decays, due to the reliability of perturbative techniques. This precision is further increased by taking ratios of processes with electrons or muons in the final state, since they are affected equally by the strong force, which does not couple directly to leptons. These ratios, for the semileptonic decays $B \to D^{(*)}\ell\bar{\nu}_\ell$ and $B \to K^{(*)}\ell^+\ell^-$, are defined as

$$R_{K^{(*)}[q^2_{min}, q^2_{max}]} = \frac{\mathcal{B}(B \to K^{(*)}\mu^+\mu^-)_{q^2 \in [q^2_{min}, q^2_{max}]}}{\mathcal{B}(B \to K^{(*)}e^+e^-)_{q^2 \in [q^2_{min}, q^2_{max}]}} \quad (4.144)$$

$$R_{D^{(*)}} = \frac{\mathcal{B}(B \to D^{(*)}\tau\bar{\nu}_\tau)}{\mathcal{B}(B \to D^{(*)}\ell\bar{\nu}_\ell)} \quad [\ell = e, \mu] \quad (4.145)$$

where $\mathcal{B}$ denotes the branching fraction for the given decay mode. $R_K$ and $R_{K^*}$ are measured over a bin size of $[q^2_{min}, q^2_{max}]$ for the squared invariant mass $q^2$ of the lepton pair, whereas $R_D$ and $R_{D^*}$ deal with the total branching ratios. The branching fractions in the ratios differ only by the leptons in the final state, hence these ratios are expected to be 1 by virtue of LU, with small deviations induced by phase space differences and QED corrections [159–163]. Should flavour anomalies be confirmed, by increasing precision in future experiments and improved theoretical analyses, it would be a clear signal of physics beyond the SM.

# Chapter 5
# The Mass of Neutrinos

In the SM neutrino masses are taken to be equal to zero, but there are no fundamental principles (like gauge invariance in the case of photons and gluons) requiring that neutrinos are massless. As seen in Chap. 1, initial concepts regarding neutrino masses, mixing, and oscillations emerged in 1957–1958 out of the ingenuity of the Italian physicist Bruno Pontecorvo [31, 33], but only in 1998 neutrino oscillation—clear evidence that neutrinos are massive particles—was observed for the first time [53]. The discovery of nonzero neutrino masses is among the most important particle physics results of the last three decades, since it provides the first evidence of particle physics outside the framework of the SM, commonly indicated as physics beyond the Standard Model (BSM).

A still unsolved consequence of the neutrino having mass is how to extend the SM to include massive neutrinos. All BSM models, like supersymmetry and GUTs, are now bound to predict massive neutrinos in agreement with data.[1] At present, it is also unknown whether neutrinos are Dirac or Majorana fermions. In the former case, neutrinos and antineutrinos are distinct and have a total of four degrees of freedom, exactly as the charged leptons and quarks do. Majorana fermions, on the other hand, are their own antiparticles, and they have just two degrees of freedom corresponding to left- and right-handed helicity. Dirac neutrinos preserve total lepton number conservation, while Majorana neutrino masses violate lepton number conservation by two units. After reviewing neutrinos in the SM in Sect. 5.1, we discuss possible mass terms in the Lagrangian and the character of massive neutrinos, Dirac versus Majorana, from Sect. 5.2 onwards.

It is perhaps fair to say that theoretical prejudice in the choice of Dirac or Majorana neutrinos, as judged by number of papers, prefers the latter possibility. One reason is that Majorana fermions are allowed by the Poincaré group, and they constitute the simplest spinorial representation. Another one is that if right-handed

---

[1] We do not discuss the generation of neutrino mass in specific BSM models, deferring this topics to other textbooks, see e.g. Ref. [164].

neutrinos exist, at the SM level, aside Dirac masses, they can have gauge-invariant Majorana masses, leading to Majorana mass eigenstates overall. Nevertheless, theoretical prejudice or popularity in the literature is not necessarily a reliable guide to the actual behaviour of nature, so one should give due consideration to both possibilities.

Neutrinos masses are much smaller than any other fermionic mass. Charged lepton masses vary widely between the families, for instance the $\tau$-lepton mass is roughly three orders of magnitude larger than the electron mass, but the smallness of the neutrino mass remains a problem even within one lepton family. Many different BSM models include an explanation of the tiny neutrino masses, depending also on whether neutrinos are considered Dirac or Majorana fermions. We highlight some possibilities from Sect. 5.7 onwards. In particular we describe the so-called 'seesaw' models, which propose a way to explain the smallness of neutrino masses introducing heavy right-handed neutrinos—the higher the mass of the new heavy neutrinos, the lower the mass of the left-handed standard ones.

The description of the studies aimed at determining the absolute values of neutrino masses is postponed to Chap. 10.

## 5.1 The Neutrino in the Standard Model

In the SM there are three generations of leptons, and as a consequence three neutrinos $\nu_e$, $\nu_\mu$ and $\nu_\tau$. This number is supported by precision studies of the $Z$-line shape. Each neutrino type $\nu_\ell$ with $\ell \in (e, \mu, \tau)$ contributes the same amount $\Gamma_\nu \equiv \Gamma(Z \to \nu_\ell \bar{\nu}_\ell)$ to the $Z^0$ total width $\Gamma_Z$. The number of different neutrino species, $N_\nu$, can then be derived from the precision measurements of the $Z^0$ total width $\Gamma_Z$ and of its partial widths into hadrons and leptons according to the equation

$$\Gamma_Z = \Gamma_{ee} + \Gamma_{\mu\mu} + \Gamma_{\tau\tau} + \Gamma_h + N_\nu \Gamma_\nu = \Gamma_\ell + \Gamma_h + N_\nu \Gamma_\nu \quad (5.1)$$

where the first three terms are the widths of decays into electrons, muons and tauons, which summed give $\Gamma_\ell$, and $\Gamma_h$ is the sum of the widths of decays into $u$, $d$, $s$, $c$ and $b$ quarks. The term $N_\nu \Gamma_\nu$ is generally indicated as invisible width, $\Gamma_{inv} \equiv N_\nu \Gamma_\nu$, since it corresponds to $Z$ boson decays into particles that are not detected, contrary to charged quarks and leptons. The simultaneous measurement of $\Gamma_Z$ and of observables related to the hadronic and leptonic widths of the $Z^0$ boson, allows one to determine the value $N_\nu$. In order to reduce the model dependence, the SM value for the ratio of the neutrino to charged leptonic partial widths

$$\left(\frac{\Gamma_\nu}{\Gamma_\ell}\right)_{SM} = 1.991 \pm 0.001 \quad (5.2)$$

## 5.1 The Neutrino in the Standard Model

is used instead of $\Gamma_\nu$ in the SM to determine the number of light neutrino types:

$$N_\nu = \frac{\Gamma_{inv}}{\Gamma_\ell} \left(\frac{\Gamma_\ell}{\Gamma_\nu}\right)_{SM}. \tag{5.3}$$

The combined result from four experiments (ALEPH, DELPHI, L3, OPAL) at the LEP electron-positron collider at CERN gave $N_\nu = 2.984 \pm 0.008$ [147]. This result supports the notion that there are 3 neutrinos participating to the weak interactions. A fourth active one is not allowed because it would contribute to the Z boson invisible decay. On the theory side, we expect the number of leptons families to equate the number of quark and lepton families due to chiral anomaly cancellation, as mentioned in Sect. 4.2. The measurement of the Z boson's invisible width, which was used to deduce the number of active light neutrinos, is among the highlights of the LEP experiment [147] and it has remained the most precise for two decades. In a bid to provide an independent and complementary test of the SM at a new energy regime, the CMS experiment at LHC (CERN, Switzerland) has now performed a precise measurement of the Z-boson invisible width—the first of its kind at a hadron collider [165]. This measurement is competitive with the LEP combined result and is currently the world's most precise single direct measurement.

In the SM, the three neutrinos are massless. We have seen in Sect. 2.14 that in the case of a massless Dirac fermion the chirality operator acquires two useful characteristics: it becomes the same as the helicity operator and commutes with the Dirac Hamiltonian. In 1929, soon after Dirac proposed his equation (1.11), the German mathematician and physicist Hermann Weyl suggested [103] that massless particles could be described by the chirality eigenstates $\psi_L$ and $\psi_R$ satisfying the massless Dirac equation

$$i\gamma^\mu \frac{\partial}{\partial x^\mu} \psi(x)_{L(R)} = i\slashed{\partial}\psi(x)_{L(R)} = 0. \tag{5.4}$$

They are the Weyl fields discussed in Sect. 2.14. Let us also remind that in Sect. 2.4.1 we have seen that the spinor representations of the proper Lorentz group are the two non-equivalent (1/2,0) and (0,1/2) representations. By definition, they act on Weyl spinors which have two components (two degrees of freedom).

At the time of Weyl proposal, the two-component fields for massless particles were rejected, because Eq. (5.4) does not conserve parity. In fact, as seen in Eq. (2.166), under the space inversion the left-handed (right-handed) fields are transformed into right-handed (left-handed) fields. The proposal for a massless two component neutrino was resumed after 1956, when large violation of parity in the $\beta$-decay and other weak processes was discovered [166–168].

If the neutrino is the two-component massless Weyl particle, it can be a left-handed $\nu_L$ having helicity equal to $-1$, or a right-handed neutrino $\nu_R$ with helicity $+1$. The antineutrino helicities are opposite. The neutrino helicity was measured in the experiment by M. Goldhaber, L. Grodzins and A.W. Sunyar in 1958 [32]. It was shown that neutrino helicity was negative, confirming the two component neutrino theory, and at the same time excluding the existence of $\nu_R$ and $\bar{\nu}_L$. As all leptons, neutrino do not participate in the strong interactions, as all neutral particles they do not couple to the photon, and only left-handed neutrinos and right-handed antineutrinos participate in charged and neutral EW currents (for $Z^0$ couplings, see Table 4.1). The success of the two-component theory of neutrino, in the sixties, led to a general belief that neutrinos were massless particles. This assumption weighted on the SM, at the time in the making, and neutrinos were assumed to be massless.

As seen in Sect. 4.1, in the SM, leptons belong to $S(2)_L \otimes U(1)_Y$ doublets of hypercharge $Y = -1/2$

$$\begin{pmatrix} \nu_{\ell L} \\ \ell_L \end{pmatrix} \tag{5.5}$$

where $\ell = e, \mu, \tau$. In the SM there are no right-handed neutrino fields and therefore Yukawa interactions of neutrino fields with the Higgs doublet do not exist (see Eq. (4.86)). This means that the Higgs mechanism in the SM gives mass to the EW gauge bosons and all fermions except the neutrinos. After spontaneous symmetry breaking of the electroweak symmetry neutrinos remain massless Weyl particles.

As shown in Sect. 4.6, the diagonalization of the charged leptons mass matrix, and the substitution of weak eigenstates with mass eigenstates, do not affect the form of the charged and neutral currents in the leptonic sector of the Lagrangian. In other terms, there is no equivalent of the CKM matrix in the leptonic sector, lepton families do not mix and lepton-flavour numbers are conserved. In the SM, with massless neutrinos, the lepton-flavour numbers $L_e$, $L_\mu$ and $L_\tau$, as well as the total lepton number, $L = L_e + L_\mu + L_\tau$, are conserved (see Sect. 4.9).

## 5.2 Dirac Mass Terms

The two-component representation discussed in Sect. 5.1 is no more appropriate for massive neutrinos. In fact, in case of massive neutrinos, the chirality eigenstates $\nu_L$ and $\nu_R$ do not coincide with eigenstates of helicity, and their helicity is no more conserved, as seen in Sect. 2.14. For massive fermions the two chiral components are combinations of the helicity eigenstates.

The SM does not include mass terms for neutrinos, but experiments have shown that neutrinos have a non vanishing, albeit tiny, mass. Several extensions of the SM have been proposed to fill this gap. The simplest possibility, considered in this section, is to include massive neutrinos in the SM as Dirac spinors, in complete analogy with the quark case. Right-handed neutrinos are added to the SM particle content. Thus the four-component massive Dirac neutrino can be viewed as composed by a pair of two-component spinors, the left-handed spinor of the SM doublet, $\nu_L$, and an independent right-handed, SM singlet, spinor that we call $N_R$. Each spinor represent two physical degrees of freedom. That is in perfect analogy with the quark sector of the SM. Let us underline the independence of $N_R$, that

## 5.2 Dirac Mass Terms

is not, for instance, obtained by CPT transformations of $v_L$. The massive, four component, Dirac spinor is defined as

$$v = v_L + N_R. \tag{5.6}$$

Separating the left handed and the right handed part of a field is indicated sometimes with the term chiral decomposition. The right-handed neutrino field is called sterile—an adjective which indicates its neutrality under weak interactions. It is opposite to active, that design a neutrino which participates in weak interactions. Therefore, it is a singlet under the $SU(2)_L \times U(1)$ gauge group. This is in agreement with experimental observations, since interactions involving right(left)-handed (anti-)neutrinos have never been detected.

The Dirac spinor corresponding to (5.6) has the right-handed component different from the charge conjugate partner $(v_L)^C$ of the $v_L$.

$$(v_L)^C = P_R v^C = (v^C)_R \equiv v_R^C. \tag{5.7}$$

We have indicated the charge conjugate of the left-handed Weyl spinor as $v_R^C$. Even if it is right handed, it is a different state than $N_R$, since charge conjugation does not change its doublet nature. Thus the charge conjugate partner of the Dirac neutrino $v$ has the following chiral decomposition

$$(v)^C = (v_L)^C + (N_R)^C = v_R^C + N_L^C. \tag{5.8}$$

In the following, for sake of simplicity, we rename

$$v_R \equiv N_R. \tag{5.9}$$

The neutrino mass term can be generated thorough an additional Yukawa term in the weak Lagrangian, analogous to the Yukawa term for up-type quarks in the Standard model, Eq. (4.84)

$$-\mathcal{L}_D = Y_\nu \bar{\ell}_L \tilde{\phi} v_R + h.c. \tag{5.10}$$

where $\ell_L$ is the lepton doublet (5.5) and $\tilde{\phi}$ is the Higgs doublet defined in Eq. (4.81). We can now repeat step by step the passages described in Sect. 4.5. Spontaneously symmetry breaking implies

$$\langle \phi \rangle = \langle 0|\phi|0 \rangle = \frac{1}{\sqrt{2}} \begin{pmatrix} 0 \\ v \end{pmatrix}. \tag{5.11}$$

In case of more than one lepton generation, $Y_\nu$ is an arbitrary, complex matrix; after symmetry breaking the Dirac mass term takes the form

$$-\mathcal{L}_D = \bar{\hat{\nu}}_L M \hat{\nu}_R + h.c. \qquad M \equiv Y_\nu \frac{v}{\sqrt{2}} \qquad (5.12)$$

where $\hat{\nu}_L$ and $\hat{\nu}_R$ refer collectively to the family states, that is $\hat{\nu}_L \equiv (\nu_{eL}, \nu_{\mu L}, \nu_{\tau L})$ and $\hat{\nu}_R \equiv (\nu_{eR}, \nu_{\mu R}, \nu_{\tau R})$.

The $M$ matrix is not necessarily hermitian, nor there is a a priori theoretical reason that it should be diagonal in the generation index. As in the quark case discussed in Sect. 4.6, $M$ can be made hermitian and diagonal by a bi-unitary transformation

$$U_L^{\nu\dagger} M U_R^\nu = m_D = \text{diag}(m_1, m_2, m_3) \qquad (5.13)$$

where $U_{L(R)}^\nu$ are unitary matrices, $m_D$ is diagonal, $m_1$, $m_2$ and $m_3$ are its real and positive eigenvalues. This transformation corresponds to the transformations of the neutrino states

$$\hat{\nu}_L = U_L^\nu \tilde{\nu}_L \qquad \hat{\nu}_R = U_R^\nu \tilde{\nu}_R \qquad (5.14)$$

where the states $\tilde{\nu} = \tilde{\nu}_L + \tilde{\nu}_R \equiv (\nu_1, \nu_2, \nu_3)$ have definite mass and are described by four component spinors. In terms of the $\tilde{\nu}$ state, the mass term in the weak Lagrangian becomes

$$-\mathcal{L}_D = \bar{\tilde{\nu}}_L U_L^{\nu\dagger} M U_R^\nu \tilde{\nu}_R + h.c. = \bar{\tilde{\nu}}_L m_D \tilde{\nu}_R + h.c. \qquad (5.15)$$

that is

$$\mathcal{L}_D = -\sum_{i=1}^{3} m_i (\bar{\nu}_{iL} \nu_{iR} + \bar{\nu}_{iR} \nu_{iL}) = -\sum_{i=1}^{3} m_i \bar{\nu}_i \nu_i. \qquad (5.16)$$

In the more generation case, flavour eigenvalues and mass eigenvalues of massive neutrinos do not coincide. In the EW interaction a neutrino is always produced, or absorbed, in a definite flavour state. But, as in the quark case, the flavour eigenstates (states with definite flavour) are not identical to the mass eigenstates (states which have definite mass) for massive neutrinos. The consequences of this mismatch will be discussed in Chap. 6.

Let us observe that the eigenvalues of the mass matrix $M$ and the ones of the Yukawa matrix are connected by the relation $m_i = y_{\nu i} v/\sqrt{2}$, where $v \sim 246$ GeV [169]. The neutrino masses from experiment are expected to be of the order of fractions of eV, and this implies that the matrix elements of $Y_\nu$ are much smaller than the corresponding Yukawa values for other leptons, for instance about $10^{-6}$ smaller than the Yukawa values for the lightest charged lepton, the electron.

If $\nu_R$ is the same type of field as right-handed components of other fermions, such smallness looks rather unnatural.

In the presence of the Dirac mass terms, neutrinos of different flavour mix into states of definite mass, and each individual lepton flavour number, $L_e$, $L_\mu$ and $L_\tau$, is violated, but the total lepton number $L$ is not.[2] In fact, the neutrino Dirac mass term and other terms of the total Lagrangian are invariant under the following phase transformations for the neutrino $\nu_i$ and the charged lepton $\ell_i$

$$\nu_{i\,L(R)} \to e^{i\alpha} \nu_{i\,L(R)} \qquad \ell_{i\,L(R)} \to e^{i\alpha} \ell_{i\,L(R)} \tag{5.17}$$

where $\alpha$ is an arbitrary constant. From invariance under this global, continuous $U(1)$ transformation, it follows, from the Nöther's theorem, a conserved charge, identified with the total lepton number, for all charged leptons and neutrinos. Such accidental $U(1)$ symmetry is also discussed in Sect. 4.9. By convention we assign a value $L = 1$ to leptons, hence $L = -1$ to antileptons. For both charged and neutral leptons $\psi_\ell$, Dirac mass term are of the type $\bar\psi_\ell \psi_\ell$ and results in transitions of (anti-)leptons to (anti-)leptons, with $\Delta L = 0$, since a $\psi_\ell$ field annihilates a lepton and create an anti-lepton, and $\bar\psi_\ell$ acts oppositely. Lepton number violating (LNV) processes are forbidden. This selection rule, that holds in the SM for charged leptons, could be lifted for neutrinos, should their nature be different from the Dirac one just presented. Then, as we will discuss in the next sections, other types of mass terms, where $\Delta L \neq 0$, make their appearance and LNV transitions for neutrinos become, in principle, possible.

## 5.3 Majorana Neutrinos

Under charge conjugation, every particle is carried into its antiparticle (see Sects. 2.7 and 2.9). The positron is the electron antiparticle; it is clearly a different particle, since it has opposite electric charge. A charged particle cannot coincide with its anti-particle, otherwise it violates electric charge conservation. The same happens if a particle carries a conserved quantum number, like the strangeness, since its antiparticle has opposite quantum number. In the SM neutral baryons cannot coincide with their antiparticles since the baryon number $B$ is conserved and baryons and anti-baryons have opposite $B$ number. On the contrary, there are neutral bosons that do not carry quantum numbers, like for instance the photon and the $\pi^0$, which transform into themselves by charge conjugation, and which are thus their own antiparticles. As seen in Sect. 1.2, in 1937 Majorana first suggested that a neutral fermion could also behave this way [20], and Racah [21] pointed out that the neutrino could be such a particle. A neutrino which is identical with its antiparticle is called a Majorana neutrino.

---

[2] In analogy with the conservation of the baryonic number in the quark sector, see Sect. 2.6.

A Majorana neutrino is described by a Dirac spinor $\nu$ that satisfies the constraint

$$\nu^C = e^{i\xi}\nu \tag{5.18}$$

where $\xi$ is an arbitrary phase (sometimes called the Majorana creation phase).

For Majorana neutrinos we cannot define a lepton quantum number $L$, which by definition would have opposite values for the lepton and its anti-lepton. This is a manifest difference with the case described in Sect. 5.2, where we have discussed a massive neutrino, indicated as a Dirac neutrino, equipped with a lepton quantum number $L$ which is not violated in EW interactions. There are other noticeable differences. In the Majorana case, a $CPT$ transformation changes only the helicity:

$$\nu(\mathbf{p}, h) \xrightarrow{CPT} \nu(\mathbf{p}, -h) \tag{5.19}$$

while in the Dirac case

$$\nu(\mathbf{p}, h) \xrightarrow{CPT} \bar{\nu}(\mathbf{p}, -h) \tag{5.20}$$

as shown in Sect. 2.13. The number of degrees of freedom for each value of the momentum is equal to the number of independent states with the same momentum. For a 4-component Dirac neutrino with positive and negative helicity $\pm h$ and fixed momentum $\mathbf{p}$ there are four possibilities

$$\nu(\mathbf{p}, h) \quad \nu(\mathbf{p}, -h) \quad \bar{\nu}(\mathbf{p}, h) \quad \bar{\nu}(\mathbf{p}, -h). \tag{5.21}$$

The Majorana particle coincides with its antiparticle, which halves the allowed four degrees of freedom for Dirac spinors. Since a Majorana neutrino is self-charge conjugated, we have only two possible states of a Majorana neutrino for each value of the momentum $\mathbf{p}$

$$\nu(\mathbf{p}, h) \quad \nu(\mathbf{p}, -h). \tag{5.22}$$

Instead of four independent components, a Majorana spinor has only two. It may seem that the massive Majorana neutrino is described by the two component Weyl spinor, but it is not, and a spinor cannot have Majorana and Weyl character at the same time. In fact, a Weyl spinor by definition has a definite chirality and its conjugate spinor, also a Weyl spinor, has opposite chirality (see Eq. (2.167)); hence the basic Eq. (5.18) does not hold. This difference disappears in the massless case, since in both Dirac and Majorana cases the left-handed and right-handed chiral components of the massless neutrino field obey the decoupled Weyl equations (5.4). However, one could argue that in the limit of zero mass the two components are

## 5.3 Majorana Neutrinos

independent in the Weyl case, while in the Majorana case they may still considered to be connected by the relation

$$\nu_R^C \equiv (\nu_L)^C. \tag{5.23}$$

In the massless case, helicity and chirality coincide and are conserved. Since only the left-handed chiral component of the neutrino field interacts weakly and the right-handed chiral component is irrelevant for interactions, the phenomenological consequences are the same in both the Dirac and Majorana descriptions of a neutrino. They have different phenomenological consequences only if the neutrino is massive. In practice one can unveil the Dirac and Majorana nature of neutrinos only by measuring some effect due to the neutrino mass, because otherwise the massless theory applies in an effective way. In addition, the mass effect must not be of kinematical nature, since the kinematical effects of Dirac and Majorana masses are the same. For example, Dirac and Majorana neutrinos cannot be distinguished by means of neutrino oscillations (see Sect. 10.3).

In principle, the alternative between massive Majorana and Dirac descriptions could be simply tested by a neutrino beam colliding on a fixed nucleus target.[3] The neutrinos in the beam are assumed generated by the charged weak interaction $W^+ \to e^+ \nu_e$. They would be described as negative helicity particles, since weak interactions only couple to left-handed states. In case of neutrinos of Dirac nature, their collision with the fixed target would create electrons, by conservation of the lepton number. If now an observer "overtakes" the neutrino by going into a reference frame which travels with a higher speed than the neutrino beam, in the direction of the neutrino velocity, he would see them as positive helicity, provided he uses the same helicity convention in the center of mass frame. In the moving frame, Dirac neutrinos become sterile, and do not produce charged leptons while colliding with the target. In the Majorana case, as we "overtake" the neutrino beam, the left-handed Majorana neutrino transforms into a right-handed Majorana neutrino, which however, because the Majorana is its own anti-particle, needs to interact as if it were a right-handed Majorana anti-neutrino, hence it is not sterile. Majorana neutrinos incident on a target at rest in the frame in which they are born would generate electrons, but when one boosts past it and the helicity flips, the interaction with a target at rest in the boosted frame would produce positrons. This is consistent with the lepton-number non-conservation induced into the SM by Majorana neutrinos. The discriminating effects are suppressed by correction factors of order $(m_\nu/E_\nu)^2$. The observation just outlined is technically unfeasible, since neutrinos studied directly so far are ultra-relativistic and cannot be overtaken. As in the case of massless neutrino, their behaviour appears almost completely insensitive to whether they are Dirac or Majorana particles. In practice, the most promising way to find if

---

[3] This "gedanken" experiment has been discussed in literature with several names ("helicity reversal paradox", "autobahn helicity paradox", "rabbit paradox"), see e.g. Ref. [170].

neutrinos are Majorana particles is the search for neutrinoless double $\beta$ decay, that we will study in Sect. 10.3.

A Majorana field can be expressed in terms of creation and destruction operators as a Dirac field, with the additional constraint that there is only one type of operator, since there is no distinction between particles and antiparticles. We can write

$$v(x) = \int \frac{d^3 p}{(2\pi)^3 2E} \sum_{h=\pm 1} \left[ a^h(p) u^h(p) e^{-ip\cdot x} + a^{h\dagger}(p) v^h(p) e^{ip\cdot x} \right] \quad (5.24)$$

where the operators $a^h(p)$ and their complex conjugates satisfy the canonical anti commutation relations. Although there is no distinction between particles and antiparticles, in practice it is customary to speak of neutrinos and antineutrinos also in the Majorana case. Majorana neutrinos with negative helicity are generally called neutrinos $v \equiv v(\mathbf{p}, -h)$, while Majorana neutrinos with positive helicity are often referred to as antineutrinos $\bar{v} \equiv v(\mathbf{p}, h)$. The weak charged leptonic interactions, expressed in one generation case in the form $\bar{\ell}_L \gamma^\mu v_L$, creates (right-handed) antineutrinos in the Dirac case, and destroys (left-handed) neutrinos. In the Majorana case, it mainly creates ultra relativistic Majorana neutrinos with positive helicity. Majorana neutrinos with negative helicity are suppressed by the ratio $m_v/E_v$.

## 5.4 Majorana Mass Terms

If neutrinos have a Majorana nature, it is not necessary to hypothesise a separate right-handed, sterile, Weyl neutrino, as we did for the massive Dirac neutrino (see Sect. 5.2). Indeed we can use the charge conjugate of the left-handed chiral component of the neutrino field in Eq. (5.23)

$$v_R^C \equiv (v_L)^C \quad (5.25)$$

which is right-handed. Then, in analogy with Eq. (5.6), we build a mass eigenstate of the form

$$\chi = v_L + v_R^C. \quad (5.26)$$

The Majorana fermion $\chi$ is self-conjugate by construction. We have

$$\chi = \chi^C \quad (5.27)$$

## 5.4 Majorana Mass Terms

corresponding to $e^{i\xi} = 1$ in Eq. (5.18). Its mass term reads

$$\mathcal{L}_{M_L} = -\frac{1}{2} m_{M_L} \overline{\chi} \chi =$$
$$= -\frac{1}{2} m_{M_L} [\overline{\nu_L} (\nu_L)^C + \overline{(\nu_L)^C} \nu_L] =$$
$$= -\frac{1}{2} m_{M_L} (\overline{\nu_L} \nu_R^C + \overline{\nu_R^C} \nu_L) \qquad (5.28)$$

in the case of one generation, in analogy to Eq. (5.16). The factor 1/2 avoids the double counting in the Eulero-Lagrange equations for the fields. It is not present in the Dirac mass term.

By using left and right projection operators we can recover the two components

$$\nu_L = P_L \chi \qquad \nu_R^C = P_R \chi. \qquad (5.29)$$

Let us remark that for any particle spinor the chirality switches if one passes to its Dirac adjoint, namely (see Eq. (2.165))

$$\nu_{L(R)} \to \overline{\nu_{L(R)}} = \overline{\nu} P_{R(L)}. \qquad (5.30)$$

Therefore, while $\nu_{L(R)}$ and its complex conjugate $(\nu_{L(R)})^C \equiv \nu_{R(L)}^C$ have opposite chirality, $\nu_{L(R)}$ and $\overline{(\nu_{L(R)})^C} = \overline{\nu_{R(L)}^C}$ have the same chirality.

The Majorana mass term $\mathcal{L}_{M_L}$ is also not consistent with the SM, since it is not invariant under the $SU(2)_L \times U(1)_Y$ SM gauge symmetries.[4] Indeed, in the SM the left-handed neutrinos are in the same $SU(2)_L$ doublet as the charged leptons, so without the breaking of $SU(2)_L$, any term that exists for neutrinos must also exist for charged leptons. For the latter, however, a Majorana mass term is forbidden because the charges of a particle and its anti-particle have opposite values, as all quantum numbers. Charged fermions cannot be self-conjugate.

Let us notice that we can express the Majorana mass term in Eq. (5.28) in a different form yet

$$\mathcal{L}_{M_L} = -\frac{1}{2} m_{M_L} (\overline{\nu_L} C \overline{\nu_L}^T - \nu_L^T C^\dagger \nu_L). \qquad (5.31)$$

In fact, from Eq. (2.127) we have

$$\overline{\nu^C} = -\nu^T C^{-1} \qquad (5.32)$$

---

[4] We will discuss further this point in Sect. 5.5.

and the same relation holds for the left and right-handed projections

$$\overline{(\nu_{L(R)})^C} = -\nu_{L(R)}{}^T C^{-1}. \tag{5.33}$$

The form of the Majorana mass term in Eq. (5.31) follows from Eqs. (2.168) and (5.33), together with the relation $C^{-1} = C^\dagger$.

Until now, we have discussed a minimal Majorana scheme, where only left-handed flavour neutrino fields $\nu_L$ and their charge conjugate states enter into the Lagrangian. However, in Sect. 5.2, we have assumed the possibility of the existence of a sterile right-handed neutrino field, which does not participate in weak interactions and it is a singlet under the $SU(2)_L \times U(1)$ gauge group. This hypothesis allows to build not only a Dirac mass term, but another, different, Majorana mass term as well. Indeed, by indicating the right-handed sterile neutrino with $\nu_R$, and taking its charge conjugate[5]

$$\nu_L^C \equiv (\nu_R)^C \tag{5.34}$$

we can define another self-conjugate neutral fermion

$$\omega = \nu_R + \nu_L^C. \tag{5.35}$$

This is again a Majorana neutrino, since

$$\omega = \omega^C. \tag{5.36}$$

By application of left and right projection operators, we recover the chiral components

$$\nu_R = P_R \omega \qquad \nu_L^C = P_L \omega. \tag{5.37}$$

The mass term for the Majorana neutrino $\omega$ reads

$$\begin{aligned}\mathscr{L}_{M_R} &= -\frac{1}{2} m_{M_R} \overline{\omega} \omega = -\frac{1}{2} m_{M_R} [\overline{\nu_R} \, (\nu_R)^C + \overline{(\nu_R)^C} \, \nu_R] = \\ &= -\frac{1}{2} m_{M_R} (\overline{\nu_R} \, \nu_L^C + \overline{\nu_L^C} \, \nu_R) = \\ &= -\frac{1}{2} m_{M_R} (\overline{\nu_R} C \overline{\nu_R}^T - \nu_R^T C^\dagger \nu_R). \end{aligned} \tag{5.38}$$

From the last expression in Eq. (5.38), it is self-evident that $\mathscr{L}_{M_R}$ satisfies the symmetries of the SM, because $\nu_R$ is a singlet of the SM gauge group. Therefore,

---

[5] Let us observe that in the Dirac case we used a different symbol for a sterile right handed neutrino, namely $N_R$. In this section, instead, we indicate the sterile neutrino with $\nu_R$.

## 5.4 Majorana Mass Terms

both Dirac $\mathscr{L}_D$ and Majorana $\mathscr{L}_{M_R}$ mass terms are allowed in the framework of the SM augmented with a right-handed chiral field $\nu_R$.

Under any global $U(1)$ transformations (5.17), that is

$$\nu_{L(R)} \to e^{i\alpha} \nu_{L(R)} \tag{5.39}$$

where $\alpha$ is an arbitrary constant, we have

$$\overline{\nu_{L(R)}} \, (\nu_{L(R)})^C + \overline{(\nu_{L(R)})^C} \, \nu_{L(R)} \to e^{-2i\alpha} \overline{\nu_{L(R)}} \, (\nu_{L(R)})^C + e^{2i\alpha} \overline{(\nu_{L(R)})^C} \, \nu_{L(R)}. \tag{5.40}$$

The mass term $\mathscr{L}_{M_R}$ is not invariant, and the same holds for the mass term $\mathscr{L}_{M_L}$. As expected, there is no conserved lepton number distinguishing leptons and antileptons. If one insists in defining it, the total lepton number $L$ is violated by both Majorana mass terms. By assigning $L = 1$ to leptons and $L = -1$ to antileptons, we can say that both Majorana mass terms, $\mathscr{L}_{M_L}$ and $\mathscr{L}_{M_R}$, viewed as an interaction, generate $\Delta L = \pm 2$ transitions. Majorana mass terms connect conjugate states with opposite chirality. They represent a transition from a left(right)-handed antineutrino into a right(left)-handed neutrino. Equivalently, they can be viewed as the creation or annihilation of two neutrinos, therefore allowing neutrinoless double beta decay, as we will see in Sect. 10.3. This is not what happens for the Dirac mass term, which transforms a neutrino into a neutrino or an antineutrino into an antineutrino, in agreement with the conservation of the lepton number $L$. A pictorial representation is given in Fig. 5.1.

In case of a generic number $n$ of families, the Majorana mass terms change since the Majorana masses $m_{M_L}$ and $m_{M_R}$ become complex $n \times n$ matrices, $M_{M_L}$ and

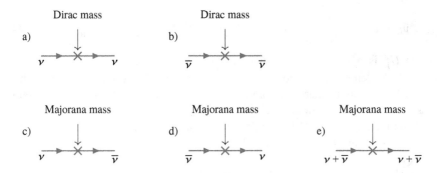

**Fig. 5.1** The effects of mass terms in the Lagrangian: Dirac terms (**a**) incoming and outcoming neutrino (**b**) incoming and outcoming anti-neutrino; Majorana terms: (**c**) incoming neutrino and outcoming anti-neutrino (**d**) incoming antineutrino and outcoming neutrino (**e**) mass eigenstate $\nu + \bar{\nu}$ sent back into itself

$M_{M_R}$. It is easy to show that each mass matrix $M_{M_{L(R)}}$ is symmetric, that is that it is equal to its transpose:

$$M_{M_L(M_R)} = M_{M_L(M_R)}^T. \tag{5.41}$$

Let us consider for example Eq. (5.31) in the case of $n$ generations

$$\mathscr{L}_{M_L} = \frac{1}{2} \sum_{i,j=1}^n v_{iL}^T C^\dagger (M_{M_L})_{ij} v_{jL} + h.c. \tag{5.42}$$

The product $v_{iL}^T C^\dagger v_{jL}$ is a scalar, and it is equal to its transpose. Then by taking its transpose we have $v_{iL}^T C^\dagger v_{jL} = (v_{iL}^T C^\dagger v_{jL})^T = v_{jL}^T C^\dagger v_{iL}$. In the last passage we have considered that the fermion fields are anticommuting, as well as the relation $C^T = -C$, given in Eq. (2.124). By substituting in Eq. (5.42), we find

$$\sum_{i,j=1}^n v_{iL}^T C^\dagger (M_{M_L})_{ij} v_{jL} = \sum_{i,j=1}^n v_{jL}^T C^\dagger (M_{M_L})_{ij} v_{iL} =$$

$$= \sum_{i,j=1}^n v_{iL}^T C^\dagger (M_{M_L})_{ji} v_{jL}. \tag{5.43}$$

Comparing the left and right sides, one can see that $M_{M_L}$ is symmetric. A complex symmetrical matrix can be diagonalised with the help of one unitary matrix [171]. The diagonalization of the mass matrix corresponds to a transformation of the Majorana neutrino states analogous to the one in Eq. (5.14) for Dirac neutrino states, namely

$$\hat{v}_L = U_L^\nu \tilde{v}_L \qquad \hat{v}_R^C \equiv (\hat{v}_L)^C = (U_L^\nu)^C (\tilde{v}_L)^C \equiv (U_L^\nu)^C \tilde{v}_R^C \tag{5.44}$$

where $U_L^\nu$ is a $n \times n$ unitary matrix. One passes from the flavour eigenstates, collectively denoted as $\hat{v}_L$, to the Majorana states $\tilde{v} = \tilde{v}_L + \tilde{v}_R^C \equiv (v_1, v_2, \ldots v_n)$ which have definite mass and are described by four component spinors. Same reasonings can be followed to diagonalize $M_{M_R}$. In this case, massive Majorana neutrino have the form of Eq. (5.35) and $n$ can be arbitrary. Instead, in the case of the mass term $M_{M_L}$, we are constrained to $n = 3$, the number of SM flavours $\hat{v}_L \equiv (v_e, v_\mu, v_\tau)$.

Finally, let us mention the following useful relation

$$\overline{v_1^C} v_2^C = \overline{v_2} v_1 \tag{5.45}$$

which follows from Eqs. (2.122) and (2.127).

## 5.5 Mass Terms and Gauge Symmetry

Let us now discuss in more detail if the mass terms for neutrinos are compatible with the SM, that is if they are invariant under the SM gauge symmetries or can be derived from gauge invariant operators through SSB.

Neutrino fields, as all particles in the SM, are classified according to the representation of their left or right handed components under the gauge group $SU(3)_C \otimes SU(2)_L \otimes U(1)_Y$. Being leptons, they are singlets under $SU(3)_C$, and therefore the dimension of their color representation is fixed to 1. Hence left or right handed neutrino components can be labelled by the pair $(I, Y)$, where $I$ is the dimension of the $SU(2)_L$ representation and $Y$ is the value of the hypercharge.

The SM neutrino $\nu_L$ is left-handed and it belongs to a $SU(2)_L$ doublet with $T_3 = -Y = 1/2$ (see Eq. (4.3)). Charge conjugation changes the sign of the hypercharge and the chirality (see Sect. 4.1), but it does not change the representation of the $SU(2)_L$ gauge group. Thus the conjugate state $\nu_R^C = (\nu_L)^C$, even if right-handed, still belongs to a $SU(2)_L$ doublet, although with opposite $T_3$ and $Y$. We label them as

$$\nu_L \sim (2, -1/2) \qquad \nu_R^C \sim (2, 1/2). \tag{5.46}$$

If we extend the SM to include a right handed neutrino $\nu_R$ that does not participate to weak interactions (sterile), we have to accommodate it in a weak isospin singlet state ($T_3 = 0$) with $Y = 0$. Its (left-handed) conjugate state $\nu_L^C = (\nu_R)^C$ will be in the same $SU(2)$ representation, and also have zero hypercharge

$$\nu_R \sim (1, 0) \qquad \nu_L^C \sim (1, 0). \tag{5.47}$$

We can write these labels for another $SU(3)_C$ singlet, the SM Higgs. We have (see Eqs. (4.77) and (4.81))

$$\phi \sim (2, 1/2) \qquad \tilde{\phi} \sim (2, -1/2). \tag{5.48}$$

As seen in Sects. 5.2 and 5.4, the introduction of a right-handed component for the neutrino state can lead to both a Dirac and a Majorana $\mathscr{L}_{M_R}$ mass term. In the former case, we can build a gauge invariant Yukawa coupling as in the case of quarks (see Eq. (5.10)), but using a neutrino doublet and a neutrino singlet, namely

$$\mathscr{L}_D = -Y_D (\bar{\nu}\, \bar{e})_L \tilde{\phi}\, \nu_R + h.c. \tag{5.49}$$

in the one-generation case. After SSB, it converts to the usual Dirac mass term $-m_\nu \bar{\nu} \nu$, where the the neutrino mass is $m_\nu = Y_D\, v/\sqrt{2}$ and $v/\sqrt{2}$ is the non-zero VEV of the SM Higgs field. In a group theoretical language, $\mathscr{L}_D$ represents a $SU(2)$ singlet, pulled out from the product of $SU(2)$ representations

$$2 \otimes 2 \otimes 1 = (3 \oplus 1) \otimes 1 = 3 \oplus 1. \tag{5.50}$$

We do not need the Higgs doublet to restore gauge invariance in the case of the Majorana $\mathscr{L}_{M_R}$ mass term (5.38). As it can be seen immediately by this formulation

$$\mathscr{L}_{M_R} = -\frac{1}{2} m_{M_R} [\overline{\nu_R}\, \nu_L^C + \overline{\nu_L^C}\, \nu_R]. \tag{5.51}$$

$\mathscr{L}_{M_R}$ is not only Lorentz invariant but also gauge invariant, because $\nu_R$ and $\nu_L^C$ are singlets of the SM gauge group.

The absence of right-handed neutrinos does not prevent the existence of mass terms for neutrinos. They are of the Majorana type as in Eq. (5.28)

$$\mathscr{L}_{M_L} = -\frac{1}{2} m_{M_L} [\overline{\nu_L}\, (\nu_L)^C + \overline{(\nu_L)^C}\, \nu_L]. \tag{5.52}$$

However, $\mathscr{L}_{M_L}$ is not gauge invariant. It breaks the symmetry between the two components of the $SU(2)$ doublet, by including only neutrinos. Moreover, it is not $U(1)_Y$ invariant; for instance we can easily see that

$$\overline{(\nu_L)^C}\, \nu_L = \overline{\nu_R^C}\, \nu_L \tag{5.53}$$

has hypercharge Y=−1. In order to be included in the SM Lagrangian, $\mathscr{L}_{M_L}$ should be derived through SSB by a gauge invariant term. In analogy to Eq. (5.49), we are tempted to use the coupling

$$(\overline{\nu}\, \overline{e})_L\, \tilde{\phi}\, \nu_R^C. \tag{5.54}$$

This four dimensional term couples $\nu_R^C = (\nu_L)^C$ to the SM left handed lepton doublet and to the Higgs field, but it is gauge invariant. In fact, it is composed by three fields that behave as doublets under $SU(2)_L$, and this composition cannot produce a $SU(2)_L$ singlet. In group theoretical terms, we express it as

$$2 \otimes 2 \otimes 2 = (3 \oplus 1) \otimes 2 = (3 \otimes 2) \oplus 2 = 4 \oplus 2 \oplus 2. \tag{5.55}$$

One possibility to induce the Majorana mass term $\mathscr{L}_{M_L}$ while preserving gauge invariance is to extend the Higgs sector of the SM. For instance, one can introduce a new Higgs scalar field $\Delta$ with hypercharge Y=+1 transforming as a triplet under $SU(2)_L$, and add to the SM Lagrangian a term indicated symbolically as

$$f_\Delta \left( \overline{(\nu_L)^C}\, \overline{(e_L)^C} \right) \otimes \Delta \otimes \begin{pmatrix} \nu_L \\ e_L \end{pmatrix} + h.c \tag{5.56}$$

where $f_\Delta$ is an arbitrary coupling. The group theoretical analysis for a term including $\Delta$ and two $SU(2)$ doublets gives

$$2 \otimes 3 \otimes 2 = (2 \otimes 3) \otimes 2 = (4 \oplus 2) \otimes 2 = (4 \otimes 2) \oplus (2 \otimes 2) = 5 \oplus 3 \oplus 3 \oplus 1. \tag{5.57}$$

which contains a $SU(2)$ singlet. After SSB, a non zero VEV $\langle\Delta\rangle$ of the new Higgs field $\Delta$ would produce a Majorana neutrino mass $m_{M_L} = f_\Delta \langle\Delta\rangle$.

## 5.6 Dirac-Majorana Mass Term

When the interacting $\nu_L$ and sterile $\nu_R$ neutrinos (and their C-conjugates) are simultaneously present, the most general mass term includes Dirac and Majorana mass terms. In the case of $n$ lepton families, one writes

$$\mathscr{L}_{DM} = \mathscr{L}_D + \mathscr{L}_{M_L} + \mathscr{L}_{M_R} = -\frac{1}{2} \overline{\nu_R} M_D \nu_L - \frac{1}{2} \overline{(\nu_L)^C} M_D^T \nu_R^C +$$
$$- \frac{1}{2} \overline{(\nu_L)^C} M_{M_L} \nu_L - \frac{1}{2} \overline{\nu_R} M_{M_R} (\nu_R)^C + h.c. \tag{5.58}$$

where $\nu = (\nu_{\alpha_1}, \nu_{\alpha_2}, \ldots, \nu_{\alpha_n})$ and $M_D$, $M_{M_L}$ and $M_{M_R}$ are $n \times n$ complex matrices. We know from Sect. 5.4 that $M_{M_L}$ and $M_{M_R}$ are symmetric, that is $M_{M_L} = M_{M_L}^T$ and $M_{M_R} = M_{M_R}^T$. In Eq. (5.58), the Dirac term (5.12) has been divided in two, and half of it has been transformed as

$$\overline{(\nu_L)^C} M_D^T (\nu_R)^C = -\nu_L^T C^{-1} M_D^T C \overline{\nu_R}^T = \overline{\nu_R} M_D \nu_L \tag{5.59}$$

taking into account the anticommuting properties of the fermionic fields, together with Eqs. (2.168) and (5.33). This transformation is functional to another way to write $\mathscr{L}_{DM}$, that is

$$\mathscr{L}_{DM} = -\frac{1}{2} \left( \overline{(\nu_L)^C} \; \overline{\nu_R} \right) \begin{pmatrix} M_{M_L} & M_D^T \\ M_D & M_{M_R} \end{pmatrix} \begin{pmatrix} \nu_L \\ (\nu_R)^C \end{pmatrix} + h.c. =$$
$$= -\frac{1}{2} \overline{(n_L)^C} M_{DM} n_L + h.c. \tag{5.60}$$

In the last passage, we have defined the $2n \times 2n$ complex, symmetric, matrix $M_{DM}$ as

$$M_{DM} \equiv \begin{pmatrix} M_{M_L} & M_D^T \\ M_D & M_{M_R} \end{pmatrix} \tag{5.61}$$

and the $2n$ column matrix of left handed chiral fields as

$$n_L \equiv \begin{pmatrix} \nu_L \\ (\nu_R)^C \end{pmatrix}. \tag{5.62}$$

Let us observe that the definition of $n_L$ is not unique. One can also interchange chiral components by taking elements of the Lagrangian's hermitian conjugate part, defining for instance $n_L \equiv ((\nu_L)^C, \nu_R)^T$. This choice just interchanges the position of $M_D$ and $M_D^T$ in the full $M_{DM}$ matrix, leaving it complex and symmetric.

The Dirac-Majorana mass term has the structure of a Majorana Mass term for two chiral neutrino fields coupled by the Dirac mass. The chiral fields are not the mass eigenstates—these are found by diagonalising the matrix $M_{DM}$, as we will see in the following.

### 5.6.1 One-Generation Case

For simplicity, we analyse the case $n = 1$, where

$$M_{DM} = \begin{pmatrix} m_{M_L} & m_D \\ m_D & m_{M_R} \end{pmatrix} \qquad n_L \equiv \begin{pmatrix} \nu_L \\ (\nu_R)^C \end{pmatrix}. \tag{5.63}$$

The chiral fields $\nu_L$ and $\nu_R$ refers to one lepton generation, and $\nu_L$ and $(\nu_R)^C$ are the left-handed fields in the flavour basis. The two flavour fields $\nu_L$ and $\nu_R$ are, respectively, active and sterile, because $\nu_L$ participates in weak interactions, whereas $\nu_R$ is a singlet of the SM gauge symmetries. The matrix $M_{DM}$ is a $2 \times 2$ symmetric matrix, whose components $m_D, m_{M_R}, m_{M_L}$ come from the Dirac and Majorana terms of Eqs. (5.12), (5.28) and (5.38). Because of their simultaneous presence in the Lagrangian, $m_D, m_{M_R}$ and $m_{M_L}$ have lost their role as mass terms and represent arbitrary parameters, which can be complex. In fact, if for instance $m_{M_R}$ and $m_D$ are chosen to be real and positive by an appropriate rephasing of the chiral fields $\nu_R$ and $\nu_L$, respectively, there is no additional freedom to cancel a possible complex phase of $m_{M_L}$. The same holds for other equivalent choices.

In order to find the neutrino fields of definite mass it is necessary to diagonalize $M_{DM}$. Since it is a symmetric matrix, it can always be done by means of a unitary matrix $U$ [171]

$$M_{DM} = (U^\dagger)^T m U^\dagger \equiv (U^\dagger)^T \begin{pmatrix} m_1 & 0 \\ 0 & m_2 \end{pmatrix} U^\dagger \tag{5.64}$$

where $m \equiv \text{diag}(m_1, m_2)$ with real and positive $m_{1,2}$. By substitution in the Dirac-Majorana mass terms of the Lagrangian Eq. (5.60), we have

$$\mathscr{L}_{DM} = -\frac{1}{2} \overline{(n_L)^C} M_{DM} n_L + h.c. = -\frac{1}{2} \overline{(n_L)^C} (U^\dagger)^T m U^\dagger n_L + h.c. =$$

$$= -\frac{1}{2} \overline{(N_L)^C} m N_L + h.c. \tag{5.65}$$

## 5.6 Dirac-Majorana Mass Term

In the last passage we have defined

$$N_L = U^\dagger n_L \equiv \begin{pmatrix} \nu_{1L} \\ \nu_{2L} \end{pmatrix} \qquad (5.66)$$

and used Eq. (5.32) to identify

$$\overline{(N_L)^C} = \overline{(U^\dagger n_L)^C} = -(U^\dagger n_L)^T C^{-1} = -n_L^T (U^\dagger)^T C^{-1} = \overline{(n_L)^C} (U^\dagger)^T. \qquad (5.67)$$

We can go on, and transform the Hermitian conjugate part according to the relations

$$(\overline{(N_L)^C} \, m \, N_L)^\dagger = -N_L^\dagger m \, (N_L^T C^{-1})^\dagger = -N_L^\dagger m \, C N_L^*$$

$$\overline{N_L} m (N_L)^C = \overline{N_L} m \, C \overline{N_L}^T = N_L^\dagger \gamma_0 \, m \, C \gamma_0^T N_L^* = -N_L^\dagger m \, C N_L^*. \qquad (5.68)$$

In the first equality we have used the relation $(C^{-1})^\dagger = C$ in Eq. (2.124), in the second one the relations $C\gamma_0^T = -\gamma^0 C$ and $\gamma_0^2 = 1$. Finally, we obtain

$$\mathscr{L}_{DM} = -\frac{1}{2} \overline{(N_L)^C} \, m \, N_L + h.c. = -\frac{1}{2} \left( \overline{(N_L)^C} \, m \, N_L + \overline{N_L} m (N_L)^C \right). \qquad (5.69)$$

We are now in the position to define the Majorana doublet field

$$N \equiv N_L + (N_L)^C = \begin{pmatrix} \nu_{1L} + (\nu_{1L})^C \\ \nu_{2L} + (\nu_{2L})^C \end{pmatrix} = \begin{pmatrix} \nu_{1L} + \nu_{1R}^C \\ \nu_{2L} + \nu_{2R}^C \end{pmatrix} \equiv \begin{pmatrix} \nu_1 \\ \nu_2 \end{pmatrix} \qquad (5.70)$$

which describes particles with defined mass. By using the previous relations and taking care of cancellations due to different chiralities, we can write

$$\mathscr{L}_{DM} = -\frac{1}{2} \left( \overline{(N_L)^C} \, m \, N_L + \overline{N_L} m (N_L)^C \right) =$$

$$= -\frac{1}{2} \overline{N} \, m \, N = -\frac{1}{2} \sum_{k=1,2} m_k \, \bar{\nu}_k \nu_k. \qquad (5.71)$$

We have just shown that in the case of one generation with both left-handed and right-handed chiral neutrino fields $\nu_L$ and $\nu_R$, the diagonalization of the most general Dirac–Majorana mass term implies that there are two different neutrino states $\nu_1$ and $\nu_2$ with real and positive masses $m_1$ and $m_2$, respectively

$$\nu_i = \nu_{iL} + (\nu_{iL})^C = \nu_{iL} + \nu_{iR}^C \qquad i \in \{1, 2\}. \qquad (5.72)$$

The massive neutrinos $\nu_1$ and $\nu_2$ are obviously Majorana particles, since they are self-conjugates.[6] The new chiral fields $\nu_{1L}$ and $\nu_{2L}$ represent their chiral components.

### 5.6.1.1 Real Dirac-Majorana Mass Matrix

In order to understand the implications of the Dirac–Majorana mass term, it is useful to consider the simplest case of a real mass matrix. When $M_{DM}$ in Eq. (5.63) is real, it can be diagonalised by an orthogonal matrix $O$

$$M_{DM} = O m' O^T = O \begin{pmatrix} m'_1 & 0 \\ 0 & m'_2 \end{pmatrix} O^T \qquad O^T O = 1 \qquad (5.73)$$

where $m' = \mathrm{diag}(m'_1, m'_2)$. Any $2 \times 2$ orthogonal matrix can be expressed in terms of a generic mixing angle $\theta$ without loss of generality

$$O = \begin{pmatrix} \cos\theta & \sin\theta \\ -\sin\theta & \cos\theta \end{pmatrix} \qquad \tan 2\theta = \frac{2 m_D}{m_{M_R} - m_{M_L}}. \qquad (5.74)$$

The eigenvalues of the mass matrix (5.73) are

$$m'_{2,1} = \frac{1}{2} \left[ m_{M_L} + m_{M_R} \pm \sqrt{(m_{M_L} - m_{M_R})^2 + 4 m_D^2} \right] \qquad (5.75)$$

which are not necessarily positive. In order to ensure positivity for the mass values we observe that a complex, unitary matrix $U$ which diagonalises the symmetric matrix $M_{DM}$ can be built from the orthogonal matrix $O$ introducing ad-hoc phases. Namely, we define

$$U^\dagger \equiv \sqrt{\rho}\, O^T \qquad \rho \equiv \begin{pmatrix} \rho_1 & 0 \\ 0 & \rho_2 \end{pmatrix} \qquad |\rho_k^2| = 1 \qquad (5.76)$$

where $\rho$ is a phase matrix. The role of the matrix $\rho$ is to change the sign of the mass eigenvalues, ensuring that the mass eigenvalues is positive, thus the only possibilities are $\rho_k = \pm 1$. The relation (5.73) can be recast in the same form than Eq. (5.64), that is

$$M_{DM} = (U^\dagger)^T m\, U^\dagger = \sqrt{\rho}\, O \begin{pmatrix} m_1 & 0 \\ 0 & m_2 \end{pmatrix} \sqrt{\rho}\, O^T \qquad U^\dagger U = 1 \qquad (5.77)$$

---

[6] In Sect. 5.6.1.1 we will see that there are special cases when the Dirac–Majorana mass term reduces to the Dirac mass term and we recover a massive Dirac neutrino.

## 5.6 Dirac-Majorana Mass Term

with $m = \mathrm{diag}(m_1, m_2)$ and

$$m'_k = m_k \rho_k. \tag{5.78}$$

An appropriate choice of $\rho_k$ gives positive mass eigenvalues.

Inverting now the formula (5.66), we have the mixing relation

$$n_L = \begin{pmatrix} \nu_L \\ (\nu_R)^C \end{pmatrix} = U N_L = U \begin{pmatrix} \nu_{1L} \\ \nu_{2L} \end{pmatrix} \tag{5.79}$$

where

$$U = \begin{pmatrix} (\sqrt{\rho_1})^* \cos\theta & (\sqrt{\rho_2})^* \sin\theta \\ -(\sqrt{\rho_1})^* \sin\theta & (\sqrt{\rho_2})^* \cos\theta \end{pmatrix}. \tag{5.80}$$

The mixing relation connects the left-handed fields $\nu_L$ and $(\nu_R)^C$ in the flavour basis and the left-handed fields $\nu_{1L}$ and $\nu_{2L}$ in the mass basis

$$\nu_L = (\sqrt{\rho_1})^* \cos\theta \, \nu_{1L} + (\sqrt{\rho_2})^* \sin\theta \, \nu_{2L}$$
$$(\nu_R)^C = -(\sqrt{\rho_1})^* \sin\theta \, \nu_{1L} + (\sqrt{\rho_2})^* \cos\theta \, \nu_{2L}. \tag{5.81}$$

The two flavour fields $\nu_L$ and $\nu_R$ are active and sterile, because only $\nu_L$ participates in weak interactions, whereas $\nu_R$ is a singlet of the gauge symmetries of the SM. Eqs. (5.81) show that the active neutrino field $\nu_L$ and the sterile field $(\nu_R)^C$ are linear combinations of the massive neutrino fields $\nu_{1L}$ and $\nu_{2L}$. Inverting Eqs. (5.81) one founds the relations of mixing in the flavour basis. They imply that oscillations between active and sterile neutrinos are possible. The possibility of sterile-active mixing, even in the one generation case, is a general result, which follows from the diagonalization of $M_{DM}$, and it is not limited to the case of a real mass matrix.

There are some special formulations of the Dirac–Majorana mass term. When $\theta = \pi/4$, the so called maximal mixing occurs. In this case $m_M \equiv m_{M_L} = m_{M_R}$, as can be seen from the expression in Eq. (5.74), and the mass eigenvalues in Eq. (5.75) reduce to

$$m'_{2,1} = m_M \pm m_D. \tag{5.82}$$

Another special case is the so-called Dirac limit, when $m_{M_L} = m_{M_R} = 0$. We have

$$m'_{2,1} = \pm m_D. \tag{5.83}$$

Since in this case the Dirac–Majorana mass term reduces to the Dirac mass term, the mass eigenstates are massive Dirac neutrinos. In the case where $m_{M_L}$ and $m_{M_R}$ are not zero, but much smaller than $m_D$, the neutrinos are indicated as pseudo-Dirac ones.

An important case is the limit where $m_{M_L} = 0$ and $m_{M_R} \gg m_D$, generating the so-called seesaw mechanism, which will be discussed in detail in Sect. 5.7.

## 5.6.2 More Generations

The previous reasonings can be replicated when there are more neutrino generations. Let us consider the Dirac-Majorana mass term in Eq. (5.58), and observe that, in the most general case, the number $n_a$ of active and $n$ sterile flavours can be different. We set $n_a = 3$ for the left-handed weak interacting components, as deduced by precision studies of the Z-line shape (see Sect. 5.1). The number of sterile neutrinos cannot be inferred by this measurement and it is not constrained either by requirement of anomaly cancellation; thus, it can be completely arbitrary. Therefore, in the most general case, the matrices $M_{M_L}$ and $M_{M_R}$ are $n_a \times n_a$ and $n \times n$ square symmetric matrices, respectively, and $M_D$ is a $n \times n_a$ rectangular complex matrix. In other terms, the Dirac-Majorana mass Lagrangian of Eq. (5.58) can be written as

$$\mathscr{L}_{DM} = \mathscr{L}_{M_L} + \mathscr{L}_{M_R} + \mathscr{L}_D =$$

$$= -\frac{1}{2} \sum_{\alpha,\beta=e,\mu,\tau} \overline{(\nu_{\alpha L})^C} (M_{M_L})_{\alpha\beta} \nu_{\beta L} - \frac{1}{2} \sum_{\alpha,\alpha'=1}^{n} \overline{\nu_{\alpha R}} (M_{M_R})_{\alpha,\alpha'} (\nu'_q)^C +$$

$$- \sum_{\alpha=1}^{n} \sum_{\beta=e,\mu,\tau} \overline{\nu_{\alpha R}} (M_D)_{\alpha\beta} \nu_{\beta L} + h.c. \qquad (5.84)$$

where we have set $n_a = 3$ and indicated the three left-handed neutrinos with their names in the flavour basis.

We can still rewrite $\mathscr{L}_{DM}$ in the form (5.60). Then the $n_a + n$ column matrix $n_L$ (5.62) will have $n_a$ components $\nu_L = (\nu_{1L}, \nu_{2L}, \ldots, \nu_{n_a L})$ components and $n$ components $(\nu_R)^C = ((\nu_{1R})^C, (\nu_{2R})^C, \ldots, (\nu_{nR})^C)$. The matrix $M_{DM}$ becomes a $(n_a + n) \times (n_a + n)$ square complex and symmetric matrix. This way the Dirac–Majorana mass term is formally written in the same form discussed in the one generation case (Sect. 5.6.1) and we can proceed to a likewise diagonalization. The diagonalization results in $n_a + n$ mass eigenstates and eigenvalues.

## 5.7 The Seesaw Mechanism

Let us start with the simplest case of one generation. The situation where the limits of the Dirac-Majorana mass matrix are $m_{M_L} = 0$ and $m_{M_R} \gg m_D$, namely

$$M_{DM} = \begin{pmatrix} 0 & m_D \\ m_D & m_{M_R} \gg m_D \end{pmatrix} \qquad (5.85)$$

has considerable phenomenological importance.

## 5.7 The Seesaw Mechanism

The first assumption, $m_{M_L} = 0$, allows to obtain all mass terms respecting the gauge symmetries of the SM. Indeed, the Majorana mass term $\mathscr{L}_{M_L}$ of Eq. (5.31) is not invariant under $SU(2)_L \times U(1)$, as detailed in Sect. 5.5. The Dirac mass term $\mathscr{L}_{M_D}$, also forbidden by the SM gauge symmetries, can nevertheless arise as a consequence of SSB, in analogy to the mass terms of charged leptons in the SM.

The second assumption, $m_{M_R} \gg m_D$, is connected to the relative weight of the remaining matrix elements. The value $m_D$, generated through the Higgs mechanism of the SM, would be proportional to the VEV of the Higgs doublet, and expected not to be much larger than the electroweak scale, namely $O(10^2)$ GeV, as it happens for the other fermions. This fact is often expressed by saying that the Dirac mass $m_D$ is protected by the symmetries of the SM. On the other hand, since $\mathscr{L}_{M_R}$ in Eq. (5.38) is already invariant under the SM gauge symmetries, no Higgs mechanism is necessary and $m_{M_R}$ can have arbitrary large values. The Majorana mass $m_{M_R}$ of the sterile neutrino field $\nu_R$ is not protected by the SM symmetries. If $m_{M_R}$ is generated by SSB at a high BSM energy scale, its value could be in principle as high as such energy scale. For example, in GUT theories, it could reach the value of the grand unification scale, which is generally believed, on the basis of coupling unification, to be about $10^{16}$ GeV [172].

In Eq. (5.85), one can also assume that the matrix $M_{DM}$ is real without loss of generality. In fact, the Majorana mass $m_{M_R}$ can be made real and positive by an appropriate rephasing of the right-handed chiral field $\nu_R$. Once the phase of $\nu_R$ is fixed, also the Dirac mass $m_D$ can be chosen to be real and positive by an appropriate rephasing of the chiral field $\nu_L$. In general, there would be no additional freedom to cancel a possible complex phase of the Majorana mass $m_{M_L}$, but in the limit $m_{M_L} = 0$ this is not necessary.

Following the procedure discussed in Sect. 5.6.1.1 for a real matrix, we find the eigenvalues of $M_{DM}$, at the lowest order in $m_D^2/m_{M_R} \ll 1$ (see Eqs. (5.75) and (5.78))

$$m_1 \simeq s\frac{m_D^2}{m_{M_R}} \quad (\rho_1 = -1)$$
$$m_2 \simeq m_{M_R} \quad (\rho_2 = +1)$$
$$\frac{1}{2}\tan 2\theta \simeq \theta \simeq \frac{m_D}{m_{M_R}}. \quad (5.86)$$

The arbitrary phases $\rho_{1,2}$ have been used to make the eigenvalues $m_1$ and $m_2$ positive. From Eq. (5.81), we have the relation among the flavour and mass eigenstates, at the lowest order in $\theta \simeq m_D/m_{M_R} \ll 1$

$$\nu_L \simeq i\,\nu_{1L} \quad (\nu_R)^C \simeq \nu_{2L}. \quad (5.87)$$

By using the definition in Eq. (5.72) we obtain the Majorana mass eigenstates $v_1$ and $v_2$

$$v_1 \simeq -i \left[ v_L - (v_L)^C \right] \qquad v_2 \simeq v_R + (v_R)^C. \qquad (5.88)$$

Summarising, the choice of the Dirac-Majorana matrix (5.85) gives rise to one extremely light Majorana particle $v_1$, that can be identified with the ordinary neutrino, plus one extremely heavy Majorana neutrino, $v_2$. It is called the seesaw mechanism, the name following from the fact that an increase of the mass of the heavy state $v_2$ naturally decreases mass of the light state $v_1$ [173–179]. The seesaw mechanism provides a very simple explanation of the smallness of neutrino mass with respect to the masses of the charged leptons and quarks, relating it to the existence of very large mass scales.

The mechanism just detailed is the most popular realisation of the seesaw mechanism, specifically addressed as type I, in order to distinguish it from other variants, as e.g. type II and III (see Sect. 5.9), where a new scalar triplet or triplet fermions are added to the Lagrangian in place of adding singlet right-handed neutrinos.

### 5.7.1 More Generations

The seesaw mechanism can be generalised to more flavours. In the mass terms, the mass parameters become non-diagonal matrices. As discussed in Sect. 5.6.2, we consider three left-handed active neutrinos and an arbitrary number $n$ of sterile right-handed neutrino fields. In the seesaw hypothesis, the mass matrices at play are a $n \times n$ square Majorana mass matrix $M_{M_R}$, a $n \times 3$ rectangular Dirac mass matrix $M_D$, and a square $(3+n) \times (3+n)$ Dirac Majorana mass matrix, which reads

$$M_{DM} = \begin{pmatrix} 0 & M_D^T \\ M_D & M_{M_R} \end{pmatrix} \qquad (5.89)$$

where $M_D^T$ is the transpose of the $M_D$ matrix, and $M_{M_R}$ is such that all its entries (and eigenvalues) are much larger than those of $M_D$. Since the Dirac mass matrix $M_D$ is generated by the SM Higgs mechanism, its elements are expected to be at the most of the order of the masses of the charged fermions, namely the order of the EW symmetry breaking scale $\approx 10^2$ GeV. Instead, the masses of right handed neutrinos are expected to emerge as a manifestation of BSM at higher scales, and need to be the heaviest for the seesaw mechanism to work out.

## 5.7 The Seesaw Mechanism

The matrix $M_{DM}$ is symmetric and therefore can be diagonalised by means of a square $(3+n) \times (3+n)$ unitary matrix $U$, that can be written in general as

$$U = \begin{pmatrix} V_{3\times 3} & S_{3\times n} \\ R_{n\times 3} & T_{n\times n} \end{pmatrix} \tag{5.90}$$

where $V_{3\times 3}$ is a $3 \times 3$ matrix, and so on. The condition of unitarity $U^\dagger U = UU^\dagger = I_{3+n}$, where $I_{3+n}$ is the $(3+n) \times (3+n)$ unit matrix, implies

$$V_{3\times 3}^\dagger V_{3\times 3} + R_{3\times n}^\dagger R_{n\times 3} = V_{3\times 3} V_{3\times 3}^\dagger + S_{3\times n} S_{n\times 3}^\dagger = \mathbb{1}_{3\times 3}. \tag{5.91}$$

Hence $V_{3\times 3}$, the flavour mixing matrix of three light Majorana neutrinos, must be non-unitary if $R_{n\times 3}$ and $S_{3\times n}$ are nonzero.

The diagonalization procedure is performed in two steps: first a block-diagonalization in which the off-diagonal block is removed, and second an individual diagonalization of the diagonal blocks. The first step gives

$$U^T M_{DM} U = \begin{pmatrix} M_{light} & 0 \\ 0 & M_{heavy} \end{pmatrix}. \tag{5.92}$$

Because of the assumption on the relative sizes of the matrix elements of $M_{M_R}$ and $M_D$, we can neglect contributions at high orders in $M_D M_{M_R}^{-1}$. Hence we cast the matrix $U$ in the form

$$U = \begin{pmatrix} \sqrt{1-\rho\rho^\dagger} & \rho \\ -\rho^\dagger & \sqrt{1-\rho^\dagger\rho} \end{pmatrix} \tag{5.93}$$

where the square-root of matrices is understood in the sense of the Taylor expansion, with the matrix $\rho^\dagger = M_{M_R}^{-1} M_D$ as expansion parameter. At the lowest order, we identify

$$\begin{cases} V = \sqrt{1-\rho\rho^\dagger} \simeq 1 - \frac{1}{2} M_D^T (M_{M_R} M_{M_R}^\dagger)^{-1} M_D \\ S = \rho = M_D^T (M_{M_R}^{-1})^\dagger \\ R = -\rho^\dagger = -M_{M_R}^{-1} M_D \\ T = \sqrt{1-\rho^\dagger\rho} \simeq 1 - \frac{1}{2} M_{M_R}^{-1} M_D M_D^T (M_{M_R}^\dagger)^{-1}. \end{cases} \tag{5.94}$$

We have omitted the subscripts indicating the number of lines and columns in the matrices $V$, $R$, $S$ and $T$ for the sake of clarity. After the diagonalization in Eq. (5.92), we find that the two matrices on the principal diagonal, at the lowest order, are

$$M_{light} \simeq -M_D^T M_{M_R}^{-1} M_D$$
$$M_{heavy} \simeq M_{M_R}. \tag{5.95}$$

The elements of the matrix $M_{light}$ are much smaller than those of $M_{heavy}$, which justifies the name given to these two matrices. They are suppressed with respect to the elements of the Dirac mass matrix $M_D$ by the small matrix factor $M_D^T M_{M_R}^{-1}$ and the actual sizes of the light neutrino masses depend on the value of this suppression. Equation (5.95) gives the original seesaw formula, today called Type I see-saw. We can easily check that non-diagonal elements are zero. One element reads

$$V^T M_D^T T + R^T M_D S + R^T M_{M_R} T \simeq$$
$$\simeq M_D^T - (\rho^\dagger)^T M_{M_R} = M_D^T - M_D^T (M_{M_R}^{-1})^T M_{M_R} = 0 \quad (5.96)$$

since the matrix $M_{M_R}$ is symmetrical. A similar reasoning holds for the other non diagonal entry, which is zero as well.

It remains to diagonalize the $n \times n$ $M_{heavy}$ matrix and the $3 \times 3$ $M_{light}$ matrix. The eigenvalues of $M_{heavy}$ are the mass values of $n$ sterile heavy neutrinos, whereas the the eigenvalues of $M_{light}$ give the light masses of the three interacting ones.[7] Both the light and the heavy neutrinos are Majorana particles.[8]

We can see the seesaw mechanism at work in two simple realisations. In the so-called quadratic seesaw, one assumes that

$$M_{M_R} = \Lambda\, I_n \quad (5.97)$$

where $\Lambda$ is the high-energy scale of new physics and $I_n$ is the $n \times n$ unit matrix. In this case, we have

$$M_{light} \simeq -\frac{M_D^T M_D}{\Lambda}. \quad (5.98)$$

Each light neutrino mass $m_i$, with $i \in \{1, 2, 3\}$, is given by

$$m_i \approx \frac{m_{Di}^2}{\Lambda} \quad (5.99)$$

where $m_{Di}^2$ is one of the three eigenvalues of the $3 \times 3$ matrix $M_D^T M_D$. We recover the same suppression formula of Eq. (5.86) in the one-generation case. The light neutrino masses scale as the squares of the masses $m_{Di}$

$$m_1 : m_2 : m_3 = m_{D1}^2 : m_{D2}^2 : m_{D3}^2. \quad (5.100)$$

Hence the name quadratic see-saw.

---

[7] Let us anticipate that since the matrices $M_{light}$ and $M_{heavy}$ are not necessarily hermitian and diagonal, the passage from flavour to mass eigenstates requires a lepton mixing matrix (see Sect. 6.1) for the light neutrino mixing and a new unitary matrix for the heavy neutrino one.

[8] Let us just mention that also in the Dirac case one can formulate a kind of seesaw mechanism that originates very light Dirac neutrinos, see for instance Ref. [180].

Another possibility is the so-called linear seesaw, where we assume $n = 3$ and the following relation of proportionality between the Dirac and the Majorana matrices

$$M_{M_R} = \frac{\Lambda}{\Lambda_D} M_D. \qquad (5.101)$$

Here $\Lambda$ is once again the high-energy scale of new physics, while $\Lambda_D$ is the energy scale of the elements of $M_D$. The matrix $M_{light}$ takes the form

$$M_{light} \simeq -\frac{\Lambda_D}{\Lambda} M_D \qquad (5.102)$$

and the light neutrino masses are given by

$$m_i \approx \frac{\Lambda_D}{\Lambda} m_{Di} \qquad (5.103)$$

where $m_{Di}$ is one of the three eigenvalues of the $3 \times 3$ matrix $M_D$. The neutrino masses are suppressed by the ratio $\Lambda_D/\Lambda \ll 1$. The name linear see-saw follows because they scale as the masses $m_{Di}$

$$m_1 : m_2 : m_3 = m_{D1} : m_{D2} : m_{D3}. \qquad (5.104)$$

Several extensions of the SM lead to a see-saw mechanism for neutrino masses, well known examples being the $SO(10)$ Grand Unified Theories [181, 182] and left-right symmetric models [176]. The see-saw mechanism in the full BSM theory can be considered effectively realised at low energies in the SM with three light Majorana neutrinos. This is an example of the approach discussed in Sect. 5.8, where the SM takes the role of a low energy effective theory.

## 5.8 The Effective Lagrangian

At present, the amount of experimental information available is still insufficient to allow a particular BSM model with massive neutrinos to be chosen over another. In this situation, one way to proceed is to probe the characteristic features of new physics using the formalism of effective field theories (EFT). This is based on the assumption that we can organise calculations by physical energy scales and that low-scale physics is largely insensitive to high-scale physics. Under these premises, we can integrate out, that is effectively eliminate, all heavy particles (or more in general degrees of freedom) with masses well above the scale associated with the problem at hand. The resulting effective Lagrangian contains only light particles with interactions approximated by local non-renormalizable terms, constrained by the symmetries of the theory, and should consistently describe all of the important low energy observables.[9]

In BSM physics, the effective Lagrangian $\mathcal{L}_{eff}$ provides a low energy description, in the SM observables, of a more complete theory that arises at a scale $\Lambda$ well above the EW scale. The effective Lagrangian is written as the SM Lagrangian enforced by a series of nonrenormalizable operators consistent with the SM gauge symmetries, namely

$$\mathcal{L}_{eff} = \mathcal{L}_{SM} + \delta\mathcal{L}^{d=5} + \delta\mathcal{L}^{d=6} + \ldots \quad (5.105)$$

$\mathcal{L}_{SM}$ contains all the renormalizable operators invariant under the SM gauge symmetry, having energy dimension $[E^d]$ with $d \leq 4$. The other gauge invariant terms $\delta\mathcal{L}^d$ contain all nonrenormalizable operators with dimension $d > 4$ built out of SM fields. They parameterise the impact of the high-energy theory after heavy fields of mass of order $\Lambda$ have been integrated out, and account for the effects of BSM physics at low energies.

The underlying assumption is that the non renormalizable operators are generated by a renormalizable high-energy theory, which must include the gauge symmetries of the SM in order to be effectively reduced to the SM at low energies. Since any term of the Lagrangian has dimension $d = 4$, an operator with dimension $d > 4$ must appear in the Lagrangian with a coupling constant proportional to $M^{4-d}$, where $M \sim O(\Lambda)$. The strong suppressing factor represented by the inverse powers of $M$ limits in practice the observability of the low-energy effects of new physics.

The effective approach is rather general; an example is the study of weak interaction processes at energies well below the EW scale. After integrating out the $W^\pm$ and $Z^0$, they decouple from the low energy theory, and we are left with an effective Fermi four point interaction, which is not renormalizable, but perfectly working at energies much lower than the EW scale. The four-point interaction corresponds to a dimension-six operator of an effective Lagrangian whose coupling, the dimensional Fermi constant, is proportional to $m_W^{-2}$, where $m_W$ is the $W$-boson mass. Historically, Fermi's theory was used for weak decay calculations even when the scales $m_W$ and $m_Z$ were not known. Effective field theory methods allow us to separate scales in a multi-scale problem, organise calculations in a systematic way, and probe the structure of the underlying theory.

Because of the scaling of the coefficients, in the effective Lagrangian approach the BSM effects are dominated by the lowest dimension operators. There is a unique lowest-dimension, non-renormalizable operator that violates the lepton number and is allowed by the SM gauge invariance and particle content. It is the gauge singlet

---

[9] One way to eliminate the heavy degrees of freedom is to adopt the Feynman path integral formalism for Green's functions, divide the integration process in two steps and integrate first over high degrees of freedom. The result can be interpreted in term of an effective action in the light degrees of freedom that brings the marks of the underlying theory at higher scales. Effective field theories have pervaded about the last six decades of research, and we refer to the vast literature on the subject for insights.

## 5.8 The Effective Lagrangian

dimension-5 operator written generically as

$$\frac{\lambda}{M}\left(\overline{(\nu_L)^C}\ \overline{(e_L)^C}\right) \otimes \phi \otimes \phi \otimes \begin{pmatrix} \nu_L \\ e_L \end{pmatrix} \quad (5.106)$$

where the field contractions are not specified.[10] Its coefficient in the effective series can be expressed as a generic complex parameter $\lambda$, suppressed by the high scale $M$. In Eq. (5.106) one can insert the SM doublet $\phi = (\phi^+, \phi^0)^T$ or $\tilde{\phi} = (\phi^0, -\phi^-)^T$ (see Eqs. (4.48) and (4.81)) to ensure zero total hypercharge.

After field contraction, the possibility to obtain an $SU(2)_L$ singlet operator is guaranteed by the group theoretical analysis, where we have

$$2 \otimes 2 \otimes 2 \otimes 2 = (4 \oplus 2 \oplus 2) \otimes 2 = 4 \otimes 2 \oplus 2 \otimes 2 \oplus 2 \otimes 2 =$$
$$= 5 \oplus 3 \oplus 3 \oplus 1 \oplus 3 \oplus 1. \quad (5.107)$$

One possible formulation of the operator of Eq. (5.106) is

$$\lambda_{ij}\left(\overline{\ell_{iL}^C}\ \tilde{\phi}^*\right)\left(\tilde{\phi}^\dagger\ \ell_{jL}\right) + h.c =$$
$$= c_{ij}\left(\overline{\ell_{iL}^C}\ \tilde{\phi}^*\right)\left(\tilde{\phi}^\dagger\ \ell_{jL}\right) + c_{ij}^*\left(\overline{\ell_{jL}}\ \tilde{\phi}\right)\left(\tilde{\phi}^T\ \ell_{iL}^C\right) \quad (5.108)$$

where $\lambda_{ij} \propto 1/M$ is the model dependent, complex coefficient, $i, j$ are flavour indices and $\ell_{iL}^T = (\nu_{iL}, e_{iL})$ are the SM lepton doublets (see Eq. (4.16)). This gauge invariant operator is known as the Weinberg operator [183]. It is non-renormalizable and it violates lepton number. Explicitly expanding the doublets $\ell_L$ and $\phi_C$ in the case of one-generation, one gets

$$\frac{\lambda}{M}\left(\overline{\nu_L^C}\phi^0 - \overline{e_L^C}\phi^+\right)\left(\phi^0 \nu_L - \phi^+ e_L\right) + h.c.. \quad (5.109)$$

It is self-evident that one cannot generate the equivalent of the Weinberg operator in the quark sector. If one replaces lepton and quark doublets $\ell_L \to q_L$ in Eq. (5.108), one produces a Lagrangian term that cannot be invariant under a $U(1)_Y$ gauge transformation, because $Y(q_L) = 1/6$ and $Y(\phi) = 1/2$. The effective operators of lowest dimension involving quark fields and violating lepton (and baryon) number are dimension-six operators, as for example $(q_L^T q_L)(q_L^T \ell_L)$. Although such six dimensional operators are suppressed by coefficients proportional to $M^{-2}$, they are interesting because they can induce proton decay. Proton decay is a generic prediction of Grand Unified Theories (GUTs), that combine quarks and leptons and include interactions that allow their transition from one to the other.

---

[10] We say one operator for convenience, but it is to be understood that there are also family indices so we really have a set of operators.

After SSB $\phi^T = (\phi^+, \phi^0) \to (0, 1/\sqrt{2}\,(v+h))$ and the Weinberg operator, with a real coefficient $\lambda$, yields the Majorana mass term

$$\frac{\lambda v^2}{2M}\left(\overline{v_L^C}v_L + \overline{v_L}v_L^C\right). \tag{5.110}$$

Since $v \simeq 246$ GeV is the value of the Higgs VEV (see Sect. 4.4.1), the Majorana neutrino masses generated by the Weinberg operator

$$m_{M_L} = \frac{\lambda}{2M}v^2 \tag{5.111}$$

are naturally much smaller than all other fermion masses because of the suppression factor $1/M$. Known the neutrino masses, the amount of this suppression could provide information on the scale of new physics at which the SM is violated. If we assume $\lambda \sim O(1)$, at least $M > 10^{14}$ GeV is required for neutrino masses $m_{M_L} < 1$ eV.

It is very suggestive that the lowest-order effect of high-energy beyond the Standard Model physics may be neutrino masses. The task now is to derive, from an underlying renormalizable (or UV complete) theory, one of the Weinberg-type operators[11] as the leading contribution to neutrino mass. This process has come to be termed "opening up the operator". Since the Weinberg operator violates lepton number we are dealing with Majorana neutrinos. In the next section we discuss this derivation at tree level in the simplest possible way, using only exotic massive fermions and scalars as the new physics. The available renormalizable interactions in the theory extending the SM are then just of Yukawa and scalar-scalar type. It is also possible generate Weinberg-type operators at loop level.

## 5.9 Opening Up the Weinberg Operator at Tree Level

Let us consider BSM physics in the EFT framework. We assume that the BSM Lagrangian respects the SM gauge symmetry while including new massive fermions and scalars of mass $O(M)$, that we indicate generically with $h$. Thus it is fair to assume that it includes Poincaré and SM gauge invariant interaction terms as $h\phi\ell_L$, $h\phi\phi$ and $h\ell_L\ell_L$, where $\phi$ and $\ell_L$ are the usual SM Higgs and left-handed lepton doublets. In the effective framework any heavy field $h$ is integrated out, decoupling from the low energy theory. As a result, the exchange of an heavy field $h$ at tree level in the full theory originates a low energy four-point interaction in the SM fields of the form (5.106). After EW symmetry breaking, this term produces effective Majorana neutrino masses $O(v^2/M)$, with $v$ being the VEV of the SM Higgs field.

---

[11] There are generalisations of the Weinberg operator including powers of $\phi^\dagger\phi$.

## 5.9 Opening Up the Weinberg Operator at Tree Level

It follows that, whatever BSM extension we are considering, it includes a low energy realisation of the seesaw mechanism in order to give mass to neutrinos.

In the case of $h\phi\ell_L$ couplings, invariance under the Poincaré group requires these $h$ to be spin 1/2 fields, while for both $h\phi\phi$ and $h\ell_L\ell_L$ interaction terms $h$ must be a scalar field. In all three cases we have

$$\phi \otimes \ell_L = 2 \otimes 2 = 1 \oplus 3$$
$$\phi \otimes \phi = 2 \otimes 2 = 1 \oplus 3$$
$$\ell_L \otimes \ell_L = 2 \otimes 2 = 1 \oplus 3. \tag{5.112}$$

Hence for $SU(2)_L$ invariance the heavy fields $h$ either transform as an $SU(2)_L$ singlet or as an $SU(2)_L$ triplet.

Invariance under the hypercharge for the coupling $h\phi\ell_L$ gives $Y = 0$ to the fermion $h$. Instead, in both $h\phi\phi$ and $h\ell_L\ell_L$ couplings, one needs a scalar field $h$ with hypercharge $Y = +1$, which transforms either as a singlet or a triplet of $SU(2)_L$. However, a singlet $h$ ($T_3 = 0$) would have an electric charge $Q = T_3 + Y = +1$. Consequently, any possible Yukawa coupling could not generate tree-level neutrino masses because any possible VEV of $h$ would break $U(1)_{em}$, being therefore forbidden by the gauge symmetry of the SM. The scalar triplet option is the only one left.

Summarising, we are left with BSM extensions including three possible sets of new particles, fermionic singlets, scalar triplets, and fermionic triplets. The addition of these new heavy particles leads to three different realisations of the seesaw mechanism, named type I, II, III, respectively. In each of them, the lepton number violating operator (5.106) is generated at tree level in the effective framework, contracted in three different ways. The large value $M$ characterises the seesaw scale. In Fig. 5.2, tree-level diagrams, which exchange the new BSM particles and can induce the effective operator of Eq. (5.106) at low-energy, are shown.

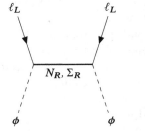 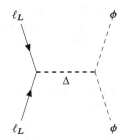

**Fig. 5.2** Tree-level diagrams inducing the effective operator of Eq. (5.106) at low-energy. These interactions correspond to the exchange of a fermionic singlet $N_R$ or triplet $\Sigma_R$ (left), and a scalar triplet $\Delta$ (right)

Type-I seesaw is the one detailed in Sect. 5.7. Fermionic singlets under $SU(2)_L$ of zero hypercharge, namely $n$ heavy right-handed Majorana neutrinos $N_{Ri}$, are added into the SM. The BSM gauge invariant Lagrangian reads

$$\mathscr{L}_{BSM} = \mathscr{L}_{SM} + \frac{i}{2}\sum_{i=1}^{n} \overline{N_{Ri}} \slashed{\partial} N_{Ri} - \frac{1}{2}\sum_{i=1}^{n} M_{Ri} \overline{(N_{Ri})^C} N_{Ri} +$$
$$- \sum_{\alpha=1}^{3}\sum_{i=1}^{n} Y_{\alpha i}^T \bar{\ell}_{L\alpha} \tilde{\Phi} N_{Ri} + h.c. \qquad (5.113)$$

where $\ell_{L\alpha}$ is the lepton doublet (5.5) and $\tilde{\Phi}$ is the Higgs doublet defined in Eq. (4.81). The first term added to the SM Lagrangian $\mathscr{L}_{SM}$ is the kinetic term for each field $N_{Ri}$ and the second one is the Majorana (heavy) mass term, where $M_R = \text{diag}(M_{R1}, M_{R2}, ... M_{Rn})$ is the $n \times n$ real and diagonal mass matrix for the right-handed neutrino fields. The large values of its matrix elements characterise the seesaw scale. The last term describes the interactions among the Higgs field, the lepton doublets and the heavy neutrinos, where $Y$ is a $n \times 3$ complex Yukawa coupling matrix. After SSB, this term originates a Dirac mass for neutrinos, as detailed in Sect. 5.2, and we have all the elements for the seesaw mechanism. According to Eq. (5.95), we can write

$$M_{heavy} \simeq M_{M_R}$$
$$M_{light} \simeq -M_D^T M_{M_R}^{-1} M_D \qquad (5.114)$$

where

$$M_D \equiv Y \frac{v}{\sqrt{2}}. \qquad (5.115)$$

In the effective framework, the five-dimensional operator (5.106) is generated at low energy by integrating out the heavy fields in tree level diagrams with exchange of a virtual $N_R$ between lepton-Higgs pairs, as in Fig. 5.2, left. More specifically, the contraction of these diagrams originates the Weinberg operator in (5.108), where each of the pairs $\bar{\ell}_L^C \otimes \phi$ and $\phi \otimes \ell_L$ forms a singlet. After the EW symmetry breaking, the Weinberg operator contributes to the effective Lagrangian with neutrino mass terms $O(v^2 Y/M)$, and the see-saw mechanism is directly implemented.

In type II seesaw, detailed in Sect. 5.10, the Weinberg operator is induced by the diagrams with exchange of the virtual heavy scalar triplet boson $\Delta$ between lepton and Higgs pairs (see Fig. 5.2, right). In type III seesaw it is induced by the diagrams with exchange of a virtual heavy Majorana triplet fermion $\Sigma_R$ between the lepton-Higgs pairs (see Fig. 5.2, left). We will detail type III seesaw mechanisms in Sect. 5.11.

## 5.10 Type II See-Saw Mechanism

Another realisation of the seesaw mechanism, mentioned in Sect. 5.9, is the so-called Type II seesaw. It provides a minimal framework to explain the neutrino masses by adding into the SM an heavy scalar Higgs which transforms as a triplet under $SU(2)_L$, with hypercharge $Y = 1$ [184–187], namely

$$\mathbf{\Delta} = \begin{pmatrix} \Delta_1 \\ \Delta_2 \\ \Delta_3 \end{pmatrix}. \tag{5.116}$$

We need now to use a few notions of group theory, introduced in Sect. 2.1. The Lie algebra of the matrix group $SU(N)$, denoted as $\mathfrak{su}(N)$, consists of all complex, $N \times N$ traceless, anti-hermitian matrices. A basis in $\mathfrak{su}(2)$ is composed by the set of matrices $T_\ell = i\tau_\ell/2$ ($\ell \in 1, 2, 3$), where $\tau_\ell$ are the Pauli matrices (see Eq. (2.57)). The commutators are

$$[T_i, T_j] = -\varepsilon_{ijk} T_k \quad i, j, k \in \{1, 2, 3\} \tag{5.117}$$

where $\varepsilon_{ijk}$ is the Levi-Civita symbol. In the basis $T_\ell$ of $\mathfrak{su}(2)$ we can write the following $3 \times 3$ traceless matrices, namely

$$\mathrm{ad}_{T_1} = \begin{pmatrix} 0 & 0 & 0 \\ 0 & 0 & 1 \\ 0 & -1 & 0 \end{pmatrix} \quad \mathrm{ad}_{T_2} = \begin{pmatrix} 0 & 0 & -1 \\ 0 & 0 & 0 \\ 1 & 0 & 0 \end{pmatrix} \quad \mathrm{ad}_{T_3} = \begin{pmatrix} 0 & 1 & 0 \\ -1 & 0 & 0 \\ 0 & 0 & 0 \end{pmatrix}. \tag{5.118}$$

These matrices follow the same commutation rules in Eq. (5.117) since ad is a Lie algebra morphism. Let us now set $T_\ell^{adj} \equiv -i\,\mathrm{ad}_{T_\ell}$. When the matrices are exponentiated, they act on the flavour Higgs triplet $\mathbf{\Delta}$, according to the $SU(2)$ transformation relations (4.12), which reads[12]

$$\begin{pmatrix} \Delta_1 \\ \Delta_2 \\ \Delta_3 \end{pmatrix} \to \exp\left[i\alpha_\ell T_\ell^{adj}\right] \begin{pmatrix} \Delta_1 \\ \Delta_2 \\ \Delta_3 \end{pmatrix} =$$

$$\exp\left[\begin{pmatrix} 0 & \alpha_3 & -\alpha_2 \\ -\alpha_3 & 0 & \alpha_1 \\ \alpha_2 & -\alpha_1 & 0 \end{pmatrix}\right] \begin{pmatrix} \Delta_1 \\ \Delta_2 \\ \Delta_3 \end{pmatrix}. \tag{5.119}$$

---

[12] Let us observe that we obtain the same transformation rule by directly setting $(T_i^{adj})_{jk} = -i\varepsilon_{ijk}$, see Eq. (2.12).

Since the weak hypercharge is the generator of the abelian group $U(1)_Y$, we can identify it with the unit diagonal matrix $Y = \mathrm{diag}(1,1,1)$ in the adjoint representation. The charge operator is given by the relation in Eq. (4.3)

$$Q = T_3^{adj} + Y. \tag{5.120}$$

Thus in the adjoint representation we have

$$Q = \begin{pmatrix} 1 & -i & 0 \\ i & 1 & 0 \\ 0 & 0 & 1 \end{pmatrix}. \tag{5.121}$$

In order to identify the charge eigenvalues of the flavour Higgs triplet $\boldsymbol{\Delta}$, we can diagonalize the charge operator $Q$ through a change of $SU(2)_L$ basis, parameterised by a unitary matrix $U$. We have

$$Q_D = U^\dagger Q U \tag{5.122}$$

where $Q_D = \mathrm{diag}(2, 1, 0)$ is the diagonal charge matrix and

$$\boldsymbol{\Delta}' = \begin{pmatrix} -\Delta^{++} \\ \Delta^+ \\ \Delta^0 \end{pmatrix} = U^\dagger \boldsymbol{\Delta} = \begin{pmatrix} -\frac{(\Delta_1 - i\Delta_2)}{\sqrt{2}} \\ \Delta_3 \\ \frac{(\Delta_1 + i\Delta_2)}{\sqrt{2}} \end{pmatrix} \tag{5.123}$$

where

$$U = \frac{1}{\sqrt{2}} \begin{pmatrix} -1 & 0 & 1 \\ -i & 0 & -i \\ 0 & \sqrt{2} & 0 \end{pmatrix}. \tag{5.124}$$

It is convenient to define a $2 \times 2$ matrix $\Delta$

$$\Delta = \boldsymbol{\tau} \cdot \boldsymbol{\Delta} = \begin{pmatrix} \Delta_3 & \Delta_1 - i\Delta_2 \\ \Delta_1 + i\Delta_2 & -\Delta_3 \end{pmatrix} = \begin{pmatrix} \Delta^+ & \sqrt{2}\Delta^{++} \\ \sqrt{2}\Delta^0 & -\Delta^+ \end{pmatrix}. \tag{5.125}$$

This way, the triplet elements are the coefficients of the $\tau$ matrices in $\Delta$ and $i\Delta$ is an element of $\mathfrak{su}(2)$.

Let us now go back to $\ell_L$, the fundamental weak doublet of chirality $-1$ for the lepton family, defined in Eq. (4.16):

$$\ell_L = \begin{pmatrix} \nu_{eL} \\ e_L \end{pmatrix} \qquad Y = -\frac{1}{2}. \tag{5.126}$$

## 5.10 Type II See-Saw Mechanism

Under $SU(2)_L$, it transforms, according the relation in Eq. (4.12), as

$$\ell_L \to \exp\left[i\alpha_\ell \frac{\tau_\ell}{2}\right]\ell_L. \tag{5.127}$$

Then we can re-write Eq. (5.119) as

$$\Delta \to \exp(\alpha_\ell \, \mathrm{ad}_{T_\ell}) \tag{5.128}$$

which according to Eq. (2.6) is equivalent to

$$\Delta \to \exp\left[i\alpha_\ell \frac{\tau_\ell}{2}\right] \Delta \exp\left[-i\alpha_\ell \frac{\tau_\ell}{2}\right]. \tag{5.129}$$

At this point, we are ready to write an additional Yukawa coupling outside the ones in the SM

$$-\mathscr{L}_Y^\Delta = \frac{1}{\sqrt{2}} \sum_{ij} Y_{ij}^\Delta \, \overline{\ell_{iR}^C} \Delta \ell_{jL} + \mathrm{h.c.} \tag{5.130}$$

Charge conjugation changes the sign of the hypercharge and the chirality (see Sect. 4.1), but it does not change the representation of the $SU(2)_L$ gauge group. Thus the conjugate states

$$(\nu_{eL})^C \equiv \nu_{eR}^C \qquad (e_L)^C \equiv e_R^C \tag{5.131}$$

even if right-handed, still belong to a $SU(2)_L$ doublet, although with opposite $T_3$ and $Y$. The conjugate lepton doublet has been defined in Eq. (4.25)

$$\ell_R^C \equiv \begin{pmatrix} -e_R^C \\ \nu_{eR}^C \end{pmatrix} \qquad Y = \frac{1}{2}. \tag{5.132}$$

We can easily see that $\mathscr{L}_Y^\Delta$ is a singlet under $SU(2)_L \times U(1)_Y$. The Higgs triplet violates the lepton number by its interactions with the lepton doublet.

The Higgs potential now involves both the Higgs triplet $\boldsymbol{\Delta}$ (Eq. (5.116)) and the SM doublet $\phi$ (Eq. (4.48)). As in the case of the SM doublet (see Eq. (4.51)), one can assume that the triplet expressed in terms charge eigenfields, $\boldsymbol{\Delta}'$ in Eq. (5.123), has also a non-zero VEV, namely

$$\langle \boldsymbol{\Delta}' \rangle = \frac{1}{\sqrt{2}} \begin{pmatrix} 0 \\ 0 \\ v_\Delta \end{pmatrix} \tag{5.133}$$

Hence after SSB the additional Lagrangian $\mathscr{L}_Y^\Delta$ generates a left Majorana mass terms for neutrinos (see Eqs. (5.28) and (5.42)). In the case of more lepton generations, labelled by $i, j$, the mass term reads

$$\mathcal{L}_{M_L}^\Delta = -\sum_{ij}[\overline{(\nu_{iL})^C}(M_{M_L})_{ij}\nu_{jL} + \overline{\nu_{iL}}(M_{M_L})_{ij}(\nu_{jL})^C] \tag{5.134}$$

where

$$(M_{M_L})_{ij} \equiv Y_{ij}^\Delta \frac{v_\Delta}{\sqrt{2}}. \tag{5.135}$$

Without right-handed neutrinos one has $M_D = M_{M_R} = 0$ in the Dirac-Majorana mass term (5.61), and the only possible neutrino mass term remains the previous Majorana mass term. The diagonalization of the mass matrix $M_{M_L}$ gives the neutrino mass eigenvalues and eigenstates.

Since the new scalar triplet participates in gauge interactions, a kinetic term for the triplet is required in the Lagrangian, which has to be hermitian and invariant under $SU(2)_L \times U(1)_Y$; it can be written as

$$\mathcal{L}_{kin}^\Delta = \text{Tr}[(D_\mu \Delta)^\dagger (D^\mu \Delta)] \tag{5.136}$$

where $D_\mu$ the covariant derivative defined in Eq. (4.1). After electroweak SSB, the masses of the gauge bosons differ from those obtained in the SM since they depend on both triplet and doublet Higgs VEVs. By a procedure analogous to the one described in Sect. 4.4, the parameter $\rho$ defined in Eq. (4.68) is found different from 1, namely

$$\rho \equiv \frac{M_W^2}{M_Z^2 \cos^2\theta_W} = 1 - \frac{2v_\Delta^2}{v + 4v_\Delta^2}. \tag{5.137}$$

Since $\rho$ is observable and its value is practically 1, one deducts that $v_\Delta \ll v$.

In presence of an Higgs triplet, the SM Lagrangian is augmented by a term which couples the scalar triplet to the Higgs doublet, and can be written as

$$-\mathcal{L}_{\phi\Delta\phi} = \mu_\Delta (\phi^\dagger \Delta \tilde{\phi} + \text{h.c.}) \tag{5.138}$$

where the SM Higgs doublet $\tilde{\phi}$ has been defined in Eq. (4.81). It is also augmented by a scalar potential $V(\phi, \Delta)$ (see e.g. Refs. [188, 189]), which includes self-interaction terms for the Higgs triplet, such as $\text{Tr}[\Delta^\dagger \Delta]$. Minimising the scalar potential, one obtains the following approximate relation

$$v_\Delta \approx -\frac{\mu_\Delta v^2}{M_\Delta^2} \tag{5.139}$$

where $M_\Delta$ is the mass value for the scalar triplet. Indeed, after the EW SSB, there are seven massive physical Higgs bosons: one CP-even neutral Higgs boson $\phi$, the scalar (neutral CP-even) and pseudoscalar (neutral CP-odd) components of $\Delta^0$, the singly and doubly charged states $\Delta^\pm$ and $\Delta^{\pm\pm}$. The Higg boson $\phi$ is SM-like and the rest of the Higgs states are $\Delta$-like, having approximately the same mass value $M_\Delta$. quartic interactions. The triplet VEV is small, $v_\Delta \ll v$, if $M_\Delta^2 \gg \mu_\Delta v^2$. The essence of the type II seesaw mechanism consists in the suppression of the triplet VEV $v_\Delta$ compared to the Higgs VEV $v$. A small values of $v_\Delta$ implies small entries in the mass matrix (5.135) and therefore small neutrino masses. The masses of neutrinos are small if the mass of the scalar triplet is very large. The type II seesaw is analogous to the usual seesaw mechanism for obtaining small Majorana neutrino masses, except that we do not need any right-handed neutrino.[13]

Summarising, the type II seesaw mechanism suppresses the VEV of the neutral component of scalar gauge-SU(2) triplet, in such a way that the left-handed neutrinos, which acquire Majorana masses from their Yukawa couplings to those neutral components, are extremely light. Just like the type I seesaw mechanism, the type II seesaw mechanism requires a very high mass scale, which now occurs in the mass terms of the scalar triplets. Those large mass terms make the scalar triplets extremely heavy and therefore the type II seesaw mechanism, like the type I seesaw mechanism, is very difficult to test experimentally. In the general case, for instance in Grand Unified Theories based on the gauge group SO(10), both type I and type II seesaw mechanisms are present [177, 190–192].

Another way of handling the heavy Higgs triplet is to integrate it out. The effective Lagrangian is induced in this case by the diagrams with exchange of a virtual $\Delta$ between lepton and Higgs pairs. As discussed in Sect. 5.9, the Weinberg operator is generated at tree level by the exchange of the neutral component of a heavy scalar multiplet transforming as a triplet under $SU(2)_L$. It corresponds to the contraction $\phi \otimes \phi$ into a triplet, and to the contraction $\bar{\ell}_L^C \otimes \ell_L$ into another triplet.

Although the theory of type-II seesaw is quite simple, it has a rich phenomenology to offer. The components of the Higgs scalar triplet couple to the charged leptons and gauge bosons and can thus mediate lepton number violating interactions at tree level. For instance, if Yukawa couplings are sufficiently large, the doubly charged component of the triplet will decay predominantly into same-sign charged leptons which is a clear signature of lepton number violation [189, 193]. This opens up attractive possibilities, but as it has not been observed yet, it also places strong limits on the couplings and masses of the triplet.

## 5.11 Type III See-Saw Mechanism

The last type of seesaw mechanism mentioned in Sect. 5.9 is the type-III one. It is based on the addition into the SM of $n$ right-handed fermionic triplets $\Sigma_{iR} = (\Sigma_{iR}^1, \Sigma_{iR}^2, \Sigma_{iR}^3)^T$, $i \in \{1, 2, .., n\}$, with zero weak hypercharge [194].

---

[13] When there is also a right-handed neutrino, the left-handed neutrino also gets a mass from the Higgs triplet in addition to the canonical seesaw mechanism.

As in the type II seesaw scenario, it is convenient to define a $2 \times 2$ matrix $\Sigma_i$ as

$$\Sigma_{iR} = \boldsymbol{\tau} \cdot \boldsymbol{\Sigma}_{iR} = \begin{pmatrix} \Sigma_{iR}^3 & \Sigma_{iR}^1 - i\Sigma_{iR}^2 \\ \Sigma_{iR}^1 + i\Sigma_{iR}^2 & -\Sigma_{iR}^3 \end{pmatrix} = \begin{pmatrix} \Sigma_{iR}^0 & \sqrt{2}\Sigma_{iR}^+ \\ \sqrt{2}\Sigma_{iR}^- & -\Sigma_{iR}^0 \end{pmatrix}. \quad (5.140)$$

In the last passage, the matrix is expressed in terms of charge eigenfields, which are related to the flavour eigenfields by

$$\Sigma_{iR}^\pm = \frac{\Sigma_{iR}^1 \mp i\Sigma_{iR}^2}{\sqrt{2}}, \qquad \Sigma_{iR}^0 = \Sigma_{iR}^3. \quad (5.141)$$

By following the same argument that led to Eq. (5.130) in the type II seesaw, one can build a Yukawa interaction term to be added to the SM Lagrangian, since in both cases the triplets pertain to the adjoint representation of $SU(2)_L$; it reads

$$-\mathscr{L}_Y^\Sigma = \frac{1}{\sqrt{2}} \sum_{\alpha=1}^{3} \sum_{i=1}^{n} Y'^T_{\alpha i} \, \overline{\ell}_{L\alpha} \, \Sigma_{iR} \, \tilde{\Phi} + h.c. \quad (5.142)$$

where $Y'$ is an arbitrary $n \times 3$ matrix. Since the new fermionic triplets participate in gauge interactions, their kinetic terms, which are gauge invariant and hermitian as the kinetic term for the scalar triplet in Eq. (5.136), are also added to the SM Lagrangian: hence we have

$$\mathscr{L}_{kin}^\Sigma = \sum_{l=1}^{n} \text{Tr}[\overline{\Sigma}_{lR} i \slashed{D} \Sigma_{lR}] \quad (5.143)$$

where $D_\mu$ the covariant derivative defined in Eq. (4.1).

The type-III seesaw mechanism is similar to the type-I one, but instead of right-handed neutrinos—which are $SU(2)_L$ singlets—it uses right-handed fermion $SU(2)_L$ triplets. Allowing lepton number violation, it is possible to build a Majorana mass term for the triplet similar to the one in Eq. (5.113), while substituting the fermionic triplet $\Sigma_{iR}$ to the right-handed Majorana neutrinos $N_{R_i}$

$$-\mathscr{L}_{M_R} = \frac{1}{2} \sum_{i,j=1}^{n} M_{Rij} \overline{(\Sigma_{iR})^C} \Sigma_{jR} + h.c. \quad (5.144)$$

where $M_R$ is a $n \times n$ matrix. To this Majorana mass term, one adds the Dirac mass term provided by Eq. (5.142) after SSB. Maintaining the substitution $N_{R_i} \to \Sigma_{iR}$, one can follow the procedure described in Sect. 5.7.1 in the type I seesaw case, and find a suppression for the neutrino mass analogous to the one in Eq. (5.114), namely

$$M_{light} = -M_D^T M_{M_R}^{-1} M_D \quad (5.145)$$

## 5.11 Type III See-Saw Mechanism

where $M_R$ is the matrix defined in Eq. (5.144) and $M_D \equiv Y'v/\sqrt{2}$.

The diagrams with exchange of a virtual $\Sigma_R$ between the lepton-Higgs pairs (see Fig. 5.2, left) induce the Weinberg operator once one integrates out the neutral component of the fermion triplet. The contractions are originated by the triplets that are composed by $\bar{\ell}_L^C \otimes \phi$ and $\phi \otimes \ell_L$.

The neutrino phenomenology of these scenarios is similar to that of the type I seesaw, but the existence of new charged fields—lepton-like states with electric charge $\pm 1$—rules out the possibility that the triplet Majorana masses M are smaller than hundreds of GeV.

# Chapter 6
# The Mixing of Neutrinos

The first step of particle physics beyond the SM has been to discover that neutrino have masses as all the other fermions, albeit much smaller. The concept of neutrino with given flavour identifies the neutral particle associated by weak interactions to the charged leptons with given flavour, namely the electron, the muon or the tau. In the SM, a neutrino with given flavour (aka a flavour eigenstate) has zero mass, and it is also a mass eigenstate (state with definite mass). The lepton number is conserved within each family. As seen in Chap. 5, in case of massive neutrinos the family lepton number is violated. Neutrinos do not carry any quantum number, with the exception of the total lepton number, which is also violated in the case of Majorana neutrinos. Thus transitions (oscillations) between neutrinos which belong to different families and have different flavour become possible.

Neutrino oscillations are a typical quantum mechanics phenomenon that can be easily described considering the wave nature of neutrinos. Let us suppose that one generates a neutrino at a source. Weak interactions act on flavour eigenstates, hence this neutrino will have a definite flavour. The neutrino field that has a given flavour does not coincide with a neutrino field that has given mass but rather it coincides with linear combinations of fields that have given mass. In other terms, in the basis of mass eigenvectors, the flavour neutrino eigenstate can be expressed as a linear combination of states of definite mass. The mass eigenstates are the free particle solutions to the wave-equation and can be taken to propagate as plane waves. If the masses are zero (or the same), the mass states remain in phase, and the state remains in the initial linear combination. If at least one mass is different from zero (and from the others), the phase between the mass states will change during the propagation. As a consequence, at the detector there will be a non zero probability to find a flavour state which was not of the original flavour.

The evidence for neutrino masses is based on the experimental observation of neutrino oscillations, revealed for the first time in fluxes of atmospheric and solar neutrinos at the end of the twentieth century. Since neutrinos cannot oscillate if they have zero mass, the observation of neutrino oscillations implies that neutrinos must have some mass, however small.

## 6.1 The Lepton Mixing Matrix

As seen in Chap. 5, neutrinos which engage in weak interactions are flavour eigenstates, and, if massive, do not coincide with the mass eigenstates. In analogy with the quark case, the passage from flavour to mass eigenstates in the Yukawa terms of the Lagrangian affects also the terms in the Lagrangian describing couplings of leptons to the charged weak bosons. Let us consider the two cases in which neutrinos have Dirac or Majorana nature.

In the case of three Dirac neutrinos, we can mimic the discussions made in Sect. 4.6, integrated by the formalism of Sect. 5.2. We use $\hat{\nu}_L$ to refer collectively to the left-handed family states participating in the weak interactions and $\tilde{\nu}$ to the Dirac mass eigenstates, namely $\hat{\nu}_L \equiv (\nu_{eL}, \nu_{\mu L}, \nu_{\tau L})$ and $\tilde{\nu} = (\nu_1, \nu_2, \nu_3)$, respectively. We use an analogous notation for charged leptons. The states that diagonalise the mass terms in the lepton sector are obtained by a field redefinition, that for left-handed neutrinos we can write as

$$\tilde{\nu}_L = U_L^{\nu\dagger} \hat{\nu}_L, \tag{6.1}$$

where the $3 \times 3$ matrix $U_L^{\nu}$ is unitary. An expression analogous to Eq. (6.1) can be written for left-handed charged leptons. Following the procedure adopted in the quark case and described in Sect. 4.6, one finds that the leptonic weak charged currents contain the unitary matrix

$$U_{\text{PMNS}} \equiv U_L^{\ell\dagger} U_L^{\nu}. \tag{6.2}$$

For instance, the charged current $J_W^{\ell+\mu}$ is written as

$$J_W^{\ell+\mu} = \frac{1}{\sqrt{2}} \overline{\hat{\nu}_L} \gamma^\mu \hat{\ell}_L = \frac{1}{\sqrt{2}} \overline{\tilde{\nu}_L} U_{\text{PMNS}}^\dagger \gamma^\mu \tilde{\ell}_L. \tag{6.3}$$

In the last formula we have the charged lepton states of definite mass $\tilde{\ell} \equiv (e, \mu, \tau)$. In the basis of mass eigenstates, it is the unitary matrix $U_{\text{PMNS}}$ that appears in the weak charged-current interactions. The lepton mixing matrix

$$U_{\text{PMNS}} = \begin{pmatrix} U_{e1} & U_{e2} & U_{e3} \\ U_{\mu 1} & U_{\mu 2} & U_{\mu 3} \\ U_{\tau 1} & U_{\tau 2} & U_{\tau 3} \end{pmatrix} \tag{6.4}$$

## 6.1 The Lepton Mixing Matrix

is called the PMNS (Pontecorvo-Maki-Nakagawa-Sakata matrix.[1]) It emphasises that flavour eigenvectors are not stationary states with definite masses, but quantum superpositions of them. The $U_{\text{PMNS}}$ matrix is analogous to the CKM matrix in the quark sector. Let us observe that there is a different convention for the CKM matrix, which connects weak isospin $T_3 = 1/2$ quarks on the left to $T_3 = -1/2$ quarks on the right, and the PMNS mixing matrix, which connects $T_3 = -1/2$ charged leptons on the left to $T_3 = 1/2$ neutrinos on the right. It is obvious that one could switch to the same convention by transforming one of the two mixing matrix to its Hermitian conjugate. In the lepton case, it is also customary to choose the basis where the mass matrix of the charged leptons is diagonal, which implies that the matrix $U_L^\ell$ is the unit matrix. It is equivalent to state that one can identify the interaction eigenstates of he charged leptons with the corresponding mass eigenstates up to a phase redefinition. In this case the lepton mixing matrix coincides with the mixing matrix of neutrinos:[2]

$$U_{\text{PMNS}} = U_L^\nu. \qquad (6.5)$$

This convention is meant to simplify the form of the charged currents, preventing the appearance of a mixing matrix in the leptonic sector, as in the massless case. In our example (6.3) we have

$$J_W^{\ell+\mu} = \frac{1}{\sqrt{2}} \overline{\hat{\nu}_L} \gamma^\mu \hat{\ell}_L = \frac{1}{\sqrt{2}} \overline{\tilde{\nu}_L} U_{\text{PMNS}}^\dagger \gamma^\mu \hat{\ell}_L \rightarrow \frac{1}{\sqrt{2}} \overline{\tilde{\nu}_L} U_L^{\nu\dagger} \gamma^\mu \hat{\ell}_L =$$

$$= \frac{1}{\sqrt{2}} \overline{\tilde{\nu}_L} \gamma^\mu \tilde{\ell}_L. \qquad (6.6)$$

Also in this case each neutrino flavour couples to the corresponding charged lepton only, as for massless neutrinos. The neutrino neutral current, reported in Eq. (4.43), does not change when expressed in terms of massive states

$$J_Z^\mu = g_L^\nu \overline{\hat{\nu}_L} \gamma^\mu \hat{\nu}_L = g_L^\nu \overline{\tilde{\nu}_L} U_L^{\nu\dagger} \gamma^\mu U_L^\nu \tilde{\nu}_L = g_L^\nu \overline{\tilde{\nu}_L} \gamma^\mu \tilde{\nu}_L \qquad (6.7)$$

where $g_L^\nu = 1/(2\cos\theta_W)$. The neutrino flavour is conserved at the vertex, as for massless neutrinos, and flavour changing is not allowed.

---

[1] As mentioned in Chap. 1, Bruno Pontecorvo was the first to consider the possibility of neutrino-antineutrino oscillations. In 1962, the year when the muon neutrino was discovered, Ziro Maki, Masami Nakagawa and Shoichi Sakata, proposed a "particle mixture theory of neutrino assuming the existence of two kinds of neutrinos" [37].

[2] Because of this convention, the term mixing matrix in literature refers to either the matrix $U_{\text{PMNS}}$ or $U_L^\nu$.

In Chap. 5 we have introduced three mass terms of the Lagrangian for Majorana fermions: $\mathscr{L}_{M_L}$, $\mathscr{L}_{M_R}$ and $\mathscr{L}_{MD}$. For each of them one can repeat the reasonings made for Dirac neutrinos, and obtain a mixing matrix. In the case of $\mathscr{L}_{M_L}$, the mass matrix is a $3 \times 3$ matrix, and we recover Eq. (6.4). The mixing among the flavour and the massive fields acts on the SM currents similarly to what observed for Dirac neutrinos. Instead, in the case of $\mathscr{L}_{M_R}$, the mass matrix is a $n \times n$ matrices, where $n$ is not constrained by the number of SM flavours. We can have $n > 3$ massive neutrinos, which are sterile and do not affect SM interactions. The case where a Dirac-Majorana mass term is present is more complex, since the possibility of mixing among active and sterile neutrinos opens up. As seen in Sect. 5.6.2, in case of $n$ right-handed neutrinos, the mass matrix can be written as $(3+n) \times (3+n)$ matrix whose diagonalization results in $3 + n$ mass eigenstates and eigenvalues. Thus Eq. (6.1) implies that the mixing matrix $U = U_L^{\nu}$ is a $3 \times (3 + n)$ rectangular matrix. The mixing matrix is not unitary, since $UU^{\dagger} \neq U^{\dagger}U$. However, if the see-saw mechanism is operative, one expects the contributions of super-heavy neutrinos to be negligible, and the matrix $U$ to be a $3 \times 3$ unitary matrix to a very good approximation.[3]

A last comment is in order. For a given weak decay, the flavour of the quarks involved can be determined experimentally, by measuring masses and charges of the hadrons involved. Then the corresponding CKM matrix elements can be determined unambiguously. Charged leptons can also be easily identified, but not neutrinos, because they are neutral and have tiny mass differences. In most cases the neutrino mass eigenstates cannot be distinguished and one has to sum over the rates associated to the production of all the three possible neutrino mass eigenstates. The same happens for antineutrinos. In a process where an interaction eigenstate $\nu_{\tilde{\alpha}}$ is produced, its overlap with mass eigenstates $\nu_i$ introduces a factor $\sum_{i=1,2,3} |(U_{\text{PMNS}})_{\tilde{\alpha} i}|^2$ in the process width. In the SM and in most of its extensions this is equal to 1 due to the unitarity of the lepton mixing matrix; then, the $U_{\text{PMNS}}$ matrix loses its role and can be ignored.

### 6.1.1 Parameter Counting

In case of neutrinos of Dirac nature, the counting of independent parameters of the lepton mixing matrix is analogous to the counting of independent parameters made for the CKM matrix in Sect. 4.7. We find the same number of parameters and we are at liberty to employ the same formalism used for the CKM matrix. Indeed, the $3 \times 3$ PMNS matrix may be parameterised by three Euler mixing angles $\theta_{12}$, $\theta_{23}$ and $\theta_{13}$ and by a CP violating phase $\delta$, which is a physical observable. In this

---

[3] We will return to the mixing with sterile neutrinos in Sect. 10.5.1.

## 6.1 The Lepton Mixing Matrix

parameterisation, the lepton mixing matrix has the same form than the CKM matrix in Eq. (4.119)

$$U_{\text{PMNS}} = \begin{pmatrix} c_{12}c_{13} & s_{12}c_{13} & s_{13}e^{-i\delta} \\ -s_{12}c_{23} - c_{12}s_{23}s_{13}e^{i\delta} & c_{12}c_{23} - s_{12}s_{23}s_{13}e^{i\delta} & s_{23}c_{13} \\ s_{12}s_{23} - c_{12}c_{23}s_{13}e^{i\delta} & -c_{12}s_{23} - s_{12}c_{23}s_{13}e^{i\delta} & c_{23}c_{13} \end{pmatrix} =$$

$$= \begin{pmatrix} 1 & 0 & 0 \\ 0 & c_{23} & s_{23} \\ 0 & -s_{23} & c_{23} \end{pmatrix} \begin{pmatrix} 1 & 0 & 0 \\ 0 & 1 & 0 \\ 0 & 0 & e^{i\delta} \end{pmatrix} \begin{pmatrix} c_{13} & 0 & s_{13} \\ 0 & 1 & 0 \\ -s_{13} & 0 & c_{13} \end{pmatrix} \begin{pmatrix} 1 & 0 & 0 \\ 0 & 1 & 0 \\ 0 & 0 & e^{-i\delta} \end{pmatrix} \begin{pmatrix} c_{12} & s_{12} & 0 \\ -s_{12} & c_{12} & 0 \\ 0 & 0 & 1 \end{pmatrix} =$$

$$= \begin{pmatrix} 1 & 0 & 0 \\ 0 & c_{23} & s_{23} \\ 0 & -s_{23} & c_{23} \end{pmatrix} \begin{pmatrix} c_{13} & 0 & s_{13}e^{-i\delta} \\ 0 & 1 & 0 \\ -s_{13}e^{i\delta} & 0 & c_{13} \end{pmatrix} \begin{pmatrix} c_{12} & s_{12} & 0 \\ -s_{12} & c_{12} & 0 \\ 0 & 0 & 1 \end{pmatrix} \quad (6.8)$$

where $c_{ij} = \cos\theta_{ij}$ and $s_{ij} = \sin\theta_{ij}$. The last expression in Eq. (6.8) decomposes the lepton mixing matrix into three matrices each characterized by a single angle, enlisting three sectors, the so called atmospheric ($\theta_{23}$), the reactor ($\theta_{13}$), the solar one ($\theta_{12}$). The reason of these names will become clear in Sects. 6.5 and 8.3. In analogy with the CKM matrix (see Sect. 4.7.2), the angles $\theta_{ij}$ are taken, without loss of generality, to lie in the first quadrant, namely $\theta_{ij} \in [0, \pi/2]$, and the phase $\delta \in [0, 2\pi]$. Values of $\delta$ different from 0 and $\pi$ imply $CP$ violation in neutrino oscillations in vacuum. If neutrinos are Majorana fermions, the counting of the phases of the lepton mixing matrix changes. A complex $n \times n$ unitary matrix has $n(n+1)/2$ independent phase parameters. In the Dirac case, one may eliminate $2n - 1$ parameters by a rephasing of the quark fields, and obtain $n_\alpha = (n-1)(n-2)/2$ independent phases (that is, 1 independent phase for 3 families). This counting excludes an overall phase which is redundant in the Dirac case, but not in the Majorana one. The overall phase is related to the total lepton number which remains a symmetry of the massive theory in the Dirac case and thus cannot be used to reduce the number of physical parameters in the mass matrix. In the Majorana case, there is no independent right-handed neutrino field, nor is lepton number a good symmetry. Therefore, one may eliminate only $n$ parameters, remaining with $n_\alpha^M = (n^2 - n)/2$ independent phases, that reduce to 3 for 3 families.

A standard parametrisation of the mixing matrix for Majorana neutrinos $U_{\text{PMNS}}$ is given by the parameterisation in terms of the product of two matrices, one with the usual CKM parameterisation that we may name $V_{\text{PMNS}}$ (Dirac case i.e. Eq. (6.8)) and the other containing the additional Majorana phases

$$U_{\text{PMNS}} = V_{\text{PMNS}}(\theta_{12}, \theta_{13}, \theta_{23}, \delta) \begin{pmatrix} e^{i\alpha_1} = 1 & 0 & 0 \\ 0 & e^{i\alpha_2} & 0 \\ 0 & 0 & e^{i\alpha_3} \end{pmatrix}. \quad (6.9)$$

Since all measurable quantities depend only on the differences of the Majorana phases, one phase can be set to zero without loss of generality. We have chosen $\alpha_1 = 0$, in agreement with usual conventions. Only two of the $\alpha_i$ phases, referred to as Majorana phases, are physically observables. The phase $\delta$ is also physical and is usually called Dirac phase.

The current picture of neutrino oscillations is characterised by large mixing angles, strikingly different from the pattern of the small mixing angles of the CKM quark mixing matrix. Two of the mixing angles are presently known fairly well: $\theta_{12} \sim 34°$ and $\theta_{13} \sim 9°$. The third mixing angle, with higher uncertainty, is close to maximal $\theta_{23} \sim 45°$ [195].

## 6.2 Neutrino Oscillations

The relation (6.1) among neutrinos participating to weak interaction $\nu_\alpha$ and mass eigenstates $\nu_i$ can be written as

$$\nu_\alpha = \sum_{i=1}^{n} U_{\alpha i} \, \nu_i \qquad (6.10)$$

where $U$ is the unitary lepton mixing matrix.[4] When calculating the neutrino oscillation probabilities, it is convenient to work with the neutrino wave functions or state vectors. The one-neutrino mass eigenstate $|\nu_i\rangle$ is generated by the Hermitian-conjugate field operator $\nu_i^\dagger(x)$ acting on the vacuum state, and likewise for the flavour states. Hence we have

$$|\nu_\alpha\rangle = \sum_{i=1}^{n} U_{\alpha i}^* |\nu_i\rangle. \qquad (6.11)$$

For simplicity, we choose a basis of orthonormal massive neutrino states, and the unitarity of the mixing matrix implies that also the flavour states are orthonormal.

In quantum field theory the evolution of states with time is regulated by the Schrödinger equation. We consider here the evolution of states in vacuum. Neutrinos are produced in charged current interactions, together with the corresponding lepton, and are described as flavour eigenstates. The time evolution of a neutrino that at the production point, at time $t = 0$, is in the flavour eigenstate $|\nu_\alpha\rangle$ is given by

$$|\nu_\alpha(t)\rangle = e^{-iHt} |\nu_\alpha\rangle \qquad (6.12)$$

where $H$ is the free total Hamiltonian. A flavour neutrino state is a precise combination of states $|\nu_i\rangle$ with masses $m_i$ described by Eq. (6.11). Mass eigenstates

---

[4] We adopt the convention (6.5) and drop the apex $\nu$ from the mixing matrix.

## 6.2 Neutrino Oscillations

can be approximated as stationary states of energy $E_i$. Hence, the time evolution becomes[5]

$$|\nu_\alpha(t)\rangle = e^{-iHt}|\nu_\alpha\rangle = \sum_{i=1}^{n} U^*_{\alpha i} e^{-iE_i t}|\nu_i\rangle. \tag{6.13}$$

In general the neutrino energies $E_i$ ($i = 1, 2, \ldots n$) are different. At $t \neq 0$, the mass eigenstate components in the original flavour state get different phases. As a result, there is a non-zero probability that the flavour measured at $t \neq 0$ is different than the flavour produced at $t = 0$. Expanding the state (6.13) in the basis of flavour neutrino states we obtain

$$|\nu_\alpha(t)\rangle = \sum_\beta A_{\nu_\alpha \to \nu_\beta}(t)|\nu_\beta\rangle. \tag{6.14}$$

where

$$A_{\nu_\alpha \to \nu_\beta}(t) \equiv \langle \nu_\beta | \nu_\alpha(t)\rangle = \sum_{k=1}^{n} U_{\beta k} e^{-iE_k t} U^*_{\alpha k}. \tag{6.15}$$

$A_{\nu_\alpha \to \nu_\beta}(t)$ is the amplitude of the $\nu_\alpha \to \nu_\beta$ transition at the time $t$. The last relation follows from Eqs. (6.11), (6.13) and the orthogonality relations of the states. The probability that a state $\nu_\alpha$ produced at $t = 0$ is revealed after the time $t$ as a state $\nu_\beta$ is defined as

$$P_{\nu_\alpha \to \nu_\beta} \equiv |A_{\nu_\alpha \to \nu_\beta}(t)|^2 = \left|\sum_{k=1}^{n} U_{\beta k} e^{-iE_k t} U^*_{\alpha k}\right|^2 =$$

$$= \sum_{k,j} U^*_{\alpha k} U_{\beta k} U_{\alpha j} U^*_{\beta j} e^{-i(E_k - E_j)t}. \tag{6.16}$$

It is obvious that if all $E_i$ are the same, because of the unitarity of the mixing matrix the initial state does not change flavour and we have $P_{\nu_\alpha \to \nu_\beta} = \delta_{\alpha\beta}$ at any time. The probabilities of the channels with $\alpha \neq \beta$ are usually called transition probabilities, whereas the probabilities of the channels with $\alpha = \beta$ are usually called

---

[5] We are using simplified expressions, neglecting the effects of the characteristics of the production and detection processes in amplitudes and treating neutrinos as plane waves, rather that as wavepackets which are localized at the production time and propagate from production to detection at a velocity close to the velocity of light. This approach, with the addition of the realistic assumptions on velocities, energy and momenta, has the advantage to reproduce the expressions for the probability of neutrino oscillations derived in the correct quantum mechanical treatment [196–198].

survival probabilities. Neutrino transitions among flavour eigenstates are referred to as oscillations.

The massive neutrino states $|\nu_i\rangle$ are eigenstates of the Hamiltonian with energy eigenvalues

$$E_i = \sqrt{p^2 + m_i^2} \simeq p + \frac{m_i^2}{2p} \tag{6.17}$$

where $p$ is $|\mathbf{p}|$. The last relation follows from the Taylor expansion in the small neutrino mass (ultrarelativistic approximation). We have also assumed that all flavour neutrinos have a definite momentum $p$, i.e. that all the massive neutrino components have the same momentum. This equal momentum assumption is motivated by the fact that all the components propagate in the same direction from source to detector. We can write

$$E_k - E_j \simeq \frac{\Delta m_{kj}^2}{2E} \qquad \Delta m_{kj}^2 \equiv m_k^2 - m_j^2 \tag{6.18}$$

where we have set $p = E$ in the denominator, neglecting the small mass contribution. As already noted, we are dealing with plane waves, even if localized particles are described by wave packets, with a velocity coinciding with the group velocity of the wave packet. Neutrinos are ultrarelativistic, and we can make the approximation $L \simeq ct \simeq t$, sometimes called light-ray approximation. Indeed, in neutrino oscillation experiments, what is measured is the distance $L$ between the neutrino source and the detector and not the propagation time $t$. We obtain

$$P_{\nu_\alpha \to \nu_\beta} = \sum_{k,j} U_{\alpha k}^* U_{\beta k} U_{\alpha j} U_{\beta j}^* \, e^{-i \frac{\Delta m_{kj}^2}{2E} L}. \tag{6.19}$$

The quartic products in (6.19) $U_{\alpha k}^* U_{\beta k} U_{\alpha j} U_{\beta j}^*$ do not depend on the specific parameterisation of the PMNS mixing matrix and are invariant under the rephasing transformations $U_{\alpha k} \to e^{i\phi_\alpha} U_{\alpha k}$ and $U_{\alpha k} \to U_{\alpha k} e^{i\phi_k}$. Therefore, given the parameterisation in Eq. (6.9), it is immediate to observe that $P_{\nu_\alpha \to \nu_\beta}$ is independent of the Majorana phases, that can be safely set to zero. The probabilities are the same in case of either Dirac or Majorana neutrinos. This has the important consequence that the analyses of transition or survival probabilities do not provide information on the Dirac or Majorana nature of neutrinos. It is expected, because flavour oscillations, although occurring only when neutrinos have mass, conserve the total lepton number.

It is also immediate to see that even if neutrino oscillations imply that neutrinos are massive, their measurement may yield information only on the values of the squared-mass differences $\Delta m_{kj}^2$, not on the absolute values of neutrino masses.

## 6.2 Neutrino Oscillations

Let us go back to the expression (6.16). Taking into account that a common phase cannot be observed, it can be transformed into the form

$$P_{\nu_\alpha \to \nu_\beta} = \left|\sum_{k=1}^{n} U_{\beta k}\, e^{-iE_k t}\, U^*_{\alpha k}\right|^2 = \left|\sum_{k=1}^{n} U_{\beta k}\, e^{-i(E_k - E_j)t}\, U^*_{\alpha k}\right|^2 \tag{6.20}$$

where $j$ is fixed. By using the unitarity relation of the $U$ matrix,

$$\sum_{k=1}^{n} U_{\beta k}\, U^*_{\alpha k} = \delta_{\alpha\beta} \tag{6.21}$$

we find the following convenient expression

$$P_{\nu_\alpha \to \nu_\beta} = \left|\sum_{k\neq j}^{n} U_{\beta k}\, e^{-i(E_k-E_j)t}\, U^*_{\alpha k} + U_{\beta j}\, U^*_{\alpha j}\right|^2$$

$$= \left|\delta_{\alpha\beta} + \sum_{k\neq j}^{n} U_{\beta k}\, U^*_{\alpha k}\left[e^{-i(E_k-E_j)t} - 1\right]\right|^2$$

$$\simeq \left|\delta_{\alpha\beta} + \sum_{k\neq j}^{n} U_{\beta k}\, U^*_{\alpha k}\left[e^{-i\frac{\Delta m^2_{kj}}{2E}L} - 1\right]\right|^2. \tag{6.22}$$

We see immediately that no mixing ($U$ equal to the unit $3 \times 3$ matrix) implies no transitions among different flavours. The same happens if the argument of the exponential becomes almost zero, that if $\Delta m^2 L/E \ll 1$ holds for all $\Delta m^2$. Flavour transitions can only be observed if at least one $\Delta m^2$ satisfies the condition

$$\Delta m^2 \gtrsim \frac{E}{L}. \tag{6.23}$$

There are other ways to present the expression for the neutrino transition probability $P_{\nu_\alpha \to \nu_\beta}$. Starting from Eq. (6.19), we can write

$$P_{\nu_\alpha \to \nu_\beta} = \sum_{k,j} U^*_{\alpha k}\, U_{\beta k}\, U_{\alpha j}\, U^*_{\beta j}\, e^{-i\frac{\Delta m^2_{kj}}{2E}L}$$

$$= \sum_{k} |U_{\alpha k}|^2\, |U_{\beta k}|^2 +$$

$$+ 2\,\mathrm{Re}\sum_{k>j} U^*_{\alpha k}\, U_{\beta k}\, U_{\alpha j}\, U^*_{\beta j}\, e^{-i\frac{\Delta m^2_{kj}}{2E}L}. \tag{6.24}$$

We have separated a constant term from the oscillating term. An incoherent average of the oscillation probability over the energy resolution of the detector or over the uncertainty of the distance $L$ only leaves the constant term

$$\langle P_{\nu_\alpha \to \nu_\beta} \rangle = \sum_k |U_{\alpha k}|^2 |U_{\beta k}|^2 \tag{6.25}$$

since the average of the exponential functions in Eq. (6.24) is zero. Let us observe that this expression has the same form than the (not averaged) incoherent transition probability, that is the transition probability when, for some reason, different massive neutrinos are produced or detected in an incoherent way. Since the oscillating term in Eq. (6.24) is produced by the interference of the different massive neutrino components of the state in Eq. (6.13), its existence depends on the coherence of the massive neutrino components and it also disappears in the incoherent transition probability.

Let us now go back to the coherent, non averaged, formulation (6.24). The equality

$$\sum_k |U_{\alpha k}|^2 |U_{\beta k}|^2 = \delta_{\alpha\beta} - 2\operatorname{Re} \sum_{k>j} U^*_{\alpha k} U_{\beta k} U_{\alpha j} U^*_{\beta j} \tag{6.26}$$

follows from the unitarity relation (6.21). Combining now (6.26) and (6.24) we have

$$P_{\nu_\alpha \to \nu_\beta} = \delta_{\alpha\beta} - 2\operatorname{Re} \sum_{k>j} U^*_{\alpha k} U_{\beta k} U_{\alpha j} U^*_{\beta j} \left(1 - e^{-i\frac{\Delta m^2_{kj}}{2E}L}\right). \tag{6.27}$$

For any two complex numbers $x$ and $y$ the following equality holds

$$\operatorname{Re} x\, y = \operatorname{Re} x \operatorname{Re} y - \operatorname{Im} x \operatorname{Im} y. \tag{6.28}$$

Then we can write the relation (6.27) as

$$P_{\nu_\alpha \to \nu_\beta} = \delta_{\alpha\beta} - 2 \sum_{k>j} \operatorname{Re}(U^*_{\alpha k} U_{\beta k} U_{\alpha j} U^*_{\beta j}) \left(1 - \cos \frac{\Delta m^2_{kj}}{2E} L\right) +$$

$$+ 2 \sum_{k>j} \operatorname{Im}(U^*_{\alpha k} U_{\beta k} U_{\alpha j} U^*_{\beta j}) \sin \frac{\Delta m^2_{kj}}{2E} L. \tag{6.29}$$

Finally, since for any angle $\theta$

$$1 - \cos\theta = 2\sin^2 \frac{\theta}{2}, \tag{6.30}$$

## 6.2 Neutrino Oscillations

we obtain the expression (6.29) as

$$P_{\nu_\alpha \to \nu_\beta} = \delta_{\alpha\beta} - 4 \sum_{k>j} \text{Re}\,(U^*_{\alpha k} U_{\beta k} U_{\alpha j} U^*_{\beta j}) \sin^2 \frac{\Delta m^2_{kj}}{4E} L +$$

$$+ 2 \sum_{k>j} \text{Im}\,(U^*_{\alpha k} U_{\beta k} U_{\alpha j} U^*_{\beta j}) \sin \frac{\Delta m^2_{kj}}{2E} L. \qquad (6.31)$$

This form underlines the oscillatory character of the transition probability.

In analogy with relation (6.11), the state describing a flavour antineutrino $|\bar{\nu}_\alpha\rangle$ can be expressed in terms of mass eigenstates antineutrinos $|\bar{\nu}_i\rangle$[6]

$$|\bar{\nu}_\alpha\rangle = \sum_{i=1}^n U_{\alpha i} |\bar{\nu}_i\rangle. \qquad (6.32)$$

Hence the amplitude of antineutrino transitions is given by

$$A_{\bar{\nu}_\alpha \to \bar{\nu}_\beta}(t) \equiv \langle \bar{\nu}_\beta | \bar{\nu}_\alpha(t) \rangle = \sum_{k=1}^n U^*_{\beta k} e^{-iE_k t} U_{\alpha k} \qquad (6.33)$$

which differs from the corresponding amplitude for neutrinos only by the exchange $U \to U^*$. Hence to find the probability of the transition among flavour antineutrinos $P_{\bar{\nu}_\alpha \to \bar{\nu}_\beta}$ one follows the same procedure described before and finds the same expression (6.22) but with the substitution $U \to U^*$. By continuing the derivation along the same path, one obtains an expression for $P_{\bar{\nu}_\alpha \to \bar{\nu}_\beta}$ that differs from formula (6.31) only for the sign of the imaginary part, that is $\text{Im} \to -\text{Im}$.

The constraints imposed by $CPT$ symmetry are

$$P_{\nu_\alpha \to \nu_\beta} = P_{\bar{\nu}_\beta \to \bar{\nu}_\alpha}. \qquad (6.34)$$

In particular, $CPT$ symmetry implies the equality of the survival probabilities of neutrinos and antineutrinos

$$P_{\nu_\alpha \to \nu_\alpha} = P_{\bar{\nu}_\alpha \to \bar{\nu}_\alpha} \qquad (6.35)$$

as well as the equality of the neutrino and antineutrino masses.

---

[6] As already discussed in Sect. 5.3 the difference between Majorana neutrinos and antineutrinos is conventional. Hence, the standard flavour neutrino states describe Dirac or Majorana neutrinos with negative helicity and the states in Eq. (6.32) describe Dirac antineutrinos with positive helicity or Majorana neutrinos with positive helicity.

## 6.3 Two-Generations Mixing

Let us consider neutrino mixing in the approximation in which only two massive neutrinos are considered. This approximation is useful not only because formulas are much simpler, but also because there are many experiments which are not sensitive to the influence of three-neutrino mixing, and their data can be safely analysed in the two-neutrino mixing framework.

For two generations, the 2 × 2 lepton mixing matrix is real and can be written without loss of generality in a form analogous to the Cabibbo matrix, where $\theta$ is the mixing angle. There is no physical phase, which implies no *CP* violation. The relation (6.10) connecting flavour and mass eigenstates becomes

$$\begin{pmatrix} \nu_\alpha \\ \nu_\beta \end{pmatrix} = \begin{pmatrix} \cos\theta & \sin\theta \\ -\sin\theta & \cos\theta \end{pmatrix} \begin{pmatrix} \nu_1 \\ \nu_2 \end{pmatrix}. \tag{6.36}$$

The mixing is said maximal when $\theta = \pi/4$ and therefore $\sin\theta = \cos\theta = \sqrt{2}/2$. When the mixing is maximal there is an equal probability to find either of the mass eigenvalues in each of the flavour eigenvalues.

The transition probability ($\alpha \neq \beta$) is given by the relation (6.22), setting $j = 1$:

$$P_{\nu_\alpha \to \nu_\beta} = \left| U_{\beta 2} U^*_{\alpha 2} \left[ \exp\left( -i \frac{\Delta m^2 L}{2E} \right) - 1 \right] \right|^2$$

$$= \frac{1}{2} \sin^2 2\theta \left( 1 - \cos \frac{\Delta m^2 L}{2E} \right) =$$

$$= \sin^2 2\theta \sin^2 \frac{\Delta m^2 L}{4E} \tag{6.37}$$

where $\Delta m^2 \equiv \Delta m^2_{21} = m_2^2 - m_1^2$ and $\sin^2 2\theta = 4|U_{\alpha 2}|^2 |U_{\beta 2}|^2 = 4\sin^2\theta \cos^2\theta$. We have used Eq. (6.30). Formula (6.37) holds in the ultrarelativistic approximation of Eq. (6.18), where $t \simeq L$ and the energy difference is proportional to the difference of the square masses $\Delta E \equiv E_2 - E_1 \simeq \Delta m^2/2E$. The periodic behaviour of the transition probability is determined by the mass difference and by the parameter angle $\theta$. There is no mixing unless both $\Delta m^2$ and $\theta$ are non-vanishing. The maximal amplitude is obtained at the point of maximal mixing, that is when $\theta = \pi/4$. For two generations, the transition probability is the same for neutrinos and antineutrinos, because there are no imaginary entries in the mixing matrix. Moreover, it is easy to check that the probabilities of direct ($\nu_\alpha \to \nu_\beta$) and inverted ($\nu_\beta \to \nu_\alpha$) channels are equal.

From the second step in Eq. (6.37), by noting that the average of the cosine function is zero, we obtain the average transition probability

$$\langle P_{\nu_\alpha \to \nu_\beta} \rangle = \frac{1}{2} \sin^2 2\theta \tag{6.38}$$

## 6.3 Two-Generations Mixing

which is the same as the incoherent transition probability, as discussed in the case of Eq. (6.25). Because experiments have finite resolution on $L$ and $E$, and a spread in beam energies, this probability average occurs when $\Delta m^2 \ll L/E$ and one loses sensitivity to $\Delta m^2$.

From Eq. (6.37) one finds that the survival probability ($\alpha = \beta$) is given by

$$P_{\nu_\alpha \to \nu_\alpha} = 1 - P_{\nu_\alpha \to \nu_\beta} = 1 - \frac{1}{2}\sin^2 2\theta \left(1 - \cos\frac{\Delta m^2 L}{2E}\right) =$$

$$= 1 - \sin^2 2\theta \sin^2 \frac{\Delta m^2 L}{4E}. \qquad (6.39)$$

Summarising, the transition and survival probabilities in Eqs. (6.37) and (6.39) can be written as

$$P_{\nu_\alpha \to \nu_\beta} = \delta_{\alpha\beta} + (-1)^{\delta_{\alpha\beta}} \sin^2 2\theta \sin^2 \frac{\Delta m^2 L}{4E}. \qquad (6.40)$$

Since the mixing matrix has no imaginary part, the same formula holds for antineutrinos. It may be useful to pass from natural units to realistic units where the energy $E$ is expressed in GeV, $\Delta m^2$ in eV$^2$ and $L$ in km. In this case, the oscillation argument becomes

$$\frac{\Delta m^2 L}{4E} \to \frac{\Delta m^2 L}{4E} \frac{c^3}{\hbar} \simeq \frac{1.27 \Delta m^2 L}{E} \qquad (6.41)$$

since $c = 1$ and $\hbar \simeq 0.1973$ eV$^2$ GeV$^{-1}$ km. Therefore, starting at 0 km with a neutrino $\nu_\alpha$, the probability of detecting a neutrino $\nu_\beta$ after $L$ km, that is $P_{\nu_\alpha \to \nu_\beta}$ in Eq. (6.37), can be written as

$$P_{\nu_\alpha \to \nu_\beta} \simeq \sin^2 2\theta \sin^2 \frac{1.27 \Delta m^2 L}{E} = \sin^2 2\theta \sin^2 \frac{\pi L}{\lambda} \qquad (6.42)$$

where we have defined the oscillation length $\lambda$ as

$$\lambda \equiv 2.47 \frac{E}{\Delta m^2}. \qquad (6.43)$$

At any distance $L + n\lambda$, $n$ being an integer number, the transition probability $P_{\nu_\alpha \to \nu_\beta}$ acquires the same value. Formula (6.42) is particularly suited to show the phenomenon of neutrino oscillations. The $\sin^2 2\theta$ factor is the oscillation amplitude. Clearly if there is no mixing angle oscillations cannot happen. For a given $\Delta m^2$, the probability of oscillation changes as one moves away from the detector, or scans over different neutrino energy. At the production point $L = 0$, the oscillation probability is zero and the corresponding survival probability is one. As neutrino propagate and their distance from the production point $L$ increases, the oscillations

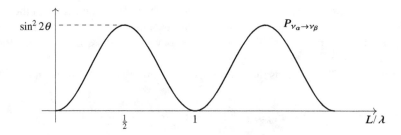

**Fig. 6.1** Transition probability $P_{\nu_\alpha \to \nu_\beta}$ as a function of $L/\lambda$

begin to switch on until $\pi L/\lambda = \pi/2$. At this point the oscillation is a maximum. If the mixing angle is such as $\sin^2(2\theta) = 0.\text{x}0$, we know that at this point x% of the initial neutrinos have oscillated away. As $L$ increases further, the oscillation dies down until, at $\pi L/\lambda = \pi$, the beam is entirely composed of the initial neutrino flavour, and so on. When $\theta = \pi/4$, that is at the maximal mixing, $sin^2 2\theta = 1$ and 100% of the neutrinos we started with will oscillate to the final neutrinos at some point on the path to the detector. In Fig. 6.1 we plot the behaviour of a generic transition probability $P_{\nu_\alpha \to \nu_\beta}$ as a function of $L/\lambda$.

We distinguish two kinds of neutrino oscillation experiments. They both start with a pure beam of known flavour. This first kind, called "disappearance" experiment, looks to see how many have disappeared. It measures the survival probability. The key issue with disappearance experiments is an accurate understanding of the incoming neutrino flux and the efficiency with which you detect events. The second type of experiment searches the number of neutrinos of a different flavour at the detector: it is an "appearance" experiment. Since in general the neutrino flavour is defined by the charged lepton which is produced in its charged current interactions,[7] an appearance measurement implies additional requirements, namely (a) energies large enough to produce the final state lepton, i.e. kinematically above the lepton mass threshold (b) a detector able to distinguish this final-state charged lepton from other final states that may occur.

Both kinds of experiments require probabilities large enough than events can be actually observed—in a reasonable amount of time. Hence experiments aim at the maximum probability, which happens when

$$\frac{L}{\lambda} = \frac{1}{2} \quad \text{that is} \quad \frac{L}{E} = 1.2 \frac{1}{\Delta m^2}. \qquad (6.44)$$

---

[7] The only way to do an "appearance" experiment with low-energy neutrinos is direct comparison between neutral-current reactions, which are sensitive to all neutrino types, and charged-current reactions, as in the solar-neutrino experiment SNO (see Sect. 7.2.5). In general, this is not easy and disappearance is the only realistic option.

In the experimental setting at colliders, we are free to either change the beam energy $E$ or the production-detection distance $L$, the so-called baseline, or both. To probe smaller and smaller masses we want to maximise $L$ and minimise $E$. One experimental difficulty to take into account, though, is that neutrino beams diverge, so the surface area of a detector placed at a distance $L$ grows roughly by $L^2$. At the same time, decreasing the neutrino energy implies that the neutrino cross-section decreases and the running time to collect a useful number of events increases.

## 6.4 CP Violation

With the discovery of neutrino oscillations, searches for $CP$ violation in the lepton sector have been actively pursued as the next, relevant goal.

As discussed in Sect. 4.8, the $CP$ transformation interchanges the operators of the Lagrangian with their Hermitian conjugates but leaves invariant the coefficients, which do not transform under $C$ or $P$. Leptons only interacts by the EW interaction and in the lepton sector the EW Lagrangian transformed under $CP$ differs from the Lagrangian itself only because of the complex coefficients of the lepton mixing matrix. When the neutrinos are massless, or degenerate in mass, there is no mixing and no $CP$ violation is possible in the lepton sector. Summarising, in analogy with the quark sector, $CP$ is violated if $U_{PMNS} \neq U^*_{PMNS}$, that is in the presence of non-zero physical phases.

As seen in Sect. 6.2, independently of the Dirac or Majorana nature of neutrinos, only the phase $\delta$ is relevant in neutrino oscillations. Majorana phases in the mixing matrix do not affect the transition probabilities for neutrinos. This can also be seen by direct substitution of the mixing matrix of Eq. (6.9) into the transition probability of Eq. (6.16); one finds the equality

$$P_{\nu_\alpha \to \nu_\beta} = \left| \sum_{k=1}^n U_{\beta k} e^{-iE_k t} U^*_{\alpha k} \right|^2 = \left| \sum_{k=1}^n V_{\beta k} e^{-iE_k t} V^*_{\alpha k} \right|^2 \tag{6.45}$$

where the matrix $V$ is defined as the PMNS one $U$ stripped by Majorana phases. The same happens for the antineutrino transition probabilities.

$CP$ violation in neutrino oscillations can be experimentally measured by comparing the probability that a neutrino, prepared in a flavour state, is detected after propagation in vacuum as a given different flavour, with the probability of an antineutrino under the same conditions. Then $CP$ asymmetry in neutrino oscillations is expressed by the inequality

$$A^{CP}_{\alpha\beta} \equiv P_{\nu_\alpha \to \nu_\beta} - P_{\bar\nu_\alpha \to \bar\nu_\beta} \neq 0 \tag{6.46}$$

regardless of their Dirac or Majorana mass term. As said before, this asymmetry is independent from Majorana phases, being affected only by possible irreducible complex phases in the $V$ matrix.

Let us consider the transition probability expressed in the form (6.31), The corresponding expression for the antineutrino differs only for the sign of the imaginary part, that is Im $\to$ $-$Im. By setting $U \to V$, we have

$$A_{\alpha\beta}^{CP} = P_{\nu_\alpha \to \nu_\beta} - P_{\bar{\nu}_\alpha \to \bar{\nu}_\beta} =$$

$$= 4 \sum_{k>j} \text{Im}\,(V_{\alpha k}^* V_{\beta k} V_{\alpha j} V_{\beta j}^*) \sin \frac{\Delta m_{kj}^2}{2E} L =$$

$$= 4 \sum_{k>j} J_{\alpha\beta}^{jk} \sin \frac{\Delta m_{kj}^2}{2E} L \qquad (6.47)$$

where $\Delta m_{kj}^2 \equiv m_k^2 - m_j^2$. In the last passage we have defined

$$J_{\alpha\beta}^{jk} \equiv \text{Im}\,(V_{\alpha k}^* V_{\beta k} V_{\alpha j} V_{\beta j}^*) \qquad (6.48)$$

which is antisymmetric under the exchange of the lower indexes $\alpha, \beta$ and the upper indexes $k, j$. Indeed, we have

$$J_{\alpha\beta}^{jk} = -J_{\beta\alpha}^{jk} \qquad (6.49)$$

$$J_{\alpha\beta}^{jk} = -J_{\alpha\beta}^{kj} \qquad (6.50)$$

$$J_{\alpha\alpha}^{jk} = \text{Im}|V_{\alpha k}|^2|V_{\alpha j}|^2 = 0 \qquad (6.51)$$

$$J_{\alpha\beta}^{jj} = \text{Im}|V_{\alpha j}|^2|V_{\beta j}|^2 = 0. \qquad (6.52)$$

For transitions between the same flavour, i.e. for $\alpha = \beta$, no $CP$ asymmetry can arise since

$$A_{\alpha\alpha}^{CP} = 0 \qquad (6.53)$$

in agreement with $J_{\alpha\alpha}^{jk} = 0$ from Eq. (6.51). It follows that the disappearance probability in Eq. (6.31) becomes

$$P_{\nu_\alpha \to \nu_\alpha} = 1 - 4 \sum_{k>j} \text{Re}\,(U_{\alpha k}^* U_{\alpha k} U_{\alpha j} U_{\alpha j}^*) \sin^2 \frac{\Delta m_{kj}^2}{4E} L$$

$$= 1 - 4 \sum_{k>j} |V_{\alpha k}|^2 |V_{\alpha j}|^2 \sin^2 \frac{\Delta m_{kj}^2}{4E} L. \qquad (6.54)$$

## 6.4 CP Violation

From the unitarity of the matrix $V$ the sums over un upper or a lower index cancel

$$\sum_j J^{jk}_{\alpha\beta} = \sum_j \text{Im}\,(V^*_{\alpha k} V_{\beta k} V_{\alpha j} V^*_{\beta j}) = \text{Im} \sum_j (V_{\alpha j} V^*_{\beta j} V^*_{\alpha k} V_{\beta k}) =$$

$$= \text{Im}(\delta_{\alpha\beta} V^*_{\alpha k} V_{\beta k}) = 0$$

$$\sum_\alpha J^{jk}_{\alpha\beta} = \sum_\alpha \text{Im}\,(V^*_{\alpha k} V_{\alpha j} V_{\beta k} V^*_{\beta j}) = \text{Im} \sum_\alpha (V_{\alpha j} V^*_{\alpha k} V^*_{\beta j} V_{\beta k}) =$$

$$= \text{Im}(\delta_{jk} V^*_{\beta j} V_{\beta k}) = 0. \tag{6.55}$$

It follows that

$$\sum_\beta A^{CP}_{\alpha\beta} = 0 \tag{6.56}$$

which expresses the conservation of probabilities

$$\sum_\beta P_{\nu_\alpha \to \nu_\beta} = 1 \qquad \sum_\beta P_{\bar{\nu}_\alpha \to \bar{\nu}_\beta} = 1. \tag{6.57}$$

Imposing CPT invariance as in Eq. (6.34) implies that $A^{CP}_{\alpha\beta}$ is antisymmetric in the indexes $\alpha$ and $\beta$

$$A^{CP}_{\alpha\beta} = -A^{CP}_{\beta\alpha}. \tag{6.58}$$

In the case of three massive neutrinos, the indexes $\alpha$ and $\beta$ run over the three flavours $e$, $\mu$, $\tau$. The six terms $\Delta_{\alpha\beta}$ for $\alpha \neq \beta$ reduce to three because of the antisymmetry of $\Delta_{\alpha\beta}$, but there is only one independent CP asymmetry. In fact, from Eq. (6.56) we have

$$A^{CP}_{\mu e} + A^{CP}_{\mu\tau} = A^{CP}_{e\mu} + A^{CP}_{e\tau} = A^{CP}_{\tau\mu} + A^{CP}_{\tau e} = 0 \tag{6.59}$$

which implies that the three CP asymmetries are connected by a cyclic relation

$$A^{CP}_{\mu e} = A^{CP}_{e\tau} = -A^{CP}_{\mu\tau}. \tag{6.60}$$

In a similar way, we can demonstrate that the parameters $J^{jk}_{\alpha\beta}$ equate $\pm$ a single object $J$. By using the first of Eqs. (6.55) and (6.52) we can write

$$J^{21}_{\alpha\beta} + J^{31}_{\alpha\beta} = J^{12}_{\alpha\beta} + J^{32}_{\alpha\beta} = J^{13}_{\alpha\beta} + J^{23}_{\alpha\beta} = 0. \tag{6.61}$$

By exploiting the antisymmetric relation (6.50), these parameters reduce to one, holding the cyclic relation

$$J^{13}_{\alpha\beta} = J^{32}_{\alpha\beta} = J^{21}_{\alpha\beta}. \tag{6.62}$$

Analogously, we can exploit the second of Eq. (6.55) and write

$$J^{jk}_{\mu e} + J^{jk}_{\tau e} = J^{jk}_{e\mu} + J^{jk}_{\tau\mu} = J^{jk}_{e\tau} + J^{jk}_{\mu\tau} = 0 \tag{6.63}$$

which gives further cyclic relations

$$J^{jk}_{e\mu} = J^{jk}_{\mu\tau} = J^{jk}_{\tau e}. \tag{6.64}$$

We can fix, for instance, by convention

$$J \equiv J^{12}_{e\mu} = \mathrm{Im}\,(V^*_{e2} V_{\mu 2} V_{e1} V^*_{\mu 1}). \tag{6.65}$$

It is evident that whatever the indexes for all non-zero elements $J^{jk}_{\alpha\beta}$ the following relation holds

$$J^{jk}_{\alpha\beta} = \pm J \tag{6.66}$$

where the sign depends on the indexes. By definition, $J$ is invariant, independent by changes of arbitrary phases and parametrisation in the $V$ matrix. By writing the lepton mixing matrix elements in the parametrisation (6.8), $J$ becomes

$$J = c_{12} c_{23} c^2_{13} s_{12} s_{23} s_{13} \sin \delta =$$
$$= \frac{1}{8} \sin(2\theta_{12}) \sin(2\theta_{13}) \sin(2\theta_{23}) \cos \theta_{13} \sin \delta. \tag{6.67}$$

According to current measurements, we have $J \simeq 0.033 \sin \delta$ [172]. It is clear that the parameter $J$ controls the magnitude of CP violation effects in neutrino oscillations in the case of three-neutrino mixing. In this respect, it is analogous to what is called the Jarlskog invariant in the quark sector $J_q$ [199], but it has the potential to be three orders of magnitude larger ($J_q \simeq 3 \times 10^{-5}$ [172]).

According to definition (6.47), we have

$$A^{CP}_{\alpha\beta} = 4 \left( J^{12}_{\alpha\beta} \sin \frac{\Delta m^2_{21}}{2E} L + J^{13}_{\alpha\beta} \sin \frac{\Delta m^2_{31}}{2E} L + J^{23}_{\alpha\beta} \sin \frac{\Delta m^2_{32}}{2E} L \right) =$$

$$= 4 J^{12}_{\alpha\beta} \left( \sin \frac{\Delta m^2_{21}}{2E} L - \sin \frac{\Delta m^2_{31}}{2E} L + \sin \frac{\Delta m^2_{32}}{2E} L \right) =$$

$$= \pm 4 J \left( - \sin \frac{\Delta m^2_{21}}{2E} L + \sin \frac{\Delta m^2_{31}}{2E} L - \sin \frac{\Delta m^2_{32}}{2E} L \right) \tag{6.68}$$

## 6.4 CP Violation

where the $\pm$ sign depends on the values of the indexes $\alpha$ and $\beta$. In particular, one has

$$\begin{aligned} A^{CP}_{e\mu} = A^{CP}_{\mu\tau} = -A^{CP}_{e\tau} = \\ = 4J \left( +\sin\frac{\Delta m^2_{21}}{2E}L + \sin\frac{\Delta m^2_{32}}{2E}L + \sin\frac{\Delta m^2_{13}}{2E}L \right) = \\ = -16J \left( \sin\frac{\Delta m^2_{21}L}{4E} \sin\frac{\Delta m^2_{32}L}{4E} \sin\frac{\Delta m^2_{13}L}{4E} \right). \end{aligned} \quad (6.69)$$

The last equality comes from the identity $\sin 2\theta + \sin 2\alpha + \sin 2\beta = -4\sin\theta\sin\alpha\sin\beta$, which holds when $\theta + \alpha + \beta = 0$, as it can easily be verified by using the standard trigonometric formulas of addition. If two or more of the masses are degenerate, the asymmetry is zero and there is no $CP$ violation.

From the expression of $J$ in Eq. (6.67) one can see that $CP$ is preserved if $\delta = 0$ or $\pi$ or at least one of the mixing angles is zero. A value of $\delta \neq 0, \pi$ indicates $CP$ violation in neutrino oscillations. One also sees that to measure a non-zero value, one needs that the conditions $\theta_{12} \neq 0, \pi/2, \theta_{23} \neq 0, \pi/2$ and $\theta_{13} \neq 0, \pi/2$ are all true.

Let us summarize the main points of this section

- $CP$ violation in oscillations cannot provide information about the Dirac or Majorana nature of neutrinos;
- since it depends on the phase $\delta$, $CP$ violation in oscillations is a genuine three (or more) flavour effect;
- it cannot show up in disappearance neutrino oscillation experiments, where one flavour converts to the same one, but only in appearance experiments, where one flavour oscillates into another one;
- $CP$ violation for three generation neutrino mixing is zero when any of these conditions is verified
  - two masses are degenerate
  - one mixing angle is zero;
  - $\delta = 0$ or $\delta = \pi$.

In 2020 the T2K collaboration has reported for the first time the strongest hint of leptonic $CP$ violation, based on an analysis of nine years of neutrino-oscillation data [200]. At present, the relatively large asymmetry $A^{CP}_{\mu e}$ found in the T2K experiment [201], which favored $CP$-violating values of $\delta$, was in slight tension with the results of the experiments NOvA [202], whose data disfavored the combinations of oscillation parameters that give rise to a large asymmetry. A joint fit for oscillations has been undertaken by the two collaborations, which finds a $1.9\sigma$ exclusion of $CP$ conservation (defined as $J = 0$) [600]. Further data are required to confirm these findings and a possible future discovery of leptonic $CP$ violation.

## 6.5 Hierarchy and Simplified Mixing

At present, oscillation data seem to be well described by three active massive neutrinos with flavour eigenstates $\nu_e$, $\nu_\mu$ and $\nu_\tau$ and mass eigenstates $\nu_1$, $\nu_2$, $\nu_3$ (with masses $m_1$, $m_2$ $m_3$).[8] In the three-mixing case, as in the simplified two-mixing one, neutrino oscillation experiments are only sensitive to mass square differences and not to the absolute values of neutrino masses. Current oscillation data indicate two squared mass differences, which differ for about two orders of magnitude, and are dubbed as the solar mass splitting $\Delta m_s^2$ and the atmospheric mass splitting $\Delta m_a^2$ [203]:

$$|\Delta m_s^2| \sim 7.4 \times 10^{-5} \text{ eV}^2 \qquad |\Delta m_a^2| \sim 2.5 \times 10^{-3} \text{ eV}^2. \qquad (6.70)$$

The two closest massive states are conventionally taken to be the states with masses $m_1$ and $m_2$. We number the massive neutrinos in such a way that $m_1 < m_2$, so that

$$\Delta m_s^2 \equiv \Delta m_{21}^2 = m_2^2 - m_1^2 > 0. \qquad (6.71)$$

The larger of the two neutrino mass squared differences is identified as

$$\Delta m_a^2 \equiv \Delta m_{31}^2 \equiv m_3^2 - m_1^2. \qquad (6.72)$$

So far the sign of $\Delta m_{31}^2$ has not been determined yet in a conclusive way. We have that

$$\Delta m_{21}^2 \ll |\Delta m_{31}^2|. \qquad (6.73)$$

It follows that the neutrino spectrum has two possible mass orderings, called hierarchies,[9] and distinguished by the sign of $\Delta m_{31}^2$:

- normal mass hierarchy ($\Delta m_{31}^2 > 0$). The lowest state in mass is $\nu_1$, with a closer heavier state $\nu_2$, and the heaviest state $\nu_3$ at some distance:

$$m_1 < m_2 \ll m_3 \qquad \left[\Delta m_{32}^2 \gg \Delta m_{21}^2 > 0\right] \qquad (6.74)$$

---

[8] Let us remark that there are experiments that do not validate this framework, as we will discuss in Sect. 10.5. In this section we limit to consider the commonly accepted three-neutrino framework.

[9] We assumes a hierarchical structure, as a vast part of the literature. However, Ref.[204] claims that a quasi-degenerate spectrum, where $m_1 \simeq m_2 \simeq m_3$, is still possible.

## 6.5 Hierarchy and Simplified Mixing

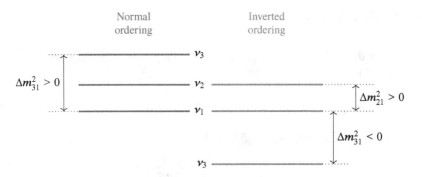

**Fig. 6.2** Normal and inverted mass ordering of neutrinos

- inverted mass hierarchy ($\Delta m_{31}^2 < 0$). The lightest state is $\nu_3$, followed at some distance by $\nu_1$, with the highest state in mass $\nu_2$ at a closer distance:

$$m_3 \ll m_1 < m_2 \qquad \left[-\Delta m_{31}^2 \gg \Delta m_{21}^2 > 0\right]. \tag{6.75}$$

The two mass orderings are illustrated in Fig. 6.2.

Oscillation data implicate a lower bound on the sum of the neutrino masses $\sum m_\nu \equiv \sum_i m_i$. Since there are two distinct mass gaps, there must be, at least, two massive neutrinos. In the normal order, one neutrino needs to have a mass above the value $\sqrt{\Delta m_{21}^2} \approx 0.009$ eV and the other one above the value $\sqrt{|\Delta m_{31}^2|} \approx 0.05$ eV. This yields

$$\sum m_\nu = m_1 + \sqrt{m_1^2 + \Delta m_{21}^2} + \sqrt{m_1^2 + \Delta m_{31}^2}$$

$$\gtrsim \sqrt{\Delta m_{21}^2} + \sqrt{\Delta m_{31}^2} \approx 0.06 \, \text{eV}. \tag{6.76}$$

In the inverted order, the lower bound is higher

$$\sum m_\nu = m_3 + \sqrt{m_3^2 + \Delta m_{21}^2 - \Delta m_{31}^2} + \sqrt{m_3^2 - \Delta m_{31}^2}$$

$$\approx m_3 + \sqrt{m_3^2 - \Delta m_{31}^2} + \sqrt{m_3^2 - \Delta m_{31}^2}$$

$$\gtrsim 2\sqrt{-\Delta m_{31}^2} \approx 0.10 \, \text{eV}. \tag{6.77}$$

## 6.5.1 Two-Neutrino Formalism

Because of the ranking (6.73), it is possible to study a simplified framework where $\Delta m_{21}^2$ is set to zero. Then we have

$$\Delta m_{32}^2 = \Delta m_{31}^2 - \Delta m_{21}^2 \simeq \Delta m_{31}^2. \tag{6.78}$$

The probability of transition in Eq. (6.22) becomes

$$P_{\nu_\alpha \to \nu_\beta} \simeq \left| \delta_{\alpha\beta} + U_{\alpha 3} U_{\beta 3}^* \left[ \exp\left( -i \frac{\Delta m_{31}^2 L}{2E} \right) - 1 \right] \right|^2. \tag{6.79}$$

The transition probabilities depend only on the largest mass squared difference $\Delta m_{13}^2$ and on the elements $U_{\alpha 3}$ that connect flavour neutrinos with the neutrino $\nu_3$. Using the parameterisation in Eq. (6.8), we obtain, for the specific oscillation transitions:

$$P(\nu_\mu \to \nu_\tau) \simeq 4|U_{33}|^2 |U_{23}|^2 \sin^2\left( \frac{\Delta m_{31}^2 L}{4E} \right) =$$

$$= \cos^4 \theta_{13} \sin^2 2\theta_{23} \sin^2\left( \frac{\Delta m_{31}^2 L}{4E} \right)$$

$$P(\nu_e \to \nu_\mu) \simeq 4|U_{13}|^2 |U_{23}|^2 \sin^2\left( \frac{\Delta m_{31}^2 L}{4E} \right) =$$

$$= \sin^2 \theta_{23} \sin^2 2\theta_{13} \sin^2\left( \frac{\Delta m_{31}^2 L}{4E} \right)$$

$$P(\nu_e \to \nu_\tau) \simeq 4|U_{13}|^2 |U_{33}|^2 \sin^2\left( \frac{\Delta m_{31}^2 L}{4E} \right) =$$

$$= \cos^2 \theta_{23} \sin^2 2\theta_{13} \sin^2\left( \frac{\Delta m_{31}^2 L}{4E} \right). \tag{6.80}$$

In the last passage of the previous relations, we have exploited the identity $\sin 2\alpha = 2 \sin \alpha \cos \alpha$ for a generic angle $\alpha$. Only two angles enter these formulas: $\theta_{23}$ and $\theta_{13}$. The transition probability can be re-written in a formula analogous to Eq. (6.37) in the two-neutrino mixing case:

$$P_{\nu_\alpha \to \nu_\beta} = \sin^2 2\theta_{\alpha\beta}^{eff} \sin^2\left( \frac{\Delta m^2 L}{4E} \right) \qquad (\alpha \neq \beta) \tag{6.81}$$

## 6.5 Hierarchy and Simplified Mixing

where $\Delta m^2 \equiv \Delta m_{31}^2$ and $\sin^2 2\theta_{\alpha\beta}^{eff} \equiv 4|U_{\alpha 3}|^2|U_{\beta 3}|^2$. The survival probability for $\nu_e$ has the form

$$P(\nu_e \to \nu_e) = 1 - P(\nu_e \to \nu_\mu) - P(\nu_e \to \nu_\tau)$$
$$= 1 - \sin^2 2\theta_{13} \sin^2\left(\frac{\Delta m_{31}^2 L}{4E}\right) \quad (6.82)$$

where we have exploited the identity $\cos^2\alpha + \sin^2\alpha = 1$. It has the same form as the survival probability in the two-neutrino mixing case, see Eq. (6.39), if we identify $(\Delta m^2, \theta)$ with $(\Delta m_{31}^2, \theta_{13})$. The survival probability for $\nu_\mu$ reads

$$P(\nu_\mu \to \nu_\mu) = 1 - P(\nu_\mu \to \nu_\tau) - P(\nu_e \to \nu_e)$$
$$= 1 - 4\sin^2\theta_{23}\cos^2\theta_{13}\,(1 - \sin^2\theta_{23}\cos^2\theta_{13})\sin^2\left(\frac{\Delta m_{31}^2 L}{4E}\right). \quad (6.83)$$

The experimental value of the angle $\theta_{13}$ is relatively small. If the angle $\theta_{13}$ is approximate to zero, the muon can oscillate only in tau neutrinos, and we are again back to the two two-neutrino mixing case, when we set $(\Delta m^2, \theta)$ to $(\Delta m_{32}^2, \theta_{23})$ (see also Eq. (6.78)). The $\nu_\tau$ survival probability can be similarly derived.

Since $\Delta m_{31}^2$ enters the probabilities only as the argument of the squared sine, it is obvious that its sign cannot be determined in this approximation. Therefore, in order to find effects that are sensitive to the mass hierarchy, one has to go beyond the limit $\Delta m_{21}^2 \simeq 0$.[10]

A $CP$ transformation exchanges neutrinos with antineutrinos, implying the following equality

$$P_{\nu_\alpha \to \nu_\beta} = P_{\bar{\nu}_\alpha \to \bar{\nu}_\beta} \quad (6.84)$$

These relations are always satisfied by the probabilities (6.80), independently from the $CP$ invariance in the lepton sector. Therefore, an experiment which is sensitive only to the largest squared-mass difference cannot probe $CP$ violation.[11]

---

[10] Another possibility is exploring neutrino oscillation in matter (see Sect. 6.6) rather than in the vacuum, where an additional phase from the interaction between neutrinos and the matter constituents makes the hierarchy determination possible.

[11] Let us observe that the two-neutrino mixing framework can be approximately realised in experimental settings where $\Delta m^2 L/4E \ll 1$, that is in short baseline experiments with $L/E \ll 4/\Delta m^2$. Thus they do not represent a convenient experimental setting to probe $CP$ violation.

Analogously, a $T$ transformation interchanges the initial and final states, implying

$$P_{\nu_\alpha \to \nu_\beta} = P_{\nu_\beta \to \nu_\alpha}$$
$$P_{\bar\nu_\alpha \to \bar\nu_\beta} = P_{\bar\nu_\beta \to \bar\nu_\alpha} \qquad (6.85)$$

which are always approximately valid as well, as expected because of $CPT$ invariance.

### 6.5.2 Electron Neutrino Disappearance

Let us observe that if we want sensitivity to the smallest mass $\Delta m_{21}^2$, we need small energies and large baselines, yielding $\Delta m_{21}^2 \sim O(1)$. In that case $\Delta m_{31}^2 L/4E \gg 1$, and the oscillating factor becomes an average

$$\sin^2\left(\frac{\Delta m_{31}^2 L}{4E}\right) \simeq \sin^2\left(\frac{\Delta m_{32}^2 L}{4E}\right) \simeq \frac{1}{2}. \qquad (6.86)$$

The disappearance probability for an electron neutrino is, from Eq. (6.54)

$$P_{\nu_e \to \nu_e} = 1 - 4 \sum_{k>j} |V_{ek}|^2 |V_{ej}|^2 \sin^2 \frac{\Delta m_{kj}^2}{4E} L$$

$$= 1 - 4|V_{e2}|^2 |V_{e1}|^2 \sin^2 \frac{\Delta m_{21}^2}{4E} L - 4|V_{e3}|^2 |V_{e1}|^2 \sin^2 \frac{\Delta m_{31}^2}{4E} L$$

$$- 4|V_{e3}|^2 |V_{e2}|^2 \sin^2 \frac{\Delta m_{32}^2}{4E} L. \qquad (6.87)$$

By using Eqs. (6.8) and (6.86) we have

$$P_{\nu_e \to \nu_e} = 1 - 4s_{12}^2 c_{12}^2 c_{13}^4 \sin^2 \frac{\Delta m_{21}^2}{4E} L - 4s_{13}^2 c_{12}^2 c_{13}^2 \sin^2 \frac{\Delta m_{31}^2}{4E} L$$

$$- 4s_{13}^2 s_{12}^2 c_{13}^2 \sin^2 \frac{\Delta m_{32}^2}{4E} L$$

$$= 1 - 4s_{12}^2 c_{12}^2 c_{13}^4 \sin^2 \frac{\Delta m_{21}^2}{4E} L - 2s_{13}^2 c_{13}^2$$

$$= s_{13}^4 + c_{13}^4 - 4s_{12}^2 c_{12}^2 c_{13}^4 \sin^2 \frac{\Delta m_{21}^2}{4E} L. \qquad (6.88)$$

In the last term we have used the equality $1 = 1^2 = (c_{13}^2 + s_{13}^2)^2 = s_{13}^4 + c_{13}^4 + 2s_{13}^2 c_{13}^2$. Finally we can write

$$P_{\nu_e \to \nu_e} = c_{13}^4 P_{\nu_e \to \nu_e}^{2\nu} + s_{13}^4 \tag{6.89}$$

where we have indicated with $P_{\nu_e \to \nu_e}^{2\nu}$ the disappearance probability in Eq. (6.40) in the two neutrino formalism referred to the 12 sector

$$P_{\nu_e \to \nu_e}^{2\nu} = 1 - \sin^2 2\theta_{12} \sin^2\left(\frac{\Delta m_{21}^2 L}{4E}\right). \tag{6.90}$$

Note that this expression is the same for anti-neutrinos, and it does not depend on either the phase $\delta$ of the lepton mixing matrix or the sign of $\Delta m_{21}^2$. In the approximation $\theta_{13} = 0$, we have, for disappearance and appearance probabilities

$$P_{\nu_e \to \nu_e} \simeq P_{\nu_e \to \nu_e}^{2\nu} = 1 - \sin^2 2\theta_{12} \sin^2\left(\frac{\Delta m_{21}^2 L}{4E}\right)$$

$$P_{\nu_e \to \nu_{\mu,\tau}} \simeq 1 - P(\nu_e \to \nu_e)$$

$$= \sin^2 2\theta_{12} \sin^2\left(\frac{\Delta m_{21}^2 L}{4E}\right). \tag{6.91}$$

Due to the long baseline, the distance Sun-Earth, at low energies and neglecting mass effects, these approximate expressions could be used in solar neutrino experiments, bringing about the adjective solar for the $\theta_{12}$ angle.

## 6.6 Neutrino Oscillations in Matter

In all expressions above we have assumed that oscillations occur in vacuum, which is not always the case. For instance solar neutrinos produced in the inner region, the core, of the Sun have to traverse the solar medium to reach the detector. The same happens to atmospheric neutrinos which traverse parts of the Earth. Here we consider oscillations of neutrino in a normal medium, composed of electrons and up and down quarks. Neutrinos and anti-neutrinos interact with matter, and we expect a different pattern of oscillation than in vacuum. Indeed, due to the difference of the weak interactions undergone by different flavour neutrinos in matter, the effective mixing can be dramatically enhanced with respect to vacuum.

The formalism of neutrino oscillations in matter was worked out between 1978 and 1986[12] [42, 206, 207]. Let us first ignore matter interaction, and set the

---

[12] For an historical account you can see for example the contribution by A. Smirnov at the International Conference on History of the Neutrino, Sept. 5–7, 2018, Paris, France [205].

formalism of oscillations in vacuum in a way that is suitable to generalisation to matter oscillations. We limit to the simpler case of two families,[13] where the mixing matrix is the orthogonal matrix $U$ (see Sect. 6.3)

$$U = \begin{pmatrix} \cos\theta & \sin\theta \\ -\sin\theta & \cos\theta \end{pmatrix}. \tag{6.92}$$

Let us take the electron and muon neutrinos as the flavour eigenstates. They are connected to the mass eigenstates $\nu_1$ and $\nu_2$ by the mixing matrix (6.36):

$$\begin{pmatrix} \nu_e \\ \nu_\mu \end{pmatrix} = \begin{pmatrix} \cos\theta & \sin\theta \\ -\sin\theta & \cos\theta \end{pmatrix} \begin{pmatrix} \nu_1 \\ \nu_2 \end{pmatrix}. \tag{6.93}$$

The mass eigenstates evolve according to the Schrödinger equation

$$i\frac{d}{dt}\begin{pmatrix} \nu_1 \\ \nu_2 \end{pmatrix} = H^D \begin{pmatrix} \nu_1 \\ \nu_2 \end{pmatrix} \tag{6.94}$$

where the Hamiltonian $H^D$ is diagonal. If the ultrarelativistic approximation (6.17) holds, we have

$$H^D = \begin{pmatrix} E_1 & 0 \\ 0 & E_2 \end{pmatrix} \simeq \begin{pmatrix} E + \frac{m_1^2}{2E} & 0 \\ 0 & E + \frac{m_2^2}{2E} \end{pmatrix} \tag{6.95}$$

by setting $E \simeq |\mathbf{p}|$. The corresponding Schrödinger equation for the flavour eigenstates is

$$i\frac{d}{dt}\begin{pmatrix} \nu_e \\ \nu_\mu \end{pmatrix} = H_V \begin{pmatrix} \nu_e \\ \nu_\mu \end{pmatrix} \tag{6.96}$$

where the Hamiltonian $H_V$ is a not-diagonal matrix defined as

$$H_V = U H^D U^T. \tag{6.97}$$

In the previous approximation $H_V$ becomes

$$H_V \simeq \left(E + \frac{m_1^2 + m_2^2}{4E}\right)\begin{pmatrix} 1 & 0 \\ 0 & 1 \end{pmatrix} + \frac{\Delta m_{21}^2}{4E}\begin{pmatrix} -\cos 2\theta & \sin 2\theta \\ \sin 2\theta & \cos 2\theta \end{pmatrix}$$

$$\simeq \frac{\Delta m_{21}^2}{4E}\begin{pmatrix} -\cos 2\theta & \sin 2\theta \\ \sin 2\theta & \cos 2\theta \end{pmatrix}. \tag{6.98}$$

---

[13] For three families see for example Sect. 16.4 of Ref. [144].

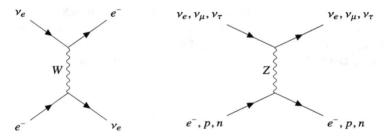

**Fig. 6.3** Feynman diagrams of tree level neutrino weak interactions mediated by charged and neutral gauge bosons

In the last line, multiples of the identity have been omitted as they introduce a constant phase factor which does not affect oscillations. Besides, we have used the trigonometric relations $\sin 2\theta = 2\cos\theta \sin\theta$ and $\cos 2\theta = \cos^2\theta - \sin^2\theta$. The angle $\theta$ is given by

$$\tan 2\theta = \frac{2H_{V12}}{H_{V22} - H_{V11}} \qquad (6.99)$$

Let us now switch to neutrinos travelling in matter. Different species of neutrinos oscillate differently while traveling through matter. On their journey, neutrinos meet practically only electrons $e^-$, protons $p$ and neutrons $n$, whose interactions affect differently $\nu_e$ and $\nu_\mu$. Consider a neutrino beam so low in energy that muons cannot be produced through scattering off nuclei and electrons, as it is the case for solar neutrinos. While both $\nu_e$ and $\nu_\mu$ undergo neutral current scattering, only $\nu_e$ can induce charged current interactions off nuclei as well as the electrons. At low energies only the elastic forward scattering is relevant (inelastic scattering can be neglected) [42]. Figure 6.3 shows the Feynman diagrams of charged and neutral current electroweak scattering. In this approximation, $\nu_e$ interacts with an usual medium according to the diagram on the left-side of Fig. 6.3, which contributes to an effective term

$$\mathcal{H}_{\text{eff}} = \frac{G_F}{\sqrt{2}} \bar{e}\gamma_\alpha(1-\gamma_5)\nu_e \, \bar{\nu}_e\gamma^\alpha(1-\gamma_5)e =$$
$$= \frac{G_F}{\sqrt{2}} \bar{e}\gamma^\alpha(1-\gamma_5)e \, \bar{\nu}_e\gamma_\alpha(1-\gamma_5)\nu_e \qquad (6.100)$$

where $G_F$ is the Fermi constant. The second form is obtained via Fierz transformation.[14] In the rest frame of the medium, when the electron is nearly at rest and the neutrino relativistic, only the $\alpha = 0$ component is relevant. In fact, the axial

---

[14] Fierz transformations are discussed in many textbooks, see for instance the appendix of [208].

current reduces to spin in the non-relativistic approximation, which is negligible for a non-relativistic, non polarised, collection of electrons, and the spatial components of the vector current give the average velocity, which is negligible as well. Since $\bar{e}\gamma^0 e = n_e$ can be interpreted as the electron density for a non polarised medium, we have

$$\mathcal{H}_{\text{eff}} = \sqrt{2} G_F n_e \, \bar{\nu}_{eL} \gamma_0 \nu_{eL}. \tag{6.101}$$

One can collectively describe interactions of neutrino and antineutrinos with matter by a non-zero potential $V$ where the neutrinos are moving. Writing down the equation of motion for a neutrino, one finds that the effect of the term (6.101) is to change its effective energy. Hence one can interpret the potential generated by charged currents as [42]

$$V_{CC} = \pm \sqrt{2} G_F n_e \tag{6.102}$$

which is written in terms of $G_F$ and the electron density in matter $n_e$. The positive (negative) sign applies to electron-neutrino (antineutrino). In matter the neutrino and anti-neutrino mixing are different because they have opposite effective potentials.

The neutral currents contribute to the potential in the same way for all flavours, while charged-current interactions with electrons interest the electron neutrino only, thus the general potential can be written as [42]

$$V = \begin{pmatrix} V_{NC} + V_{CC} & 0 \\ 0 & V_{NC} \end{pmatrix}. \tag{6.103}$$

The evolution of neutrino beams is now governed by the Schrödinger equation

$$i \frac{d}{dt} \begin{pmatrix} \nu_e \\ \nu_\mu \end{pmatrix} = (H_V + V) \begin{pmatrix} \nu_e \\ \nu_\mu \end{pmatrix} = H_M \begin{pmatrix} \nu_e \\ \nu_\mu \end{pmatrix} \tag{6.104}$$

where $H_M$ is the effective Hamiltonian:

$$H_M \equiv H_V + V$$
$$\simeq \left( E + \frac{m_1^2 + m_2^2}{4E} + V_{NC} + \frac{\sqrt{2} G_F n_e}{2} \right) \begin{pmatrix} 1 & 0 \\ 0 & 1 \end{pmatrix} +$$
$$+ \begin{pmatrix} -\frac{\Delta m_{21}^2}{4E} \cos 2\theta + \frac{\sqrt{2} G_F n_e}{2} & \frac{\Delta m_{21}^2}{4E} \sin 2\theta \\ \frac{\Delta m_{21}^2}{4E} \sin 2\theta & \frac{\Delta m_{21}^2}{4E} \cos 2\theta - \frac{\sqrt{2} G_F n_e}{2} \end{pmatrix}$$
$$\simeq \frac{\Delta m^2}{4E} \begin{pmatrix} -\cos 2\theta + \xi & \sin 2\theta \\ \sin 2\theta & \cos 2\theta - \xi \end{pmatrix}. \tag{6.105}$$

## 6.6 Neutrino Oscillations in Matter

In the last passage we have again omitted multiples of the identity and set $\Delta m^2 = \Delta m_{21}^2 = m_2^2 - m_1^2$, since we are considering two families only. The effect of matter is described by the adimensional quantity $\xi$

$$\xi \equiv \frac{2V_{CC}E}{\Delta m^2} = \frac{2\sqrt{2}\,G_F\,n_e\,E}{\Delta m^2}. \qquad (6.106)$$

The same relation (6.105) holds for antineutrinos with the exchange $\xi \to -\xi$.

Now the Hamiltonian in matter $H_M$ can be reduced to an expression formally equivalent to vacuum Hamiltonian $H_V$, given in Eq. (6.98), if we introduce an effective mixing angle in matter $\theta_M$ and an effective difference of squared masses $\Delta m_M^2$ such that

$$H_M \simeq \frac{\Delta m_M^2}{4E} \begin{pmatrix} -\cos 2\theta_M & \sin 2\theta_M \\ \sin 2\theta_M & \cos 2\theta_M \end{pmatrix} \qquad (6.107)$$

By equating Eqs. (6.105) and (6.107), we obtain the expression for the effective mixing parameters in matter

$$\Delta m_M^2 = C\,\Delta m^2 \qquad \sin 2\theta_M = \frac{\sin 2\theta}{C} \qquad (6.108)$$

where

$$C = \sqrt{(\cos 2\theta - \xi)^2 + \sin^2 2\theta}. \qquad (6.109)$$

Moreover, the angle $\theta_M$ satisfies the relation

$$\tan 2\theta_M = \frac{2H_{M12}}{H_{M22} - H_{M11}} = \frac{\sin 2\theta}{\cos 2\theta - \xi} = \frac{\tan 2\theta}{1 - \frac{\xi}{\cos 2\theta}}. \qquad (6.110)$$

We can diagonalize $H_M$ to obtain the mixing matrix and mass eigenstates in matter via a rotation matrix, similar to that for vacuum (6.97)

$$H_M^D \equiv U_M^T H_M U_M \qquad (6.111)$$

where

$$U_M = \begin{pmatrix} -\cos\theta_M & \sin\theta_M \\ \sin\theta_M & \cos\theta_M \end{pmatrix}. \qquad (6.112)$$

The eigenvalues of $H_M$ turn out to be

$$E_{M1,2} \simeq \mp\frac{\Delta m^2}{4E}\sqrt{(\xi - \cos 2\theta)^2 + \sin^2 2\theta} = \mp\frac{\Delta m_M^2}{4E}. \qquad (6.113)$$

The eigenvalues for the complete $H_M$ in the first lines of Eq. (6.105) are recovered by adding $E + (m_1^2 + m_2^2)/4E + V_{NC} + \sqrt{2}G_F n_e/2$ to the previous values.

## 6.6.1 Uniform Density

We have just seen that the Hamiltonian in matter, $H_M$, is formally identical to its vacuum counterpart, $H_V$ except that the vacuum parameters $\Delta m^2$ and $\theta$ are replaced, respectively, by $\Delta m_M^2$ and $\theta_M$. Let us consider the case where the electron density $n_e$ in Eq. (6.102) is constant, as is sometimes a good approximation for neutrinos travelling through small portions of Earth. When $n_e$ is constant, we can proceed in complete analogy to the vacuum case. The value for the transition amplitude is

$$A_{\nu_\alpha \to \nu_\beta}(t) = \sum_{k=1,2} U_{M\beta k}\, e^{-iE_{Mk}t}\, U_{M\alpha k}^T \qquad \alpha, \beta \in (e, \mu) \qquad (6.114)$$

in analogy to Eq. (6.15) and observing that $U_M^* = U_M^T$. The transition probability $P_{\nu_\alpha \to \nu_\beta}$ has the same formal expression in matter than in vacuum (see Eq. (6.37))

$$P_{\nu_e \to \nu_\mu} = \sin^2 2\theta_M \sin^2 \frac{\Delta m_M^2 L}{4E} \qquad (6.115)$$

given the substitutions $\theta \to \theta_M$, $\Delta m^2 \to \Delta m_M^2$, and assuming $n_e$ constant.

The net result of the previous considerations, is that, inside matter, the oscillations are governed by an effective mixing angle different from the mixing angle in vacuum. As can be seen from Eq. (6.110), the mixing pattern is affected by the values of the new parameters $\xi$. Independently of the value of the mixing angle in the vacuum, $\theta$, there is an effective maximal mixing, corresponding to $\theta_M = \frac{\pi}{4}$, that is to $\sin \theta_M = \cos \theta_M = \frac{1}{\sqrt{2}}$, when the equality

$$\xi = \cos 2\theta \qquad (6.116)$$

holds. It corresponds to the density value of matter

$$n_e^R \equiv \frac{\Delta m^2 \cos 2\theta}{2\sqrt{2}G_F E}. \qquad (6.117)$$

This is the so-called MSW (Mikheyev-Smirnov-Wolfenstein) resonance effect [42, 207]. The probability of neutrino flavour transition in matter can be significantly enhanced, irrespectively of the value of the mixing angle $\theta$. Even if the mixing angle $\theta$ is tiny and the vacuum oscillation probability Eq. (6.37) is very small, suppressed by a small value of $\sin^2 2\theta$, the effective mixing is maximal if the

## 6.6 Neutrino Oscillations in Matter

resonance condition (6.117) holds, and the transition probability $P_{\nu_e \to \nu_\mu}$ in matter (Eq. (6.115)) is amplified, since $\sin 2\theta_M$ attains its maximum possible value, unity. The oscillation length in matter corresponding to $\lambda$ in Eq. (6.43) is obtained by substituting $\Delta m \to \Delta m_M^2$. We find

$$\lambda_M = \frac{\sin 2\theta_M}{\sin 2\theta} \lambda. \tag{6.118}$$

At the resonance, $\sin 2\theta_M = 1$, the oscillation length in matter is scaled by the factor $1/\sin 2\theta$ and it can also be significantly different from the oscillation length in vacuum.

Note that at the resonance the energy eigenvalues in Eq. (6.113) become $E_{M1,2} = \mp \Delta m^2 \sin 2\theta / 4E$, and the effective squared-mass difference in matter $\Delta m_M^2$ (Eq. (6.108)) attains its minimum value, that is $\Delta m_M^2 = \Delta m^2 \sin 2\theta$.

Summarising, at the resonance

(a) the flavour mixing is maximal,
(b) the level splitting is minimal,

independently of vacuum mixing.

The condition for the resonance enhancement implied by Eq. (6.117) is a positive density, that is

$$\Delta m^2 \cos 2\theta = (m_2^2 - m_1^2)(\cos^2 \theta - \sin^2 \theta) > 0. \tag{6.119}$$

Since we have set $\Delta m^2 = \Delta m_{21}^2 = m_2^2 - m_1^2$, if $\nu_2$ is heavier than $\nu_1$, one needs $\cos^2 \theta > \sin^2 \theta$, that is $\theta$ to be in the first octant. That means, for example, that there is no resonance for $\theta = \frac{\pi}{4}$. In more compact terms, the resonance condition is

$$\xi = \cos 2\theta > 0 \qquad (\theta < \pi/4). \tag{6.120}$$

Given we have adopted the standard convention $\Delta m_{21}^2 > 0$ (see Sect. 6.5), that is always the case.[15] Since, according to Eq. (6.93), we have

$$\nu_1 = \cos\theta\, \nu_e - \sin\theta\, \nu_\mu$$
$$\nu_2 = \sin\theta\, \nu_e + \cos\theta\, \nu_\mu, \tag{6.121}$$

it follows that for the resonance enhancement of neutrino oscillations in matter to be possible, the lower-mass of the two mass eigenstates must have have a larger $\nu_e$ component. Let us observe that the exchange $\xi \to -\xi$ for antineutrinos implies that, under the same assumption, the resonance condition for antineutrinos is $\cos 2\theta < 0$

---

[15] If $\Delta m^2 < 0$, one needs $\cos^2 \theta < \sin^2 \theta$.

($\theta > \pi/4$). Therefore, fixed the sign of $\Delta m^2_{21}$, either neutrinos or antineutrinos (but not both) can experience the resonantly enhanced oscillations in matter.

The low density limit corresponds to $\xi \to 0$, and we can see from Eq. (6.110) that the vacuum case is recovered, that is $\theta_M \to \theta$. In the opposite case, the very high density limit, the effective mixing parameters become

$$\Delta m_M^2 \to \xi \, \Delta m^2 \to 2 V_{CC} E$$

$$\sin 2\theta_M \to \frac{\sin 2\theta}{\xi} = \frac{\sin 2\theta \, \Delta m^2}{2\sqrt{2}\, G_F n_e E} \to 0. \tag{6.122}$$

Large values of $\xi$ imply $\xi \gg \cos 2\theta$, thus $\tan 2\theta_M$ becomes negative and tends to zero, that is $\theta_M \to \pi/2$, as can be seen from Eq. (6.110) under the assumption $\sin 2\theta > 0$.[16]

### 6.6.2 Variable Density

In general, one deals with the more realistic case of neutrinos travelling in matter of non-uniform density. An example are solar neutrinos, that are created near the centre of the Sun and propagate in the solar medium, until they leave its surface and start their journey in what can be considered vacuum to a reasonably good approximation. In the Sun, the matter density is very high in the inner region, the core, decreasing towards the surface. Other examples are supernova neutrinos, that have to cross material with a rapidly varying electron density, and atmospheric neutrinos crossing the Earth, which has a higher matter density near the center than near the surface. In the general case, the Schrödinger evolution equation (6.104) does not allow an analytic solution and has to be solved numerically. However, there is an important particular case in which one can get an approximate analytic solution, the case of slowly (adiabatically) varying matter density.

The evolution equation (6.104), with the matter effective Hamiltonian $H_M$ in Eq. (6.107), reads

$$i\frac{d}{dx}\begin{pmatrix} \nu_e \\ \nu_\mu \end{pmatrix} = H_M \begin{pmatrix} \nu_e \\ \nu_\mu \end{pmatrix} = \frac{\Delta m_M^2}{4E}\begin{pmatrix} -\cos 2\theta_M & \sin 2\theta_M \\ \sin 2\theta_M & \cos 2\theta_M \end{pmatrix}\begin{pmatrix} \nu_e \\ \nu_\mu \end{pmatrix}. \tag{6.123}$$

We have set $c = 1$ and $t = x$. This description still holds in the case of variable matter density, with the difference that, due to its dependence on the density, the mixing angle changes in the course of propagation, that is

$$\theta_M = \theta_M(n_e(x)). \tag{6.124}$$

---

[16] For antineutrinos, $\tan 2\theta_M$ become positive for large values of $\xi$ and its zero limit correspond to $\theta_M = 0$.

## 6.6 Neutrino Oscillations in Matter

As seen in Sect. 6.6, the matter effective Hamiltonian $H_M$ is diagonalised by the orthogonal matrix $U_M$ in Eq. (6.112)

$$H_M^D = U_M^T H_M U_M \qquad (6.125)$$

where

$$\begin{pmatrix} \nu_e \\ \nu_\mu \end{pmatrix} = U_M \begin{pmatrix} \nu_1 \\ \nu_2 \end{pmatrix} \qquad U_M = \begin{pmatrix} -\cos\theta_M & \sin\theta_M \\ \sin\theta_M & \cos\theta_M \end{pmatrix}. \qquad (6.126)$$

The evolution equation (6.123) becomes

$$i\frac{d}{dx}\begin{pmatrix} \nu_e \\ \nu_\mu \end{pmatrix} = i U_M \frac{d}{dx}\begin{pmatrix} \nu_1 \\ \nu_2 \end{pmatrix} + i \frac{dU_M}{dx}\begin{pmatrix} \nu_1 \\ \nu_2 \end{pmatrix} = H_M \begin{pmatrix} \nu_e \\ \nu_\mu \end{pmatrix}. \qquad (6.127)$$

In terms of mass eigenstates, it becomes

$$i\frac{d}{dx}\begin{pmatrix} \nu_1 \\ \nu_2 \end{pmatrix} = U_M^T H_M U_M \begin{pmatrix} \nu_1 \\ \nu_2 \end{pmatrix} - i U_M^T \frac{dU_M}{dx}\begin{pmatrix} \nu_1 \\ \nu_2 \end{pmatrix}$$

$$= H_M^D \begin{pmatrix} \nu_1 \\ \nu_2 \end{pmatrix} - \begin{pmatrix} 0 & -i\frac{d\theta_M}{dx} \\ i\frac{d\theta_M}{dx} & 0 \end{pmatrix}\begin{pmatrix} \nu_1 \\ \nu_2 \end{pmatrix}. \qquad (6.128)$$

If the matter density is constant, $d\theta_M/dx = 0$, we recover the results of Sect. 6.6. The last line of Eq. (6.128) can be written as

$$\begin{pmatrix} E_{M1} & -i\frac{d\theta_M}{dx} \\ i\frac{d\theta_M}{dx} & E_{M2} \end{pmatrix} = \frac{1}{2}(E_{M1} + E_{M2})\begin{pmatrix} 1 & 0 \\ 0 & 1 \end{pmatrix} + \begin{pmatrix} -\frac{1}{2}\Delta E_M & -i\frac{d\theta_M}{dx} \\ i\frac{d\theta_M}{dx} & \frac{1}{2}\Delta E_M \end{pmatrix} \qquad (6.129)$$

where $\Delta E_M \equiv E_{M2} - E_{M1}$. We argued before that the terms proportional to the unit matrix do not affect transition probabilities, so the only physical parameter in the diagonal elements is $\Delta E_M$. If the matter density is not constant, it is necessary to take into account the effect of the off-diagonal terms proportional to $d\theta_M/dx$ in the evolution equation.

We have

$$\frac{d\theta_M}{dx} = \frac{1}{2}\frac{\sin 2\theta_M}{\Delta m_M^2}\Delta m^2 \frac{d\xi}{dx} \qquad (6.130)$$

which follows by deriving the equation on the right in (6.108), using the definition of C and applying the relations (6.110) and the equation on the left in (6.108).

The evolution equations of Eq. (6.128) constitute a system of coupled equations: the instantaneous mass eigenstates $\nu_i$ mix in the evolution and are not energy eigenstates. The importance of this effect is controlled by the relative size of the off-diagonal piece with respect to the diagonal one. If the off-diagonal terms are much smaller than the difference between the diagonal term, the mass eigenstates $\nu_i$ behave approximately as energy eigenstates and they do not mix in the evolution. This is the adiabatic transition approximation. In order to quantify the amount of such transitions, one may introduce the so-called adiabaticity parameter

$$\gamma = \frac{\Delta m_M^2}{4E \left|\frac{d\theta_M}{dx}\right|}. \tag{6.131}$$

In the adiabatic limit, the effective Hamiltonian is approximatively diagonal, as in the constant density case, that is

$$\left|\frac{d\theta_M}{dx}\right| \ll \frac{\Delta E_M}{2} = \frac{\Delta m_M^2}{4E} \tag{6.132}$$

which corresponds to a large $\gamma$ in all points of the neutrino trajectory. The last equality follows from Eq. (6.113). If the evolution is adiabatic, the transitions between the effective mass eigenstates are negligible, the states evolve independently and the effect of the evolution is a simple phase factor. In the adiabatic approximation, the value for the transition amplitude is ($\alpha, \beta \in (e, \mu)$)

$$A_{\nu_\alpha \to \nu_\beta}(t) = \sum_{k=1,2} U_{M\beta k}(x)\, e^{-i \int_{x_0}^{x} E_{Mk}(x')dx'}\, U_{M\alpha k}^T(x_0) \tag{6.133}$$

in analogy to Eq. (6.15) and to Eq. (6.114), observing that $U_M^* = U_M^T$, and adding the dependence on $x$ of the matrices $U_M$ and $H_M$, as well as of the eigenvalues $E_{Mi}$.

The probability of the $\nu_\alpha \to \nu_\beta$ transition is defined as in Eq. (6.16)

$$P_{\nu_\alpha \to \nu_\beta} = |A_{\nu_\alpha \to \nu_\beta}(t)|^2 = \left|\sum_{k=1,2} U_{M\beta k}(x)\, e^{-i \int_{x_0}^{x} E_{Mk}(x')dx'}\, U_{M\alpha k}^T(x_0)\right|^2 =$$

$$= \sum_{k=1,2} |U_{M\beta k}(x)|^2\, |U_{M\alpha k}(x_0)|^2 +$$

$$+ 2 \sum_{k<i} U_{M\beta k}(x)\, U_{M\beta i}(x)^T\, e^{-i \int_{x_0}^{x} (E_{Mk}(x') - E_{Mi}(x'))dx'}\, U_{\alpha k}(x_0)^T\, U_{\alpha i}(x_0).$$

$$\tag{6.134}$$

Assuming an averaging over neutrino energy and the region in which neutrinos are produced, the last terms disappears and we have

$$\langle P_{\nu_\alpha \to \nu_\beta} \rangle = \sum_{k=1,2} |U_{M\beta k}(x)|^2\, |U_{M\alpha k}(x_0)|^2. \tag{6.135}$$

## 6.6 Neutrino Oscillations in Matter

Thus, in the adiabatic approximation, the averaged transition probability is determined by the elements of the mixing matrix in matter at the initial and final points. In the case of two neutrino flavours $U_M$ is given by Eq. (6.112), adding the dependence of the effective mixing angle on $x$.

The average survival probability reads

$$\langle P_{\nu_\alpha \to \nu_\alpha} \rangle = \cos^2 \theta_M(x) \cos^2 \theta_M(x_0) + \sin^2 \theta_M(x) \sin^2 \theta_M(x_0) =$$
$$= \frac{1}{2}(1 + \cos 2\theta_M(x) \cos 2\theta_M(x_0)). \tag{6.136}$$

When neutrinos are produced in stars, such as solar neutrinos or supernova neutrinos, they traverse a medium with a smooth density variation, the plasma which composes stars. The effective mixing angle depends on the density as in Eq. (6.124) and the adiabatic condition in Eq. (6.132) is generally satisfied. Besides, neutrinos are detected very far away from stars, practically in vacuum flavour,[17] and the final effective mixing angle $\theta_M(x)$ is equal to the vacuum mixing angle $\theta$. In this case the average survival probability becomes

$$\langle P_{\nu_\alpha \to \nu_\alpha} \rangle = \frac{1}{2}(1 + \cos 2\theta \cos 2\theta_M(x_0)) \tag{6.137}$$

which is independent of the source–detector distance.

It is easy to see that if the neutrino passes the point $\bar{x}$ where the resonance condition is satisfied, a large effect of disappearance will be observed. As seen in Eq. (6.120), the resonance condition for neutrinos implies $\xi(\bar{x}) = \cos 2\theta > 0$ ($\theta < \pi/4$), in the standard convention of $\Delta m^2 > 0$. The parameter $\xi(x)$ is connected to the density by the relation (6.106). If at the production point $x_0$ the density is larger than at the resonance point we have $\xi(x_0) > \xi(\bar{x}) = \cos 2\theta$. It follows from Eq. (6.110) that $\tan 2\theta_M(x_0) < 0$, assumed that $\sin 2\theta(x_0) > 0$ (see (6.119) and below). Since Eq. (6.108) ensures that $\sin 2\theta_M(x_0) > 0$, we have $\cos 2\theta_M(x_0) < 0$. From Eq. (6.136) we can conclude that

$$\langle P_{\nu_\alpha \to \nu_\alpha} \rangle < \frac{1}{2}. \tag{6.138}$$

In the limit of very high density at the production point $x_0$, $\xi(x_0) \gg \cos 2\theta$ and $\theta_M(x_0) \simeq \pi/2$. The averaged survival probability in Eq. (6.137), at the observation point where the density of matter is very low, becomes

$$\langle P_{\nu_\alpha \to \nu_\alpha} \rangle \simeq \frac{1}{2}(1 - \cos 2\theta) = \sin^2 \theta. \tag{6.139}$$

---

[17] The density of the actual detector is irrelevant with respect to the medium in which the neutrino propagates before reaching the detector, the size of the detector being too small to induce significant flavour transitions.

This is the average survival probability of electron neutrino in vacuum. At small $\theta$ the survival probability is close to zero; almost all neutrinos change flavour.

In the non-adiabatic case, formula (6.136) generalises to the so-called Parke formula [209]

$$\langle P_{\nu_\alpha \to \nu_\alpha} \rangle = \frac{1}{2} + \left(\frac{1}{2} - P_{12}\right) \cos 2\theta_M(x) \cos 2\theta_M(x_0) \qquad (6.140)$$

where $P_{12}$ is the transition probability of jumping from one eigenstate to another.

# Chapter 7
# Natural Neutrino Sources

In our Universe, neutrinos are produced everywhere: in stars, in supernovae, in gamma-ray bursts, in cosmic ray interactions in the atmosphere, in decays of radioactive elements (mostly Uranium and Thorium) on the Earth. The existence of Big Bang relic neutrinos—exact analogues of the Big Bang relic photons comprising the cosmic microwave background radiation—is a basic prediction of standard Big Bang cosmology. Neutrino energies range from μeV for neutrinos left over from the Big Bang, right up to PeV for neutrinos from galactic sources. In this chapter neutrinos from natural sources are surveyed.

In Sect. 7.1 solar neutrinos, along with the standard solar model, are introduced, and in Sect. 7.2 a few significant experiments aimed to detect solar neutrinos are outlined. We move on to neutrinos produced in supernovae (Sect. 7.3) and atmospheric neutrinos (Sect. 7.4), discussing their production mechanisms and detection. In Sect. 7.5 and in Sect. 7.6 we delve into cosmic neutrinos and high energy astrophysical neutrinos. We examine cutting-edge experiments like IceCube in Sect. 7.7. Since the mid-twentieth century, geoneutrinos, which are electron antineutrinos generated from $\beta$ decay emitters within our planet, have been proposed as valuable tools for investigating the Earth's interior. They are the focal point of Sect. 7.8.

## 7.1 The Standard Solar Model

The Sun is an intense source of photons and electron neutrinos, produced by fusion reactions at its center. Electron neutrinos are emitted with variable energy up to about 15 MeV. Photons easily interact with the solar medium, giving rise to other photons of decreasing energy and increasing number. On the contrary, since neutrino interactions with matter is extremely weak, practically all neutrinos almost freely escape from the Sun after they are produced, and reach us directly. The only change during their way inside the solar medium is the change in the rate of oscillation, according to the general analysis of neutrino oscillations in matter of Sect. 6.6. Hence the detection of solar neutrinos offers direct evidence for thermonuclear reactions in the center of the Sun.

The Sun density is very high near the inner region, the core, about $10^5$ kg m$^{-3}$, and decreases gradually towards the surface. The gravitational force first compresses the hydrogen gas which is in the core, raising its temperature (about $1.5 \times 10^7$ K) and giving the protons the necessary energy to overcome electrical repulsion and start the fusion process. At these temperatures the core is in a plasma state, that is a mix of positively charged ions and electrons. The Sun shines by steadily fusing protons, that is nuclei of hydrogen ($^1$H or H), into Helium-4 ($^4$He) nuclei. This process is known as the main sequence in the life of a star. The overall reaction can be represented symbolically by the relation

$$4p \rightarrow (^4\text{He})^{++} + 2e^+ + 2\nu_e. \tag{7.1}$$

The nucleus of Helium-4 is an $\alpha$ particle, that is a bound state of two protons and two neutrons. Each neutron is produced, alongside with an electron neutrino and a positron, in the $\beta^+$ decay of a proton. The two positrons immediately annihilate with two electrons. Therefore the energy generation process is

$$4p + 2e^- \rightarrow (^4\text{He})^{++} + 2\nu_e + Q_H. \tag{7.2}$$

In formula (7.2) it is indicated the energy $Q_H$ released in the form of photons (heat) during of the overall conversion. The energy is also released in the form of kinetic energy of the neutrinos $\langle E_{k\nu} \rangle$ (the kinetic energy of the Helium-4 nucleus is negligible because of its large mass). Then the $Q$-value, which is the amount of energy released, as in this case, or absorbed during a nuclear reaction, amounts to $Q = Q_H + 2\langle E_{k\nu} \rangle$. By expressing the $Q$-value in term of proton, electron and $^4$He masses, one can obtain its value, that is $Q = 4m_p + 2m_e - m_{^4\text{He}} \sim 26.7$ MeV. The largest part of the energy is carried away by photons: the two neutrinos emitted from this reaction take only 2% of the total energy for every four protons burned, namely $2\langle E_{k\nu} \rangle \sim 0.6$ MeV.

Solar neutrinos produced by nuclear fusion processes in the core of the sun yield exclusively electron neutrinos. One has to observe that the probability of four protons meeting at a point is negligibly small even at the large densities existing in the solar core. Almost all reactions involve collisions of only two protons. Hence making helium from four protons involves a sequence of steps, which is not unique; there are different series of nuclear reactions up to the task.

The two paths more likely to occur in stellar cores are the so-called proton-proton ($pp$) chain and the carbon cycle. In the $pp$-chain two protons start the process by combining to form the deuteron and further protons are added later on. In the carbon cycle the four protons are successively absorbed in a series of nuclei, starting and ending with carbon, which is most abundant among the heavy elements in stars. Both in the carbon cycle and the $pp$-chain, the net process is the same

## 7.1 The Standard Solar Model

as in Eq. (7.1), namely the fusion of four protons to form an $\alpha$ particle with the emission of two positrons and two neutrinos. Each conversion of four protons to an $\alpha$ particle is known as a termination of the chain of energy-generating reactions that accomplishes the nuclear fusion. Which path is taken is determined in large part by the temperature at the star's core. This temperature essentially determines the pressure which opposes the gravitational collapse, that is the contraction of the star because of its own gravity. For stars with masses no greater than the Sun's mass, the central temperature does not exceed $1.6 \times 10^7$ K, and hydrogen burning proceeds predominantly via the $pp$ chain. For stars more massive than the Sun, with higher central temperature, hydrogen burning occurs primarily through the carbon cycle.[1]

A standard solar model (SSM) aims to describe the observed luminosity[2] and radius of the Sun, the observed heavy-element-to-hydrogen ratio at its surface, and neutrino fluxes. The concept behind SSMs is that of a well-defined framework within which a physical description of the Sun can be constructed and predictions be made. An influential role in neutrino physics has been played by the SSM elaborated in the sixties by Bahcall and collaborators [210]. The current SSMs predict that in the Sun the main reaction chain, the $pp$-chain, produces about 99% of the solar energy (and most of the solar neutrinos), leaving only 1% for the CNO cycle. As can be seen in Fig. 7.1, the first basic elementary reaction of the $pp$-chain is the fusion of two protons into deuterium

$$p + p \to {}^2\text{H} + e^+ + \nu_e. \tag{7.3}$$

The (7.3) reaction is the principal source of neutrinos from the Sun, but these neutrinos are quite difficult to measure since they have low energies, below 0.42 MeV. As illustrated in Fig. 7.1, the reaction (7.3) is followed by different sequences of nuclear reactions, the branches or sub-chains. Despite some heavier elements appearing at intermediate stages, the common outcome is to burn hydrogen only to helium. More energetic neutrinos are emitted along the way. The branch denominated $pp$II gives the so-called "Beryllium neutrinos" through the reaction

$$^7\text{Be} + e^- \to {}^7\text{Li} + \nu_e. \tag{7.4}$$

This reaction produces a two-body final state, comprising about 90% of the time $^7$Li in the ground state and a mono-energetic neutrino of 0.86 MeV. In the remaining 10%, this reaction induces an excited lithium state and the emission of neutrinos with an energy of 0.384 MeV. Thus mono-energetic neutrinos are produced in both

---

[1] Roughly speaking, a larger charge of a nucleus implies a stronger Coulomb barrier to overcome for the proton capture, and a higher temperature is needed.

[2] In astronomy, luminosity is the total amount of electromagnetic energy emitted per unit of time by a star, galaxy, or other astronomical objects. In IS units, luminosity is measured in joules per second, or watts.

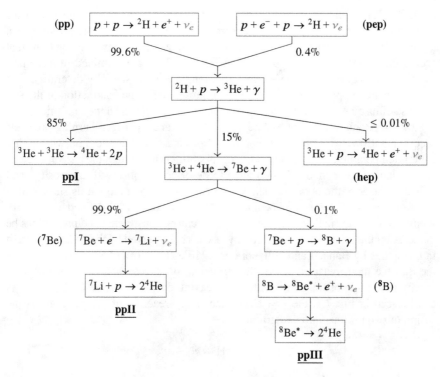

**Fig. 7.1** The proton-proton (pp) chain of reactions and its three branches (ppI, ppII, ppIII). Each branch is identified at the end reaction. The *pep* and *hep*-chain are also shown. The relative percentages of the branches [211] are indicated

cases. The so-called *pp*III branch gives "Boron neutrinos"

$$^8B \to 2\,^4He + e^+ + \nu_e. \tag{7.5}$$

They have the highest energy of the main branches of the *pp* chain, with a spectrum reaching about 14 MeV and as such are the least difficult to detect. In Fig. 7.1 it is also reported another reaction initiating the chain, the *pep* (proton-electron-proton) reaction, a three body reaction involving two protons and an electron, which produces deuterium and one neutrino. While the primary *pp* fusion occurs with a probability of 99.6%, the *pep* process occurs with the remaining probability of 0.4%. It has a lower frequency ratio, but the neutrinos produced in this reaction (*pep* neutrinos) are mono-energetic, producing a sharp energy line of 1.44 MeV. Detection of solar neutrinos from this reaction were reported by the Borexino collaboration in 2012 [212]. In addition to the main *pp/pep*, $^7$Be and $^8$B neutrinos, it is shown a fourth source of neutrinos from a weak side-branch of the *pp* chain,

## 7.1 The Standard Solar Model

the ppIV branch, also indicated as *hep* or $^3$He+$p$, that is the branch ending with the reaction

$$^3\text{He} + p \rightarrow {}^4\text{He} + e^+ + \nu_e. \tag{7.6}$$

These neutrinos (*hep* neutrinos) have a very low flux, but are the most energetic produced by the Sun, with a maximum energy of about 19 MeV. They populate energy bins above the $^8$B neutrino endpoint.

The carbon cycle is also called CNO (carbon-nitrogen-oxygen) cycle, since it utilises the presence of C, N and O as catalysers for the process of synthesising helium from hydrogen. It relies on the fact that these elements are fairly abundant in the universe, and are present in a star even before it starts nuclear burning. Catalysts enable the reaction to take place, but are not themselves consumed or created by it. The two main channels of the CNO cycle are the CN and NO cycle (or sub-cycles), which we report in Fig. 7.2. It can be seen that in the CN cycle, $^{12}$C combines with a proton giving $^{13}$N, which $\beta$ decays into $^{13}$C. This $^{13}$C combines with a proton giving $^{14}$N, which in turn combines with a proton giving $^{15}$O. The $\beta$ decays of $^{15}$O produces $^{15}$N. Finally, $^{15}$N combines with a proton giving back the original $^{12}$C and a newly synthesised $^4$He. The net result is that four protons fuse to produce one alpha particle, two positrons and two electron neutrinos, namely the reaction (7.1), using carbon, nitrogen and oxygen isotopes as a catalyst. An alternative route, the NO cycle, starts at the combination of $^{15}$N and a proton, resulting in the much less likely production of $^{16}$O, and continues as described by the arrows in the lower half of Fig. 7.2. Also in this case the overall result is the reaction (7.1).

The CNO solar neutrino flux is sensitive to the abundance of heavy elements in the Sun (metallicity), an experimental input parameter in solar models. A precise CNO solar neutrino flux measurement has therefore the potential to discriminate

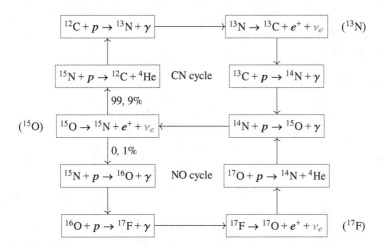

**Fig. 7.2** Main channels of the CNO cycle

between competing models and to shed light on the inner workings of heavy stars. Since no net production of these elements occurs, even at high central temperatures, the CNO cycle could not have occurred in metal-free primordial high-mass stars.

About 600 million tons of hydrogen are burned every second to supply the solar luminosity. The dominant $pp$ flux is directly proportional to the total solar photon luminosity, which is very well known. The energy release from the fusion processes and the total number of neutrinos emitted by the Sun are correlated; we have seen that for each about 27 MeV of energy received, we count 2 neutrinos. A rough estimate of the total neutrino flux yields

$$\phi_\nu \approx 2 \frac{\mathscr{L}_S}{Q} \approx 6.4 \times 10^{10} \, \frac{1}{\text{cm}^2 \, \text{s}} \tag{7.7}$$

where the factor 2 takes account of the fact that two neutrinos are produced from hydrogen fusion, $Q \simeq 26.7$ MeV is the $Q$-value for each $4p$ fusion process (Eq. (7.1)), and $\mathscr{L}_S \simeq 8.534 \times 10^{11}$ MeV cm$^{-2}$ s$^{-1}$ is the apparent solar luminosity [127]. These many solar neutrinos are passing through our body and the Earth.

In spite of this extremely large flux, the detection of solar neutrinos is difficult and requires large detectors because of the small neutrino interaction cross-section. In any experiment, one major problem is the identification and the drastic reduction of the backgrounds, namely of those phenomena that can simulate the events being sought (the signal). By working in deep underground laboratories one can shield the detectors from cosmic rays, whose interactions in the detector represent one of the main backgrounds in most neutrino experiments.

Several underground laboratories searching for solar neutrinos are located at latitudes around 35°–45° in the Northern Hemispere. Therefore, during the night solar neutrinos come from below, passing through the Earth on their way to the detector, while during the day they come from above, crossing the atmosphere and only the thin part of the Earth crust above the detector. This difference is important, since the matter density in these two regions is different, and may show up as a day-night difference of the detected neutrino flux.

Solar neutrinos have been at first detected by means of radio-chemical techniques, employing $^{37}$Cl and $^{71}$Ga, then by means of water and heavy water Cherenkov detectors, which having a threshold of about 3 MeV as a minimum can only observe Boron neutrinos. More recently, experiments have employed ultra-pure liquid scintillators, which having a threshold around 40 keV allow all neutrino components to be detected, including those produced during the CNO cycle [213]. Since the first observations, physicists were puzzled by the discrepancy between solar neutrino measurements and the expectations based upon SSM flux calculations. The measured flux appeared to be substantially lower. Such deficit of observed solar electron neutrinos with respect to the SSM prediction became known as the *solar neutrino problem* or *puzzle*. It can be understood if the electron neutrinos produced in the Sun would get transformed into other flavours that the experiments could not detect.

## 7.2 Solar Neutrino Experiments

The first evidence of nonzero neutrino mass was obtained from the neutrino oscillation interpretation of a deficit in the solar neutrino rate, observed since the end of the 1960s. Solar neutrino experiments detect neutrinos generated in the core of the Sun, whose energy range is about $(0.2 - 19) \times 10^{-3}$ GeV. Since the Sun-Earth distance is roughly $L \simeq 1.5 \times 10^8$ km, on the grounds of the assumptions made in Sect. 6.3 one expects that they are sensitive to extremely small values of $\Delta m^2$, of order $\Delta m^2 \gtrsim 10^{-12}$ eV$^2$.

Owing to the low neutrino energies, $\nu_\mu$ and $\nu_\tau$, from electron neutrino oscillations, cannot be observed in charged current interactions, which should produce the corresponding charged leptons.[3] Hence only electron neutrinos can be revealed and solar neutrino observations are in general disappearance measurements. A notable exception is the SNO experiment, which was sensitive to fluxes of all active neutrinos measured through neutral charge interactions, acquiring in this respect the status of an appearance experiment.

The deficit of solar electron neutrinos was first observed by the Homestake experiment [214]. The Homestake experiment was to measure the flux of neutrinos from our Sun's core and prove the validity of the fusion model developed over the previous three decades. The deficit was confirmed by many other following experiments, chiefly GALLEX [215] at Laboratori Nazionali del Gran Sasso (LNGS) in Italy, SAGE [216] at the Baksan Laboratory in Russia, and Kamiokande at the Kamioka observatory in Japan [217].

Kamiokande (see Sect. 7.2.3) was a pioneering experiment, running from 1983 to 1996. It was upgraded to Kamiokande-II (1986–90) and to Kamiokande-III (1990–95). It was based on water as detector medium, in which neutrino detection was made by means of Cherenkov radiation produced by electrons/positrons, either emitted by inverse $\beta$ decay on protons or by elastic scattering on electrons. The detector of Kamiokande employed 3000 tons of pure water. It was located at about 1 km underground in the lead and zinc Kamioka mine. The main goal of Kamiokande was to find the decays of protons, but eventually it produced very important results on solar, supernova (SN1987A), and atmospheric neutrinos. It represented the first generation of experiments based on a water Cherenkov detector, which will be described in more detail in Sect. 7.2.2.

A second generation started with the Super-Kamiokande (also referred as Super-K or SK) Cherenkov detector, operating since 1996 in the same location than Kamiokande. The Super-Kamiokande is an upright self-supporting cylinder, 39.3 m in diameter and 41.4 m in height, made of welded stainless steel, separated in two optically insulated sections: the Inner Detector (ID) and the Outer Detector (OD). It contains more than 50 kton of ultra-pure water and it is surrounded by more than 12,000 photomultiplier tubes (PMTs). In the last years, gadolinium (Gd) sulphate

---

[3] Muons and tauons have a mass of about 100 MeV and 1.7 GeV, respectively.

was dissolved in the water, reaching a concentration of Gd of 0.03%, in order to enhance neutron detection efficiency. SK has confirmed the solar neutrino deficit with real-time neutrino imaging of the Sun, measured the solar neutrino energy spectrum above 5 MeV for the first time, recorded the first unambiguous evidence of atmospheric neutrino oscillations, set the world's highest lower limits on partial lifetimes for proton decay modes [218]. It is worth mentioning here that the SK detector is also used as the far detector of the T2K experiment, an experiment made by sending a muon neutrino beam produced with the J-PARK accelerator and directed to SK.

There is also a future project called Hyper-Kamiokande (HK), which will employ 258 kton of pure water—a third generation of water Cherenkov detectors. It will be buried at about 600 m and it will be the largest water Cherenkov detector in the world. Hyper-Kamiokande relies on the same detection principles as SK, and it is designed in a similar way, but with some improvements to overcome the new challenges it brings: a larger detector to improve statistics and improved photo-sensors for better efficiency among all. More details on this experiment will be given in Sect. 8.2.4.

Heavy water ($D_2O$) as a detection medium was used at the Sudbury Neutrino Observatory (SNO), that was sited 2 km underground in a mine near Sudbury in Canada [219, 220], the current location of SNOLab. This experiment was initiated in 1984 to provide a definitive answer to the solar neutrino problem and stopped data-taking, after three experimental phases, in 2006.[4] The main results from SNO for solar neutrinos had shown that electron neutrinos from $^8B$ in the solar core changed their flavour in the transit to Earth, proving beyond doubts that neutrino oscillations explain the solar neutrino problem.

Summarising, after the Homestake experiment, which first had detected solar neutrinos in 1968, the experiments that have given clear and unambiguous evidence of neutrino oscillations started in the end of the previous century. They were the Super-Kamiokande experiment in Japan (still running in 2024) and SNO in Canada (completed in 2006). The use of water and heavy water allowed the clear identification of neutrino oscillations both in solar neutrinos (SNO) and in atmospheric neutrinos (SK) by means of excellent detectors, clever tricks, and a little luck.[5] The two main teams in Japan and Canada came from different areas of fundamental physics. The history of these experiments is very well described by the 2016 Nobel winners McDonald and Kajita in their lectures [221, 222].

---

[4] As a historical note, SNO was not the first heavy-water solar-neutrino experiment. In 1965, Tom Jenkins, along with other members of what was then the Case Institute of Technology, began the construction of a 2 ton small heavy-water Cherenkov detector in the Morton salt mine in Ohio. This experiment finished in 1968 and served to place an upper limit on the solar $^8B$ flux.

[5] The fact that the atmospheric neutrino oscillation length at GeV scale is comparable with the Earth diameter was crucial to obtain the results.

## 7.2.1 The Homestake Experiment

The first detection of solar neutrinos was announced in 1968 by the Homestake experiment [40, 223]. Electron neutrinos were detected observing electron neutrino capture by an isotope of chlorine

$$\nu_e + {}^{37}\text{Cl} \rightarrow e^- + {}^{37}\text{Ar} \qquad (\nu_e + n \rightarrow e^- + p). \qquad (7.8)$$

This detection method, known as the radiochemical method, was first suggested by B. Pontecorvo [224]. It exploits a detection target which, upon absorption of a neutrino, is converted into a radioactive element. The product nuclei are first accumulated during exposure of large targets, then the reaction product is separated from the target with radiochemical techniques and subsequently detected by radioactive decay. Only the energy threshold of the selected reaction restricts the minimum energy for neutrino detection. A critical condition in radiochemical experiments is the reduction of production background. For this, both cosmic radiation and natural radioactivity must be reduced far below the normal environmental level, since secondary protons can mimic neutrino capture via $(p,n)$ reactions in both cases. The background is reduced by using material extremely free of radioactive components, and shielding the cosmic rays by going underground. All radiochemical solar neutrino experiments are located in deep mines or in mountain tunnels.

The Homestake experiment, headed by astrophysicists Raymond Davis, was to measure the flux of solar neutrinos and prove the validity of the fusion model. It was built in the sixties, in the Homestake Gold Mine in Lead (South Dakota, USA) at about 1.6 km depth. The chlorine-argon reaction occurred in a tank containing approximately 380,000 litres of perchloroethylene ($C_2Cl_4$), a common colourless liquid widely used for dry cleaning of fabrics. The produced $^{37}\text{Ar}$ is radioactive, with half-life $\tau_{1/2}^{Ar}$ of about 35 days.

In nuclear physics, the half-life $\tau_{1/2}$ is the average time required for one-half the nuclei to decay. In general, starting at $t = 0$ with $N_0$ unstable nuclei, the number of nuclei remaining at time $t$ is given by

$$N(t) = N_0 \left(\frac{1}{2}\right)^{t/\tau_{1/2}}. \qquad (7.9)$$

Then

$$\frac{1}{N}\frac{dN}{dt} = -\frac{\ln 2}{\tau_{1/2}}. \qquad (7.10)$$

Let us compare it with the lifetime $\tau$ and the decay constant (or decay rate or decay width) $\Gamma$. Starting at $t = 0$ with $N_0$ unstable nuclei, the number of nuclei remaining at time $t$ can be expressed as

$$N(t) = N_0 e^{-\Gamma t} = N_0 e^{-\frac{t}{\tau}} \qquad \tau \equiv \frac{1}{\Gamma}. \qquad (7.11)$$

Differentiating, one has

$$\frac{dN}{dt} = -\Gamma N(t) \qquad (7.12)$$

Therefore, one obtains

$$\frac{\Gamma}{\ln 2} = \frac{1}{\tau_{1/2}}. \qquad (7.13)$$

After an exposure of the detector (two to three times $\tau_{1/2}^{Ar}$), the reaction products were chemically extracted and introduced into low-background proportional counters, where they were counted observing the energy spectrum of the electrons from $^{37}$Ar decays. The radio-chemical technique is powerful but limited, because it does not measure the energy of the incoming neutrinos nor their direction. It simply counts the integrated neutrino flux above a kinematic threshold, 0.814 MeV for $^{37}$Cl neutrino capture. Homestake could therefore measure only the energetic but small component of the whole solar neutrino flux due the beryllium and boron neutrinos.

The Homestake experiment became operational in 1967 and the final results were published in 1994 by R. Davis [223]. Even after the first run, it was clear the solar neutrino flux was lower than predicted by SSMs. It was the beginning of the solar neutrino problem. The uncertainties both on the experimental side (chiefly, the efficiency in collecting and counting the $^{37}$Ar atoms) and on the theoretical side (the precise determination of the expected flux produced by the complex nuclear fusion chain operating in the Sun) did not allow to draw firm conclusions from the observed discrepancy, although the results gave a hint that a real problem existed.

Homestake was a true pioneering experiment. It was the first to observe a deficit in the flux of neutrinos from the Sun, which since then has consistently remained the same, namely a factor of about 1/3 for $^8$B neutrinos. Now we know that electron neutrinos are transformed into other neutrinos on the flight from the Sun to the detector and the observed deficit is fully explained by the fact that only electron neutrinos can contribute to the radiochemical reaction (solar neutrino energies are well below the value of the muon mass).

### 7.2.2 The Cherenkov Detector

A Cherenkov detector allows the detection of neutrinos in real time by observing the tracks of the ultrarelativistic charged leptons produced by neutrino interactions. It consists essentially of a medium where Cherenkov radiation is produced (usually called the radiator) and a system of photodetectors to detect Cherenkov photons. By definition, if a medium has an index of refraction $n$, the phase velocity of light travelling in that medium is $c/n$. The Vavilov-Cherenkov (commonly called just Cherenkov) radiation occurs when a charged particle moves in a dielectric medium

## 7.2 Solar Neutrino Experiments

with velocity $v > c/n$ [225, 226]. In other terms, when the particle velocity, that is not affected by the refractive index, is higher than the *local* speed of light. Cherenkov radiation does not require that the charge is accelerated and the basic mechanism is very similar to that of sound shock waves in gases.

According to classical physics, a charged particle moving through a dielectric medium polarises the molecules and arrange the dipoles of the medium, setting up a macroscopic polarization. After the particle has passed, the polarised molecules rapidly revert to their ground state, emitting electromagnetic radiation. According to the Huygens principle, the emitted electromagnetic waves move out spherically at the phase velocity of the medium. Usually, the dipole are disposed symmetrically along the direction of the particle, the electromagnetic waves interfere destructively and the perturbation can relax while the particle moves on. There is no resultant field at large distances and therefore no radiation. But if its velocity exceeds the velocity of light in the medium, the polarization is asymmetric along that direction, in front and at the rear of the particle, and the perturbation can no longer decay since the reaction time of the medium is not high enough. The electromagnetic waves add up constructively, leading to a resulting field and to a coherent radiation, the Cherenkov radiation. This is not emitted in all directions, but at angle $\theta$ with respect to the particle direction. At each point of the particle's trajectory, after the time $\Delta T$, the particle travels a length $\beta c \, \Delta T$ and the light, moving at inferior speed, travels a smaller distance $c/n \, \Delta T$, creating a coherent wavefront.

This is illustrated in Fig. 7.3a. Some points along the particle's trajectory reached at different $\Delta T$ are indicated: $A$, $P_1$, $P_2$. They are vertices of similar rectangular triangles, where one side is on the particle's trajectory, the other one on the direction of the emitted light, and the third one along the direction (line BC) of the wavefront.

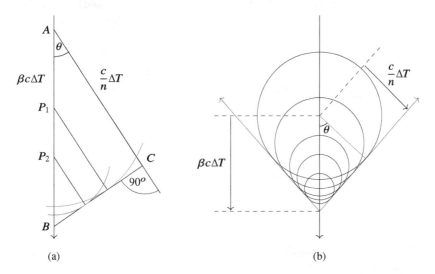

**Fig. 7.3** Formation of a coherent wavefront of Cherenkov radiation

The angle $\theta$ between the particle's trajectory and the direction of the emitted light is given by the relation

$$\cos\theta = \frac{c/n}{\beta c} = \frac{1}{n\beta}. \tag{7.14}$$

Let us underline that it is not the particle that emits light, but the bound particles of the immediately surrounding (dielectric) medium. The wavefront is actually spherical, since the wavelets originate all around the track of the particle, forming a coherent cone (see Fig. 7.3b).

The signature of the Cherenkov effect is a cone of emission in the direction of particle motion. The axis of the light cone is just the track of the running charged lepton. If the media is transparent the Cherenkov light can be detected.[6] The intensity of Cherenkov radiation increases with decreasing wavelength. That is, it is weighted towards the blue end of the visible spectrum, which generally gives the Cherenkov radiation a characteristic blue appearance. One can observe the rings of light produced by the light cone. The emitted light can be converted into an electric signal by photomultiplier tubes in optical contact with the Cherenkov medium. Photomultiplier tubes (PMTs) are vacuum tubes with an internal photocathode, exposed to light through a transparent glass window. If an incident photon extracts an electron (photoelectron) from the photocathode due to photoelectric effect, it is accelerated and multiplied by a set of dynodes[7] and finally collected by the anode where it generates an electric signal which is digitized and stored by the front-end electronics.

One characteristics of a Cherenkov detector is that a minimum particle velocity is required in a given medium in order to generate any Cherenkov light, which may represent a disadvantage. One can translate the minimum (threshold) velocity $v_{th} = 1/n$ into the threshold on momentum $p_{th}$ for a particle with rest mass $m_0$

$$\beta = \frac{v}{c} \implies \beta_{th} = \frac{1}{n} \implies$$

$$p_{th} = m_0 \gamma_{th} \beta_{th} c = \frac{m_0 \beta_{th} c}{\sqrt{1-\beta_{th}^2}} \implies p_{th} = \frac{m_0 c}{\sqrt{n^2-1}} \tag{7.15}$$

On the other side, there is the advantage that very large target masses can be instrumented at comparatively modest cost, and that long muon tracks produced by charged current muon neutrino interactions can be disentangled from electromagnetic showers produced by electron neutrinos, allowing a separation of muons and electrons samples.

---

[6] Cherenkov effect is not only optical, radio or microwave Cherenkov emission (also called Askarian effect) in dense dielectric materials (salt, sand, ice, etc.) is also possible.

[7] A dynode is an electrode in vacuum that emits electrons when hit by an electron with sufficient kinetic energy.

### 7.2.3 The Kamiokande Experiment

The solar neutrino problem called for additional experiments and detection techniques. During the period 1987–1990 the Kamiokande experiment (Kamioka Nuclear Decay Experiment), led by Masatoshi Koshiba and based on a water Cherenkov detector, measured the solar neutrino flux. The Kamiokande experiment was located in an underground laboratory built in the Kamioka mine in central Japan (Gifu Prefecture), about 220 km north-west of Tokyo, at the depth of about 1 km below the top of Mount Ikeno. The mine was one of two adjacent mines (Mozuki mine) owned by the Kamioka Mining and Smelting Company, which produced zinc and lead. Kamiokande, whose name with time was extended to mean also the Kamioka Neutrino Detection Experiment, was decommissioned in 1997.

Kamiokande was originally designed to search for proton decay, but by the end of 1986 it was modified to detect solar neutrinos. It identified solar neutrinos by the Cherenkov light produced by the recoiling electrons in the elastic scattering

$$\nu_x + e^- \rightarrow \nu_x + e^- \tag{7.16}$$

occurring inside a water Cherenkov detector made of about 3000 tons of pure water. This scattering takes place with all active neutrinos, although the cross sections for muon or tau neutrinos are about 15% smaller than the ones for electron neutrinos. Because of its high energy threshold (∼7.5 MeV), Kamiokande was sensitive to Boron neutrinos only. The observed rate was once again smaller than the rate predicted by the Standard Solar Model. Kamiokande also detected a neutrino burst from the supernova, SN1987A, confirming theories about the physics of supernova explosions [46]. Because of all these important achievements, Masatoshi Koshiba was awarded, together with Raymond Davis,[8] 1/2 of the 2002 Nobel prize.

The Kamiokande data on the solar neutrino deficit were confirmed by Super-Kamiokande, a greatly improved version of the Kamiokande experiment, located in another cavern of the Kamioka mine. It started to operate in 1996, using a Cherenkov detector with much more (50,000 tons) pure water and better PMT (photomultiplier tube) coverage. The high level of light collection, combined with an advanced water purification system that ensured good water clarity and low levels of radioactive background, allowed the experiment to lower the threshold for solar neutrino detection to 5 MeV (and lower, in later upgrades of the detector). Due to their energy thresholds, both Kamiokande and Super-Kamiokande measured Boron neutrinos. By measuring the directions of the electrons hit by the neutrinos in the elastic scattering, Super-Kamiokande also established that neutrinos were coming from the Sun, becoming the first neutrino telescope. After so many experimental evidences of the neutrino deficit, it remained to be established if it was due to a flaw in the solar models or to an anomalous behaviour of neutrinos. The answer could

---

[8] Raymond Davis led the Homestake experiment, see Sect. 7.2.1.

come only by a measurement of the $pp$ neutrino flux (see Eq. (7.3)), that could be estimated in an almost model independent way by the solar luminosity.

SK also carried out analyses of atmospheric neutrinos, with results that were fundamental to establish neutrino oscillations (see Sect. 7.4.2).

### 7.2.4 The GALLEX and SAGE Experiments

After SK confirmed the solar neutrino problem, two radiochemical experiments were built to measure the $pp$ neutrino flux, GALLEX (Gallium Experiment) [227] in Italy and SAGE (Soviet–American Gallium Experiment) [216] in Russia. They were both underground and both employed gallium as the target. From 1990 to 2000, they measured the process

$$\nu_e + {}^{71}\text{Ga} \rightarrow e^- + {}^{71}\text{Ge} \tag{7.17}$$

with an energy threshold of 233 keV. They were thus sensitive to low-energy $pp$ neutrinos, whose end point energy is 423 keV, and provided a feasible mean to measure low-energy solar neutrinos. The produced $^{71}$Ge is radioactive with half-life of about 12 days. The two experiments follow the radiochemical techniques, sketched in Sect. 7.2.1 for the Homestake experiment.

These two experiments are characterised by different procedures for the chemical separation of germanium from gallium. The SAGE experiment used metallic gallium (which becomes a liquid at just above room temperature); germanium was extracted from the Ga metal into an aqueous solution by an oxidation reaction [228]. The GALLEX experiment used gallium in the form of a $GaCl_3$. The germanium extraction started by sweeping gaseous nitrogen through the process tank, which carry out the Germanium chloride ($GeCl_4$) originating from the solar neutrinos.

The GALLEX experiment published the first results in 1992: they observed deficits in the flux of neutrinos from the Sun with a factor of about 1/2 [229]. SAGE soon confirmed this value. GALLEX ended in 1997, becoming (in an improved version) GNO, which ended in 2003. In 2014, the SAGE experiment was modified to become part of the neutrino oscillation experiment BEST (Baksan Experiment on Sterile Transitions) in the same laboratory, which is still running. GALLEX and SAGE gave a big boost to the interpretation of the neutrino deficit in terms of oscillations, although they were not able to completely prove it. Being able to measure solar neutrinos at low energy threshold, where the flux is dominated by $pp$ neutrinos, they provided a much more solid evidence that the problem was relying on neutrino physics and not on our poor understanding of the solar core temperature.

From the 1990s, further controls and refining precisions, as well as an impressive set of elio-seismological measurements, brought to a growing confidence on the SSM and to the conclusion that the neutrino deficit had its root in the neutrinos behaviour. Apparently, electron neutrinos were disappearing in large fractions on

## 7.2 Solar Neutrino Experiments

their way from the solar core to the Earth, and the most probable and natural hypothesis was flavour oscillations.

### 7.2.5 The SNO Experiment

The SNO (Sudbury Neutrino Observatory) experiment has been the key one to establish solar neutrino oscillations. With the aid of a crucial reservoir of heavy water[9] available by the Canadian government and by means of a very well conceived set of experimental phases, SNO was able to measure precisely both the flux of electron neutrinos and, at the same time, the total flux of all neutrino species.

The detector, which used 1000 tons of heavy water, contained by a 12 m diameter acrylic vessel, was located in the Vale's Creighton mine near Sudbury (Ontario, Canada), about 2 km deep underground. Neutrinos reacted with the heavy water to produce Cherenkov radiation, which was detected by an array of 9600 photomultiplier tubes [230]. Observations started in 1999 and ended in 2006.

SNO was designed to observe neutrinos in three different reactions, which depends on neutrino flavour type differently:

- the inverse $\beta$ decay on a deuterium nucleus

$$\nu_e + {}^2H \rightarrow p + p + e^- \tag{7.18}$$

which occurs through a charged-current (CC) weak interaction. It is possible for electron neutrinos only, since the incoming solar neutrinos do not have enough energy to produce a muon or tau in the final state. This reaction has an energy threshold of 1.4 MeV, and the electron can be identified by the Cherenkov light it yields in the liquid. It could be used to measure the flux of solar electron neutrinos.
- the NC inelastic scattering on a deuterium nucleus, mediated by a $Z^0$ boson,

$$\nu_x + {}^2H \rightarrow p + n + \nu_x \tag{7.19}$$

which is possible with the very same cross section for all neutrino flavours $\nu_x \in \{\nu_e, \nu_\mu, \nu_\tau\}$. The net result is just to break apart the deuterium nucleus; the liberated neutron is eventually captured by another nucleus emitting $\gamma$ rays. The gamma rays will scatter electrons which produce detectable light via the Cherenkov process. This reaction was used to measure the neutrino flux independent of the neutrino flavour.

---

[9] The heavy water, that is deuterium oxide ($D_2O$ or ${}^2H_2O$), is a form of water where the hydrogen has been substituted with its isotope deuterium (${}^2H$ or D).

- elastic scattering (ES) on electrons:

$$\nu_x + e^- \to \nu_x + e^- \tag{7.20}$$

which is possible for neutrinos $\nu_x$ of any flavour, but with different cross sections. In fact, it takes place by $Z^0$ exchange for all neutrinos $\nu_x$ and, in addition, by $W$ exchange for electron neutrinos. The final state electrons are identified via the Cherenkov emission. This reaction was precisely measured by SK as well.

The experiment was sensitive in the higher part of the spectrum, namely to Boron electron neutrinos. The SNO collaboration found that their flux was only one third of the *total registered* flux of solar neutrinos [54, 231]. The reduction of the flux with respect to SSM predictions was in line with previous results, but the SNO experiment demonstrated that not only did electron neutrinos disappeared but other neutrino flavours appeared instead. The final result for the flux of electron neutrinos $\phi$ was [221, 232]:

$$\phi = (5.54 \pm 0.33|_{\text{stat}} \pm 0.36|_{\text{syst}}) \times 10^6 \, \text{cm}^{-2} \, \text{s}^{-1}, \tag{7.21}$$

where the first and the second error are statistical and systematic, respectively. This value is in excellent agreement with the SSM prediction, taking into account the effect of solar matter on oscillations. SNO also identified unambiguously the oscillation parameters relevant for solar neutrinos, namely $\Delta m_{21}^2$ and $\sin^2 \theta_{12}$, pointing to a large value of the mixing angle.

The SNO experiment, with its seminal NC measurement, had solved the solar neutrino problem, definitively establishing that the total flux of neutrinos from the Sun is consistent with theoretical expectations and that neutrinos do oscillate from electron flavour to either muon or tau flavours during their path to the Earth.

The importance of the SNO results was recognised by the 2015 Nobel prize awarded for an half to the SNO leader Arthur McDonald "for the discovery of neutrino oscillations, which shows that neutrinos have mass".

### 7.2.6 The KamLAND Experiment

Solar neutrino oscillations were confirmed by the disappearance experiment KamLAND [55, 233] (Kamioka Liquid Scintillator Anti-Neutrino Detector). Although it started operation in 2002, one could classify KamLAND among "early" solar neutrino experiments, since it has been the first experiment to explore with a terrestrial beam the region of neutrino oscillation parameters that is relevant for solar neutrino oscillations. The experiment was located in the cavern of the Kamioka Mine (Japan) that originally held Kamiokande, and offered sensitivity to electron antineutrinos produced by more than 40 nuclear reactors at various commercial nuclear power plants. They had different power and were located at a different distance from the detector. Nuclear reactors are copious sources of electron

## 7.2 Solar Neutrino Experiments

antineutrinos produced in the $\beta$-decays of neutron-rich nuclei. In the KamLAND detector, 1 kiloton of high-purity liquid scintillator is used, surrounded by almost 2000 photomultipliers. Reactor $\bar{\nu}_e$ are detected through the observation of the inverse $\beta$ decay in the detector:

$$\bar{\nu}_e + p \rightarrow n + e^+. \tag{7.22}$$

The positron carries most of the energy of the incoming antineutrino from the reaction and is annihilated quickly. It emits scintillation lights when it goes through the liquid scintillator, providing a prompt signal. On the other hand, the neutron collides with protons and it is eventually absorbed by a proton to form a deuterium nucleus

$$n + p \rightarrow {}^2H + \gamma, \tag{7.23}$$

emitting a mono-energetic $\gamma$-ray (delayed signal). The signature of the event in the KamLAND experiment (and in other reactor neutrino experiments[10]) is a coincidence between the prompt signal and the delayed signal.

In Eq. (6.43) we have defined the oscillation length $\lambda$ as

$$\lambda \equiv 2.47 \frac{E}{\Delta m^2}. \tag{7.24}$$

The flux-weighted average distance of the reactors to KamLAND is about 180 km, and the average energy of the reactor antineutrinos is about 3.6 MeV. These values for $\lambda$ and $E$ constitute an appropriate setting to study neutrino oscillations driven by the solar neutrino mass-squared difference, which is in the $\Delta m^2 = \Delta m^2_{12} \approx 10^{-5}$ range. The KamLAND experiment was able to probe directly the oscillation pattern identified by previous solar neutrino experiments and particularly by SK and SNO. It observed signals from electron antineutrinos, provided a precise measurement of the mass splitting and a very clear oscillation pattern, which was crucial to exclude non-standard solutions and establishing robustness of large angle oscillations driven by $\theta_{12}$.

In 2011 the experiment underwent a major upgrade and became part of the KamLAND-Zen project (see Sect. 10.3.3).

### 7.2.7 The Borexino Experiment

The Borexino experiment has been the first to measure low-energy (less than 1 MeV) solar neutrino events in real time. The Borexino detector is a large volume (mass of 278 tons) organic liquid scintillator, surrounded by more than

---

[10] Experiments using antineutrinos from reactors will be discussed in more detail in Sect. 8.1.

two thousand photomultipliers. The measurement of the scintillation light allows to determine the energy of the electrons. Compared with water Cherenkov detectors, liquid scintillators have a lower energy threshold, but at the same time, since the scintillation photons are emitted isotropically, there is no information about the direction of the electrons.

The Borexino experiment, located at Laboratori Nazionali del Gran Sasso has collected data from 2007 to 2021. Solar neutrinos of all flavours were detected by means of the elastic scattering

$$\nu_x + e \to \nu_x + e. \qquad (7.25)$$

Electron anti-neutrinos were detected by means of their inverse beta decay on protons or carbon nuclei. The electron (positron) recoil energy was converted into scintillation light which was then collected by the photomultipliers.

The aim of the experiment was to build a detector capable to determine independently and by direct counting each solar neutrino component of the *pp* fusion chain and also those from the CNO cycle. Because the energy threshold in the Borexino experiment had to be low, the major requirement was an extremely low radioactive contamination of the scintillator. This result was achieved by developing the most ultra-pure detector ever built. Thanks to this extreme purity (several orders of magnitude better than competing detectors), the low energy threshold (about 40 keV, well below *pp* neutrinos energy), and the demonstrated capability to identify each solar neutrino component independently by spectral analysis [234, 235], including those of CNO cycle [213], Borexino has completed the job. It has confirmed the oscillation pattern of solar neutrinos and directly also the MSW effect by proving convincingly that the survival probability of electron neutrinos originating from the Sun core has a clear energy dependence not accounted for by pure vacuum oscillations. The Borexino collaboration has shown "Comprehensive measurement of *pp*-chain solar neutrinos" [234], the "First simultaneous precision spectroscopy of *pp*, $^7$Be, and *pep* solar neutrinos with Borexino Phase-II" [235] and the "Experimental evidence of neutrinos produced in the CNO fusion cycle in the Sun", a first direct observation, with a high statistical significance [236].

## 7.3 Supernova Neutrinos

A supernova (SN) is the name given to the explosion of a star at the end of its life, one of the most spectacular phenomena in astrophysics. The observed luminosity of the star suddenly increases to typical values of the order of $10^8$ $L_\odot$,[11] amounting to a total energy budget in radiation of about $10^{49}$ erg, and it may remain very bright

---

[11] Convenient astronomical units are the mass, radius and luminosity of the Sun, $M_\odot \approx 1.988 \times 10^{30}$ kg, $R_\odot \approx 6.957 \times 10^8$ m and $L_\odot \approx 3.828 \times 10^{26}$ W, respectively, and the radius of the Earth, $R_\oplus \approx 6.378 \times 10^6$ m.

## 7.3 Supernova Neutrinos

for a few months. The name supernova comes from Tycho Brahe's manuscript in latin "De nova et nullius aevi memoria prius visa stella" (Concerning the Star, new and never before seen in the life or memory of anyone), describing a supernova visible to the naked eye in 1572. Each supernova is designated with the SN prefix followed by the year of discovery, which is followed, in years with more than one supernova, by an upper-case letter A through Z for the first 26 supernovae (SNe) or the lower-case letters aa, ab, and so on, for the following supernovae discovered in the same year. A supernova is a rare events, estimate to happen on average a few times per century in our galaxy. After thirty-three years from Tycho's Nova, another supernova appeared in the Milky Way, the SN1604, also known as the Kepler's Nova, which was again observed by the naked eye. The next supernova was observed in 1885 in Andromeda galaxy, and all subsequent ones were also observed in galaxies different from our galaxy. The development of increasingly advanced telescopes and technologies over the past century has brought the discoveries of more than 36,000 recorded supernovae [237].

When supernovae were first classified, it was done by looking at their spectra. If there were no hydrogen lines, it was a Type I supernova. If there were hydrogen lines, it was a Type II. It was later realised that Type I could be further divided into subclasses: type Ia (showing silicon (Si) absorption), type Ib (signs of helium (He)) and type Ic (neither Si or He). As the sample of supernovae has increased, so too has our ability to understand and model these events, leading to two primary SN explosion paths: thermonuclear or gravitational core collapse. The former case, also called thermal runaway, can occur when two stars orbit near each other, and one or both of those stars is a white dwarf; the latter one arises in the final stage of single stars with high mass, at least about 8 $M_\odot$.

In a star, neutrinos appear in a number of reactions and freely escape from the stellar matter, producing a powerful mechanism of their cooling. One way to produce neutrinos is through $\beta^\pm$ decays, as it happens when the sun is in the main sequence, steadily fusing hydrogen to helium (Eq. (7.1)). There are other reactions where neutrinos are emitted, for instance pair annihilation ($e^+ + e^- \to \bar{\nu}_e + \nu_e$) or nucleon-nucleon bremsstrahlung ($n + n \to n + n + \bar{\nu} + \nu$). They reactions are known as thermal processes, while $\beta^\pm$ decays are called weak. These names are a little misleading: neutrinos can only be produced by weak reactions and, in a star, the rate of emission is always very sensitive to the temperature. The number of neutrinos emitted from thermal processes during the main sequence is small, but once nuclear fusion reactions start transforming three helium-4 nuclei (alpha particles) into carbon (triple-alpha process), they increase significantly. However, independently of their abundance, thermal neutrinos tend to peak at energies of about 1 MeV, which means low-threshold detectors are needed to see them. The weak processes are capable of emitting neutrinos up to energies of about 10 MeV, which makes them much easier to detect.

In this section we discuss how a star becomes a SN and how neutrinos are produced in the star during this transition. We focus on core-collapse SNe, where neutrino emission is more copious. The observation of the burst of neutrinos from supernova SN1987A in the Large Magellanic Cloud just outside our Milky

Way galaxy confirmed the basic picture of core-collapse supernovae. A total of about two dozen events were recorded, with data that was too sparse for detailed quantitative tests [238–240]. A new generation of neutrino experiments awaits the next supernova burst, expecting to be able to record a larger number of neutrino events.

### 7.3.1 Formation of a Supernova

Stars evolve through dynamics between gravity and pressure, and their final state is determined in large part by their mass at birth. White dwarfs represent the final evolutionary stage of middle size stars, with mass between approximately 0.8 and 8 $M_\odot$. The first step of the star towards this ultimate state is to reach the main sequence and start fusing H into He (see Sect. 7.1). This fusion reaction is an exothermic process and energy is released until the supply of hydrogen in the star begins to exhaust. The outward radiation pressure necessary to resist the inward gravitational pressure can then be created through fusion of helium, and so on, creating and burning consecutively heavier elements. Higher and higher temperatures are needed to fuse heavier elements. In the core of middle size stars, hydrogen and helium are burned into carbon C and oxygen O, but no further burning takes place since temperatures are too low. When the star is about to run out of nuclear fuel, it loses mass by shedding its outer layers until only a dense C-O core remains—a white dwarf. The white dwarf is typically half as massive as the Sun, yet only slightly bigger than Earth. That means that a white dwarf is one of the densest collections of matter, surpassed only by neutron stars. In both cases, there is no internal heat source, and the star is held up by degeneracy pressure. The significance of degeneracy pressure comes about naturally because the Heisenberg uncertainty principle ensures that, at very high densities, when the inter-particle spacing becomes small, the particles of the gas must possess large momenta according to the relation $\Delta p \Delta x \sim \hbar$. These large quantum mechanical momenta provide the pressure holding up the core. This effect becomes important for electrons at much larger inter-particle spacings than the protons and neutrons, because of their smaller masses, which corresponds at smaller momenta for equal velocities.

Without any nuclear energy source, except for residual H burning in some cases, a white dwarf simply cools down for the rest of its life. There is, however, another possibility for a white dwarf within a binary system. When the distance between the stars is small enough, the white dwarf within a binary system can attract matter from the companion star due to its very strong gravity. The white dwarf gradually becomes more massive—this process is known as accretion. During the accretion, the mass of the white dwarf increases until it reaches a limit of about 1.4 $M_\odot$ (the so-called Chandrasekhar limit), above which electron degeneracy pressure is no longer sufficient to prevent a collapse due to the star's own gravitational self-attraction. Eventually, the increased pressure on the core is sufficient to reignite and produce a chain of runaway fusion reactions that creates more energy than that which holds the star together gravitationally, causing the star to explode, and form a SN. This is the so-called thermonuclear (or thermal runaway) mechanism. The explosion does not leave any compact remnant nor produce significant neutrino emission through fusion processes.

The wave of thermonuclear fusion ripping through the white dwarf excites the synthesis of iron-peak elements (Ni, Co, Fe) in the dense inner regions and of intermediate mass elements (Si, S, Ca, Mg, O) where burning is incomplete. Sometimes unburned material (C,O) is left near the outer layers [241]. Though the explosion provides the kinetic energy of the SN, this is not what we see as the SN. The lightcurve is powered by the radioactive decay of $^{56}$Ni (half-life 6.1 days) to $^{56}$Co, and ultimately to $^{56}$Fe (half-life 77 days). Gamma rays produced in the decays are thermalized, and at peak light about 85% of the light output of the SN is in the optical, peaking at 4000 Å, with the remainder mostly radiated in the near-ultraviolet and the near-infrared [242]. These SNe, also known as thermonuclear SNe, are classified as type Ia. Type Ia supernovae happen in all types of galaxies with no preference for star-forming regions, consistent with their origin from an old or intermediate age stellar population.

The other types of SNe (type II, Ib and Ic) happen only in star-forming regions where young massive stars are found. They are believed to originate from a different mechanism, the gravitational core collapse. It arises in the final stage of single stars with high mass, at least about 8 $M_\odot$. In the stellar core, nuclear fusion initially burns hydrogen nuclei into helium nuclei. A massive star can reach the high temperatures requested in the subsequent burning stages, that can produce carbon (C), oxygen (O), neon (Ne), silicon (Si) and iron (Fe). Each time one fuel runs out, the star contracts, heats up and then burns the next one. With the passing of each stage, the centre of the star grows hotter and more dense. After helium burning, the evolution is greatly accelerated by neutrino losses. Besides, electron-positron annihilation creates occasionally a neutrino-antineutrino pair, which escape the star with ease and force the burning to go faster to replenish the loss. Although the fusion of hydrogen and helium takes millions of years, the last burning phase—silicon burning—lasts only two weeks. At the end of the fusion processes, a massive star consists of concentric shells that are the relics of its previous burning phases (hydrogen, helium, carbon, neon, oxygen, silicon). An inert iron core builds up while the fuel of lighter nuclei is consumed.

Iron lies near the peak of the nuclear binding energy curve (Nickel-62 is most tightly bound) and can no longer act as a fuel, since nuclear fusion of iron is endothermic, requiring energy, rather than producing it. The core is supported against gravity mainly by electron degeneracy pressure. As the mass of the core builds, the degenerate electron gas cannot exert enough pressure to support the star against gravitational collapse. The star becomes unstable with respect to gravity, and eventually succumbs to a dramatic gravitational collapse within less than a second, after a life of about 10–40 million years. The inner core becomes extremely dense and so rigid that it bounces back producing a violent shock wave, the starting point of a sequence of events that ultimately triggers a supernova explosion. In the SN explosion parts of the star's heavy-element core and of its outer shells are ejected into the interstellar space. The collapse continues until the inner core transforms into a black hole, or until it is repelled by neutron degeneracy pressure, transforming the central core into a neutron star. Neutrons in a degenerate neutron gas are spaced much more closely than electrons in an electron-degenerate gas because the more massive neutron has a shorter wavelength at a given energy. The optical SN outburst commences when the explosion wave, generated in the optically obscured stellar center, eventually reaches the surface layers of the star, a few hours after the onset. That implies that in order to get direct and practically immediate information one has to rely on the observations of neutrinos generated in the core collapse and of gravitational waves which are emitted when the collapse does not proceed perfectly symmetrically because of rotation, violent turbulent mass motions, and anisotropic neutrino emission. Moreover, neutrinos dominate the energetics of core-collapse supernovae. Only about 1% of the gravitational binding energy released in the formation process of the compact remnant ends up as kinetic energy of the expanding ejecta, whereas the remaining is radiated away in neutrinos.

### 7.3.2 Core-Collapse Supernovae

In the core-collapse supernova (CCSN), the process of SN explosion can be broadly described as a sequence of phases: the infall, the neutronization burst, the accretion phase and the cooling phase.

As seen in Sect. 7.3.1, an inert stellar core made of iron-group elements builds up as a consequence of nuclear fusion of lighter elements. The infall stage starts when the inner core approaches the Chandrasekhar mass, that is the mass above which electron degeneracy pressure is no longer sufficient to balance the star's own gravitational self-attraction and the collapse begins. During the collapse, the inner core is compressed and heated. As $\alpha$ particles in iron are bound by about 2 MeV per nucleon, sufficient heating can induce the photo-disintegration

$$\gamma + {}^{56}\text{Fe} \rightarrow 13\alpha + 4n. \tag{7.26}$$

Then it takes about 7 MeV to release nucleons from $\alpha$ particles. At the same time, electrons start to be captured on nuclei and free protons, emitting electron neutrinos through the process of electron capture

$$e^- + p \rightarrow n + \nu_e. \tag{7.27}$$

During the infall phase, lasting about 100 ms, the inner core increases its composition in neutrons (neutronization). In the early stages of the infall, when the density of the iron core is not too high, the electron neutrinos readily escape, carrying away energy and lepton number, since their mean free path is longer than the radius of the core. The mean free path $\lambda$, which represents the most probable distance

## 7.3 Supernova Neutrinos

before interaction, is connected to the density, namely $\lambda \approx 1/(n\sigma)$, where $\sigma$ is the neutrino cross section and $n$ is the density of scattering centres, for instance heavy nuclei. One can define the region until the last scattering surface, from where neutrinos can escape, the so-called neutrinosphere. When the inner core density increases and reaches about $10^{12}$ g/cm$^3$, neutrinos are trapped, and further loss of lepton number from the core (deleptonization) ceases. From that moment on, the electron neutrinos produced by ongoing electron captures—now dominantly on free protons—are swept inward and move with the infalling matter.

Both the electron capture and the nuclear excitation and disassociation take energy out of the electron gas, which is the star's only source of support. This means that the collapse continues until 3–4 times nuclear density, namely about $10^{14}$ g cm$^{-3}$, is reached, and at that point the pressure of degenerate non-relativistic nucleons abruptly stops the collapse. The inner core takes the form of a proto-neutron star (radius of about 10 km), halts abruptly, overshoots and springs back a little ("bounce") like a spherical piston into the still infalling outer core. The clash of incoming matter with outgoing matter, at the moment of the bounce, generates a supersonic shock wave at the edge of the inner core, headed towards the exterior of the star.

As the shock propagates, the energy is dissipated by the photo-dissociation of nuclei into protons and neutron. This means that the material behind the shock wave is mainly composed of free nucleons. At the same time neutrinos are produced by electron capture on the free protons left in the wake of the shock. These neutrinos pile up behind the shock, until the shock reaches the electron neutrino neutrinosphere (radius a few tens of km), corresponding to an area at lower densities, about $10^{11}$ g/cm$^3$, a few milliseconds after the bounce, and the electron neutrinos behind the shock are released in the subsequent few milliseconds. This is the so-called the neutronization burst (or shock breakout or prompt electron neutrino burst or deleptonization burst), when a total energy of about $3 \times 10^{51}$ erg, which is similar to the optical and kinetic energy release, is radiated in milliseconds. This brief emission of electron neutrinos amounts to about 1% of the total SN luminosity, that is the energy carried per unit of time by neutrinos and antineutrinos.

The hot core of the proto-neutron star produces neutrinos of all flavours by thermal processes:

- pair annihilation

$$e^+ + e^- \rightarrow \bar{\nu}_e + \nu_e \tag{7.28}$$

- electron-nucleon bremsstrahlung

$$e^\pm + n \rightarrow e^\pm + n + \bar{\nu} + \nu \tag{7.29}$$

- nucleon-nucleon bremsstrahlung

$$n + n \rightarrow n + n + \bar{\nu} + \nu \tag{7.30}$$

- plasmon[12] decay

$$\tilde{\gamma} \to \bar{\nu} + \nu \qquad (7.31)$$

- photo-annihilation

$$e^{\pm} + \gamma \to e^{\pm} + \bar{\nu} + \nu. \qquad (7.32)$$

Since the medium is composed of protons, neutrons, and electrons, and the neutrino energy does not allow creation of muons and taus, electron neutrinos and antineutrinos interact with nuclear matter via both charged- and neutral-current reactions, while the $\nu_\mu$, $\bar{\nu}_\mu$, $\nu_\tau$, $\bar{\nu}_\tau$, usually indicated collectively as $\nu_x$, experience only neutral current scattering. Since neutrino interactions depend on flavour and energy, there are different energy-dependent neutrinospheres for different flavour neutrinos. Neutrinos that interact more strongly have neutrinospheres at higher radii:[13] $R_{\nu_e} > R_{\bar{\nu}_e} > R_{\nu_x}$. Although the electron anti-neutrino emission dominates during the burst, the emission of $\nu_x$ begins to rise during the burst as thermal emission processes become important in the shock-heated matter. In other terms, after the burst, each neutrinosphere produces a thermal flux of the corresponding neutrino flavour.

The shock propagates through the dense matter of the outer core, which is still collapsing, abruptly decelerating it. Therefore, below the shock the dense matter falls much more slowly on the surface of the proto-neutron star, accreting it. We are in the so-called accretion stage. In the so-called delayed SN explosion scenario [243], the propagating shock, weakened by energy dissipation, stalls and lies at a radius of about 100–300 km, well outside of the neutrinospheres. While the shock is stalled, matter continues to accrete on the protoneutron star passing through the shock. The hot material behind the shock, composed mainly of free nucleons, electrons and photons, is heated by the accretion and by subsequent compression. Both electron neutrinos and antineutrinos are produced in large numbers by charged-current processes in the hot mantle of the protoneutron star. The mass of this mantle grows continuously, because it is fed by the accretion flow of the collapsing stellar matter that falls through the stalled shock. In the denser core region, the high densities and temperatures allow the production of neutrinos and antineutrinos of all flavours through thermal processes. Since the stalled shock is out of the neutrinosphere, these neutrinos can free-stream out of the star.

A supernova explosion can be achieved only if the shock is revived by some mechanism that is able to renew its energy. The mechanism which is currently thought to give the most important contribution is the energy deposition by the huge

---

[12] A plasmon is a quanta of electromagnetic field in a plasma which, in contrast to ordinary photons in vacuum, can be not only transverse, but longitudinal as well.

[13] Since the mantle of the proto-neutron star is neutron-rich, we expect a smaller cross section for $\bar{\nu}_e$ than $\nu_e$, and a smaller radius for the corresponding neutrinosphere.

## 7.3 Supernova Neutrinos

neutrino flux produced thermally in the proto-neutron star in the matter behind the shock. The efficiency of the neutrino energy deposition is helped by (essentially three-dimensional) convective effects. If the shock is revived, the SN explosion is produced, and the external layers of the star are expelled into space. The SN explosion marks the end of the accretion phase, about 0.5 s after the bounce. In this time span, the thermal emission of neutrinos of all flavours carries away around 10–20% of the available energy. After the SN explosion, the star loses energy by emitting neutrinos of all flavours and the cooling stage (about 10 s) starts, until a neutron star or a black hole is formed. In this phase, the remaining 80–90% of the SN energy is released, mostly in neutrinos and antineutrinos of all flavours. Later on, the shock wave sweeping the mantle of the star continues to accelerate and heat the star's external layers. Eventually (after a few hours), the shock breaks out, originating the explosion which is visible in light.

Summarising, during the stellar collapse that leads to the explosion of a SN, neutrino emission occurs in different stages, whose general features are qualitatively similar for most models:

- neutronization, in which a fusion reaction between electrons and protons produces neutrons and electron neutrinos. This process produces a peak in the $\nu_e$ luminosity curve as a function of time that lasts for about 25 ms. During neutronization, electron neutrinos are trapped behind the shock wave formed by the collapse and are only released when the matter density becomes sufficiently low;
- accretion, in which the matter from the collapsing star is attracted to the newly formed neutron star. Electron neutrinos and antineutrinos are produced in large numbers by charged-current processes in the mantle, while thermal neutrinos and antineutrinos of all flavours are produced in the core;
- cooling, which occurs after the SN explosion. The proto-neutron star releases all types of neutrinos and antineutrinos for about 10 s. The luminosities of all kinds of neutrinos and antineutrinos become similar (within about 10%) and decline with time in parallel.

The probability of having core-collapse SN events in our galaxy, the Milky Way, over the coming decade is unfortunately small, since these events are estimated to be a few per century—in contrast with the observable universe, where a core-collapse SN event is estimated to occur every second (see e.g. [244, 245]). In the unlucky situation where we do not observe a SN in our galaxy in the near future, the large number of core-collapse SNe in the universe may nevertheless provide us with the chance to detect the neutrinos emitted from the ensemble of SNe at cosmological distances. This cumulative flux of neutrinos and antineutrinos, emitted by all core-collapse supernovae in the causally-reachable universe, is called the diffuse supernova neutrino background (DSNB). It will appear isotropic and time-independent in feasible observations. Given the large number of SNe per year, we expect a nonzero rate of the DSNB, even when the large distance of the SN reduces the probability of detecting a single neutrino to practically zero.

### 7.3.3 Detection

When Supernova SN1987A was detected in the Large Magellanic Cloud, a satellite galaxy of our Milky Way at a distance from the Earth of about 50 kpc (165,000 lyr), it became possible for the first time to measure the neutrino emission from a nascent neutron star, turning SN1987A into the best studied of all SNe. The light of the SN1987A was first optically observed in the sky of the Southern Hemisphere; it was only after news of this supernova was disseminated that neutrino-sensitive experiments, in the Northern Hemisphere, thought to look at data from the detectors that were relative to a few hours earlier. Totally a few tenths of electron antineutrinos, with average energy around 10 MeV, out of the $\sim 10^{57}$ expected, were detected by three experiments:[14] Kamiokande-II [46, 248], IMB [47] and the Baksan Underground Scintillator Telescope (BUST) [48].

The BUST experiment is located in the Baksan Neutrino Observatory (BNO), an observatory in the Baksan Valley (North Caucasus, Russia) dominated by one of a series of peaks of Europe's highest mountain, Mount Elbrus. It has been detecting neutrinos since the late 1970s. Its detector consists of oil-based liquid scintillator and the total target mass is about 0.3 kt (kiloton), an order of magnitude smaller than the Kamiokande-II and IMB detectors.

Both Kamiokande-II and IMB utilised water Cherenkov detectors. The dominant signals came from the charged-current, IBD process

$$\bar{\nu}_e + p \to n + e^+. \tag{7.33}$$

The Kamiokande-II detector's energy threshold for $\bar{\nu}_e$ was about 8 MeV. In total 12 neutrino events were observed, and the signals lasted about 13 s. With a high energy threshold of 20 MeV, the IMB experiment totally recorded 8 neutrino events. The BUST experiments reported 5 events, that were originally believed to arrive 25 s later than the first IBM event; eventually the difference was reconciled in the uncertainty of time measurements.

Because of the poor statistics, the energy spectra of SN neutrinos was poorly determined, but other significative constraints were derived. One can set an inferior limit on the neutrino lifetime, expected longer than the propagation time on the distance Large Magellanic Cloud–Earth, assuming that the number of SN neutrinos or antineutrinos is not significantly reduced by neutrino decays. Neutrinos, with the average energy of 10–15 MeV, are ultra-relativistic. Moreover, since neutrinos

---

[14] A fourth detector, the Liquid Scintillation Detector (LSD), a scintillator detector with 90 tons of active mass, located in the Mont Blanc Underground Neutrino Observatory, in a tunnel underneath Mont Blanc in France, had also claimed neutrino observation [246]. However, its results are controversial, mostly because they do not fit in the framework of actual SNe theories, according to which a gravitational stellar collapse must occur in a very short time, of the order of a few seconds or even less (see for instance Ref. [247]). The LSD events were detected about 5 hours before the simultaneous observations by Kamiokande-II and IBM.

## 7.3 Supernova Neutrinos

are massive, their arrival times must be different from the arrival time of massless neutrinos, and limits can be set on their masses. Constraints on neutrino magnetic moments can be derived from analyses that take into account the fact that possible spin flips in the supernova converting left-handed neutrinos into right-handed ones, with no standard weak interactions, would cause their free escape, and consequently an energy loss and a reduction of the duration of neutrino signals. Other information on neutrino mixing was obtained from SN1987A data considering the effect of vacuum oscillations or MSW resonant transitions on the fluxes of different flavours.

Future experiments have been built and planned for high-statistics observations of neutrino bursts from future SNe. Current and future SN neutrino detectors are much larger than the detectors in operation during 1987. When the next galactic supernova occurs, detectors will be ready to seize the moment and record a number of events as large as $\sim 10^4$. Some of these experiments are based on water Cherenkov detectors. Because water is hydrogen-rich, these detectors are primarily sensitive to inverse beta decay interactions, $\bar{\nu}_e + p \rightarrow n + e^+$, revealing the Cherenkov light of the final-state positron. Elastic scattering on electrons $\stackrel{(-)}{\nu_x} + e^- \rightarrow \stackrel{(-)}{\nu_x} + e^-$ is another common interaction channel, which can be particularly useful for reconstructing the direction of neutrinos because the Cherenkov light cone indicates the direction of the scattered electron.

Water Cherenkov detectors come in two varieties. The first involves storing water in a spherical or cylindrical vessel and instrumenting the vessel walls with the PMTs. Examples of this type of detector include Super-Kamiokande and the future Hyper-Kamiokande. The second variety involves a natural body of water or ice as the interaction medium and instrumenting it with an array of long strings with optical modules attached. Examples include IceCube at the South Pole, KM3NeT in the Mediterranean Sea, and the Baikal Deep Underwater Neutrino Telescope (BDUNT) in the Baikal lake in Russia (see Sect. 7.7).

The inverse beta decay of SN neutrinos can also be observed in a scintillation detector. Scintillator detectors use organic hydrocarbons as an interaction medium, most commonly in liquid form. The passage of charged particles through a scintillating material excites its molecules which then isotropically emit photons. Liquid scintillators are rich in hydrocarbons, so, much like Cherenkov detectors, inverse beta decay is the dominant interaction mode for supernova neutrinos. Examples of this type of detector include BUST, Borexino, KamLAND, SNO+[15] and JUNO, that will be discussed in Sect. 8.1.3.

---

[15] As seen in Sect. 7.2.5, the SNO detector took data up to 2006. Then it was substituted by the SNO+ detector [249], that was initially emptied by its target heavy water and filled with light water. For the second operating phase, or scintillator phase, the inner detector volume was replaced with 780 tonnes of organic liquid scintillator. This increased the light yield by a factor of 50 with respect to water. In April 2021, SNO+ completed liquid scintillator filling. Finally, for the third operating phase, or tellurium phase, the scintillator will be loaded with 3.9 tonnes of natural tellurium.

The large statistics collected by the above detectors will not only improve our knowledge of neutrino properties, but will also test our understanding of SNe physics.

## 7.4 Atmospheric Neutrinos

In 1911 the Austrian physicists Victor Hess started a series of balloon flights, equipped with electroscopes, that measured the ionization produced by radiation. He found that the ionization rate at first decreased with altitude, but then started to increase up to a height of 5.3 km, the greatest height he reached. High in the atmosphere, Hess had discovered a natural source of high-energy particles: cosmic rays. For his discovery, he shared the 1936 Nobel Prize in Physics.[16]

Nowadays, with the term (primary) cosmic rays one generally indicates the flow of charged elementary particles and nuclei originated outside the solar system.[17] The energy spectrum of these particles extends from GeV up to more than $10^{11}$ GeV, although their flux decreases rapidly with the increasing energy. In the GeV (per nucleon) energy region, where the flux is largest (tens of particles per square metre per second), the cosmic-ray particles are mostly protons, about 5% of helium nuclei and a still smaller fraction of heavier nuclei.

The primary cosmic rays spectrum decreases with energy from the peak value in the GeV range following a nearly pure, single power law

$$\frac{dN}{dE} \propto E^{-\gamma} \qquad (7.34)$$

of spectral index $\gamma \sim 2.7$. Such approximate behaviour maintains strikingly stable over the whole range, which covers more than 10 orders of magnitude, with few deviations. The most important ones can be classified as:

- the steepening at $\sim 10^{15}$ eV, where the spectral index rises from 2.7 to 3.0—the "knee";
- the steepening at $\sim 10^{17}$ eV, where the spectral index rises to 3.3—the "second knee" or "dip";
- the flattening at $\sim 10^{18}$, where the spectral index falls back to 2.7—the "ankle";
- the apparent extinction of the cosmic-ray flux, starting at an energy of $\sim 10^{19}$ eV—the "suppression";

---

[16] The other 1936 Nobel laureate was Carl David Anderson "for his discovery of the positron".

[17] Sometimes in literature the term cosmic rays includes also particles associated with solar flares. The term cosmic rays was coined in the 1020s by Robert Millikan, who was convinced of their electromagnetic nature, and it stuck, despite the fact that over time it became clear that they were in fact charged particles, as the intensity of the radiation correlated with the strength of the local Earth magnetic field.

## 7.4 Atmospheric Neutrinos

- a more recently revealed feature between the "ankle" and the "suppression", where the spectral index rises to 3.0, at an energy of $\sim 10^{19}$ eV—the "instep" [250].

These spectral features have been commonly named after the features of the human leg, because of shape similarity. Their detailed understanding remains an open question. It is generally assumed that "knee" is the maximum energy that protons can be accelerated to by the dominant class of galactic cosmic-ray sources, whereas the "ankle" marks the transition from galactic to extragalactic cosmic rays. Although the sources of both galactic and extragalactic cosmic rays, together with the question on how they are accelerated, remain topics of debate, supernova remnants (SNRs)[18] are commonly considered to be the dominant source of galactic cosmic rays, and Active Galactic Nuclei (AGN)[19] of extragalactic ones.

### 7.4.1 Production

The (primary) cosmic ray particles, once entered into the Earth's atmosphere, interact with the nuclei in the atmosphere and produce a shower of (secondary) particles.[20] Typically, in these interactions, many pions, and, less abundantly, kaons are produced. Neutral pions decay into two photons, feeding the electromagnetic part of the shower. The charged pions practically only decay into muons and muon (anti-)neutrinos. If the muons are not sufficiently energetic to reach ground level before decaying, also positrons (electrons) and electron (anti-)neutrinos can be produced. The decay chains are (see Fig. 7.4):

$$\pi^+ \to \mu^+ + \nu_\mu \to e^+ + \nu_e + \bar{\nu}_\mu + \nu_\mu$$
$$\pi^- \to \mu^- + \bar{\nu}_\mu \to e^- + \bar{\nu}_e + \nu_\mu + \bar{\nu}_\mu. \tag{7.35}$$

---

[18] Astronomers have long assumed that supernovae are the sources of the cosmic rays. With the discovery that supernova remnants were emitting radio synchrotron radiation, which requires the presence of relativistic electrons, it was suggested that relativistic protons and other accelerated atomic nuclei are also present. These results pointed toward supernova remnants, rather than supernovae, as locations of particle acceleration, albeit with the energy having been provided by the supernova explosion. Other possible, although not considered dominant, sources are a pulsar wind nebula, that is the nebula surrounding the pulsar (a fast rotating neutron stars), and binary systems.

[19] AGN are nuclei of active Galaxies with a central supermassive black hole, emitting non-thermal radiation at multiple wavelengths by accreting matter from the surrounding environment. The production/acceleration of cosmic rays could be achieved also in other sources as gamma-ray bursts (a sudden release of $\gamma$-rays), magnetars (a type of neutron star with an extremely powerful magnetic field), and so on.

[20] If one does not restrict to cosmic ray interactions in the atmosphere, as we do here, but considers the astrophysical origin of cosmic rays, one can denote as primary cosmic rays the particles accelerated at astrophysical sources and as secondary cosmic rays the particles produced in interaction of the primaries with interstellar gas. If it is not clear from the context, one can use the more specific adjectives galactic secondary or atmospheric secondary.

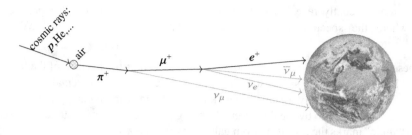

**Fig. 7.4** A typical neutrino production process induced by cosmic ray interactions in the Earth's atmosphere

The typical height of the neutrino production is 15 km above the ground.

Neutrinos created by cosmic-ray interactions in the Earth's atmosphere are called atmospheric neutrinos. Relations as Eq. (7.35) indicate that the fluxes of muon, electron neutrinos and antineutrinos are strictly related to each other. For instance, let us consider only pion decays and assume that all muons decay in flight before reaching the Earth's surface. Then the probability of finding muon neutrinos and antineutrinos is the same, and the $\pi^+/\pi^-$ production rate is the same than the $\mu^+/\mu^-$ flux ratio and the $\nu_e/\bar{\nu}_e$ flux ratio. Moreover, by counting the number of neutrinos and antineutrinos produced for electron and muon flavours, one estimate a 1:2 production ratio.

Atmospheric neutrinos originating from the decays of charged pions and kaons are indicated as "conventional". On the other hand, above TeV energies, also heavier hadrons, mostly containing a quark charm ($D$ mesons or charmed baryons), can be produced in the interaction of cosmic rays with the atmosphere. The charmed hadrons immediately decay, due to their extremely short lifetimes, on the order of $10^{-12}$ s. Neutrinos from semileptonic decays of these charmed hadrons are called "prompt" neutrinos. Because in general the production of an heavy particle is disfavoured by the phase space, we expect a smaller production cross-section for prompt neutrinos.

As their decay lengths increase with energy,[21] pions and kaons are apt to loose energy through interactions with other particles in the atmosphere before they decay. The resulting conventional neutrino flux rapidly decreases with energy, with a steeper energy spectrum ($\sim E^{-3.7}$), while the flux of prompt neutrinos, which derives from speedy decays, has a spectral shape following approximately the shape of the cosmic rays ($\sim E^{-2.7}$), and becomes dominant in the high-energy regions, above a few tens TeV. There are a few good reasons to distinguish these two components of atmospheric neutrino fluxes: their spectrum, angular distribution, and flavour composition are different. For instance, a small component of tau neutrinos is expected in prompt decays, through $D_s \rightarrow \tau \nu_\tau$ decay. Although experimental

---

[21] Increasing energy will increase the kinetic energy, and therefore velocity, and the Lorentz factor for highly relativistic particles.

## 7.4 Atmospheric Neutrinos

limits constraints have been established [251], as of today a definitive measurement of the prompt atmospheric neutrino flux has yet to be made.

The calculation of atmospheric neutrino flux requires the estimate of the primary cosmic-ray flux at the top of the atmosphere, the knowledge of the mechanism of propagations and interactions of cosmic rays in the air to produce mesons, and the knowledge of how mesons and muons decays while propagating through the air and Earth. The detailed computation of the prompt neutrino fluxes, in particular, is not straightforward, since it requires the knowledge of the production cross sections for heavy quarks, which create the heavy hadrons through fragmentation. The production of charm quarks, for instance, takes place mostly via gluon fusion, that is via $g + g \to c\bar{c}$, where $g$ is a gluon and $c$ and $\bar{c}$ are the charm quark and its antiquark, respectively.

The cosmic ray fluxes in interstellar space are to a very good approximation constant in time and isotropic in direction, as cosmic rays come from outside the solar system. However, at the top of the atmosphere, the cosmic ray flux differs from that of interstellar space because of solar wind[22] and geomagnetic fields.

The solar wind deflects the primary cosmic rays away from Earth and decelerates them, causing the so-called modulation of the cosmic ray flux. This effect is correlated with solar activity, in particular with the eleven-year solar cycle, and it is therefore largely time-dependent. It is most prominent for energies below 10 GeV.

One effect of geomagnetic fields is to bend the trajectories of charged particles in primary cosmic rays. The trajectory of a charged particle depends on its momentum and charge, whose ratio is called the magnetic rigidity. Particles with different momenta and charge, but identical rigidity, show the same dynamics in a magnetic field. In a uniform magnetic field $B$, the radius of the circular motion of a charged particle, the so-called Larmor radius (or gyroradius or cyclotron radius), is directly proportional to the rigidity and inversely proportional to $B$. Thus a small rigidity implies a small Larmor radius, that is a larger bending of the particle trajectory, which signifies that low energy cosmic rays are affected more strongly by geomagnetic effects. The deflection induced by the the geomagnetic fields potentially prevents primary low-energy cosmic rays from reaching the Earth, which would reflect into a suppression of the low-energy atmospheric neutrino flux. This is a threshold effect process known as (rigidity) cutoff, and depends on the latitude of the detector. The cutoff is generally lower near the poles, and therefore the low energy flux is higher for detectors located near the poles, than those near the equator. Indeed the charged particles approaching the earth near the poles travel almost along the direction of the magnetic lines of force, but the charged particles that approach at the equator have to travel in a perpendicular direction to the field and are deflected away.

---

[22] The solar wind is the stream of charged particles released from the upper atmosphere of the Sun.

## 7.4.2 Detection

Atmospheric neutrinos were first detected in 1965. Two experiments that were carried out in extremely deep mines in India [38] and South Africa [39] successfully observed muons produced by atmospheric muon neutrino interactions. The rock overburden allowed to shield the experiments from the background muons produced by cosmic rays travelling in the atmosphere.

In the early 1980s several proton decay experiments started, motivated by proton decay predictions of Grand Unified Theories (GUT), which were proposed in the 1970s. These experiments did not observe any convincing signal of proton decays. However, these experiments observed hundreds of atmospheric neutrino interactions in the detectors. Atmospheric neutrino interactions are a serious background for proton decay searches, since a neutrino arrives at the detector without showing any evidence for incidence and may interact with a nucleon, producing visible secondary particles. Studying these neutrino interactions became necessary to separate the proton decay signal and atmospheric neutrino background.

Starting from the late 1980s, the Kamiokande [252] and the IMB experiments [253, 254], both built with the purpose of searching for proton decays, became to observe a clear deficit (about 1/2) in atmospheric muon neutrinos. This deficit was called the "atmospheric neutrino anomaly". Both experiments were based on water Cherenkov technology (see Sect. 7.2.2), which allowed the determination of energy, incoming direction and also the electron or muon type of the incoming neutrinos. This muon neutrino deficit was so large that it did not seem possible to ascribe it to statistical effects. Several scenarios were considered, included some unknown systematics. The possibility of neutrino oscillations was ventured, but the statistical evidence was insufficient to support this conclusion.

A determining experiment in the study of atmospheric neutrinos and their oscillations was the Super-Kamiokande experiment, already described in Sect. 7.2. It was able not only to improve the statistics, but also study the complete angular distribution of the upward-going and downward-going neutrinos entering the detector, for both muon and electron types. Oscillations of atmospheric neutrinos can be revealed in a model-independent way by measuring the fluxes of upward-going and downward-going high-energy atmospheric neutrinos. If there were no neutrino oscillations, these neutrino fluxes should be equal to each other. Since the downward-going neutrinos are produced in the atmosphere above the detector, they travel a distance of some tens of km, whereas upward-going neutrinos, coming from the other side of the globe, which has a diameter of about 13 thousands of km, cover much larger distances. In the presence of oscillations, upward-going neutrinos have time enough to oscillate into other flavours. That causes a suppression of the upward-going flux with respect to the downward going one. In 1988 the Super-Kamiokande experiment showed a clear evidence of a nonzero flux

## 7.4 Atmospheric Neutrinos

asymmetry, particularly significant in the case of muon neutrinos.[23] [255] With the confirmed disappearance of muon neutrinos, attention turned to the possible oscillation channel. Data favoured the $\nu_\mu \to \nu_\tau$ oscillation. In the two-flavour neutrino oscillation framework, the probability that a neutrino of energy E with a flavour state $\nu_\mu$ will later be observed in the $\nu_\tau$ flavour eigenstate after traveling a distance L in vacuum is:

$$P_{\nu_\mu \to \nu_\tau} = \sin^2 2\theta \sin^2 \frac{1.27 \Delta m^2 L}{E} = \sin^2 2\theta \sin^2 \frac{\pi L}{\lambda} \qquad (7.36)$$

where $\theta$ is the mixing angle between the mass eigenstates and the flavour eigenstates, $\Delta m^2$ is the difference of the squares of the masses of the mass eigenstates and the oscillation length is $\lambda = 2.47 E / \Delta m^2$ (see Eq. (6.42)). In 2018, SK has shown the change of muon neutrinos to tau neutrinos during their propagation with significance level of 4.6 sigma [256]. Strong argument against $\nu_\mu \to \nu_e$ came also from the non-observation of any effect in CHOOZ [257] and Palo Verde [258] long-baseline reactor experiments. In Eq. (7.36), the neutrino oscillation length can be as long as $10^4$ km. By setting as the typical neutrino energy the value that maximises its flux ($\sim 1$ GeV), one finds that atmospheric neutrino experiments are sensitive to values of $\Delta m^2$ larger than $10^{-4}$ eV$^2$ (see Sect. 6.5).

The SK result was a game changer. Neutrino oscillation physics was real and the community started to take it seriously,[24] particularly with the approvals of dedicated experiments aimed at exploring the sector relevant to atmospheric oscillations by means of neutrino beams produced in laboratories. They were built in Japan (K2K and later T2K), Europe (OPERA and ICARUS) and USA (NOνA). We will discuss them in Chap. 8. These experiments have thoroughly confirmed SK results and contributed to the precise determination of oscillation parameters. In 2015, the SK leader Takaaki Kajita was awarded one half of the Nobel Prize in Physics for "the discovery of neutrino oscillations, which shows that neutrinos have mass".[25]

The study of atmospheric neutrinos can be used to investigate the Earth's structure [259]. Since the neutrino-nucleon cross section increases with energy [260], at very high (TeV-PeV) energies the Earth is not a fully transparent medium for neutrinos and there is a non-negligible probability for atmospheric neutrinos to be absorbed. The attenuation of the neutrino flux, as measured by the signals in large Cherenkov detectors, provides information about the nucleon matter density of the Earth [261] (neutrino absorption tomography).

On the other hand, at lower (MeV to GeV) energy, atmospheric neutrinos crossing the Earth undergo flavour oscillations which are modified due to coherent

---

[23] As can be seen in Eq. (6.80), for a small angle $\theta_{13}$ the probability of the transition $\nu_\mu \to \nu_e$ is suppressed with respect to $\nu_\mu \to \nu_\tau$.

[24] Until the 1998 SK discovery and actually even a few years later, most of the scientific community was skeptical about the interpretation of (atmospheric and solar) deficits as neutrino oscillations—the SNO results finally convinced 'all'.

[25] The other half went to Arthur McDonald, leader of the SNO Collaboration.

forward scattering on electrons (see Sect. 6.6). The signature of these matter effects in the angular, energy and flavour distributions of neutrinos detected at the surface may therefore provide sensitivity to the electron density $N_e$ in the different layers of matter traversed. The ratio of electron density $N_e$ to mass density scales with the average ratio of the atomic number to the mass number $(Z/A)$, which depends on the chemical and isotopic composition of the medium itself. Thus the matter effect on neutrino oscillations may lead to constraints on the compositional models of the Earth's interior [262] (neutrino oscillation tomography).

## 7.5 Cosmic Neutrinos

So far we have discussed neutrinos generated by nuclear reactions inside the Sun, supernova neutrinos and neutrinos produced by cosmic rays interacting with nuclei in the Earth's atmosphere. But there are other neutrinos, generically coming from further afield in our Universe. Together with Supernova neutrinos, they are indicated as cosmic or astrophysical neutrinos. Astrophysical neutrinos are particularly prized as information carriers because their tiny interaction rate allows them to pass through clouds of gas and dust that would otherwise hide distant astrophysical objects. Since there are a large number of experiments where in principle we can learn on neutrino properties, we can afford to use cosmic neutrinos to acquire informations on their astrophysical sources, and talk therefore of neutrino astronomy.

Astrophysical neutrinos can be subdivided into categories according to their production mechanisms, energy and detection techniques. In Sect. 7.5.1 we discuss neutrinos from the cosmic neutrino background (CNB). They are neutrinos that were abundantly produced in the early hot stages of the Universe, and nowadays exist as a remnant background, analogous to the cosmic microwave background. These neutrinos, also called relic neutrinos, have the lowest possible energy in the neutrino energy spectrum, and have not been directly detected yet. In Sect. 7.6, we discuss instead neutrinos in the highest part of the energy spectrum and in Sect. 7.7 the experiments particularly designed to detect them. Among them stands a one of a kind experiment, the IceCube experiment located at the South Pole. Since 2013, it has reported detections of more than 100 cosmic neutrinos in the energy range $10^{14}$–$10^{16}$ eV, an achievement that has marked the beginning of galactic and extra-galactic neutrino astronomy.

### 7.5.1 Cosmic Neutrino Background

The formulation of the Big Bang model began in the 1940s [263–266]. The classical hot Big Bang cosmology is based on general relativity and it provides a description

## 7.5 Cosmic Neutrinos

of the evolution of the Universe from about the Planck time[26] (approximately $\sim 10^{-43}$ s after the Big Bang), up to now, approximately 13 billions of years later. It assumes that our Universe evolved from an isotropic initial state which was very hot (temperature $T \sim 10^{19}$ GeV[27] at the Planck time) and dense, enough so as to allow for the nucleosynthetic processing of hydrogen. Its subsequent expansion went through successive stages of lower energy density, which correspond to lower temperature.

The fundamental set of observations that support the Big Bang model are: (1) the isotropic Hubble expansion of the Universe; (2) the existence of the cosmic microwave background (CMB) radiation; (3) the abundance pattern of the lightest elements (hydrogen, deuterium, helium, and lithium) seen in the most primitive samples of the cosmos.

Early in the twentieth century the Universe was thought to be static: Hubble's law provided key evidence for the expansion of the Universe, thus supporting the Big Bang model.

The CMB is a relic background radiation with a temperature of order a few K, observed for the first time 16 years after its prediction in the Big Bang model [267]. At the very beginning, radiation materialised into pairs of particles and antiparticles, while, in subsequent cooler stages, nuclei could form and survive. In this context, the CMB is the electromagnetic radiation left over when electrons and protons combine to form neutral hydrogen (the so called recombination phase). Following this process, the absence of charged particles around them allows photons to be basically free of interactions, decouple from the hydrogen gas and henceforth propagate freely in the expanding Universe (photon decoupling). It was the observation of this radiation that singled out the Big Bang model as the prime candidate to describe our Universe, confirmed by subsequent work on Big Bang Nucleosynthesis (BBN).

BBN (also known as primordial nucleosynthesis) predicts the abundances of the light elements in the Universe. Before the 1960s, it was generally assumed that all elements in our Universe were produced either in star interiors or during supernova explosions, but quantitative estimates, especially of the helium abundance in our Universe, were in contrast with experimental observations. This challenge has now been met successfully for the light nuclei, that, according to BBN, are originated in the first stages of our Universe after the Big Bang. After the Big Bang, when the Universe expanded and hence cooled, nuclei could form (and survive)

---

[26] Before the Planck time (the Planck Era) we believe that the basic forces of nature, included gravity, were unified. Since we lack a theory explaining the unification between quantum forces and gravity, only highly speculative statements are possible as descriptions of the Planck era. Gravity became distinct after the Planck time.

[27] Given the Planck mass $m_P = (hc/G)^{1/2} \sim 10^{-8}$ kg, where $G$ is the gravitational constant, and the Planck time $t_P = (Gh/c^5)^{1/2} \sim 10^{-43}$ s, one can define the Planck energy $E_P = m_P c^2 \sim 10^{-44}$ J and the Planck temperature $T = E_P/k \sim 10^{32}$ K ($k$ is the Boltzmann constant). In natural units $c = k = 1$, hence temperature and energy have the same dimensions and measure units.

from the protons and neutrons available.[28] One has to remark that the Big Bang nucleosynthesis stop at lithium; the production of heavier elements takes place in the stars. By merging two alpha particles ($^4$He nuclei), it is possible to form the $^8$Be isotope, which has a half-life of about $7 \times 10^{-17}$ s and decays back into two alpha particles. Extreme temperatures and pressures, available in stellar cores, allow to create the $^8$Be isotope faster than it decays. The process where the $^8$Be isotope, originated by two alpha particles, interacts with another alpha particle to form $^{12}$C is the so called triple-alpha process. Once enough carbon is generated in the stars, a chain of reactions, called the alpha ladder, produces heavier isotopes.

The Big Bang model predicts, besides CMB, a relic neutrino background pervading the Universe. It has been named Cosmic Neutrino Background (C$\nu$B or CNB). Within this model, neutrinos were originally kept in thermal equilibrium with the primordial plasma of protons, neutrons and electrons by the weak interactions. The reactions with nucleons were negligible, because of the smaller number density of the non-relativistic nucleons compared to the density of relativistic electrons and positrons. Hence interactions occurred mainly with electrons and positrons, namely $e^+e^- \leftrightarrow \bar{\nu}\nu$, $e^-\nu \leftrightarrow e^-\nu$, $e^+\bar{\nu} \leftrightarrow e^+\bar{\nu}$. The interaction rate $\Gamma$ for each neutrino can be written as

$$\Gamma = n <\sigma v>, \tag{7.37}$$

where the angle brackets denote thermal averaging, $v \sim 1$ is the neutrino velocity, $n$ is the number density of the relativistic electrons and positrons, and $\sigma$ is the cross-section for neutrino interactions. The number density $n$ depends on the temperature $T$ as

$$n \sim T^3. \tag{7.38}$$

Such behaviour is expected on the grounds of dimensional analysis, since in the ultrarelativistic limit the only dimensional scale is the temperature $T$. We could argue, based on the dimensional reasonings, that the cross-section for neutrino interactions goes like $\sigma \sim T^{-2}$, but after the electroweak symmetry breaking the new scale, the Fermi's constant $G_F \sim 10^{-5}$ GeV, actually determines the weak interaction strength. Hence we have

$$\sigma \sim G_F^2 T^2. \tag{7.39}$$

It follows that

$$\Gamma \sim G_F^2 T^5. \tag{7.40}$$

---

[28] Nuclear reactions took place approximately during the time interval $t \in [0.01, 100]$ s. The BBN predicts the abundances of most of the universe's light elements, as the isotopes helium-4 and helium-3 ($^4$He, $^3$He), the lithium isotope lithium-7 ($^7$Li), the hydrogen isotope deuterium ($^2$H or D).

## 7.5 Cosmic Neutrinos

The key to understanding the decoupling of particles is understanding the competition between the interaction rate of particles and the expansion rate of the Universe. Particles maintain equilibrium as long as the former is larger than the latter and decouple (freeze out) when it is less or equal. The expansion rate of the Universe, or Hubble expansion rate, is given by the Hubble parameter $H$ which in the early radiation-dominated epoch can be estimated to be

$$H \sim \frac{T^2}{m_P} \tag{7.41}$$

where $m_P$ is the Planck mass $m_P \sim 1.22 \times 10^{19}$ GeV. We can see that the neutrino interaction rate decreases more quickly with the decrease of temperature due to the expansion of the Universe than the rate of expansion of the Universe itself.

When the neutrino interaction rate drops below the Hubble expansion rate, neutrinos decouple. That occurs when $\Gamma \sim H$, that is when

$$T \sim (G_F^2 m_P)^{-\frac{1}{3}} \sim 1 \text{ MeV} \tag{7.42}$$

This temperature is commonly referred to as the neutrino decoupling or freeze-out temperature. It corresponds to an age of about a second, approximately, that is much earlier than the epoch which characterises the CMB relic photons. At lower temperatures, primordial neutrinos stop interacting with the plasma, and just circulate freely—the Universe becomes transparent to them.

Neutrino decoupling leaves behind a C$\nu$B, analogous to the CMB radiation of visible photons. Its present temperature is expected around 1.9 K (see for instance Ref. [127]), which corresponds approximately to energies in the $\sim 100\,\mu$eV range. While the C$\nu$B has not yet been detected directly, it has been indirectly confirmed by the accurate agreement of predictions and observations of the primordial abundance of light elements, the power spectrum of CMB anisotropies and the large scale clustering of cosmological structures. The direct detection of relic neutrinos by experiments on Earth, however, is very challenging. This is mainly because their interaction cross-section is tiny as a result of the very low neutrino energies. Existing neutrino experiments have detection thresholds that are many orders of magnitude above the predicted C$\nu$B energy. Thus any experiment wishing to observe directly relic neutrinos therefore requires suitable new techniques of neutrino detection. Several direct detection experiments have been proposed, for instance the Ptolemy experiment (PonTecorvo Observatory for Light, Early-universe, Massive-neutrino Yield[29]) [268]. It is based on the concept of neutrino capture on $\beta$-decay nuclei, sometimes indicated as enhanced or stimulated beta decay emission. Let us consider a nucleus $N_i$ with atomic weight and number $A$ and $Z$, naturally undergoing

---

[29] Note that originally PT stayed for Princeton Tritium, since the experiment was originally planned at Princeton.

β-decay to the daughter nucleus $N_f$. Schematically, we have (see Eq. (1.25))

$$(A, Z) \to (A, Z+1) + e^- + \bar{\nu}_e. \tag{7.43}$$

By crossing channel, we have neutrino capture process, namely

$$\nu_e + (A, Z) \to (A, Z+1) + e^-. \tag{7.44}$$

This reaction has the same observable final states as its beta decay counterpart, whose energy balance, namely the difference of the nucleus masses $m_i$ and $m_f$, is always positive. Thus it has no energy threshold on the value of the incoming neutrino energy $E_\nu$ and is always energetically allowed. It can occur with neutrinos at arbitrary low energies. In the limit of zero neutrino mass $m_\nu$ and energy, the electron in the final state has the same energy of the beta decay endpoint energy $m_i - m_f$. Releasing this assumption, they become different by a value of at least 2 $m_\nu$. Indeed, in the process (7.44) the electron kinetic energy is $m_i - m_f + E_\nu$, which has $m_i - m_f + m_\nu$ as minimum value. The electron in β-decay (7.43) has at most an energy $m_i - m_f - m_\nu$, neglecting nucleus recoil energy. Because the process (7.44) has a two body final state, the neutrino capture would be signalled by a peak in the electron spectrum, shifted of about $2m_\nu$ above the β-decay endpoint. The proposal of using the process (7.44) to measure the cosmological relic neutrino background relies on its threshold-less nature and its characteristic signature [269, 270]. The proposed cosmic neutrino detector Ptolemy [268] uses tritium atoms as targets due to tritium sizeable half-life time ($\tau_{1/2} \sim 12$ yr), availability, low amount of energy released during the β-decay ($Q = m_{3H} - m_{3He} \sim 18.6$ keV) and high neutrino capture cross-section ($\sim 10^{-44}$ cm$^2$). The process of neutrino capture on tritium is given by

$$\nu_e + {}^3H \to {}^3He^+ + e^-. \tag{7.45}$$

The signal to background ratio depends crucially on the energy resolution at the β-decay endpoint.

## 7.6 Neutrinos at High Energy

In 2013 the IceCube experiment[30] reported the first detection of neutrinos with energy ranging between 10 TeV and 100 PeV [66]. The twenty-eight events included the highest energy neutrinos ever observed. Since 2013, IceCube has detected more events and accumulated more statistics on high-energy (HE, TeV to 100 PeV) neutrinos. They have flavours, directions, and energies inconsistent with

---

[30] The IceCube experiment and similar ones are detailed in Sect. 7.7.

## 7.6 Neutrinos at High Energy

those expected from the atmospheric muon and neutrino backgrounds. In particular, their energy spectrum departs significantly from that of the atmospheric neutrinos. Atmospheric neutrinos have a well-measured energy distribution that follows a steep power law, while the high-energy neutrino flux discovered by IceCube displays a much flatter power law. It diverges significantly from the atmospheric neutrino background at energies beyond ~100 TeV. Moreover, the apparent isotropy of the neutrino arrival directions favours an extragalactic origin for the neutrino flux, potentially created by a large population of distant sources.[31] In 2018, HE neutrinos were detected by another experiment, ANTARES (Astronomy with a Neutrino Telescope and Abyss environmental Research) [272]. The observation of HE neutrinos prompts the need to address questions regarding the identification of astrophysical sources and to delve deeper into understanding neutrino interactions at the highest energies.

Neutrinos with energies in the GeV range and higher cannot be generated through transitions within a nucleus or through thermal processes inside stable, collapsing, or exploding stars, which typically produce neutrinos with maximum energies in the range of a few tenths of MeV. The only uncontroversial way to make them is via highly relativistic charged particles colliding with either target particles or photons. An ample supply of such relativistic particles is provided by cosmic rays. Cosmic rays can generate neutrinos by their interactions in their source, its surroundings, or along the cosmic-ray path to Earth. It is generally assumed that the observed HE neutrinos derive for the most part from cosmic ray collisions with the gas or with the radiation present in the environment of the sources where the cosmic rays get accelerated. A very high energy proton in cosmic rays collides with a photon ($p\gamma$) or another proton ($pp$) and typically produce a nucleon and a pion, or a pion plus other hadronic states (indicated with $X$), namely

$$p + \gamma \rightarrow p + \pi^0$$
$$p + \gamma \rightarrow n + \pi^+$$
$$p + p \rightarrow \pi + X. \qquad (7.46)$$

Pion decays produce both neutrinos and gamma rays, through the processes

$$\pi^0 \rightarrow \gamma\gamma$$
$$\pi^\pm \rightarrow \mu^\pm + \nu_\mu. \qquad (7.47)$$

The generated muons undergo decay processes too:

$$\mu^- \rightarrow e^- \bar{\nu}_e \nu_\mu$$
$$\mu^+ \rightarrow e^+ \nu_e \bar{\nu}_\mu. \qquad (7.48)$$

---

[31] Details on the high-energy neutrino fluxes can be found, for example, in Ref. [271].

As seen in Eq. (7.47), the neutrino sources are expected to also emit gamma rays, establishing a significant link between HE neutrino and gamma-ray astronomy.

These production mechanisms mirror those responsible for generating atmospheric neutrinos (see Sect. 7.4.1). Atmospheric neutrinos are mostly products of the decay chain of charged mesons. As these mesons traverse the dense layers of Earth's atmosphere, they engage in collisions with air molecules, losing energy and momentum before they have the chance to decay. Consequently, the fluxes of atmospheric neutrinos at high energies experience significant attenuation. In contrast, if astrophysical neutrinos originate from sources with sparse gas or radiation targets, allowing mesons to decay before encountering further interactions, the suppression of the astrophysical neutrino flux at high energies might be less severe than that observed in atmospheric neutrinos. This diminished suppression could facilitate the observation of astrophysical neutrino fluxes above the atmospheric background, typically at energies exceeding a few tens of TeV.

To date, there are only two plausible evidences for neutrino point sources, the blazar[32] known as TXS 0506+056 [67] and the AGN of the NGC 1068 galaxy[33] [273]. These observations have put us at the doorstep of extragalactic neutrino astronomy.

The procedure that led to the identification of TXS 0506+056 is an interesting example of multi-messenger astronomy, a relatively new field which aims at the study of astronomical sources using different types of *messenger* particles: photons, neutrinos, cosmic rays and gravitational waves. On 22 September 2017, the IceCube collaboration detected a track event induced by a $\sim$300 TeV muon neutrino. The detection generated an automatic alert that caused related searches from the direction of the event by many experiments. The ANTARES experiment did not find neutrino candidates in a $\pm 1$ day period around the event time. However, on September 28, the Fermi Large Area Telescope (Fermi-LAT) Collaboration reported that the arrival direction of the neutrino was aligned with the coordinates of a known gamma-ray source, the active galaxy TXS 0506+056 (an object classified as a blazar), that was in a particularly active state, or, said in another way, was "flaring" with a gamma-ray flux that had increased in recent months. Fermi-LAT is a telescope aboard the Fermi Gamma-ray Space Telescope, sensitive to $\gamma$-rays with energies from 20 MeV to greater than 300 GeV. Since August 2008, it has operated continuously, primarily in an all-sky survey mode. Prompted by the Fermi-LAT detection, the Major Atmospheric Gamma Imaging Cherenkov (MAGIC) Telescopes performed additional observations starting 28 September and observed a significant photon flux of energies up to 400 GeV from the direction of the blazar. MAGIC consists of two 17 m telescopes, located at the Roque de los Muchachos

---

[32] Let us remind here that a blazar is an AGN characterised by a relativistic jet (a jet composed of ionized matter traveling at nearly the speed of light) pointing almost directly towards the Earth.

[33] NGC 1068, also known as Messier 77 (M77), from the discoverer, the French astronomer Charles Messier, is a barred spiral galaxy located in the constellation Cetus. It is one of the brightest and closest examples of a Seyfert galaxy. Its central supermassive black hole is hidden by a dense torus of gas and dust.

## 7.6 Neutrinos at High Energy

Observatory on the Canary Island of La Palma (Spain). Several other experiments completed the previous studies with observations of compatible emissions in the radio, optical and X-ray range. Given where to look, IceCube searched its archival neutrino data up to and including October 2017 for evidence of neutrino emission at the location of TXS0506+056. Evidence was found for a neutrino emission in excess to the background in a burst lasting 110 days, between September 2014 and March 2015, independent of and before the 2017 flaring. This burst dominated the integrated flux from the source over the last 9.5 years for which there were data, leaving the 2017 flare as a second subdominant feature [274].

While the blazar TXS 0506+056 lies at a distance of nearly five billion light-years, the NGC 1068 galaxy is significantly closer, positioned approximately 46 million light-years away from us. In November 2022, the IceCube collaboration presented the findings of a ten-year investigation into neutrinos originating from astrophysical $\gamma$-ray sources. Their results revealed an excess in the neutrino flux compared to the atmospheric background in the vicinity of NGC 1068 [273].

Efforts have also been made to search for ultra-high-energy (UHE, $>100$ PeV) neutrinos, with no results so far. The energy spectrum of cosmic rays has been measured up to energies of $10^{20}$ eV [275]. Charged particles are accelerated to extreme energies in powerful astrophysical objects, the identification of which remains an ongoing pursuit. As a result of their interactions during the journey to Earth, UHE neutrinos, indicated as cosmogenic neutrinos, can be produced.

Cosmic ray interactions with photons in the cosmic microwave background are an anticipated source of cosmogenic neutrinos. These neutrinos mainly arise as decay products of pions created through these interactions. The flux of extragalactic protons in cosmic rays can be attenuated by their interactions with the background CMB photons $\gamma_b$ via the channels

$$p + \gamma_b \to \pi^0 + p$$
$$p + \gamma_b \to \pi^+ + n \qquad (7.49)$$

above a threshold of about 400 EeV ($10^{18}$ eV = 1 EeV), the Greisen-Zatsepin-Kuzmin (GZK) cut-off [276, 277]. Also heavier nuclei are attenuated by photo-disintegration of nuclei by CMB photons at similar energies. Cosmogenic neutrinos, or GZK neutrinos, with energies predominantly in the EeV range, originate from the secondary pions, as described in Eqs. (7.47) and (7.48). The $\beta$-decay of nucleons and nuclei from photo-disintegration can also lead to neutrino production. However, while neutrinos produced from pion decay have energies that are a few percent of the parent pions, those produced from $\beta$-decay carry less than one part per thousand of the parent nucleon's energy.

The detection of cosmogenic neutrinos is yet to achieve, and, dependently on the model estimates of their flux, not necessarily in the reach of present and/or future neutrino telescopes.

## 7.7 Cubic-Kilometer Detectors

As we know, neutrinos interactions with matter are extremely feeble. This characteristics makes cosmic neutrinos the ideal astronomical messenger, since they may reach us unscathed from cosmic distances. It also make them very difficult to detect. Huge volumes are required to collect cosmic neutrinos in statistically significant numbers, and, by the 1970s, it was clear that huge meant a size of detectors of at least 1 km$^3$.

Early efforts concentrated on transforming large volumes of natural water into Cherenkov detectors. Water provides not only the target for neutrinos, which create charged particles that in turn generate Cherenkov light, it also shields the downward-moving muons from cosmic-ray interactions in the atmosphere. The first attempt was the project of a Deep Underwater Muon and Neutrino Detector [278] (DUMAND) to be built in the sea off the main island of Hawaii. It started informally in 1973 and was funded as a feasibility study in 1979. DUMAND was cancelled by the US Department of Energy in 1995, prior to starting full deployment, but left an incredibly rich legacy of ideas and technical principles.

By the end of 1979, international politics and in particular the Soviet–Afghan war forced the separation of the US and Russian scientists in DUMAND. The latter decided to push ahead with the smaller Baikal Neutrino Telescope project, that started operating in Lake Baikal (Russia) in 1996 [279, 280]. Lake Baikal is the deepest freshwater lake on Earth, with its largest depth at nearly 1700 m, and it is famous for its clean and transparent water. In late Winter it is covered by a thick ice layer which allowed deploying underwater equipment without any use of ships.

The first telescope on the scale envisaged by the DUMAND collaboration was not realised transforming a large volume of water, but a large volume of natural, transparent, Antarctic ice into a particle detector. It was the Antarctic Muon and Neutrino Detector Array (AMANDA), in operation from 2000 to 2005 [281]. It has represented a proof of concept for the world's largest neutrino detector, the kilometer-scale IceCube.

At the end of the 80s, with ongoing activities in Hawaii and at Lake Baikal and the first ideas on a telescope in polar ice, the exploration of the Mediterranean Sea as a site for an underwater neutrino telescope was natural. The Mediterranean endeavours split in three different locations.

In 2004 the NESTOR collaboration [282] installed, at a site near Pylos (Greece) and at 4 km depth, a single detector prototype, which operated for about one month and ended due to a failure of the cable to shore.

Since 1998, the NEMO (NEutrino Mediterranean Observatory) collaboration carried out research activities aimed at developing and validating key technologies for a km$^3$-scale underwater neutrino telescope. The first project, NEMO Phase-1, dates back to 2007, when a detector prototype was deployed at a depth of 2 km within a test site situated approximately 20 km off the coast of Catania, Italy. After a thorough assessment, another site at a depth of 3.5 km, located about 80 km off Capo Passero on the South-Eastern coast of Sicily (Italy), was identified as the optimal

## 7.7 Cubic-Kilometer Detectors

location for the installation of the underwater neutrino telescope. In 2013 NEMO Phase-2 was deployed at the site, where it continuously took data until 2014 [283].

The ANTARES detector, located 2.5 km under the Mediterranean Sea off the coast of Toulon (France), has been fully operative from 2008 to February 2022 [284]. With an instrumented volume at a few percent of 1 km$^3$, ANTARES has reached roughly the same sensitivity as AMANDA, and it has demonstrated the feasibility of neutrino detection in the deep sea.

ANTARES and NEMO have been superseded by KM3NeT [285], another neutrino telescope in the Mediterranean Sea, with two main components: ORCA, dedicated to the study of neutrino properties, and ARCA, optimised for high-energy neutrino astrophysics. In its final configuration, ARCA will consist of two blocks, each with 115 detector units (or strings) with 18 optical modules per string, which combined will instrument a volume of about 1 km$^3$. As of 2021, six detector units were operational in the ORCA site (off the coast of Toulon, France, close to the ANTARES site), and six were operational at the ARCA site, the former NEMO site, near Portopalo di Capo Passero [286].

GVD (Gigaton Volume Detector) [287] is a current effort to build a detector of cubic kilometer scale in Lake Baikal, Russia, following the operation of previous detectors at the site. The first phase, GVD-1, was deployed in 2021 and covers a volume of 0.4 km$^3$. Both GVD and KM3Net have been planned with a field of view complementary to that of IceCube.

### 7.7.1 The IceCube Experiment

The IceCube experiment [288], also indicated as IceCube Neutrino Observatory, began full operations in 2011, when the detector took its first set of data as a completed instrument. It is located at the Amundsen-Scott South Pole Station in the continent of Antarctica. IceCube observes the Cherenkov light emitted by charged particles produced in neutrino interactions in 1 km$^3$ of highly transparent and sterile Antarctic ice. The detector is the ice itself, beneath the surface at a depth of about 1.5 km under the surface. The in-ice component of IceCube consists of 5160 digital optical modules (DOMs), each with a photomultiplier tube of about 25 cm and associated electronics. The DOMs are attached to vertical cables called strings, and arrayed over a cubic kilometer from about 1.5–2.5 km depth. Each string holds 60 DOMs, whose vertical separation is 17 m. The strings are deployed on a hexagonal grid with 125 m spacing, but eight of them at the centre of the array are deployed more compactly, with a horizontal separation of about 70 m and a vertical DOM spacing of 7 m. This denser configuration forms the DeepCore sub-detector, which lowers the neutrino energy threshold to about 10 GeV, creating the opportunity to study neutrino oscillations. The DeepCore, together with IceTop, an air-shower array on the ice surface for cosmic-ray studies,[34] significantly enhances the capabilities of the IceCube observatory, making it a multipurpose facility. IceCube can be compared to another running Cherenkov detector of high dimensions, Super-Kamiokande; the detection techniques are based on the same principles, but the Icecube detector is much bigger, roughly 1 Gton in mass, and the energy threshold, around 10 GeV, is much higher.

IceCube observes astrophysical neutrinos in two ways. One approach selects neutrinos that interact inside the detector, while the other one selects neutrinos from the Northern Hemisphere sky (upgoing events), where the Earth serves as a filter to weed out muons produced by cosmic rays in the Earth's atmosphere.

In the Icecube detector, an incoming neutrino may undergo charged-current (CC) interactions with a nucleon $N$ of the detector medium, mediated by the exchange of a $W^{\pm}$ boson, or neutral-current (NC) interactions, via the exchange of a $Z^0$ boson. In both cases a particle shower $X$ is produced, but in the former case the shower is accompanied by a charged lepton with the same flavour as the incoming neutrino, namely we have $\nu_\ell + N \to \ell^- + X$ and $\nu_\ell + N \to \nu_\ell + X$ for CC and NC interactions, respectively. An additional channel is associated with the resonant production of $W^-$ in $\bar{\nu}_e e^-$ interactions, the Glashow resonance [289], which dominates over neutrino-nucleon interactions at a neutrino energy of 6.3 PeV. In 2021 IceCube has detected an event consistent with the production of a $W^-$ Glashow resonance initiated by an astrophysically produced ∼6.3 PeV $\bar{\nu}_e$ interacting with an electron in the ice of the detector [290].

While observing the Northern Hemisphere sky, the IceCube detector detects through-going, upgoing, track-like events which originate from outside the instrumented volume, increasing the effective volume for neutrino detection. Neutrinos from the Northern Sky reach the IceCube detector travelling through the Earth, which absorbs a fraction of them. This fraction is a function of cross section, which can be obtained by calculating a ratio of downgoing events (from Southern Sky) and upgoing events (from Northern Sky), since neutrinos from the Southern Hemisphere sky do not travel through the Earth and their number is not reduced. At the highest energies (PeV scale), neutrinos are effectively stopped by the Earth. In fact, at high momenta and energy transferred, neutrinos interact with individual nucleons into the nuclei via charged-current and neutral-current deep inelastic scattering, which will be discussed in Sect. 9.2.3. The cross-sections for these processes rise with increasing energy to the point that the Earth can stop neutrinos in their tracks— it becomes opaque to them.[35] This implies that one can actively search for PeV neutrinos looking at the Southern Sky, once the muon background is under control.

---

[34] IceTop is built as a veto and calibration detector for IceCube, but it also detects air showers from primary cosmic rays in the 300 TeV to 1 EeV energy range. The surface array measures the cosmic-ray arrival directions in the Southern Hemisphere as well as the flux and composition of cosmic rays.

[35] This SM prediction was confirmed by the IceCube experiment in 2017 [291].

IceCube analyses have confirmed the existence of astrophysical neutrinos from our galaxy as well as cosmic neutrinos from sources outside our Milky Way. Let us consider

$$P(\nu_\mu \to \nu_\mu) \simeq 1 - 4|U_{\mu 3}|^2(1 - |U_{\mu 3}|^2)\sin\left(\frac{\Delta m_{32}^2}{4}\frac{D}{E_\nu}\right), \quad (7.50)$$

where $U_{\mu 3} = \sin\theta_{23}\cos\theta_{23}$ is one of the lepton mixing matrix; $\theta_{12}$, $\Delta m_{21}^2$ and $\delta$ have no impact on the *present* data.

In 2018 the IceCube Collaboration has published its measurement of detection and reconstruction of neutrinos produced by the interaction of cosmic rays in Earth's atmosphere at energies as low as $\sim 5$ GeV. Interactions of cosmic rays in the atmosphere provide a large flux of neutrinos traveling distances from $D \sim 20$ km (vertically *down*-going) to $D \sim 1.3 \cdot 10^4$ km (vertically *up*-going) to a detector near the Earth's surface. A low energy threshold permits to measure the muon neutrino disappearance over a range of baseline up to the diameters of our Earth and thus to probe extensively the ratio of $D/E_\nu$. Assuming normal neutrino mass ordering, these analyses used neutrinos with reconstructed energies from 5.6 to 56 GeV, measuring [292]:

$$\Delta m_{32}^2 \simeq (2.31 \pm 0.13) \cdot 10^{-3} \text{ eV}^2 \quad (7.51)$$

$$\sin^2\theta_{23} \simeq 0.51 \pm 0.09. \quad (7.52)$$

These two values are consistent with those from the T2K, OPERA, MINOS and NOvA experiments.

## 7.8 Geoneutrinos

At the end of Sect. 7.4.2, we have seen how atmospheric neutrinos can be used as probe of the Earth's structure. An independent method to study the matter composition deep within the Earth can be provided by geoneutrinos. Geoneutrinos (geo-$\bar\nu$) are electron anti-neutrinos produced in $\beta$ decays of long-lived radioactive elements naturally present in the Earth. As seen in Sect. 1.3, the $\beta$ decay is the transition

$$(A, Z) \to (A, Z+1) + e^- + \bar\nu \quad (7.53)$$

where $A$ and $Z$ are the mass and atomic number of the parent nucleus. The energy $Q_0$ released in the reaction corresponds to mass difference between the parent and the daughter nuclei, and it is distributed between the electron and the antineutrino. This reaction produce energy that is converted into the heat, called radiogenic. In each decay, the emitted radiogenic heat is in a well-known ratio to the number of

**Table 7.1** Decays chains for HPEs [293]

| Decay | $E_{max}$ [MeV] | $\tau_{1/2}$ [$10^9$ yr] | $Q_0$ [MeV] |
|---|---|---|---|
| $^{238}$U $\to$ $^{206}$Pb $+ 8\,^4$He $+ 6\,e^- + 6\,\bar{\nu}_e$ | 3.27 | 4.47 | 51.7 |
| $^{235}$U $\to$ $^{207}$Pb $+ 7\,^4$He $+ 4\,e^- + 4\,\bar{\nu}_e$ | 1.23 | 0.71 | 46.4 |
| $^{232}$Th $\to$ $^{208}$Pb $+ 6\,^4$He $+ 4\,e^- + 4\,\bar{\nu}_e$ | 2.25 | 14.0 | 42.7 |
| $^{40}$K $\to$ $^{40}$Ca $+ e^- + \bar{\nu}_e$ | 1.31 | 1.28 | 1.3 |

emitted geoneutrinos [293]. Another source of heat is primordial heat accumulated during the formation of the Earth. It is currently believed that radioactive decays are responsible for roughly a half of the total heat flow of the Earth, with the other half coming from the secular cooling of the Earth.[36] By measuring the geo-neutrino flux and spectrum, it is possible to reveal the distribution of long-lived radioactivity in the Earth and to assess the radiogenic heat. This knowledge is critical in understanding complex processes driven by this release of energy, as the Earth's dynamic processes of plate tectonics, mantle convection, and the geodynamo, the mechanism generating the Earth's magnetic field.

The radiogenic heat arises mainly from the decays of isotopes with half-lives $\tau_{1/2}$ comparable to, or longer than Earth's age ($\sim 10^9$ years). There are several of them, with different abundances and life-times, but the major contribution to the Earth's radioactivity comes from decays of radioactive elements in chains of decays started with $^{40}$K, $^{238}$U, $^{232}$Th and $^{235}$U. These isotopes are labeled as heat-producing elements (HPEs). In Table 7.1 we list the decay chains of HPEs and their $\tau_{1/2}$, together with anti-neutrino maximum energy and the $Q_0$ of the corresponding $\beta$ decay.

Geoneutrinos were first discussed in the 1960s [295, 296], but to make their detection feasible several decades of progress on understanding neutrino propagation and developing low background neutrino detectors were necessary. Evidence of a signal originating from geo-neutrinos was reported for the first time in 2005 [62] by the KamLAND experiment discussed in Sect. 7.2.6. The detection of geo-neutrinos is realised through inverse $\beta$ decay, the classical reaction for antineutrinos detection

$$\bar{\nu}_e + p \to n + e^+. \quad (7.54)$$

The threshold for a scattering process $A + \bar{\nu}_e \to \sum_X X$ can be calculated by imposing that the squared center-of-mass energy $s = (p_A + p_{\bar{\nu}_e})^2$, where $p_A$ and $p_{\bar{\nu}_e}$ are the 4-momenta of the initial states, is larger than the square of the sum of the masses $(\sum_X m_X)^2$ of the particles in the final state. In this case $A = p$. In the laboratory frame, where $p$ is at rest, the four momentum of the proton is $p_p = (m_p, \mathbf{0})$. Thus one has $s = (p_p + p_{\bar{\nu}_e})^2 \simeq 2E_{\bar{\nu}_e} m_p + m_p^2$, where we have

---

[36] See Ref. [294] and references therein.

## 7.8 Geoneutrinos

neglected $m_{\bar{\nu}_e}^2$, and the threshold energy of the anti neutrino is

$$E_{\bar{\nu}_e} \geq E_{\bar{\nu}_e}^{thr} = \frac{(m_n + m_e)^2 - m_p^2}{2m_p} \qquad (7.55)$$

which corresponds to about 1.8 MeV. From Table 7.1 we observe that potassium and uranium-235 neutrinos are below the threshold energy so liquid scintillator detectors observe only uranium-238 and thorium neutrinos. The interaction cross-sections, which scale with their energy, for the detectable geoneutrinos (order of MeV) are on the order of $10^{-46}$ m$^2$ [297]. Geoneutrino flux at the detector's location can be calculated by integrating the antineutrino fluxes from all possible points of origin in the Earth and summing over all contributing $\beta$-decays. Its value is expected to be of order $10^6$ cm$^{-2}$ s$^{-1}$ [298]. The main source of background in geoneutrino detection is the production of electron anti-neutrinos by nearby nuclear power plants. Their energy spectrum extends well beyond the end point of the geoneutrino spectrum, which is 3.27 MeV, as given by Table 7.1. As a consequence, in the geoneutrino energy window (1.8–3.27 MeV), there is an overlap between geo-neutrino and reactor antineutrino signals.

All the data on the geoneutrinos obtained until now are provided by two experiments with large volume liquid scintillator detectors, KamLAND, already mentioned, and Borexino (see Sect. 7.2.7) [299]. Other planned detectors are making geoneutrinos part of their program. The SNO+ detector, mentioned at the end of Sect. 7.3, is a large liquid scintillator detector that can also observe geoneutrinos. JUNO, that will be discussed in Sect. 8.1.3, is a multipurpose neutrino detector planned to be constructed in China. Its main purpose is to measure reactor neutrino oscillations to further improve Daya Bay achievements. JUNO will primarily challenge the problem of the mass hierarchy of neutrinos at high statistical significance (3–4 $\sigma$ in about 6 years of data taking), but there are plans to use the JUNO detector also for geoneutrino studies.

Several experiments dedicated to the measurements of geoneutrinos have been devised and proposed, in various locations, in order to disentangle the contributions due to the crust and to the mantle of the Earth.

# Chapter 8
# Oscillations at Reactors and Accelerators

On Earth, intense sources of neutrino beams can be produced by man in nuclear power plants and accelerators. A neutrino oscillation experiment analyses the neutrino beam at a near site downstream of beam production, and make additional measurements at a far site placed at a distance; the baseline represents the distance $L$ between these two sites. Depending of the value of $L$, we distinguish between Long Baseline (LBL) and Short Baseline (SBL) neutrino oscillation experiments. The baseline scales with the $\Delta m^2$ sensitivity sought according to the relation for the oscillation length, see e.g. Eq. (6.43). LBL experiments are expected to be sensitive to small mass differences $\Delta m^2 \lesssim 10^{-2}$ eV$^2$, and SBL experiments to larger ones. A neutrino is produced and detected only through weak interactions and so it is in a definite flavour state both when it is produced and when it is detected. If a pure neutrino beam of known flavour is available, one naturally faces two possibilities: (1) measuring how many neutrinos of that flavour have disappeared (disappearance experiments) (2) measuring how many neutrinos of a different flavour are detected (appearance experiments). In other terms, at the distance $L$, disappearance experiments measure the survival probability of the neutrino flavour produced at the source, while appearance experiments measure the probability of detecting a neutrino flavour not produced at the source.

In Sect. 8.1 we discuss reactor neutrino oscillation experiments. Reactors are incredibly useful for studying neutrinos because they produce abundant fluxes (around $10^{20}$ $\bar{\nu}_e$/sec per GW, depending on reactor fuel and age [300]) which come in only one flavour: electron anti-neutrinos. They have variable energy that reaches as a maximum about 12 MeV. Reactor neutrino oscillation experiments use electron anti-neutrinos fluxes produces by usual, commercial, nuclear reactors and build detectors which are placed nearby, at different baselines. In 1956 a nuclear reactor was the source of the very first neutrinos ever detected (the experiment of Cowan and Reines [28]), already mentioned in Sect. 1.3.

Another way to study neutrino oscillations in an effective way is by using particle accelerators which produce high-intensity neutrino beams with hadron decays.

Experiments based on accelerators realise the concept of a neutrino beam where not only the neutrino travel distance is fixed but the energy can be controlled by beamline design at a certain level. They are discussed in Sect. 8.2, while the present status of oscillation parameters is discussed in Sect. 8.3.

## 8.1 Neutrinos at Reactors

In nuclear reactors, nuclear fission produces isotopes rich in neutrons, which are naturally unstable, and decay to stable isotopes via $\beta$ decays

$$(A, Z) \rightarrow (A, Z+1) + e^- + \bar{\nu}_e. \tag{8.1}$$

Fluxes of electron antineutrinos are produced isotropically in the $\beta$ decays of the neutron-rich heavy nuclei that are fission daughters. The nuclear fission is the phenomenon in which heavy nuclei are split into fragments, referred as fission fragments, or daughters. In 1939 it was observed for the first time that barium is produced when an uranium sample is exposed to neutrons [301–303]. In Fig. 8.1 we depict the $^{235}_{92}$U nuclear fission reaction

$$^{235}_{92}\text{U} + n \rightarrow {}^{140}_{56}\text{Ba} + {}^{90}_{36}\text{Kr} + 3n. \tag{8.2}$$

Each of the two fission fragments undergoes a cascade of $\beta$-decays, each time moving up in the periodic table and simultaneously producing an electrons and a $\bar{\nu}_e$.

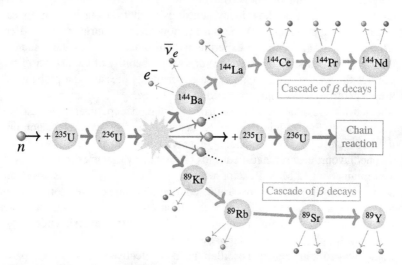

**Fig. 8.1** A representation of the nuclear fission reaction $n + {}^{235}_{92}\text{U}$, with $\beta$ decays cascades from fission fragments, originating reactor electron anti-neutrinos

## 8.1 Neutrinos at Reactors

The exact $\bar{\nu}_e$ spectrum and flux depends on thermal power and fuel of the reactor. The thermal power is the thermal energy released per second from the reactor core when operating at reference power, ignoring any inefficiencies in electrical energy extraction. Different fuels produce different combinations of neutron-rich isotopes following fission, each with their own neutrino emission spectra and decay branches. Assuming an average energy of about 200 MeV released per fission and 6 neutrinos produced along the $\beta$-decay chain of the fission products, one expects about $10^{20}$ $\bar{\nu}_e$/s emitted in a $4\pi$ solid angle for each gigawatt of thermal energy (GWth) [300].

In essentially all reactor neutrino oscillation studies, electron antineutrinos are detected by interactions with free protons via the inverse $\beta$ decay reaction (IBD)

$$\bar{\nu}_e + p \to n + e^+ \tag{8.3}$$

which has a neutrino energy threshold of 1.8 MeV, as seen in Eq. (7.55). A positron and a neutron are emitted simultaneously during each interaction event. Many experiments use organic liquid scintillators, as target and detector, which are composed mostly of hydrogen and carbon, both of which are poor absorbers of gamma rays. The interaction event causes a specific signature, in a liquid scintillator detector, namely two consecutive flashes of light. The first flash originates when the positron is annihilated by an electron in the detector (prompt signal). The neutron, after about 200 μs of migrations inside the detector, slows down and can be captured by protons, generating a deuterium nucleus with the emission of a photon with the characteristic energy of 2.2 MeV. This gives the (delayed) second flash. Therefore, it is possible to perform an efficient background rejection through positron and neutron signal time coincidence.

In order to increase the probability of absorbing low-energy neutrons, a percentage of another substance, like cadmium, lithium or gadolinium, can be added to the liquid scintillator (doping). For instance, gadolinium (Gd) has lower capture times with respect to the proton (about 28 μs) while emitting photons of higher energy (about 8 MeV). Thus, depending on whether hydrogen or gadolinium capture the neutron, the average time of the delayed signal is different. The idea to detect not only the positron but also the neutron dates back to the first experiment which successfully detected neutrinos directly, as we discuss in Sect. 8.1.1.

The electron antineutrinos have energies of a few MeV on average; they can reach about 12 MeV at the most, and peak around 0.3 MeV. These energies can be obtained by measuring the positron energy spectrum as

$$E_{\bar{\nu}_e} = E_{e^+} + m_n - m_p = E_{e^+} + 1.3 \,\text{MeV} \tag{8.4}$$

neglecting the small neutron recoil energy ($\sim 20$ keV). The fact that the principal observables are the number and energy of the positrons is an attractive feature for measurements of neutrino oscillations, that require knowledge of the antineutrino energy.

The detected event rate as a function of energy, namely the spectrum, is given by the convolution of the IBD interaction cross section with the emitted antineutrino

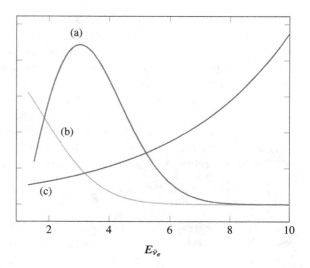

**Fig. 8.2** Shape of the IBD electron antineutrino spectrum (a) as a combination of its flux (b) and cross section (c) (oscillations neglected)

flux. The IBD cross section increases with energy whereas the emitted antineutrino flux decreases with energy, giving to the detected antineutrino spectrum the typical shape shown in Fig. 8.2. Each individual fuel isotope has a different flux curve, albeit always decreasing with energy. In addition to the $\bar{\nu}_e$ produced by fission, contribution from $\bar{\nu}_e$ produced by activation[1] of fuel and structural material is expected. A reactor $\bar{\nu}_e$ spectrum thus results from the overlapping of hundreds $\bar{\nu}_e$ spectra of fission products and activated elements, depending on the reactor features.

It is worth noting that as of today, the reactor neutrino flux itself does not seem to be fully understood, due to an approximate 6% deficit of the measured reactor neutrino flux in some experiments (see e.g. Daya Bay [304, 305] or Double Chooz [306] experiments) compared to the predicted one (reactor anti-neutrino flux anomaly or RAA). However, recent revisions of the flux prediction offer possible explanations to the deviations from the prediction, leading to a much better agreement [307, 308]. Another anomaly concerns the neutrino measured spectra, which presents an excess at about 5 MeV, called the "bump", (see e.g. Daya Bay [305, 309]) with respect to present models (shape anomaly).

Muon and tau neutrinos can be detected through charged current interactions when they produce charged particles, namely muon (mass $\simeq$ 106 MeV) or tau (mass $\simeq$ 1777 MeV), with neutrino energy sufficient to produce them. Since the peak energy of electron antineutrinos is rather low, in the few-MeV range, only reactions producing positrons are possible, and oscillations experiments based on reactors can only be of $\bar{\nu}_e$ disappearance type. Thus $CP$-invariance violation, which is only present in the appearance channels, cannot be measured directly with reactor neutrinos.

---

[1] Neutron activation is the process in which neutron radiation induces radioactivity in materials.

8.1 Neutrinos at Reactors

Reactor experiments which focus on oscillations use the IBD process just discussed to detect anti-neutrinos. However anti-neutrinos from reactor can also be detected using elastic scattering on nuclei, which we discuss in Chap. 9. These processes allow to investigate the electromagnetic properties of neutrinos, generally embodied in an effective interaction vertex, and, in particular, to set limit on the neutrino magnetic moment [310]. A particularly advantageous process is the coherent elastic neutrino-nucleus scattering. At low neutrino energies, the ones that are relevant at reactors, its cross section is two orders of magnitude larger than the IBD, without having any intrinsic energy threshold (see Sect. 9.4.1).

## 8.1.1 The First Direct Neutrino Detection

When Bethe and Peierls in 1934 [26] calculated for the first time the neutrino cross section, its extremely small value ($\sigma \approx 10^{-44}$ cm$^2$ at $E(\bar{\nu}) = 2$ MeV) seemed to preclude the possibility of ever directly detecting neutrinos. But the advent of very intense sources of neutrinos, namely fission bombs and fission reactors, changed that prospect. Clyde L. Cowan and Frederick Reines were both at Los Alamos Scientific Laboratory, USA, a laboratory which had been founded during World War II to coordinate the scientific research on the first nuclear weapons.[2] In late 1951, they began "Project Poltergeist," the first experiment in neutrino physics. The experiment at Hanford [311], in the U.S. state of Washington, was the first one to make an attempt to directly detect neutrinos. It used a fission reactor and a detector employing a liquid scintillator, which could both act as a target for the inverse $\beta$ decay reaction $\bar{\nu}+p \rightarrow e^+ +n$ and detect the emitted positrons via their annihilation to gamma rays. No neutrino signal was found, owing to high backgrounds. The detectors were then moved to Los Alamos and put underground, which confirmed that the backgrounds were from cosmic rays. As summarised by Cowan [312]: "The lesson of the work was clear: it is easy to shield out the noise men make, but impossible to shut out the cosmos. Neutrons and gamma rays from the reactor, which we had feared most, were stopped in our thick walls of paraffin, borax and lead, but the cosmic ray mesons penetrated gleefully, generating backgrounds in our equipment as they passed or stopped in it. We did record neutrino-like signals but the cosmic rays with their neutron secondaries generated in our shields were 10 times more abundant than were the neutrino signals. We felt we had the neutrino by the coattails, but our evidence would not stand up in court".

---

[2] The laboratory is located in the high desert of northern New Mexico, about 60 km from the state capital Santa Fe. During the war (1943–1945) the exact location of the laboratory (directed by the US theoretical physicist J. Robert Oppenheimer) as well its real purpose, were kept secret. After the war, Oppenheimer retired from the directorship. In 1947, the laboratory was officially baptised Los Alamos Scientific Laboratory (LASL). In 1981, it became a national laboratory and its name changed to Los Alamos National Laboratory (LANL).

The next attempt at neutrino measurement used the large flux of anti-neutrinos emitted from a nuclear site at Savannah River Plant (NC, USA), which had five fission reactors. Cowan and Reines built a nearby detector composed by two large, flat plastic tanks, each containing 200 litres of fluid. The fluid in the tanks consisted of water containing cadmium chloride. The target tanks were sandwiched between three much larger scintillation detectors. Each detector contained about 1400 litres of liquid scintillator solutions that was viewed by 110 photomultiplier tubes. The whole apparatus was located at about 11 m from the reactor and about 12 m underground [28, 29, 311].

Antineutrinos from the reactor were detected after interacting with protons in the water of the target tanks, according to inverse $\beta$ decay (1.37). Aa a consequence of the interaction, the antineutrino and the proton disappeared, and a positron and a neutron were simultaneously produced. The emitted positron rapidly annihilated with atomic electrons to produce two gamma rays in opposite directions. The scintillator produced flashes of visible light in response to the gamma photons, and that light was detected by photomultiplier tubes. Their measured cross-section confirmed the tiny order of magnitude of the theoretical prediction for inverse $\beta$ decay predicted by Bethe and Peierls already in 1934 [26]. Despite the great intensity of the neutrinos the reactor delivered, a very low counting speed was expected. In order to reduce the risk of erroneous interpretation, the experiment was designed to detect also the neutron emitted together with the positron in the detection event. The emitted neutron lost velocity in the water and was eventually captured by a cadmium nucleus (see Fig. 8.3). Cadmium is a highly effective neutron absorber and finds use in control rods for nuclear reactors because of that. When absorbing a neutron, $^{108}Cd$ produce is an excited state of $^{109}Cd$ which subsequently emits gamma rays

$$n + {}^{108}Cd \rightarrow {}^{109}Cd^* \rightarrow {}^{109}Cd + \gamma. \tag{8.5}$$

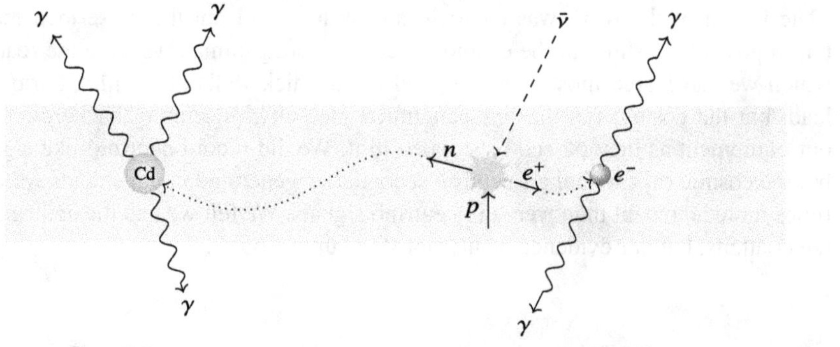

**Fig. 8.3** Schematic picture of a layer of water with added cadmium chloride which acts as a target in the neutrino detector of Reines and Cowan. An antineutrino and a proton interact, producing a positron and a neutron. The positron annihilates almost immediately with an electron of the medium, producing two back-to-back (because of momentum conservation) gamma rays, which are revealed in the adjacent scintillator tanks. This event is followed, a few microseconds later, by a burst of gamma rays in the same scintillator tanks, due to neutron capture by cadmium

## 8.1 Neutrinos at Reactors

The gamma from this process reached the detectors a few microseconds later than the gamma pair from the positron annihilation. The experimental arrangement provided on the whole a distinctive signature for the neutrino reaction—two gamma coincidence plus delayed gamma rays. The "sandwich" structure of the detector allowed to deduce the spatial origin of the event and to distinguish signals coming from inverse beta decay from spurious ones, which would most likely trigger the detectors in a random combination.

The team members stayed in Savannah River for over five months, accumulating data at a rate of about three neutrinos per hour. As additional confirmation that they were seeing neutrino events from the detection scheme described above, they also took data when the reactor was off. Their goal was to demonstrate that there was a difference in the neutrino-like signal, and precisely that it was much larger when the reactor was on than off, thus indicating that it was caused by the flux of antineutrinos coming from the reactor.

The first results in 1954 of a positive signal of neutrino detection were followed in 1956 and 1958 by more precise experiments [28, 29, 311], which also yielded a measurement of the IBD cross section $\simeq 11 \times 10^{-44}$ cm$^2$, in agreement with the theoretical prediction.

### 8.1.2 Early Oscillation Experiments

The early reactor experiments, until the 1990s, had very short baselines (VSBL) ranging from $L \sim 10$ m to $L \sim 100$ m. A short distance between the reactor core and the detector allowed to profit the most from a large flux. A large number (more that 20) of such experiments were performed, an example being the experiment of Cowan and Reines [28] (see Sect. 8.1.1), with a 11 m baseline. Another example is the VSBL experiment built at Institut Laue-Langevin (ILL), in Grenoble, France. It was taking data in the 1980s, featuring an 8.79 m baseline [313, 314]. These were all experiments based on one detector, in which the signal at the baseline $L$ was compared with the spectrum of the electron antineutrinos emitted by the reactor. The knowledge of that spectrum and its uncertainties was, therefore, an essential ingredient.

Until 2011, there had been a 3% deficit of the measured reactor neutrino flux compared to the predicted one in VSBL reactor neutrino experiments. In 2011 the reactor neutrino flux was reevaluated (Huber-Mueller model) [300, 315] which resulted in a 5.7% deficit between the neutrino flux observed by past experiments and the expected one. This discrepancy, called the Reactor Antineutrino Anomaly (RAA), could be interpreted as a new oscillation from active to eV-scale sterile neutrinos, implying the existence of a fourth non-standard (sterile) neutrino. To cast light on this, many experiments have seen the light, located at close distance to reactors in order to probe relatively large mass gaps (see Eq. (6.44)) and rely on high statistics. We discuss sterile neutrinos and related experiments in Sect. 10.5.2, here we just mention some of them: DANSS [316], NEOS [317], PROSPECT [318] and STEREO [319]. The only experiment so far which claimed an observation of sterile neutrinos is Neutrino-4 [320], at $\sim 3\sigma$, with a best-fit values of $\Delta m_{14}^2 \approx 7$ eV$^2$ and $\sin^2(2\theta_{14}) \approx 0.4$. On the contrary, the STEREO experiment has rejected the hypothesis of a light sterile neutrino [319]. It is worth noting that recent theoretical improvements in predictions are showing a better agreement with data, which disfavour the hypothesis of a light sterile neutrino as the explanation for RAA, and supports previous model deficiencies as the origin [307, 308].

In the late 1990s, the era of blind exploration had came to an end. The phenomenon of neutrino oscillations was fairly established and investigations using atmospheric and solar neutrinos progressed towards constraining the mixing parameters. In contrast to the CKM matrix in quark mixing, where all three mixing angles are very small, the mixing angles $\theta_{23}$ and $\theta_{12}$ in the neutrino-mixing matrix appeared to be large, as indicated by atmospheric and solar neutrino experiments, respectively. The measurement of the remaining mixing angle $\theta_{13}$ was recognised as the next major milestone.

As seen in Eq. (6.82), in the simplified three flavours description which exploits the neutrino squared mass difference hierarchy, the survival probability for $\bar{\nu}_e$ is the same than for $\nu_e$, and it reads

$$P(\bar{\nu}_e \to \bar{\nu}_e) = 1 - \sin^2 2\theta_{13} \sin^2\left(\frac{\Delta m_{31}^2 L}{4E}\right). \qquad (8.6)$$

This approximation holds with the average reactor neutrino energy (a few MeV) and a baseline less than 1 km, giving a clean window into the mixing angle $\theta_{13}$ and the $\Delta m_{31}^2$ squared mass difference (if $\theta_{13}$ is not too small). At larger baselines, the full survival probability has to be considered, since terms including $\Delta m_{21}^2 L/4E$ cannot be neglected any longer. Let us also observe that the disappearance probability does not depend on the Dirac $CP$ phase $\delta$. The cleanest way to measure the mixing angle $\theta_{13}$ is through kilometre-baseline reactor neutrino oscillation experiments. In fact, according to Eq. (8.6), a non-zero $\theta_{13}$ causes a deficit of the antineutrino flux at a source-detector distance $L$ of about 1–2 km. The reactor neutrino energy $E$ is typically around a few MeV and $L \simeq 1$ km gives a sensitivity to $E/L \simeq \Delta m^2 \gtrsim 10^{-3}$ eV$^2$ (see Eq. (6.44)). Furthermore, at this value of $L$ the modification of the oscillation probability induced by the matter effect can be neglected in first approximation.

The first reactor experiments to extend the baseline to a distance of about one kilometre were the Palo Verde experiment, taking data between 1998 and 2000 [258], and the CHOOZ experiment [321], taking data from 1997 to 1998. The Palo Verde detector was located in an underground bunker, under 12 m of rock, at a distance of 890 m from two reactors and 750 m from a third reactor. The reactors form the Palo Verde Nuclear Generating Station in the Arizona desert, which actually generates the largest amount of electricity in the United States per year. The CHOOZ nuclear power plant of Electricité de France consists of two reactors and it is located in the village of the same name, in the northeast of France,

(Ardennes region), close to the border with Belgium. The CHOOZ detector was positioned 1050 m away from the double unit of CHOOZ nuclear reactors and about 100 m underground, in an old tunnel, to be shielded by cosmic rays.

Neither Palo Verde or CHOOZ experiments was able to observe the deficit caused by $\theta_{13}$ oscillation. This null results, combined with the measured values of $\theta_{23}$ and $\theta_{12}$, motivated many phenomenological speculations of neutrino-mixing patterns such as bimaximal and tribimaximal mixing [322, 323]. In most of these theories, $\theta_{13}$ is either zero or very small.

### 8.1.3 Oscillation Experiments Since the Turn of the Century

The importance of knowing the precise value of $\theta_{13}$ prompted the construction of second-generation kilometre-baseline reactor experiments in the twenty-first century, namely Double Chooz [324], RENO [325] and Daya Bay [326]. A common technology used in both the first- and second-generation experiments is the gadolinium-loaded liquid scintillator as the antineutrino detection target. The choice of a Gd-doping was to maximise the neutron capture efficiency; gadolinium has the highest thermal neutron cross section. First generation reactor neutrino experiments were using only one "far" detector to extract their results. The unoscillated neutrino spectrum was extracted using the reactor predictions taking into account the reactor operation conditions. Addition of "near" detectors at baselines of a few hundred metres is the most significant improvement of the second-generation experiments over the previous ones. Since the reactor neutrino source led to the largest systematic uncertainties in the previous reactor neutrino experiments, this new set-up provided a great improvement in the search for a small mixing angle. The relative near-far measurement between nearly identical antineutrino detectors allows for the cancellation of detector-related systematic uncertainties. All the three experiments have a similar detector design. At each site, neutrino detector modules are shielded to reduce background and veto cosmic rays, e.g. they are submerged in a water pool. Each neutrino detector module consists of three nested cylindrical volumes separated by two concentric acrylic vessels and contained in an outmost vessel made of stainless steel. Serving as the target for the inverse beta-decay reaction, the innermost volume holds Gd-loaded liquid scintillator. The target region is nested within an additional multi-ton liquid scintillating region which has no Gd, so it does not serve as the primary antineutrino target. Its purpose is to increase the detector efficiency by absorbing gamma rays which happen to escape the target region ("$\gamma$ catcher"). This region is further surrounded by an inactive zone of mineral oil which shields the scintillating regions from external radiation. Photomultiplier tubes (PMTs) mounted within this region detect the scintillation light emitted by particle interactions within the detector. These three volumes and the PMTs constitute the inner detector (ID). The exact dimension and the target mass of the detector module are different in each experiment.

Double Chooz was the continuation of CHOOZ, being installed in 2009 near the Chooz nuclear power plant. It had the simplest configuration with two detectors situated at about 395 m and 1050 m from double unit of reactors. It started physics data taking in 2011 and ended it in early 2018. Its main result was to provide the first hint for the non-zero value of $\theta_{13}$ mixing angle in reactor experiments, yielding $\sin^2 2\theta_{13} = 0.105 \pm 0.014$ [306, 327].

The RENO (Reactor Experiment for Neutrino Oscillations) experiment, started in 2011, uses anti-neutrinos emitted from six reactors at the Hanbit Nuclear Power Plant (previously Yong Gwang, from the nearby city) in Korea. It was the first reactor experiment to take data with two identical near and far detectors in operation. They are located at 294 m and 1383 m, respectively, from the center of the reactor arrays. The present result is $\sin^2 2\theta_{13} = 0.087 \pm 0.008|_{\text{stat.}} \pm 0.014|_{\text{syst.}}$ [328].

The Daya Bay reactor experiment has taken data from 2011 to 2015 and was located in Daya Bay, near Dapeng city in the Guangdong province of China, 50 km north-east Hong Kong, at the cluster of six nuclear reactors, distributed in pair over three sites. It was the first to report simultaneous measurements of reactor antineutrinos at multiple km baselines. Daya Bay has a total of eight detectors, with four detectors placed at the far site and two detectors each at the two near sites. Daya Bay's far detectors lie between about 1500 and 1900 m from the sites' reactors whilst the near detectors have baselines from about 360 to 560 m. The detectors were located underground in experimental halls built beneath mountains that provide shielding from cosmic ray background. Based on 3158 days of operation in 2022 its analyses led to: $\sin^2 2\theta_{13} = 0.0851 \pm 0.0024$ and $\Delta m_{32}^2 = (2.466 \pm 0.060) [-(2.571 \pm 0.060) \cdot 10^{-3} \, (\text{eV})^2$ with normal [inverted] mass ordering [329].

Another major milestone suggested in the late 1990s was to build reactor experiments at a long baseline of about 100 km, corresponding, at a few MeV energies, to the squared mass measured by solar neutrino experiments, namely $\Delta m_s^2 \equiv \Delta m_{21}^2 \simeq 10^{-4} \, \text{eV}^2$. The goal was to demonstrate the validity of the oscillation interpretation of the solar neutrino observations also for electron antineutrinos at a terrestrial experiment and without the matter effects. Going to such a distance from the reactor core represents an obvious challenge since the electron antineutrino flux is reduced by the factor of million with respect, for instance, to $L \simeq 100$ m. The detector must be much larger, much better shielded against cosmic rays, with backgrounds reduced as much as possible. Possibly, a large array of reactors contributing in unison should be available. The unique combination of all of that was realised in the KamLAND experiment. At that time there were about 40 working power reactors in Japan, and many of them were situated approximately at a circle with radius about 180 km centred around Kamioka, where the water Cherenkov Kamiokande detector (see Sect. 7.2.3) was located. When the Kamiokande experiment was decommissioned an underground site became available in Kamioka. The KamLAND experiment had a liquid scintillator detector with unprecedented radiopurity, detecting neutrino by the IBD reaction. We have already discussed the KamLAND experiment in more detail in Sect. 7.2.6. KamLAND gave a clear visual picture of oscillations and provided a precise measurement of the mass splitting. Its results showed, for the first time, that the disappearance probability of the reactor electron antineutrinos indeed changes periodically as the function of $L/E$. Though unrelated to nuclear reactors, let us recall that another first for KamLAND was the observation of geoneutrinos, as seen in Sect. 7.8.

The Daya Bay experiment paved the way for China to build a successor, the Jiangmen Underground Neutrino Observatory (JUNO) [330]. The JUNO detector, which is in the final phase of its construction, consists of 20 kton liquid scintillator viewed by more than 40,000 photomultiplier tubes. The energy resolution is designed to be less than 3% at 1 MeV. As the Daya Bay detector, it will measure a deficit of reactor $\bar{\nu}_e$ via IBD, but it has a much longer baseline, since it is located at equal distance of about 53 km to the Yangjiang power plant and Taishan power plant. Due to the absence of high mountains in the relevant area, the JUNO detector will be deployed in an underground laboratory under the Dashi hill with a vertical overburden around 700 m. Besides, the JUNO detector will observe not only antineutrinos from the reactors, but neutrinos/antineutrinos from terrestrial and extra-terrestrial sources, including supernova burst neutrinos, diffuse supernova neutrino background, geoneutrinos, atmospheric neutrinos, and solar neutrinos. The JUNO main physics goal is to determine the neutrino mass ordering—the question of whether the third neutrino mass eigenstate is the most or least massive of the three [331].

## 8.2 Accelerator Experiments

In 1959, Bruno Pontecorvo suggested that one could use neutrinos emitted in the decay of charged pions, to check if they were different from the ones produced in beta decays, the only ones known at the time [332, 333]. In 1960 Mel Schwartz proposed the first realistic scheme of a collimated beam of neutrinos, from pions and kaons decays, intense enough to make neutrino interactions detectable, with the purpose to investigate weak interactions at high energies [334].

The same year, a new synchrotron accelerator, the AGS (Alternating Gradient Synchrotron), located at the Brookhaven National Laboratory in Long Island (US), had just completed its commissioning, that is the ensemble of all the necessary actions for checking and tuning an accelerator in order to meet its specifications. This new accelerator, operating at what at the time was record proton energies (up to 30 GeV), opened the possibility of studying neutrino interactions at the GeV scale. Neutrinos were produced during the decay in flight of charged pions. In a first step, protons circulating in the AGS and accelerated to high velocities were brought to strike an internal target made of the metal beryllium (Be). When a proton traveling near the speed of light hits a target, the proton's energy is used to produce a jet of hadrons. They are predominantly pions and kaons, because they have lower masses. At AGS, high-velocity charged pions were produced in a forward-directed pion beam. The charged pions decay mainly[3] through the channels $\pi^+ \to \mu^+ + \nu_\mu$ and $\pi^- \to \mu^- + \bar{\nu}_\mu$, when allowed to travel a path of free flight, which was set at 21 m. Thus, a beam of high-energy neutrinos was produced, forward-directed towards the detector. The beam still contained quantities of leftover pions and muons. All particles other than neutrinos were eliminated from the beam by placing a wall of steel, made from pieces of old warships and thick 13.5 m, before the detector. The detector was a spark chamber, whose design had only recently been significantly improved [335].

A spark chamber consists of a sealed box filled with helium or neon gas (or a mixture of both) and sectioned by a stack of metal plates. When a charged particle moves through the box, it ionises the gas between the plates, leaving a trail of electrons ejected from the atoms of the gas along its trajectory. The passage of the charged particle is detected by two scintillator counters, usually above and below the chamber, whose output is fed to a coincidence circuit. The transmitted signal triggers a high voltage pulse generator, which supplies a voltage of several kilovolts between each pair of neighbouring plates before the disappearance of the trail of electrons. A spark is then produced along the trajectory of the charged particle. The traversing path of the charged particle is thus revealed by the array or line of sparks, which may be seen through the chamber.

For the neutrinos to have a fair chance to interact in the detector, and produce the charged particles to be detected, the detector had to be massive. A 10 ton spark chamber was built, composed of 90 plates of aluminium, each 2.5 cm thick, and spaces between the plates were filled with neon gas. Neutrinos and antineutrinos interacted with protons and neutrons through the reactions $\nu_\ell + n \to \ell^- + p$ and $\bar{\nu}_\ell + p \to \ell^+ + n$. The detector was arranged so that the nature of the charged lepton $\ell^\pm$ causing the spark tracks could be recognised. If the neutrinos in the beam were identical to the neutrinos known from beta decay, that is electron neutrinos, the interaction in the detector should produce $\ell = e$. Instead, the experiments performed at AGS in 1962 [36] observed only $\ell = \mu$ in the above reactions, with no electrons or positrons. This brought to the conclusion that these neutrinos, produced by pion decays together with muons, were different from the neutrinos produced in beta decay– indeed, a new kind of neutrino $\nu_\mu \neq \nu_e$ [36].

The second ever neutrino beam was built brand-new at CERN (Switzerland), from 1961 to 1963. With respect to the beam at Brookhaven, it realised two important progresses: the extraction of the proton beam and the focusing of the secondary mesons. The proton beam was extracted from the Proton Synchrotron (PS), CERN's first synchrotron, which accelerated protons to approximately the same energy than AGS, around 28 GeV. The AGS and PS were the biggest machines of their time, with a diameter of about 200 m. At PS, the extracted proton beam was sent to an external target, instead of having the target positioned inside a straight section of the accelerator, as at AGS. With an external target, the trajectories of

---

[3] The pion decays into electrons and antineutrino or positron and neutrino are disfavoured with respect to the decay into muons because of helicity.

## 8.2 Accelerator Experiments

the charged mesons produced in the target are not disturbed by the peripheral magnetic field of the accelerator magnets, and there is space to surround the target with magnets, in order to focus them toward the experiment. The focusing device, inducing magnetic fields that pushed all charged particles of the same sign closer on the same axis, became known as magnetic horn, from its shape. It was invented by the Dutch physicist Simon van der Meer[4] in 1961 [336]. The focusing increases the intensity of the neutrino beam. These two measures led to an increase in the neutrino flux to the point that it was decided to put not only a spark chamber, but also a heavy liquid bubble chamber on the neutrino beam, since it seemed that the neutrino flux was sufficiently large to produce events even in this relatively low mass device (0.75 ton).

The bubble chamber is one of the earliest imaging detectors, being invented by the American physicist Donald Arthur Glaser in 1952.[5] A conventional bubble chamber consists of a sealed container filled with a liquefied gas. The chamber is designed such that pressure inside can be quickly changed. The idea is to momentarily superheat the liquid, that is to heat it to a temperature higher than its boiling point, without let it boil, when the particles are expected to pass through it. This is accomplished by suddenly lowering the pressure, with a rapid expansion, which decreases the boiling point, thus converting the liquefied gas into a superheated liquid. When particles cross it, they produce tracks of localised electron-ion pairs. The energy delivered to the liquid during this process produces tiny bubbles along the particle's track. The whole chamber is then illuminated and photographed by a high-definition camera. The photograph is analysed offline for particle identification and measurements. A disadvantage is the when the bubbles form, the chamber must be recompressed to stop their growth and photograph the track. This limits the rate at which events can be collected. The spark chamber improved on the bubble chamber as interactions could be captured much more rapidly, even if they did not capture the same amount of details (the resolution) of the bubble chambers.

The results of the CERN experiment [337] confirmed the existence of the muon neutrinos, with higher statistics and more details than the AGS experiment. Several other aspects of weak interactions were also explored, for example the possibility that neutrinos from kaon parents were different from those from pion parents. No evidence was found for this idea, as no evidence was found for the existence of the $W$ intermediate boson, already postulated to be the carrier of the weak force [338]. It is evident that lack of evidence in the first case followed from a wrong hypothesis, while in the second case from the insufficient available energy, since the $W$ boson has mass around 80 GeV.

Since these first experiments with accelerator neutrinos, numerous others have been conducted in various laboratories worldwide, all employing the method for

---

[4] Simon van der Meer shared the Nobel Prize in Physics in 1984 with Carlo Rubbia for contributions to the CERN experiment leading to the observation of the $W$ and $Z$ bosons.

[5] For this invention he was awarded the 1960 Nobel prize in physics.

producing neutrino beams, summarized in the following. Neutrino beams have to be produced as secondary beams, because no direct, strongly focused, high-energy neutrino source is available. A proton beam, extracted by a particle accelerator,[6] strikes a fixed target, producing secondaries, such as pions and kaons.[7] Those secondaries leave the target, boosted in the forward direction. By using beam optical devices (dipole or quadrupole magnets or magnetic horns), secondaries of a certain charge sign are focused into a long decay tunnel. The mesons, permitted to drift in free space, decay to neutrino tertiaries, mainly according to the decays $\pi^\pm \to \mu\nu_\mu$, $K^\pm \to \mu\nu_\mu$ or $K_L \to \pi\mu\nu_\mu$. At the end of the decay tunnel, shielding, often referred to as the "beam stop", "beam dump" or "muon filter," removes all particles in the beam except for the neutrinos (or anti-neutrinos), which continue on to the experiment.

The neutrino beams obtained this way, by pion decay in flight (DIF), are known as conventional. They are composed mainly by muon neutrinos, with a percentage of electron neutrinos.[8] In order to give a rough estimate of their relative proportions, let us assume that only positively charged particles are collected by the focussing device. Hence, the main source of neutrinos is the decay $\pi^+ \to \mu^+ + \nu_\mu$. Electron neutrinos come from the subsequent decay $\mu^+ \to \bar{\nu}_\mu + e^+ + \nu_e$. The $\pi^+$ and $\mu^+$ lifetimes are respectively around 26 ns and 2.2 μs. At high energies $\pi^+$ and $\mu^+$ are both relativistic and travel at the same speed in the beam pipe, where they can decay before being absorbed in the dump. Thus, the number of muon decays to pion decays, and consequently the number of electron neutrinos $N_{\nu_e}$ to muon neutrinos $N_{\nu_\mu}$, is roughly the same as the ratio of their lifetimes, i.e. $N_{\nu_e}/N_{\nu_\mu} \simeq 0.01$. The $\pi^+$ can also decay directly into electron neutrinos, as $\pi^+ \to e^+ + \nu_e$ decay, although with a very low branching ratio, at the level of $10^{-4}$. Additional electron neutrinos are due to $K^+$ mesons decays. The $K^+$ are produced by the primary proton interactions at a rate which is about 10% that of $\pi^+$ mesons. The main $K^+$ decay channels are $K^+ \to \mu^+ + \nu_\mu$ decays (∼65%), which add to the main $\nu_\mu$ beam, but there is a percentage (∼5%) of $K^+ \to \pi^+ + \pi^0 + \nu_e$ decays, which increases the $\nu_e$ component of the neutrino beam by about $0.1 \times 0.05 = 0.005$.

Today, modern facilities operate at energies and luminosities many orders of magnitude greater than the pioneering colliders of the early 1960s. Detectors have progressed, becoming more complex and able to detect many more particles at a time, backing the advances in collider accelerator technology and beam physics.

Experiments with conventional neutrino beams can be further classified in three categories: (1) Wide Band (WB) beam, where a system of magnetic horns focuses particles with the appropriate charge towards the beam axis. A high-intensity neutrino beam with a wide energy spectrum (spanning one or two orders of magnitude) is produced, convenient for investigating new oscillation signals in

---

[6] Extracted beams are the norm in today's experiments.

[7] The commonly used luminosity unit is protons on target (POT).

[8] Obviously, if the oppositely signed charged mesons are focused, the beam is composed mainly by $\bar{\nu}_\mu$, with a percentage of $\bar{\nu}_e$.

a wide range of values of $\Delta m^2$. A disadvantage is the difficult prediction of the absolute neutrino energy spectrum and composition. (2) Narrow Band (NB) beam, obtained with the selection of secondaries (pions and kaons) of a certain charge and momentum range, via a system of quadrupole and dipole magnets, before the decay tunnel. The main advantage of such a beam is a narrow energy spectrum, which may be convenient for precise measurements of $\Delta m^2$. On the other side, the neutrino flux intensity is reduced, comparing with a WB beam obtained from the same proton beam. (3) Off-axis (OA) beam, which use a high-intensity WB beam with the detector shifted by a small angle with respect to the axis of the beam, where the neutrino energy is almost monochromatic. The off-axis technique was first proposed in 1995 for a long-baseline experiment at Brookhaven that was never realised [339].

## 8.2.1 Non Conventional Neutrino Beams

Experiments have also used neutrinos from the decay at rest of pions (DAR or stopped pions), instead than DIF pions, to get neutrino beams with a well known spectrum in the few tens of MeV range. They are obtained when protons hit a target large enough that pions come at rest before decaying. Then, the bulk of the negative pions are strong absorbed by the target before they are able to decay, and most of the negative muon produced from the $\pi^-$ decay are captured from the atomic orbit, a process which does not give rise to $\bar{\nu}_e$. On the contrary, positive pions and their daughter $\mu^+$ decay and produce neutrinos. Let us underline that the target for the primary proton beam, where the neutrino parent particles emerge, is the medium for absorbing or stopping the hadrons. No drift space is provided for hadrons to decay in.

The flavour composition and the spectrum of DIF and DAR neutrino beams are remarkably different. In the latter case, neutrinos are produced isotropically and their spectrum shows a prompt monochromatic component of muon neutrinos at about 30 MeV, due to the sequence $\pi^+ \to \mu^+ \nu_\mu$, which has decay time of 26 ns, and delayed $\nu_e, \bar{\nu}_\mu$ components, coming from the muon decay at rest $\mu^+ \to e^+ + \nu_e + \bar{\nu}_\mu$ (decay time 22 μs). Experiments at stopped pion neutrino sources were for instance LSND [340] at LANL in US and Karmen [341] at ISIS, Rutherford Appleton Laboratory (RAL), in UK, both powered by 800 MeV proton accelerators of respectively 0.1 mA, 0.2 mA intensity. We will discuss both in Sect. 10.5.2. Neutrino cross sections on different nuclei in this energy range are relevant for understanding core-collapse dynamics and nucleosynthesis in supernovae.

In the past several decades, a number of intense accelerator-based neutron sources called spallation sources have started to operate. A spallation source consists of a high-powered accelerator that brings high energy (greater than 0.5 GeV) protons (ions) into a heavy metal target, such as mercury or tungsten. These metals "spall off" free neutrons in response to the impact. As accelerators can be readily pulsed, spallation sources are generally pulsed neutron sources, unlike most reactors that generate neutrons constantly. Neutrons are employed for various applications, including neutron-scattering experiments and a variety of interdisciplinary uses. As a by-product of the proton interacting at the target, a copious number of pions are produced, which decay at rest originating an intense source of pulsed neutrinos.

Another technique to form neutrino beams consists in dumping an high energy proton beam in a dense block of heavy material extending also in the transverse direction. In this way, most of the known long lived particles, as high energy pions and kaons, will interact before decaying, and will not produce high energy neutrinos. We can say they are absorbed in the dump. Then a beam of neutrinos can originate only by the decay of short lived particles, like for instance charmed mesons, with proper lifetimes of the order of $10^{-13}$ s, which have no time to interact before decaying. The charmed heavy hadrons decay promptly with practically equal branching ratios into electrons and muons, emitting equal fluxes of electron and muon neutrinos with energies of several tens of GeV on average (up to above 100 GeV). Neutrinos produced in this way are referred to as prompt, and these experiments as beam dump experiments. They were proposed soon after the discovery of the $J/\Psi$ in 1974, aiming at the search of the production of short lived particles in proton-nucleon interactions, by looking at a signal of prompt neutrinos. The first beam dump experiments were made in the 1970s, at the 400 GeV Super Proton Synchrotron (SPS) at CERN and at the 70 GeV U-70 proton synchrotron at the Institute for High Energy Physics in Protvino (near Serpukhov, Russia) [342–345]. Among the opportunities offered by beam dump experiments there is the possibility to detect tau neutrino interactions. The beam dump experiment DONUT (Detector for direct observation of tau neutrinos) carried on at Fermilab gave the first direct observation of tau neutrino interactions [52], as mentioned in Sect. 1.4. In the usual neutrino beams tau neutrinos and antineutrinos are practically absent. In the DONUT experiment, an 800 GeV beam of protons from the Tevatron collided into the beam dump, a large block of tungsten, to produce the charm particle $D_s$ (a meson comprised of charm and strange quarks) which decayed to an anti-tau neutrino and a tau lepton. The latter then decayed to a tau neutrino, which passed through 36 m of shielding to a detector consisting of emulsion targets followed by a spectrometer. In the emulsion targets a tau neutrino can interact with a nucleon to produce a tau detectable by the various levels of the spectrometer.

Since when neutrino oscillations were experimentally established, there has been interest to produce neutrino beams in different, not conventional, and possibly more efficiently and precise ways; several proposals have been put forward, like Muon Storage Ring and Beta-Beams. Muon storage rings [346], and their evolution, the so-called Neutrino Factories, exploit the idea of neutrinos from muons. The name Beta-Beam refers to the production of a pure beam of electron neutrinos or antineutrinos through the beta decay of accelerated radioactive ions circulating in a storage ring [347].

## 8.2.2  First Generation LBL Experiments

The proposals for the first generation of long-baseline accelerator experiments were motivated by the results on atmospheric neutrinos in the 1990s, already mentioned in Sect. 7.4.2, which produced clear evidence for neutrino oscillations. At the time there was a widespread uncertainty in the value of the squared mass difference $\Delta m^2$, estimated in the range $10^{-3}$–$10^{-1}$ (eV)$^2$ [255, 348]. Increasing the length of the oscillation baseline improves the sensitivity to smaller mass splittings. Beam energies of a few GeV allow good sensitivity to $E/L \simeq \Delta m^2 \gtrsim 10^{-3}$ (eV)$^2$ with distances ranging between 100 and 1000 km. Indeed, the main reason to locate a detector hundreds of kilometres from the neutrino beam target is to study neutrino oscillations.

Several laboratories around the world started the LBL program—the High Energy Accelerator Research Organisation (KEK) in Tsukuba (Japan), Fermilab at Batavia, near Chicago (USA); in Europe, CERN at Geneva (Switzerland) and Laboratori Nazionali del Gran Sasso (LNGS) near L'Aquila (Italy).

### 8.2.2.1  K2K

The first LBL accelerator was K2K (KEK to Kamioka) [349, 350]. It ran from 1999 until 2004, using the 12 GeV proton synchrotron accelerator at KEK. The proton beam was bent to the direction of the Kamioka underground laboratory, and hit an aluminium target producing mainly positive pions. The positive pions decayed into positive muons and $\nu_\mu$ during their flight in a 200-m decay pipe. The $\nu_\mu$ beam was monitored by a system based on 1 kton water Cherenkov neutrino detector located 300 m from the target. A neutrino beam is in general neither pure nor very stable (small contamination of wrong muon sign and electron type are unavoidable from various pion and kaon decay channels) and therefore a detector placed a few hundred meters down the proton target can be used to monitor the beam and determine its energy spectrum and flavour composition. This is indicated as the near detector, while the detector used to assess the impact of neutrino oscillations is referred to as the far detector. At K2K the far detector was located 250 km away. It was the Super-Kamiokande detector, a 50 kton water Cherenkov neutrino detector, already described in Sect. 7.2. The configuration with a far and a near detector, the latter being a smaller version of the former, allows for a large cancellation of detector-related systematic uncertainties. A GPS based system provided the synchronisation between the beam extraction and the far detector.

The attempted oscillation measurements were muon neutrinos disappearance and electron muon neutrino appearance. The energy, 1.3 GeV of average, was below threshold for $\tau$ production ($E_\nu \approx 3.5$ GeV), so no search for $\tau$ appearance was possible. In a 2-neutrino formalism, appropriate for the sensitivity of the period, the oscillation modes were determined by (see Eqs. (6.40) and (6.78)):

$$P_{\nu_\mu \to \nu_\mu} = 1 - \sin^2 2\theta_{23} \sin^2 \frac{\Delta m_{32}^2 L}{4E}$$

$$P_{\nu_\mu \to \nu_e} = \sin^2 2\theta_{23} \sin^2 \frac{\Delta m_{32}^2 L}{4E} \quad (8.7)$$

where $\theta_{23}$ and $\Delta m_{32}^2$ are the mixing angle and squared mass difference neutrinos between second and third generation neutrinos, respectively. Because of the relatively small value of the mixing angle $\theta_{13}$, the muon neutrino oscillates primarily in the tau neutrino. Oscillations into electron neutrinos occur at a sub-dominant level (see Eqs. (6.80)). With neutrinos below tau production threshold, this is observed as disappearance of muon neutrinos.

The K2K was the first accelerator experiment to confirm the prediction of the neutrino oscillation discovered by the observation of atmospheric neutrinos. No evidence for an electron neutrino appearance signal was found. The final results of disappearance analyses set bounds on the $\nu_\mu \to \nu_e$ oscillation parameters. In the full data sample collected, corresponding to $9.2 \times 10^{19}$ protons on target (POT), about 112 neutrino candidate events (within statistical errors) were observed at the far detector, against an expectation of about 158 events without oscillation. The probability that the observations were explained without neutrino oscillation was found to be 0.0015% (4.3$\sigma$) [351]. An upper limit of $\sin^2 2\theta_{23} < 0.13$ at 90% confidence level was set, at the best fit value $\Delta m_{32}^2 = 2.8 \times 10^{-3}$ eV$^2$ [351, 352].

### 8.2.2.2 MINOS

The first long baseline neutrino oscillation experiment at Fermilab was MINOS (Main Injector Neutrino Oscillation Search) [353, 354], which ran from 2003 until 2012. An upgrade, MINOS+, operated from 2013 to 2016. The experiment MINOS had as its main motivation precision measurements of the atmospheric oscillation parameters via $\nu_\mu$ disappearance, by greatly improving on the statistics of the collected data sample. The Fermilab Main Injector facility produced protons at energy of 120 GeV, which were led to strike a graphite target. A shower of hadrons was produced, consisting primarily of pions but with a significant kaon component at higher energies. They passed through two magnetic horns, which focussed either positive or negative hadrons (depending on the direction of the electric current through the horns), then down a 675 m long, helium filled pipe, in which they decayed to produce a $\nu_\mu$ beam, with a small $\nu_e$ component due to the decay $\mu^+ \to \bar{\nu}_\mu + e^+ + \nu_e$. The neutrino beam (NuMI beam) was designed with the flexibility to change the neutrino energy by tuning the momentum range of the secondaries focused. That could be obtained by adjusting the current through the focusing horns and the relative positions of the horns and target. The energy range

## 8.2 Accelerator Experiments

of the NuMI beam was 1–10 GeV. The experiment was designed in mid-1990s, a time where, as already mentioned, the range of possible values of squared mass difference was quite broad. By the time the beam line was completed and MINOS started taking data in 2005, instead, data pointed to the lower bound of such range, namely $10^{-3}$, and MINOS took data essentially only in the low energy configuration of the beam.

The MINOS near detector was a steel-scintillator calorimeter located at Fermilab, at about 1 km from the source of the neutrinos and 100 m underground, and measured the neutrino spectrum before oscillation. The far detector was larger, but otherwise similarly constructed, and it was located 735 km away, in the Soudan mine (705 m underground) in northern Minnesota. MINOS detectors were magnetised, allowing the charges of particles to be identified, and hence $\nu_\mu$ and $\bar{\nu}_\mu$ interactions to be distinguished on an event by event basis.

The first analysis from MINOS were made in 2006 in the two flavour approximation [355], and was followed by updated analyses in 2008 [356] and 2011 [357]. In 2011, the results with $7.25 \times 10^{20}$ POT for the atmospheric mass splitting and the angle $\sin^2(2\theta_{23})$ gave: $|\Delta m_{32}^2| = \left(2.32^{+0.12}_{-0.08}\right) \times 10^{-3}$ (eV)$^2$ and $\sin^2(2\theta_{23}) > 0.90$ with 90% confidence level [357]. An initial claim of $CPT$ violation through a possible difference of disappearance of neutrinos and anti-neutrinos was excluded in 2012 [358]. The final MINOS two-flavour fit considered the full $10.7 \times 10^{20}$ POT in the neutrino-dominated beam mode, a sample more than an order of magnitude larger than K2K, as well as $3.4 \times 10^{20}$ POT in the antineutrino-enhanced beam [359]. The MINOS far detector was also a very effective detector of neutrinos produced in the atmosphere. Since it was switched on in 2013, it recorded 37.9 Kton-years of atmospheric neutrinos [359].

The MINOS and MINOS+ experiments were also able to search for the subdominant appearance of electron neutrinos in the muon neutrino beam. This channel, being subdominant, must always been considered in the case of three neutrino flavours, and the main aim of this search is the measurement of the angle $\theta_{13}$ (see Eq. (6.80)) [354]. These analyses found the value of $\theta_{13}$ to be greater than zero, although their significance did not rise above the ones of other experiments, as for instance T2K [360]. Let us also remind that when electron neutrino traverse matter, they can have charged-current interactions with the electrons in the matter which give rise to a change in the oscillation pattern, as seen in Sect. 6.6. In LBS neutrino experiments aiming at precision the MSW effect cannot be neglected in the $\theta_{13}$ analysis [354]. The final results, from a full three flavour oscillation analysis, using the complete set of beam and atmospheric data taken at MINOS and MINOS+ experiments, yielded $|\Delta m_{32}^2| = 2.40^{+0.08}_{-0.09} \left[2.45^{+0.07}_{-0.08}\right] \times 10^{-3}$ (eV)$^2$ and $\sin^2\theta_{23} = 0.43^{+0.20}_{-0.04} \left[0.42^{+0.07}_{-0.03}\right]$ at 68% confidence level, for normal [inverted] order [361].

### 8.2.2.3 ICARUS and OPERA

In Europe, projects involving LBL accelerators started with the design in the late 1990s of a new neutrino beam at CERN directed toward the underground LNGS laboratory, located 730 km away. This neutrino beam was appropriately named CNGS (CERN Neutrinos to Gran Sasso). Protons at energy of 400 GeV were extracted from the SPS accelerator at CERN and transported along a 840 m long beam-line to a carbon target where kaons and pions were produced. The positively charged pions and kaons were selected by energy and guided in the direction of LNGS. They decayed into muons and muon-neutrinos in a vacuum tube of length 1 km and diameter 2.5 m. At LNGS two experiments, ICARUS (Imaging Cosmic and Rare Underground Signals) and OPERA (Oscillation Project with Emulsion Racking Apparatus), were ready to receive the CNGS neutrino beam. The CNGS beam was commissioned in 2007 and operated until 2012 [362]. One major limitation with respect to the competing facilities was the absence of a near detector, which was not possible to add in any practical way.

The ICARUS experiment [363] performed sensitive searches for anomalous electron neutrino appearance, as well as non-accelerator (atmospheric) physics, which contributed to constrain the allowed parameter space. The detector was a 760-ton liquid argon time projection chamber (LAr-TPC). Liquid argon is an attractive material for particle detection. Energy deposition in the argon produces scintillation light and ionization electrons; in the presence of a moderate electric field, a significant fraction of the electrons escape recombination, making them available for detection as a charge signal. That the liquid has a relative density of $1.4\,g/cm^3$ and that the free electrons can be drifted many meters with minimal dispersion makes it particularly attractive for use in massive time-projection chambers (TPC). The time-projection chamber was invented in the 1970s to detect and identify particles by recreating their tracks. Its basic layout consists of a cylindrical or square field cage that is filled with a gaseous (or liquid) detection medium. Charged particles produce tracks of ionization electrons that drift in a uniform electric field created between a cathode and an anode (see Fig. 8.4) with an average velocity known as drift velocity. At the anode plane, the electrons can be detected on the readout plane which is segmented in the directions perpendicular to the drift direction. By also knowing the arrival times of the drifted electrons, the track of the initial ionizing particle can be determined in three dimensions. The arrival time is the difference between the electron signal at the anode and the time of the neutrino interaction. The latter is obtained from the timing of the scintillation light emitted in the medium in the interaction and recorded by photomultipliers. The LAr-TPC detector combines the capabilities of the bubble chambers to provide a three dimensional imaging of charged particle with the excellent energy measurement of huge electronic detectors. The ICARUS experiment proved for the first time that LAr-TPC technology could be implemented at large scales. After the three-year physics run (2010–2012), the detector underwent a significant overhauling at CERN and was transferred to Fermilab. Liquid argon has been the choice of the Fermilab program investigating the possibility of additional, sterile, neutrinos (see Sect. 10.5.2).

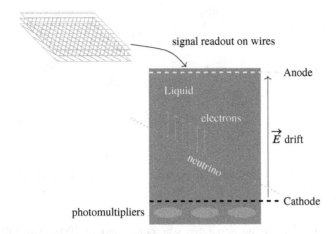

**Fig. 8.4** Schematic view of a time-projection chamber (TPC)

While the main goal of the LBS programmes in Japan and US was the muon neutrino disappearance, the OPERA experiment [364] was designed to perform the first detection of neutrino oscillations in direct appearance mode in the $\nu_\mu \to \nu_\tau$ channel, i.e. observing explicitly the appearance of $\nu_\tau$ by detecting the $\tau$ leptons produced by current charged events in the detector. The neutrino beam energy ($E_\nu \simeq 17$ GeV) was well above the kinematic threshold for the production of the tau lepton. In order to met the challenge to detect the short-lived (1 mm) $\tau$ leptons, the experiment exploited the nuclear emulsion technique, which, in 1947, had allowed, for the first time, the observation of the pion while studying cosmic rays in the Earth's atmosphere. Charged particles leave traces on nuclear emulsions observable with optical microscopes after some chemical processing, similarly to the now obsolete photographic plates.

In the OPERA experiment the main target was a hybrid emulsion/electronic apparatus made of "bricks" ($12 \times 10 \times 7\,\text{cm}^3$ of volume). Inside the bricks emulsion plates were interleaved with lead sheets, 1 mm thick, in order to reconcile the precision of space measurements (of one thousandth of a millimeter) with the need to obtain a large number of interactions (thanks to the high atomic weight of lead). The interaction vertex of $\nu_\tau$ candidates were at first roughly located by means of standard tracking techniques that identified the right brick. Then, offline, the brick was opened and the primary interaction vertex located by careful scanning of the emulsion foils. The overall mass of the lead was of 1250 tons, and the whole experiment weighted more than 3500 tons. The apparatus also included robotic machines for the construction of the bricks and their extraction during the data taking. Thanks to the use of emulsions, OPERA was the only experiment capable to observe all three types of neutrino, distinguishing the lepton produced in the interactions: the tau as an unstable particle decaying into a space of 1 mm, the muon as an extremely penetrating particle, and the electron as a particle producing a shower of other particles.

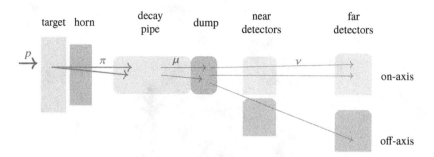

**Fig. 8.5** Schematic view of a generic LBL experiment, with near and far detectors in on- and off-axis configurations

The construction of the apparatus was completed in 2008. The full data set was collected between 2008 and 2012 and it was based on $1.8 \times 10^{20}$ POT. Like others long-baseline experiments, OPERA used a GPS based system for timing and synchronisation with the beam which allowed an experimental measurement of the neutrino velocity. In 2011 the announcement of faster then light neutrinos by the OPERA collaboration [365] made international headlines, until further cross-checks established that this result was due to an unfortunate experimental mistake [366].[9]

The OPERA experiment has not found any candidate for $\nu_\mu \to \nu_e$, setting the limit $\sin^2(2\theta_{13}) < 0.43$ (90% C.L.) [368]. On the other hand, it has observed 5 $\nu_\tau$ neutrino candidates, the first candidate event being observed in 2010 [369]. In a subsequent analysis based on looser selection experimental criteria, a sample of ten events was collected [370]. OPERA gave the first measurement of $\Delta m_{23}^2$ in appearance mode which, assuming maximal mixing ($\sin^2(2\theta_{23}) = 1$), yields the value [371] $\Delta m_{23}^2 = (2.7^{+0.7}_{-0.6}) \times 10^{-3}$ (eV)$^2$.

### 8.2.3 Off-Axis Technique

An experimental technique that started to be employed in LBL experiments, since about a couple of decades, is the off-axis (OA) technique mentioned in Sect. 8.2. In a LBL experiment, after the protons impinge on the target, the secondary hadron beam is created. In an off-axis experiment, the detector is moved a few degrees off with respect to the neutrino beam, that is with respect to the target, the secondary beam optics and the decay tunnel (see Fig. 8.5). The OA technique allows to obtain a flux of neutrinos with a narrow energy spectrum, as we will see in the following.

---

[9] It has to be noted that the ICARUS experiment, on the same CNGS beam line, had not detected faster-than-light neutrinos [367].

## 8.2 Accelerator Experiments

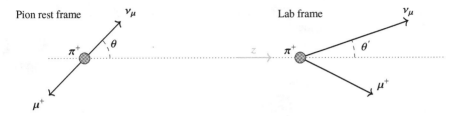

**Fig. 8.6** Kinematics of charged pion decay into leptons in the pion rest frame (left) and lab frame (right)

Let us consider the decay $\pi^+ \to \mu^+ \nu_\mu$, and set the $z$ axis along the pion direction.[10] Because of four-momentum conservation we have

$$p_\mu = p_\pi - p_{\nu_\mu} \tag{8.8}$$

which squared yields

$$m_\mu^2 = m_\pi^2 + m_{\nu_\mu}^2 - 2 p_\pi \cdot p_{\nu_\mu}. \tag{8.9}$$

We neglect the muon neutrino mass since $m_{\nu_\mu}^2 \ll m_\pi^2$ and obtain, in the rest frame of the pion

$$m_\mu^2 \simeq m_\pi^2 - 2 m_\pi E_{\nu_\mu} \tag{8.10}$$

which gives the neutrino energy

$$E_{\nu_\mu} \simeq \frac{m_\pi^2 - m_\mu^2}{2 m_\pi} \approx 30 \text{ MeV}. \tag{8.11}$$

The pion has spin zero, so the decay is isotropic in the rest frame of the pion. We identify the decay plane with the $x$-$z$ plane, and call $\theta$ the angle between the neutrino direction and the $z$ axis (see Fig. 8.6, left). Thus the neutrino 4-momentum is

$$p_{\nu_\mu} = (E_{\nu_\mu}, E_{\nu_\mu} \sin\theta, 0, E_{\nu_\mu} \cos\theta) \tag{8.12}$$

in the pion rest frame.

Now let us consider a reference frame where the pion moves with velocity $\beta_\pi = v_\pi$ on the positive verse of the zeta axis (see Fig. 8.6, right). This is a frame which moves with uniform velocity $-\beta_\pi$ with respect to the frame where the pion is at rest, that is along the negative verse of the $z$ axis. We call this frame the laboratory frame or lab frame for short. If the two frames coincide at a fixed initial time ($t = 0$), the

---

[10] The same kinematic reasonings holds for $\pi^- \to \mu^- \bar{\nu}_\mu$.

energy and momentum for each particle viewed in the two frames are given by the Lorentz transformations, mentioned in Sect. 2.4

$$\begin{pmatrix} E' \\ p'_z \end{pmatrix} = \begin{pmatrix} \gamma_\pi & \gamma_\pi \beta_\pi \\ \gamma_\pi \beta_\pi & \gamma_\pi \end{pmatrix} \begin{pmatrix} E \\ p_z \end{pmatrix} \tag{8.13}$$

where the primed energy and momentum refer to the lab frame and $\gamma_\pi \equiv (1 - \beta_\pi^2)^{-1/2}$ is the Lorentz boost. The momenta along the $x$ and $y$ direction do not change. In the frame where the pion is at rest we have $E_\pi = m_\pi$ and $\mathbf{p}_\pi = 0$. In the lab frame according to Eq. (8.13) it turns into $E'_\pi = \gamma_\pi m_\pi$, yielding

$$\gamma_\pi = E'_\pi / m_\pi \tag{8.14}$$

The neutrino 4-momentum in the lab frame is

$$p'_{\nu_\mu} = (E'_{\nu_\mu}, E'_{\nu_\mu} \sin\theta', 0, E'_{\nu_\mu} \cos\theta') \tag{8.15}$$

where the $\theta'$ angle is the angle between $z$ and the direction of the neutrino, as indicated in Fig. 8.6, on the right. By applying the Lorentz transforming Eq. (8.13) to the 4-momentum $p_{\nu_\mu}$ in Eq. (8.12), we obtain

$$p'_{\nu_\mu} = (\gamma_\pi E_{\nu_\mu}(1 + \beta_\pi \cos\theta), E_{\nu_\mu} \sin\theta, 0, \gamma_\pi E_{\nu_\mu}(\beta_\pi + \cos\theta)). \tag{8.16}$$

From Eqs. (8.15) and (8.16) it follows that

$$\tan\theta' = \frac{\mathbf{p}'_{\nu_\mu x}}{\mathbf{p}'_{\nu_\mu z}} = \frac{E_{\nu_\mu} \sin\theta}{\gamma_\pi E_{\nu_\mu}(\beta_\pi + \cos\theta)}. \tag{8.17}$$

If one wants to express the neutrino energy in terms of the lab angle $\theta'$, one has to equate the four-momenta in Eqs. (8.15) and (8.16), consider the first and the last equalities, and cancel $\cos\theta$ among them. It yields

$$E'_{\nu_\mu} = \frac{E_{\nu_\mu}}{\gamma_\pi (1 - \beta_\pi \cos\theta')}. \tag{8.18}$$

In neutrino off-axis experiments, pions are usually relativistic, having $E_\pi \gg m_\pi$. It follows that $\gamma_\pi \gg 1$ and $\beta_\pi \approx 1$. Then Eq. (8.17) gives

$$\tan\theta' \approx \frac{E_{\nu_\mu} \sin\theta}{\gamma_\pi E_{\nu_\mu}(1 + \cos\theta)} \approx \frac{E_{\nu_\mu} \sin\theta}{E'_{\nu_\mu}}. \tag{8.19}$$

## 8.2 Accelerator Experiments

Let us now make the approximation of small $\theta'$ angles. In this approximation $\tan\theta' \simeq \theta'$. Since $\sin\theta$ cannot exceed unity, $\theta'$ cannot exceeds the value $\theta'_{max} \simeq E_{\nu_\mu}/E'_{\nu_\mu}$. Using both approximations, small $\theta'$ and $\gamma_\pi \gg 1$, Eq. (8.18) becomes

$$E'_{\nu_\mu} \simeq \frac{E_{\nu_\mu}}{\gamma_\pi\left[1 - \beta_\pi(1 - \theta'^2/2)\right]} = \frac{\gamma_\pi(1+\beta_\pi)E_{\nu_\mu}}{1 + \gamma_\pi^2\theta'^2\beta_\pi(1+\beta_\pi)/2}$$

$$\simeq \frac{2\gamma_\pi E_{\nu_\mu}}{1 + \gamma_\pi^2\theta'^2} = \frac{m_\pi^2 - m_\mu^2}{m_\pi^2}\frac{E'_\pi}{1+\gamma_\pi^2\theta'^2}. \quad (8.20)$$

The first and second formula in the first line come from setting $\cos\theta' \simeq 1 - \theta'^2/2$, multiplying numerator and denominator by $\gamma_\pi(1+\beta_\pi)$ and using the definition $\gamma_\pi^2(1-\beta_\pi^2) = 1$. The two formulas on the second line are obtained by assuming $\beta_\pi \approx 1$ and by substituting to the neutrino energy in the rest frame its value (8.11) and to $\gamma_\pi$ in the numerator the value of Eq. (8.14). The final result shows that in the case $\theta' = 0$ the neutrino energy is proportional to the pion energy, but if there is a small $\theta' \neq 0$ the neutrino energy is no longer proportional to the pion energy, because of the dependence on the pion energy in the denominator due to $\gamma_\pi$ (see Fig. 8.7). If one substitutes the values for the pion and muon masses in (8.20), and uses Eq. (8.14) in the denominator one obtains

$$E'_{\nu_\mu} \simeq 0.43\frac{E'_\pi}{1+\gamma_\pi^2\theta'^2} = 0.43\,m_\pi^2\frac{E'_\pi}{m_\pi^2 + E'^2_\pi\theta'^2}. \quad (8.21)$$

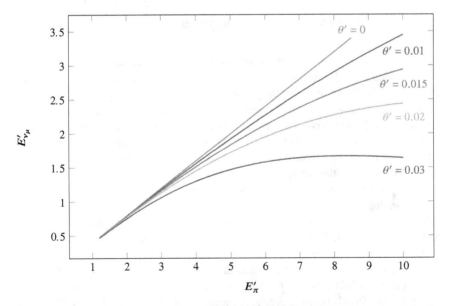

**Fig. 8.7** Qualitative plots of neutrino energy as function of the pion energy and the angle $\theta'$ in the laboratory frame

By differentiating with respect to pion energy, one finds

$$\frac{dE'_{\nu_\mu}}{dE'_\pi} = 0.43\, m_\pi^2 \frac{E'_\pi}{(m_\pi^2 + E'^2_\pi \theta'^2)^2} (m_\pi^2 - E'^2_\pi \theta'^2) \tag{8.22}$$

which is zero when $E'_\pi \theta' = m_\pi$. It means that, given a value $E'_\pi$, the curves shown in Fig. 8.7 peak at a specific value of the angle, namely $\theta' \equiv 1/\gamma_\pi = m_\pi/E'_\pi$. The peak neutrino energy $E'^{max}_{\nu_\mu}$ as a function of the angle is obtained by substituting $E'_\pi = m_\pi/\theta'$ in Eq. 8.21, which yields $E'^{max}_{\nu_\mu} \simeq 0.043\, m_\pi/2\theta'$, namely

$$E'^{max}_{\nu_\mu}(\text{GeV}) \simeq \frac{0.03}{\theta'}. \tag{8.23}$$

Summarising, if the detector is placed a few degrees off the beam axis, the neutrino energy is no longer proportional to the pion energy but rather almost all pions in a broad energy range give neutrinos in a narrow energy interval around $E'^{max}_{\nu_\mu}$, which depends only from the chosen off-axis angle. Two LBL experiments, the past T2K [372] experiment in Japan and the still running NOvA [373] experiment in US, have employed off-axis detectors. Given the baseline $L$, the choice of a suitable off-axis angle makes possible to adjust the beam peak energy $E$ in order to tune $L/E$ at the maximum of neutrino oscillations. Further details on T2K and NOvA experiments will be given in Sect. 8.2.4.

### 8.2.4 New Generation LBL Experiments

The K2K and KamLAND experiments had confirmed that neutrino oscillations were the explanation for the surprising observations of Super-K and SNO. But all of these experiments had seen only the disappearance of neutrinos, and it was therefore time for an experiment to observe the appearance of neutrinos. The T2K experiment in Japan was built to observe the appearance of electron neutrinos in a muon-neutrino beam. The objective was met in 2011, when the T2K collaboration reported the appearance of electron neutrinos from a muon-neutrino beam, which was the first significant neutrino-flavour appearance signal [372].

After the first generation of long-baseline experiments reached their goals, the priority in designing new experiments became studying the electron-muon neutrino sub-dominant oscillations, $CP$ violation in neutrino oscillations and the ordering of neutrino masses. The NOvA experiment in US, which is currently running, and the two forthcoming experiments T2HK and DUNE, are poised to address many of the remaining unknowns related to neutrino masses and their mixing. New methods and techniques have been developed which greatly improve the experimental sensitivity, e.g. using an off-axis beam to precisely measure $\sin^2 2\theta_{13}$ (T2K, NOvA, T2HK), or

8.2 Accelerator Experiments

unfolding $CP$ violation from matter effects (DUNE). The experiments T2HK and DUNE will also benefit of new, very high intensity, beams.

### 8.2.4.1 T2K

In Japan the T2K (Tokai to Kamioka) experiment [372] started in 2010 and ended its 10th data-taking run in 2020, with plans for major upgrades. It was the first long baseline neutrino oscillation experiment proposed and approved to look explicitly for the electron neutrino appearance from the muon neutrino, thereby measuring $\theta_{13}$. It is the successor of the K2K experiment and it uses the same far detector than K2K, the 50 kton water Cherenkov detector Super-Kamiokande. K2K and T2K are conceptually similar, but T2K benefits of higher beam power and higher precision in the near detectors. To achieve higher beam power, T2K uses the new accelerator complex at the Japan Proton Accelerator Research Complex (J-Parc) in Tokai Village on the east coast of Japan, about 70 km from KEK. It accelerates protons to 30 GeV, namely 2.5 times more than the PS at K2K. Protons impinge on a graphite target, and interactions in the target produce hadrons, which are focused using magnetic horns. The polarity of the magnetic field produced by the horns is reversible, allowing for the selection of positively (negatively) charged hadrons which then decay into a beam dominated by muon neutrinos (anti-neutrinos). In 2014, T2K began to operate with a muon antineutrino beam. Differences between the oscillations of neutrinos and antineutrinos could unveil $CP$ violation in the neutrino sector.

T2K adopts the off-axis method (see Sect. 8.2.3) to generate a narrow-band neutrino beam. The neutrino beam is purposely directed at an angle with respect to the baseline connecting the proton target and the far detector. The off-axis angle is set at $2.5^0$ so that the narrow-band muon-neutrino beam generated toward the far detector has a peak energy at $E \simeq 0.6$ GeV, which maximizes the effect of the neutrino oscillation at the T2K baseline of 295 km, namely the distance between J-PARC and Super-Kamiokande. The near detector site at about 280 m from the production target houses on-axis and off-axis detectors. The on-axis detector, named INGRID, measures the neutrino beam direction and profile. The suite of off-axis detectors (ND280) measures the muon neutrino flux and energy spectrum, and intrinsic electron neutrino contamination in the beam in the direction of the far detector.

The T2K collaboration has shown analyses of neutrino and antineutrino oscillations both in appearance and disappearance channels. A first important result by T2K was published in 2011 [372] by observing an indication of $\nu_\mu \to \nu_e$ appearance. It represented the first indication of a non-zero value of $\theta_{13}$. Soon after this result the experiment had to stop for more than one year due to the earthquake-tsunami that occurred in Japan in March 2011. In 2014 the T2K collaboration has provided electron neutrino appearance with $7.3\sigma$ experimental uncertainty and an independent measurement of $\theta_{13}$ [374]. In 2018 T2K had detected 89 $\nu_e$ and 7 $\bar{\nu}_e$ [375]. These counts should have been closer to 68 and 9, respectively, if $CP$

symmetry had been unbroken. The discrepancy indicated $CP$ violation within a 95% confidence interval [375]. That was a strong hint for $\delta \neq 0$ in the lepton sector, as detailed at the end of Sect. 6.4. In 2020 the T2K collaboration suggested the constraint $\pi < \delta < 2\pi$ [200]. More recently, it has provided a best fit value of $\delta = -1.97^{+0.97}_{-0.70}$ and excluded CP-conserving values of $\delta$ of 0 and $\pi$ at the 90% confidence level [376].

Let us now review T2K results on $\Delta m^2_{32}$ and $\sin^2 \theta_{23}$ given by measurements of muon neutrino and antineutrino disappearance, following the increase of protons on target (POT). The 2017 results were based on $22.5 \times 10^{20}$ POT, yielding the values (in the normal [inverted] order):

$$\Delta m^2_{32} = (2.54 \pm 0.08) \, [(-2.51 \pm 0.08)] \times 10^{-3} \, (\text{eV})^2 \quad (8.24)$$

$$\sin^2 \theta_{23} = 0.55^{+0.05}_{-0.09} \, [0.55^{+0.05}_{-0.08}]. \quad (8.25)$$

Muon neutrino and antineutrino disappearance probabilities are identical in the standard three-flavour neutrino oscillation framework, but $CPT$ violation and non-standard interactions can violate this symmetry. Thus the T2K collaboration has reported measurements independently for neutrinos and antineutrinos. In 2020, with $14.9 \, [16.4] \times 10^{20}$ POT for $\nu \, [\bar{\nu}]$, they found, by assuming normal neutrino mass ordering,

$$\Delta m^2_{32} = (2.47^{+0.08}_{-0.09}) \, [(2.50^{+0.18}_{-0.13})] \times 10^{-3} \, (\text{eV})^2 \quad (8.26)$$

$$\sin^2 \theta_{23} = 0.51^{+0.06}_{-0.07} \, [0.43^{+0.21}_{-0.05}], \quad (8.27)$$

for neutrinos [anti-neutrinos]. No significant difference between results for $\nu_\mu$ and $\bar{\nu}_\mu$ was observed, consistent with the standard neutrino oscillation picture [377]. In 2023, given $1.97 \, [1.63] \times 10^{21}$ POT for $\nu \, [\bar{\nu}]$, the results have been [378]

$$\Delta m^2_{32} = (2.48^{+0.05}_{-0.06}) \, [(2.53^{+0.10}_{-0.11})] \times 10^{-3} \, (\text{eV})^2 \quad (8.28)$$

$$\sin^2 \theta_{23} = 0.47^{+0.11}_{-0.02} \, [0.45^{+0.16}_{-0.04}] \quad (8.29)$$

No significant deviation was observed between the neutrinos and antineutrino case.

### 8.2.4.2 NO$\nu$A

The NO$\nu$A experiment [373], which has recorded its first neutrinos in 2014, exploits the Fermilab NuMI neutrino beam-line, the same, although upgraded, of the MINOS experiment. The far detector is situated off-axis at 810 km in northern Minnesota, not very far from the MINOS detector. The off-axis location, about $0.8^0$, of the detector maximises the number of detected 2 GeV neutrinos, the energy at which oscillations are expected to be maximum with the distance of 810 km. The near detector is located underground at Fermilab, approximately 1 km from the

## 8.2 Accelerator Experiments

production target. Both detectors are liquid scintillator detectors, and the far detector is the largest neutrino detector ever operated on surface, measuring $15 \times 15 \times 60$ m. It is made of 14 kton (about 344,000 cells) of extruded, highly reflective plastic (PVC) cells filled with liquid scintillator. Each cell in the far detector measures about 4 cm wide, 6 cm deep and 15 m long, that is it extends the full width of the detector. The cells are disposed in alternating planes in order to provide two orthogonal two-dimensional views of particle trajectories. The fine segmentation allows to separate electron showers initiated by electron neutrino charged current interactions from muons. Muons produced in charged-current $\nu_\mu$ interactions leave long straight tracks of detector activity that can span hundreds of cells. Electrons, in contrast, create more compact electromagnetic showers with well-characterised longitudinal and transverse profiles. This discrimination is important since the experiment focus on $\nu_e$ appearance and the measurement of the $\theta_{13}$ mixing angle. The near detector, about 300 ton, is functionally identical to the far detector, and measures the rate, energy spectrum and flavour composition of the neutrino beam prior to significant flavour oscillations.

Based on $8.85 \times 10^{20}$ protons on target (POT), in the time interval 2014–2017, the NOνA Collaboration has measured appearance of $\nu_\mu \to \nu_e$ and disappearance of $\nu_\mu \to \nu_\mu$. Joint analyses of these two processes have given the values $\Delta m^2_{32} \simeq (2.35 - 2.52) \times 10^{-3}$ (eV)$^2$, $\sin^2 \theta_{23} \simeq (0.43 - 0.51)$ or $(0.52 - 0.60)$ and $\delta \simeq (0 - 0.12\,\pi)$ or $(0.91\,\pi - 2)$ [379]. By using $12.33 \times 10^{20}$ POT, the NOνA experiment has recorded 27 $\bar{\nu}_\mu \to \bar{\nu}_e$ candidates [380]. This new antineutrino data has been combined with NOνA neutrino data [379], leading to $|\Delta m^2_{32}| = (2.48^{+0.11}_{-0.06}) \times 10^{-3}$(eV)$^2$ and $\sin^2 \theta = 0.56^{+0.04}_{-0.03}$ in the normal order. NOνA's latest measurements of neutrino oscillation parameters use data recorded between 2014 and 2020 and corresponding to $13.6 \times 10^{20}$ POT of neutrino beam and $12.5 \times 10^{20}$ POT of antineutrino beam. The resulting best-fit oscillation parameters (and their $1\sigma$ allowed ranges) are [202]:

$$\Delta m^2_{32} = (2.41 \pm 0.07) \times 10^{-3} \text{ (eV)}^2$$
$$\sin^2 \theta_{23} = 0.57^{+0.03}_{-0.04}$$
$$\delta = \left(0.82^{+0.27}_{-0.87}\right) \pi \tag{8.30}$$

Let us observe that there are differences between the current best fit values in T2K and NOνA analyses of $CP$ violation, but that there are regions of overlap in their $\sin^2 \theta_{23} \times \delta$ contours [202].

### 8.2.4.3 HK, T2HK and DUNE

Two beams of very high intensity (beam power exceeding 1 MW) will drive the field of LBL accelerator neutrino physics in the next decade, one produced at the upgraded J-PARC neutrino beamline, and the other produced at the Fermilab Long

Baseline Neutrino Facility (LBNF), serving, respectively, the Hyper-Kamiokande (HK) experiment and the Deep Underground Neutrino Experiment (DUNE). Both experiments are designed to address $CP$ violation, perform precision measurements of neutrino oscillation parameters, determine the mass hierarchy using atmospheric neutrinos (HK) or matter effects in beam (DUNE). Other science goals are search for proton decay as well as detection and measurement of the electron neutrino flux from core-collapse supernovae within our galaxy (should any occur).

The HK experiment, whose construction has been approved in early 2020, is expected to start data-taking around 2027 [381]. The HK detector represents the successor of the highly successful SK underground water Cherenkov detector. The detector tank will be cylindrical, 71 m high and with a diameter of 68 m, filled with pure water. Its effective mass for the physics analysis will amount to about 190 kton, namely 8.4 larger than that of SK. The cavern that will house the HK detector has a rock overburden of 650 m, below the peak of Mount Nijugo. In October 2023, the excavation of the dome section was completed. With 69 m of diameter and 21 m of height it is one of the largest human-made underground space.

HK will study the properties of neutrinos from various sources. It will provide unique sensitivity to core-collapse supernova neutrinos, and, due to an unprecedented statistical power, it could be able to measure short-period flux variations in solar neutrinos, realising a real-time monitoring of the solar core. Using atmospheric neutrinos, HK should have the sensitivity to determine the neutrino mass ordering due to the matter effects in the Earth.

HK will increase the existing sensitivity to proton decay. The most sensitive decay channel is $p \rightarrow e^+\pi^0$, as depicted in Fig. 8.8. This channel is dominant in a number of Grand Unified Theories, and represents a nearly model independent reaction mediated by the exchange of a new heavy gauge boson with a mass at the grand unification scale. Free protons decay back-to-back into a positron and a neutral pion, then neutral pions promptly decay into two photons in 98.8% of the times. In the final state one expects three well visible Cherenkov rings and

**Fig. 8.8** The $p \rightarrow e^+\pi^0$ decay, with indicated the three Cherenkov rings

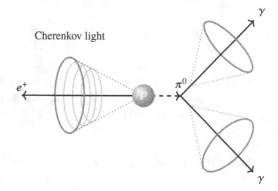

## 8.2 Accelerator Experiments

well defined kinematics.[11] It is to be noted that each photon does not directly produce Cherenkov light, since it is not charged, but it can convert, in matter, into an electron and a positron, emitted in the photon direction, which in turn radiate photons, which again can produce $e^+ e^-$ pairs, and so on. This process continues until the energies of the $e^+ e^-$ pair produced drop below a critical energy where they lose energy by atomic collisions rather than radiating photons. The cascade of secondary electrons, positrons and photons, initiated by the interaction with matter, is indicated as electromagnetic shower.[12] The nearly parallel electrons and positrons in the shower can emit Cherenkov light, which combines to make a Cherenkov ring.

In addition to operating as a standalone experiment, HK will serve as the far detector of a proposed long-baseline neutrino experiment, also known as T2HK (Tokai to Hyper-Kamiokande) [382]. It will employ a design similar to that of the T2K experiment, with similar off-axes angle and baseline. The beam, with 2.6 times beam power, will travel about 295 km, from J-PARC to HK. The main focus of this experiment will be $CP$ violation in neutrinos.

In the planned DUNE experiment[13] [383, 384] the new, high-intensity, on-axis, wide-band neutrino beam from Fermilab will be directed towards a far detector housed about 1.5 km underground at the Sanford Underground Research Facility (SURF) in the Homestake Mine (South Dakota), with a baseline of 1300 km. The detector will be a massive liquid argon time-projection chamber (LAr-TPC). The DUNE near detector, located at a distance of 574 m from the neutrino source, will be used to characterise the neutrino beam flux and flavour composition.

### 8.2.5 SBL Accelerator Experiments

The earliest accelerator experiments investigating neutrino oscillations were short-baseline experiments. They exploited, often with minimal modifications, neutrino beams and detectors designed and optimised for other physics topics. Accelerator SBL experiments have a typical source-detector distance ranging from tens of meters to about 1 km, and beam energy between MeV and GeV, which gives a sensitivity of $\Delta m^2 \gtrsim 1$ eV$^2$. Various experiments in the late 1970s and 1980s carried out oscillation searches that set only limits on $\nu_\mu$ ($\bar{\nu}_\mu$) disappearance and on $\nu_e$ ($\bar{\nu}_e$), $\nu_\tau$ ($\bar{\nu}_\tau$) appearance in $\nu_\mu$ ($\bar{\nu}_\mu$) beams.

---

[11] The HK target material is water. All the protons in an H$_2$O molecule are assumed to decay with equal probability. There are two protons coming from hydrogen and eight protons from oxygen. A proton coming from an hydrogen atom can be considered as a free proton decaying at rest. So, the momenta of the decay particles are uniquely determined by two-body kinematics. In the case of a proton coming from an oxygen atom, we have a bound state; nuclear effects and pion final state interactions complicate this simple scheme.

[12] An electromagnetic shower can also be initiated by the interaction with matter of the positron in the final state of the $p \to e^+ \pi^0$ decay.

[13] It is expected to start in 2029.

In the 1990s two SBL experiments, CHORUS (CERN Hybrid Oscillation Search Apparatus) [385] and NOMAD (Neutrino Oscillation MAgnetic Detector) [386], were built at CERN. They shared a wide band muon neutrino beam with an average energy of 24 GeV. Protons were accelerated by Super Proton Synchrotron (SPS). The detectors were located in the West Area Neutrino Facility (WANF) of CERN, at about 600 m distance from the beam production. Both experiments were motivated by the search for the $\nu_\mu - \nu_\tau$ oscillations, but adopted different search strategies.

In CHORUS the main active target consisted of four blocks of nuclear emulsions. CHORUS pioneered the use of nuclear emulsions in $\nu_\tau$ appearance searches. The neutrino beam, composed mainly by muon neutrinos, interacted with nucleons by charged current $\nu_\mu$ interactions. In the event of oscillations, one expected charged current $\nu_\tau$ interactions of tau neutrinos with nucleons $N$, namely $\nu_\tau + N \rightarrow \tau^+ + X$. As already remarked in the case of DONUT and OPERA experiments, the main advantage of emulsions is the excellent spatial resolution, of a few μm, necessary to detect the short-living tau leptons. Given their lifetime and the average energy of the CERN SPS neutrino beam, the tau leptons travel about 1 mm before decaying. Data taking at CHORUS lasted from 1994 to 1997.

The NOMAD experiment was built to perform a short-baseline search of $\nu_\mu \rightarrow \nu_\tau$ and $\nu_\mu \rightarrow \nu_e$ oscillations in appearance mode. The detector was built over a period of 4 years starting in 1991 and was in operation during the next 4 years, from 1995 to 1998. It was completely electronic, composed of a set of drift chambers used as target and tracking medium. Neutrinos were interacting in the chamber walls, mostly composed of carbon, while the chambers were used for precise charged particle tracking. The detector was located in a dipole magnet, formerly used by the UA1 experiment, giving a field of 0.4 T. The spatial resolution of the NOMAD detector, though good, was not sufficient to resolve the short tracks of tau leptons. Instead, the decaying tau leptons were identified through the kinematics of their decay products. The NOMAD experiment pioneered $\nu_\tau$ appearance searches based on kinematical criteria applied to the final state of neutrino interactions.

Both these experiments assumed values of squared mass differences $\Delta m^2 \gtrsim 1$ eV$^2$, much larger than the values measured later on, a common prejudice in theoretical models at the time. Besides, large mixing angles were forecasted in the lepton sector, in analogy with the CKM mixing matrix, overlooking the (still weak) evidence for the atmospheric neutrino deficit, which suggested a different scenario. Both CHORUS and NOMAD did not observe $\nu_\mu - \nu_\tau$ oscillations, providing only limits on oscillation parameters.

Let us observe that short-baseline searches for neutrino oscillation with mass squared differences in the 0.1–10 eV$^2$ range are still an active research topic today, motivated by experimental hints suggesting neutrino oscillation beyond the PMNS paradigm. Fermilab is running a program on sterile neutrino searches with three experiments at short baselines, SBND at 110 m from the neutrino source, MicroBooNE at 470 m and ICARUS at 600 m. These experiments follow, and are largely motivated by, an excess of electron events first observed by the experiment LSND (Liquid Scintillator Neutrino Detector), that used neutrinos from the decay at rest of pions (DAR) [340]. This excess was not confirmed by another experiment operating with a similar set-up, the KARMEN (KArlsruhe Rutherford Medium Energy Neutrino) experiment [341, 387], while compatible results were reported by the MiniBooNE SBL experiment [388]. We discuss searches for sterile neutrinos in Sect. 10.5.2.

## 8.3 Oscillation Parameters Determination

As seen in Chap. 6, with three families of leptons (without sterile neutrinos), one can parameterise the neutrino mixing by the three mixing angles $\theta_{12}$, $\theta_{13}$, $\theta_{23}$ and the phase $\delta$ of the lepton mixing matrix. In case of Majorana neutrinos, there are additional phases, which play no role in neutrino oscillations.

Because of the observed hierarchy in the squared mass differences (6.78), neutrino oscillations can be approximately described using only the larger squared mass $\Delta m^2_{32} \simeq \Delta m^2_{31}$. Adding the approximation $\sin \theta_{13} \simeq 0$ and $\cos \theta_{13} \simeq 1$, based on the experimental result $\theta_{13} \ll \theta_{23}$, it is easy to see that the primary oscillation of a beam of muon neutrinos is into tau neutrinos (see Eq. (6.80)). By using Eqs. (6.83), we observe that the survival probability of a muon beam depends only on $\theta_{23}$

$$P(\nu_\mu \to \nu_\mu) \simeq 1 - \sin^2 2\theta_{23} \sin^2 \left( \frac{\Delta m^2_{32} L}{4E} \right). \tag{8.31}$$

Hence one can approximately interpret the results of oscillation experiments with muon neutrinos assuming only two-neutrino states. In this case, the mixing matrix depends on a single mixing angle $\theta_{23}$, and no $CP$ violation effect in oscillations is possible. The values for $\theta_{23}$ and $|\Delta m_{32}|$ can be measured in atmospheric neutrino experiments, such as Super-K, and accelerator LBL disappearance experiments, for instance K2K, MINOS, T2K and NOvA. In particular, the Super-K experiment in 1998 was the first compelling evidence of neutrino flavour oscillation [389]. There are by now solid indications of a large value for the angle $\theta_{23}$, but with a deviation from the maximal value, $\theta_{23} = \pi/4$, which would have implied $\sin^2 2\theta_{23} = 1$. A recent measurement at NOvA has given [202]

$$\sin^2 \theta_{23} = 0.57^{+0.03}_{-0.04}. \tag{8.32}$$

In Chap. 7 we have discussed solar neutrinos. Since they have an energy range of about $(0.2-19) \times 10^{-3}$ GeV and travel a Sun-Earth distance of roughly $L \simeq 1.5 \times 10^8$ km, analyses of solar neutrino oscillation data are sensitive to extremely small values of $\Delta m^2$. At the same time, the low energy range prevents the production of muon and tau leptons in charged current interactions with $\nu_\mu$ and $\nu_\tau$ from electron neutrino oscillations. Hence only electron neutrinos can be revealed and solar neutrino observations are in general disappearance measurements. As observed in Sect. 6.5.2, where a two-neutrino scenario and the approximation of oscillations in vacuum were adopted, the analyses of solar neutrino data can yield the values of $\theta_{12}$ and $\Delta m_{12}$. A recent result gives [203]

$$\sin^2 \theta_{12} = 0.310^{+0.013}_{-0.012}. \tag{8.33}$$

Given that both $\theta_{12}$ and $\theta_{23}$ are large, it seems natural to expect a sizable third mixing angle $\theta_{13}$, but it turns out it is not so. The mixing angle $\theta_{13}$ is particularly important since even a small non-zero value can potentially introduce $CP$ violation via the term $\delta$ in the lepton mixing matrix (6.8).

There are at least two ways to access $\theta_{13}$. Long baseline accelerator neutrino experiments are capable to measure the angle $\theta_{13}$ by observing muon neutrino to electron neutrino transitions. As can be seen by the simplified expression in Eq. (6.80), this oscillation mode will only occur if $\theta_{13}$ is non-zero, but also depend on other parameters, as the angle $\theta_{23}$. Prior to 2012, neutrino experiments had only been able to set limits on the mixing angle $\theta_{13}$. First indications for non zero value of $\theta_{13}$ came from MINOS [390] and T2K [372] experiments.

A more direct way to measure $\theta_{13}$ is to search for the disappearance of electron antineutrinos from a reactor. In Eq. (6.84) we have seen that in the two-neutrino approximation, survival and transition probabilities are the same for neutrino and antineutrinos. As discussed in Sect. 8.1, the most common way to detect reactor $\bar{\nu}_e$ is via the distinct signal provided by inverse beta decay. Given that reactor $\bar{\nu}_e$ energies extend only up to about 10 MeV, there is not enough energy to produce muon or tau leptons from the interactions of $\bar{\nu}_\mu$ and $\bar{\nu}_\tau$ coming from $\bar{\nu}_e$ oscillations. Therefore, $\bar{\nu}_e$ oscillations into $\bar{\nu}_\mu$ and $\bar{\nu}_\tau$ cannot be observed through the detection of the corresponding charged leptons, and can be revealed only by $\bar{\nu}_e$ disappearance. The survival probability depends directly on $\theta_{13}$, as seen from the approximate relation (6.82). The angle $\theta_{13}$ is often indicated as reactor angle.

A reactor experiments of this kind is the long baseline reactor experiment Double Chooz, which gave also early indications of an angle $\theta_{13}$ different from zero [327], reporting in 2012 that $\theta_{13} = 0$ hypothesis was disfavoured at $1.7\sigma$. These hints of a non-zero $\theta_{13}$ were followed the same year by the results of the Daya Bay SBL reactor neutrino experiment, which reported the discovery of a non-zero $\theta_{13}$ with $5.1\sigma$ significance [391], by RENO SBL reactor results, which confirmed DayaBay's finding with a $4.9\sigma$ significance [392], and, a year later, by new Daya Bay results, which increased the significance to $7.7\sigma$ using a larger data set [393]. A non-zero $\theta_{13}$ was firmly established. Final results by the Daya Bay experiment yield [329]

$$\sin^2 2\theta_{13} = 0.0851 \pm 0.0024 \tag{8.34}$$

in a three-neutrino framework, equivalent to $\theta_{13} \sim 8.5^0$.

In the past, the results of neutrino oscillation experiments were interpreted assuming two-neutrino states, but with accurate measurement of $\theta_{13}$ and the improved results of all experiments, we have now entered the era of global three flavours analysis. A global fit aims to combine all the experiments where the

## 8.3 Oscillation Parameters Determination

**Table 8.1** Three-flavour oscillation parameters from a fit to global data by Nufit Collaboration in Ref. [394], to which we refer for details. The numbers in the first (last) columns are obtained assuming normal (inverted) ordering, i.e. relative to the respective local minimum. Note that $\Delta m_{3\ell}^2 \equiv \Delta m_{31}^2 > 0$ for normal ordering and $\Delta m_{3\ell}^2 \equiv \Delta m_{32}^2 < 0$ for inverted ordering

| | Normal ordering (best fit) | | Inverted ordering ($\Delta\chi^2 = 2.3$) | |
|---|---|---|---|---|
| | bfp $\pm 1\sigma$ | $3\sigma$ range | bfp $\pm 1\sigma$ | $3\sigma$ range |
| $\sin^2\theta_{12}$ | $0.303^{+0.012}_{-0.011}$ | $0.270 \to 0.341$ | $0.303^{+0.012}_{-0.011}$ | $0.270 \to 0.341$ |
| $\theta_{12}/°$ | $33.44^{+0.77}_{-0.74}$ | $31.27 \to 35.86$ | $33.45^{+0.78}_{-0.75}$ | $31.31 \to 35.74$ |
| $\sin^2\theta_{23}$ | $0.572^{+0.018}_{-0.023}$ | $0.406 \to 0.620$ | $0.578^{+0.016}_{-0.021}$ | $0.419 \to 0.623$ |
| $\theta_{23}/°$ | $49.1^{+1.0}_{-1.3}$ | $39.6 \to 51.9$ | $49.5^{+0.9}_{-1.2}$ | $39.9 \to 52.1$ |
| $\sin^2\theta_{13}$ | $0.02203^{+0.00056}_{-0.00059}$ | $0.02020 \to 0.02391$ | $0.02219^{+0.00060}_{-0.00057}$ | $0.02047 \to 0.02396$ |
| $\theta_{13}/°$ | $8.54^{+0.11}_{-0.12}$ | $8.19 \to 8.89$ | $8.57^{+0.12}_{-0.11}$ | $8.23 \to 8.90$ |
| $\delta_{CP}/°$ | $197^{+42}_{-25}$ | $108 \to 404$ | $286^{+27}_{-32}$ | $192 \to 360$ |
| $\dfrac{\Delta m_{21}^2}{10^{-5}\,\mathrm{eV}^2}$ | $7.41^{+0.21}_{-0.20}$ | $6.82 \to 8.03$ | $7.41^{+0.21}_{-0.20}$ | $6.82 \to 8.03$ |
| $\dfrac{\Delta m_{3\ell}^2}{10^{-3}\,\mathrm{eV}^2}$ | $+2.511^{+0.028}_{-0.027}$ | $+2.428 \to +2.597$ | $-2.498^{+0.032}_{-0.025}$ | $-2.581 \to -2.408$ |

neutrino flavour oscillations are relevant to obtain a complete description of the neutrino evolution.

Different analyses find consistent results for squared mass differences and oscillation angles. The latter ones have been measured with at most 4% uncertainties. We report some results form a recent global fit [394] in Table 8.1 with both mass orderings. We see that while there is indication of a sizeable $CP$ violation, the evidence is not conclusive and $CP$ conservation (for $\delta \sim 180^0$) is still allowed at a confidence level of $1$–$2\sigma$. Data are consistent with mixing angles and mass splittings being the same for neutrinos and anti-neutrinos, in agreement with $CPT$ conserving oscillation mechanism.

# Chapter 9
# Neutrino Cross Sections

In the SM, neutrinos only interact with matter via the weak interaction, as seen in Sect. 4.3. Neutrinos partake of both charged (CC) and neutral (NC) current forms of the weak interaction. In CC mediated scattering of neutrinos, a charged $W$ gauge boson couples an incoming a neutrino (anti-neutrino) and an outgoing charged lepton (anti-lepton) at tree level. For instance, muon neutrinos scatter on a target of electrons by the process

$$\nu_\mu + e^- \to \mu^- + \nu_e. \tag{9.1}$$

The flavour of the charged lepton in the final state tags the flavour of the (coupled) neutral lepton in the initial state. Besides, because the lepton flavour number is conserved in the SM, its charge determines if it is a neutrino or an antineutrino. As already remarked in the previous sections, observing the CC interactions represents the common way to identify interacting $\nu_\mu$ and $\nu_\tau$ neutrinos, with the drawback that their energies need to be above the muon and tau production thresholds, respectively. In neutral current (NC) interactions, neutrinos exchange $Z^0$ bosons at tree level. Since neutrinos have no charge, photons cannot mediate their interactions. Both an incoming and an outgoing neutrino are in the process, as for instance in the $\nu_\mu$-electron elastic scattering

$$\nu_\mu + e^- \to \nu_\mu + e^- \tag{9.2}$$

The NC interaction is available to all neutrino flavours at all energies and it does not provide a method of determining the flavour of the incoming neutrino. For this reason, NC interactions can often represent a difficult background while searching for a specific neutrino flavour.

In all modern neutrino experiments, the target of neutrino beams are nuclei. Neutrino-nucleus interaction could be treated as an incoherent sum of interactions with free nucleons. This would be the simplest description, since it would rely only on the SM, which governs the basic reactions and decays, with radiative corrections that can be accurately calculated to many orders. Unfortunately, a realistic description of neutrino-nucleus interactions is significantly more complicated because of nuclear effects. The SM does not provide enough information on initial state conditions of the nucleons in the nuclei, nuclear corrections, final state interactions, and so on; hence a variety of theoretical models have stepped in.

Since the description of the nuclear and hadronic effects generally depends on energy, it is conventional to separate the analysis of neutrino-nucleus interactions for low, intermediate and high energy regions of the neutrino sources. However, one has to keep in mind that these distinctions are only approximate, and there is no clear-cut general separation.

At low energies, the interaction length is greater than the nuclear diameter. The nucleus appears to have no structure and the neutrino interacts with it as whole. Thus the initial and final states are specific nuclear levels and we are lead to consider exclusive scattering to specific bound (or resonance) nuclear states. Up to about 1 MeV, significative processes are the neutral current exchange where a neutrino interacts coherently with the nucleus (coherent scattering), observed for the first time in 2017 [395], and neutrino capture on radioactive nuclei, already mentioned in Sect. 7.5.1, not yet observed.

Above 1 MeV the interaction length reduces and neutrinos can interact with nucleons. The intermediate region up to tens of GeV is very difficult to model because the nuclear effects may become important.

At high energies, of order of 100 GeV, the interaction length is an order of magnitude lower than 1 fm and thus nuclear effects are depressed. One is typically interested in the inclusive scattering, summing over all possible nuclear final states. At highest energies, the neutrinos can probe individual quarks ignoring the presence of the nucleons—this is called the deep inelastic scattering (DIS) energy region.

In this section, we discuss neutrino interactions in the SM and cross sections, which are a measure of the probability of an interaction occurring. The number of neutrinos you expect in any detector is proportional to the cross section of the neutrino interactions, as well as to the neutrino flux and the density of the target. We assume a prior basic knowledge of the formalism and computations in particle physics,[1] and we focus on the applications to neutrino interactions.

We start describing the simplest interaction, with occurs with charged leptons (Sect. 9.1), and continue describing neutrino interactions with nucleons (Sect. 9.2), which present some analogies with the neutrino-lepton case. The Gargamelle neutrino experiment, revealing the first evidence for weak neutral currents, is described in Sect. 9.3. Neutrino-nucleus scattering is discussed in Sect. 9.4.

---

[1] See e.g. standard particle physics textbooks as Refs. [396, 397].

## 9.1 Neutrino-Electron Scattering

The simplest interaction of neutrinos with matter is the one with electrons, which at the lowest order in the weak interaction perturbation theory involves only free leptons. In the SM, neutrinos and leptons can interact via charged currents (CC) and neutral currents (NC). As seen in Sect. 4.2, the corresponding Lagrangians are

$$\mathcal{L}_{CC} = g\,(W_\mu^+ J_W^{+\mu} + W_\mu^- J_W^{-\mu})$$
$$\mathcal{L}_{NC} = g\,Z_\mu^0 J_Z^\mu \tag{9.3}$$

where $W_\mu^\pm$ and $Z_\mu^0$ represent the heavy gauge boson fields and $g$ is the weak coupling constant. To simplify the notation, in this section we have omitted the superscript $\ell$ from the leptonic currents. The leptonic charged weak current, $J_W^{+\mu}$, is defined as

$$J_W^{+\mu} = \frac{1}{\sqrt{2}} \sum_{\ell=e,\mu,\tau} \bar{\nu}_{\ell L}\gamma^\mu \ell_L. \tag{9.4}$$

The leptonic neutral current is defined as

$$J_Z^\mu = \frac{1}{\cos\theta_W} \sum_{\ell=e,\mu,\tau} g_L^\nu \bar{\nu}_{\ell L}\gamma^\mu \nu_{\ell L} + g_L^e \bar{\ell}_L \gamma^\mu \ell_L + g_R^e \bar{\ell}_R \gamma^\mu \ell_R. \tag{9.5}$$

For each left handed and right handed coupling of the fermion $f$, with charge $Q_f$ and value of third component of the weak isospin $T_3$, one has $g_L^f = T_3 - Q_f \sin^2\theta_W$ and $g_R^f = -Q_f \sin^2\theta_W$, with $\theta_W$ being the weak mixing angle. Hence, for neutrino and electrons one has $g_L^\nu = 1/2$, $g_L^e = -1/2 + \sin^2\theta_W$ and $g_R^e = \sin^2\theta_W$.

As mentioned in Sect. 5.8, at low neutrino energies, the effects of the $W_\mu^\pm$ and $Z_\mu^0$ propagators can be neglected, and weak processes can be described by effective Fermi Lagrangians. To be more precise, when the energy scales involved in a process, that is the transferred momentum squared and the masses of the particles, are much smaller than the masses $M_W$ and $M_Z$ of the W and Z gauge bosons (order of 100 GeV), the gauge boson propagators in momentum space can be approximated by the couplings

$$i\frac{g^{\mu\nu}}{M_Z^2} \text{ (Z boson)}, \quad i\frac{g^{\mu\nu}}{M_W^2} \text{ (W boson)} \tag{9.6}$$

and the boson lines in the Feynman diagrams can be contracted to a point. This leads to effective current–current Lagrangians describing NC and a CC four-fermion interactions

$$\mathscr{L}_{NC}^{eff} = -\frac{g^2}{M_Z^2} J_{Z\mu} J_Z^\mu$$

$$\mathscr{L}_{CC}^{eff} = -\frac{g^2}{M_W^2} J_{W\mu}^+ J_W^{-\mu}. \tag{9.7}$$

### 9.1.1 Elastic and QE Amplitudes

In neutrino interactions, one can distinguish between elastic and quasi elastic (QE) scattering, the difference being that in the former process the initial and final state are the same, while in the latter one the target is not broken, but its kind is changed.[2]

The elastic scattering of neutrinos (or anti-neutrinos) of flavour $\ell = e, \mu, \tau$ with electrons occurs through the process

$$\nu_\ell(\bar{\nu}_\ell) + e^- \to \nu_\ell(\bar{\nu}_\ell) + e^-. \tag{9.8}$$

It does not have an energy threshold, since the final state is the same as the initial state. The only effect of an elastic scattering process is a redistribution of the total energy and momentum between the two participating particles.

Let us discuss separately NC and CC interactions, starting with a pure NC exchange. We consider the elastic process (9.8) with $\ell = \mu$ or $\tau$. When electrons scatters with $\nu_\mu$ or $\nu_\tau$, only NC interactions contribute to the tree level processes. The corresponding tree level NC Feynman diagram is represented by the diagram in Fig. 9.1 (a) for neutrinos and (b) for antineutrinos. In the case of NC interaction, the electron has both left-handed and right-handed couplings. At tree level in perturbation theory, the matrix element of the effective NC Lagrangian describing four-fermion interactions is (see Eq. (9.7))

$$\mathcal{M}_{NC} = -\frac{g^2}{M_Z^2} J_Z'^\mu J_{Z\mu} \tag{9.9}$$

---

[2] In general, the term quasi-elastic designates a limiting case of inelastic scattering where the energy transfer is small compared to the incident energy of the scattered particles. It was originally coined in nuclear physics to distinguish collisions associated with small energy losses from the ones showing substantial amounts of energy and mass transfer.

## 9.1 Neutrino-Electron Scattering

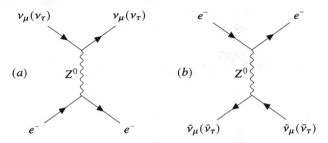

**Fig. 9.1** Feynman tree-level diagrams for NC scattering: (a) $\nu_\mu(\nu_\tau) + e^- \to \nu_\mu(\nu_\tau) + e^-$ (b) $\bar{\nu}_\mu(\bar{\nu}_\tau) + e^- \to \bar{\nu}_\mu(\bar{\nu}_\tau) + e^-$

with the currents which are appropriate for the NC scattering, namely

$$J_Z^{\prime\mu} = \frac{1}{\cos\theta_W} g_L^\nu \bar{\nu}_{\ell L} \gamma^\mu \nu_{\ell L} \qquad (9.10)$$

$$J_{Z\mu} = \frac{1}{\cos\theta_W} g_L^e \bar{e}_L \gamma_\mu e_L + g_R^e \bar{e}_R \gamma_\mu e_R. \qquad (9.11)$$

By exploiting the relation $M_W/M_Z = \cos\theta_W$, one can write

$$\mathcal{M}_{NC} = -4\sqrt{2} G_F \left(g_L^\nu \bar{\nu}_{\ell L} \gamma^\mu \nu_{\ell L}\right) \left(g_L^e \bar{e}_L \gamma_\mu e_L + g_R^e \bar{e}_R \gamma_\mu e_R\right) \qquad (9.12)$$

in terms of the Fermi coupling constant $G_F = g^2/(4\sqrt{2}M_W^2) \simeq 1.2 \times 10^{-5}$ GeV$^{-2}$, defined in Eq. (4.69).

$$\mathcal{M}_{NC} = -\sqrt{2} G_F \left[\bar{\nu}_\ell \gamma^\mu (g_V^\nu - g_A^\nu \gamma^5) \nu_\ell\right] \left[\bar{e}\gamma_\mu (g_V^e - g_A^e \gamma^5) e\right]$$
$$= -\frac{G_F}{\sqrt{2}} \left[\bar{\nu}_\ell \gamma^\mu (1 - \gamma^5) \nu_\ell\right] \left[\bar{e}\gamma_\mu (g_V^e - g_A^e \gamma^5) e\right]. \qquad (9.13)$$

The vector and axial-vector coupling constants are defined as

$$g_V^\nu = g_L^\nu + g_R^\nu = g_A^\nu = g_L^\nu - g_R^\nu = \frac{1}{2} \quad (g_R^\nu = 0)$$

$$g_V^e = g_L^e + g_R^e = -\frac{1}{2} + 2\sin^2\theta_W$$

$$g_A^e = g_L^e - g_R^e = -\frac{1}{2}. \qquad (9.14)$$

By modifying the couplings, it is easy to extend this formulation to describe neutrino-quark interactions: for any fermion $f$ one can write

$$g_V^f = g_L^f + g_R^f = \left(T_3^f - 2Q_f \sin^2\theta_W\right)$$
$$g_A^f = g_L^f - g_R^f = T_3^f. \tag{9.15}$$

We have, for instance, $g_V^u = \frac{1}{2} - 4/3 \sin^2\theta_W$ at tree level, and so on.

Now let us pass to analyse a pure CC interaction. The prototype process is the QE interaction

$$\nu_\ell + e \to \nu_e + \ell \quad (\ell = \mu \text{ or } \tau) \tag{9.16}$$

which is sometimes called inverse muon or inverse tau decay. It is similar in character to the elastic scattering, but the flavour of the final charged lepton changes. Hence there is an energy threshold of the initial $\nu_\mu$ ($\nu_\tau$) to allow the production of a muon ($\tau$). In the effective framework, the corresponding tree-level amplitude $\mathcal{M}_{CC}$ can be calculated from the expressions (9.7) and reads

$$\mathcal{M}_{CC} = -4\sqrt{2}\, G_F J_W^{\prime+\mu} J_W^{-\mu} \tag{9.17}$$

where

$$J_W^{+\mu} = \frac{1}{\sqrt{2}} \bar{\nu}_{\ell L} \gamma^\mu \ell_L$$

$$J_{W\mu}^{\prime+} = \frac{1}{\sqrt{2}} \bar{\nu}_{eL} \gamma_\mu e_L. \tag{9.18}$$

Eventually, we have

$$\mathcal{M}_{CC} = -2\sqrt{2}\, G_F \left[\bar{\ell}_L \gamma^\mu \nu_{\ell L}\right] \left[\bar{\nu}_{eL} \gamma_\mu e_L\right]$$
$$= -\frac{G_F}{\sqrt{2}} \left[\bar{\ell}\gamma^\mu(1-\gamma^5)\nu_\ell\right] \left[\bar{\nu}_e \gamma_\mu(1-\gamma^5)e\right] \tag{9.19}$$

which is casted in a form similar to the one in Eq. (9.13).

We can now consider the elastic scattering

$$\nu_e + e \to \nu_e + e \tag{9.20}$$

where both CC and NC amplitudes contribute to the tree level process, as shown in Fig. 9.2. We can calculate the contributions to the effective amplitude coming from the CC and NC current, along the lines described before, and add the two contributions. The total right handed coupling only comes from NC currents, because there is no right handed weak CC coupling. There are two left handed

## 9.1 Neutrino-Electron Scattering

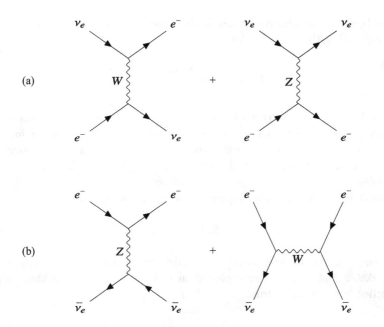

**Fig. 9.2** Feynman tree-level diagram for CC and NC scatterings: (a) $\nu_e + e^- \to \nu_e + e^-$ (b) $\bar{\nu}_e + e^- \to \bar{\nu}_e + e^-$

couplings, from NC and CC currents, and we add the associated amplitudes. In terms of vector and axial-vector coupling constants we have

$$\mathcal{M}_{NC+CC} = -\frac{G_F}{\sqrt{2}} \left\{ \left[ \bar{\nu}_e \gamma^\mu (1-\gamma^5) \nu_e \right] \left[ \bar{e} \gamma_\mu (g_V^e - g_A^e \gamma^5) e \right] + \right.$$
$$\left. - \left[ \bar{e} \gamma^\mu (1-\gamma^5) \nu_e \right] \left[ \bar{\nu}_e \gamma_\mu (1-\gamma^5) e \right] \right\}. \quad (9.21)$$

The pairing of the fermions in the CC contribution can be rearranged in the same form of the NC contribution with Fierz transformations. The result is equivalent to take Eq. (9.13) and shift $g_V^f \to g_V^f + 1$ and $g_A^f \to g_A^f + 1$, namely

$$\mathcal{M}_{NC+CC} = -\frac{G_F}{\sqrt{2}} \left[ \bar{\nu}_e \gamma^\mu (1-\gamma^5) \nu_e \right] \left[ \bar{e} \gamma_\mu ((1+g_V^e) - (1+g_A^e) \gamma^5) e \right]. \quad (9.22)$$

We observe that the interference between CC and NC contributions is constructive. In this elastic scattering all kinematic variables are accessible provided one can identify electrons and measure their energy. Assuming that incoming neutrino direction can be considered constant, neglecting the neutrino mass, and applying four-momentum conservation for a neutrino with four-momentum $p_\nu = (E_\nu, 0, 0, p) \simeq$

$(p, 0, 0, p)$ and an electron at rest in the laboratory frame with four-momentum $p_e = (m_e, 0, 0, 0)$, one gets [398]

$$1 - \cos \theta_e = \frac{m_e(1 - E_k^e/E_\nu)}{E_e} \tag{9.23}$$

where $E_k^e = E_e - m_e$ is the electron kinetic energy and $\theta_e$ is the angle of the outgoing electron with respect to the neutrino beam. As a consequence, the energy and direction of the electron provide the value of $E_\nu$. From a different standpoint, this same equation shows that this scattering is highly directional in nature. The outgoing electron is emitted at very small angles with respect to the incoming neutrino direction. Indeed, by Taylor expanding around $\theta_e = 0$ one finds

$$E_e \theta_e^2 \leq 2m_e. \tag{9.24}$$

The outgoing electron closely follows the direction of the incoming neutrino. This remarkable feature has been exploited in various experiments to determine the direction from which neutrinos arrive.

### 9.1.2 QE Scattering Cross Sections

Let us consider the inverse muon (or tau) decay in Eq. (9.16), that is the neutrino-electron CCQE scattering

$$\nu_\ell + e^- \to \ell^- + \nu_e \quad (\ell = \mu \text{ or } \tau). \tag{9.25}$$

Figure 9.3 shows the tree-level Feynman diagram for the inverse muon decay, together with lepton momenta assignments. The diagram for the inverse tau decay is the same, except for the substitution $(\nu_\mu, \mu^-) \to (\nu_\tau, \tau^-)$. Analogously to the inverse $\beta$ decay, this process can take place only if the neutrino energy is above a

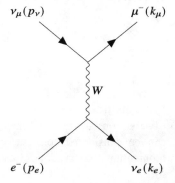

**Fig. 9.3** Diagram of 2-body scattering between an incoming muon neutrino with 4-momentum $p_\nu$ and an electron with 4-momentum $p_e$

## 9.1 Neutrino-Electron Scattering

minimum value (threshold) (see Sect. 7.8). For a generic neutrino scattering process $A + \nu \to \sum_X X$, the threshold can be calculated by imposing that the squared center-of-mass energy $s = (p_A + p_\nu)^2$, where $p_A$ and $p_\nu$ are the 4-momenta of the initial states, is larger than the square of the sum of the masses $(\sum_X m_X)^2$ of the particles in the final state. In the scattering (9.25), if the electron is at rest, the squared center-of-mass energy becomes

$$s = 2E_\nu m_e + m_e^2 \tag{9.26}$$

neglecting $m_\nu$. The threshold condition $s \geq m_\ell^2$ corresponds to a threshold energy for the initial neutrino

$$E_\nu \geq E_\nu^{thr} \equiv \frac{(m_\ell^2 - m_e^2)}{2m_e}. \tag{9.27}$$

The value of the threshold energy $E_\nu^{thr}$ is about 11 GeV for the inverse muon decay and about 3 orders of magnitude more for the inverse tau decay.

When the 4-momentum of the intermediate boson is much smaller than its mass (i.e. $|q^2| \ll M_W^2$) the propagator effects can be ignored, and the tree-level amplitude can be written in the general form of Eqs. (9.17) and (9.19), namely

$$\mathcal{M}_{CC} = -\frac{4G_F}{\sqrt{2}} J'_{\ell\mu} J^{\mu\dagger}_\ell = -\frac{G_F}{\sqrt{2}} [\bar{\nu}_e \gamma_\mu (1-\gamma_5)e][\bar{\ell}\gamma^\mu(1-\gamma_5)\nu_\ell] \tag{9.28}$$

where we have slightly modified the charged currents in (9.18), defining[3]

$$J^\mu_\ell = \bar{\nu}_\ell \gamma^\mu (1-\gamma_5)\ell$$
$$J'_{\ell\mu} = \bar{\nu}_e \gamma_\mu (1-\gamma_5)e. \tag{9.29}$$

In unpolarised cross sections, no information about spins is maintained. To allow for scattering in all possible spin configurations, we sum over final polarisations and average over initial spins. The electron has two helicity states. Once we neglect its small mass, the neutrino has only one helicity state, since the massless neutrino only interacts in the SM as left-handed. Hence averaging over final spins in the process (9.25) corresponds to divide by 1/2. The unpolarised and averaged squared matrix element is Lorentz invariant and reads

$$\overline{\mathcal{M}^2} \equiv \frac{1}{2} \sum_{spin} |\mathcal{M}_{CC}|^2. \tag{9.30}$$

---

[3] This will come handy in lepton-hadron scattering, when we deal with currents composed by more Lorentz structures that just vector and axial-vector ones.

Eventually one finds[4]

$$\overline{\mathcal{M}^2} = \frac{G_F^2}{4} L_{\mu\nu} L'^{\mu\nu} = 64 G_F^2 (p_\nu \cdot p_e)(k_\nu \cdot k_\ell) =$$
$$= 16 G_F^2 (s - m_e^2)(s - m_\ell^2) \qquad (9.31)$$

where $p_\nu$, $p_e$, $k_\nu$ and $k_\ell$ are the 4-momenta for the initial neutrino, initial electron, final neutrino and final lepton $\ell$ (muon or tau), respectively. We have introduced the lepton tensors

$$L_{\mu\nu}^{(l)} \equiv J_{\ell\mu}^{(l)} J_{\ell\nu}^{(l)\dagger}. \qquad (9.32)$$

The small neutrino masses have been neglected, and we have indicated with $m_e$ and $m_\ell$ the masses of the initial electron and final lepton $\ell$. By 4-momentum conservation we have

$$s = (p_\nu + p_e)^2 = (k_\nu + k_\ell)^2. \qquad (9.33)$$

As it is well-known, cross sections represent the intrinsic scattering probability. The interaction typically occurs when an initial neutrino beam hits a target, which contains the scattering centres of interest, in this case electrons. If $R_i$ is the interaction rate, namely the number of interactions per unit time, $\Phi_b$ is the incoming beam flux, namely the number of incident particles per unit area and unit time, and $N_t$ is the number of target particles, the total cross section $\sigma$ is defined as

$$\sigma \equiv \frac{R_i}{\Phi_b N_t}. \qquad (9.34)$$

The cross section can be thought of as the effective cross-sectional area of the target particles for the interaction to occur. It has the dimensions of a surface. In nuclear physics one uses as a unit the barn, defined as $b = 10^{-28}$ m². Since the diameter of a nucleus with atomic mass around 100 (e.g. uranium) is in the range of 10 fm, a barn is of the order of its geometrical section. In subnuclear physics the cross sections are smaller, so one uses submultiples of the barn. In order to pass to natural units ($\hbar = c = 1$), one can use the relations: 1 mb = 2.5 GeV$^{-2}$ and 1 GeV$^{-2}$ = 389 $\mu$b.

The differential interaction cross section is a Lorentz invariant quantity, and it can be written symbolically as

$$d\sigma = \frac{\overline{\mathcal{M}^2}}{F} \, d\text{Lips} \qquad (9.35)$$

---

[4] For a derivation see for instance Refs. [396, 399, 400].

## 9.1 Neutrino-Electron Scattering

in terms of Lorentz invariant expressions. dLips is the Lorentz invariant phase space factor[5]

$$\text{dLips} = (2\pi)^4 \delta^4(k_\nu + k_\ell - p_\nu - p_e) \frac{d^3 k_\nu}{(2\pi)^3 2k_\nu^0} \frac{d^3 k_\ell}{(2\pi)^3 2k_\ell^0} \qquad (9.36)$$

and the Lorentz invariant flux $F$ of the initial beam is ($m_\nu$ is the neutrino mass):

$$F = 4[(p_\nu \cdot p_e)^2 - m_\nu^2 m_e^2]^{1/2}. \qquad (9.37)$$

Aside from the centre of mass energy, other standard variables invariant under Lorentz transformations are

$$Q^2 = -q^2 = (p_\nu - k_\mu)^2 \quad \text{(4-momentum transfer)}$$

$$y = \frac{p_e \cdot q}{p_e \cdot p_\nu} \quad \text{(inelasticity)} \qquad (9.38)$$

where $q^\mu \equiv k^\mu - p^\nu$. Lorentz invariant expressions apply to any rest frame. In the collisions between an incoming neutrino and a (stationary) target lepton, the Lorentz invariant differential cross-section can be explicitly expressed as

$$\frac{d\sigma}{dq^2} = \frac{1}{16\pi} \frac{|\mathcal{M}'^2|}{[s - (m_e + m_\nu)^2][s - (m_e - m_\nu)^2]}. \qquad (9.39)$$

Neglecting neutrino masses, it becomes

$$\frac{d\sigma}{dq^2} = \frac{1}{16\pi} \frac{|\mathcal{M}'^2|}{(s - m_e^2)^2}. \qquad (9.40)$$

In the laboratory frame, the incoming and outcoming 4-momenta are

$$p_\nu = (E_\nu, \mathbf{p}_\nu) \qquad k_\mu = (E_\mu, \mathbf{k}_\mu)$$
$$p_e = (m_e, 0) \qquad k_e = (E_e, \mathbf{k}_e). \qquad (9.41)$$

It is possible to highlight a particular dependence of the differential cross-sections by making use of the appropriate Jacobian. For example, to determine the cross-section as a function of the muon's scattering angle, $\theta_\mu$, the Jacobian is given by:

$$\frac{dq^2}{d\cos\theta_\mu} = 2|\mathbf{p}_\nu||\mathbf{k}_\mu|, \qquad (9.42)$$

---

[5] It is easy to demonstrate that $d^3 k / k^0$ is a Lorenz invariant quantity, see for instance Ref. [396].

The dimensionless inelasticity parameter $y$ reflects the kinetic energy of the outgoing lepton, and it is

$$y = \frac{E_\ell - \frac{(m_\ell^2 + m_e^2)}{2m_e}}{E_\nu} \qquad 0 \leq y \leq y_{\max} = 1 - \frac{m_\ell^2}{2m_e E_\nu + m_e^2}. \qquad (9.43)$$

The Jacobian written in terms of the inelasticity parameter $y$ is given by:

$$\frac{dq^2}{dy} = 2m_e E_\nu. \qquad (9.44)$$

In the laboratory frame, by making use of the previous formulas, we can easily see that the differential cross-section with respect to the fractional energy imparted to the outgoing lepton is

$$\frac{d\sigma(\nu_\ell e \to \nu_\ell \ell)}{dy} = \frac{2m_e G_F^2 E_\nu}{\pi}\left(1 - \frac{(m_\ell^2 - m_e^2)}{2m_e E_\nu}\right), \qquad (9.45)$$

When $E_\nu \gg E_{\text{thresh}}$, integration of the above expression over the isotropic angular distribution gives the total neutrino cross section

$$\sigma(\nu_\ell e \to \nu_\ell \ell) \simeq \frac{2m_e G_F^2 E_\nu}{\pi} = \frac{G_F^2 s}{\pi} \qquad (9.46)$$

which grows linearly with energy.

Because of the different available spin states, the equivalent expression for the inverse lepton decay of anti-neutrinos:

$$\bar{\nu}_e + e \to \bar{\nu}_\ell + \ell \quad (\ell = \mu \text{ or } \tau), \qquad (9.47)$$

has a different dependence on $y$ than its neutrino counterpart, although the matrix elements are equivalent.

$$\frac{d\sigma(\bar{\nu}_e e \to \bar{\nu}_\ell \ell)}{dy} = \frac{2m_e G_F^2 E_\nu}{\pi}\left((1-y)^2 - \frac{(m_\ell^2 - m_e^2)(1-y)}{2m_e E_\nu}\right). \qquad (9.48)$$

Upon integration, the total cross-section is approximately a factor of 3 lower than the neutrino-cross-section, namely

$$\sigma(\bar{\nu}_e e \to \bar{\nu}_\ell \ell) \simeq \frac{2m_e G_F^2 E_\nu}{3\pi} = \frac{G_F^2 s}{3\pi}. \qquad (9.49)$$

The suppression comes entirely from helicity considerations.

## 9.2 Neutrino-Nucleon Interactions

Let us now turn to the elastic process (9.8)

$$\nu_\ell(\bar{\nu}_\ell) + e \to \nu_\ell(\bar{\nu}_\ell) + e \quad (\ell = \mu \text{ or } \tau) \tag{9.50}$$

mediated by a pure neutral current interaction. The matrix element is different, since it includes both left-handed and right-handed leptonic couplings. By repeating the above reasonings one obtains

$$\frac{d\sigma(\nu_\ell e \to \nu_\ell e)}{dy} = \frac{m_e G_F^2 E_\nu}{2\pi} \left( (g_V + g_A)^2 + (g_V - g_A)^2 (1-y)^2 \right.$$
$$\left. - (g_V^2 - g_A^2)\frac{m_e y}{E_\nu} \right),$$

$$\frac{d\sigma(\bar{\nu}_\ell e \to \bar{\nu}_\ell e)}{dy} = \frac{m_e G_F^2 E_\nu}{2\pi} \left( (g_V - g_A)^2 + (g_V + g_A)^2 (1-y)^2 \right.$$
$$\left. - (g_V^2 - g_A^2)\frac{m_e y}{E_\nu} \right),$$

where $g_V \equiv (2g_L^\nu g_V^\ell)$ and $g_A \equiv (2g_L^\nu g_A^\ell)$.

## 9.2 Neutrino-Nucleon Interactions

At very low energies, the nucleus appears to have no structure and the neutrino interacts with it as whole. Rising to intermediate neutrino energies, neutrinos can scatter off an entire nucleon in the nucleus of the target material in a quasi elastically manner. Roughly speaking, given a nucleon size of about 1 fm $\simeq 1/200$ MeV$^{-1}$, a momentum transfer $\gtrsim 200$ MeV justifies assuming the interaction with the entire nucleon.

Neutrino interactions with a free nucleon have the same basic characteristics as those of lepton scattering, though the manner in which one builds the hadronic current is different. Neutrinos can interact elastically with a single proton or neutron by neutral electroweak currents. Elastic neutrino-nucleon interactions are

$$\nu_\ell + n(p) \to \nu_\ell + n(p)$$
$$\bar{\nu}_\ell + n(p) \to \bar{\nu}_\ell + n(p). \tag{9.51}$$

Neutrinos can interact with a single proton or neutron also by charged electroweak currents. These processes, namely

$$\nu_\ell + n \to \ell^- + p$$
$$\bar{\nu}_\ell + p \to \ell^+ + n \tag{9.52}$$

are referred to as quasi-elastic (QE) or CC quasi-elastic (CCQE) scatterings. They will be discussed in Sect. 9.2.1.

Inelastic scattering processes start when the neutrino energy becomes high enough to pass the threshold for the production of a single pion. In inelastic processes the final state does not have the same composition of the initial state, namely one lepton and one nucleon. A single pion can be produced via decay of resonance excitations or non-resonant interactions. As the neutrino energy increases, neutrino interactions can also induce production of multiple pions, strange mesons and hyperons. The single pion production channels make up the largest fraction of the neutrino-nucleus cross section in the 1–3 GeV range. We discuss resonant (RES) single pion production in neutrino-nucleon scattering in Sect. 9.2.2.

In lepton-hadron scattering, when the wavelength of the exchanged virtual boson, of 4-momentum $q$, is much smaller than the size of the nucleon ($|q^2| \gg 0.71$ GeV$^2$) and significant energy is transferred, the lepton can begin to resolve the internal structure of the target. A lepton scattering off an individual quark inside the nucleon is a well known process called deep inelastic scattering (DIS). It has played an important role to validate our knowledge on both the nucleon structure and the strong interaction. We recall some DIS features for neutrino-nucleon scattering in Sect. 9.2.3. Closely related to DIS are the Bjorken scaling and the parton model, briefly underlined in Sect. 9.2.4.

When we compare charged lepton-nucleon scatterings with neutrino-nucleon ones, several differences stand up. Electron (or muons or tauons) interactions at lowest order are mediated only by neutral currents ($\gamma$ and $Z$), while a neutrino interaction can also be mediated by charged currents ($W^\pm$), carrying an added vector-axial contribution. The energy distribution of a neutrino beam is broad in several experiments, and the incoming neutrino energy (and, therefore, the momentum and energy transfer) are generally not well known experimentally and must be reconstructed from observations of the final state. At accelerator experiments, the reason stems from the way neutrino beams are produced. A high-current proton beam is fired into a thick target to produce many secondary particles such as charged pions and kaons, which are then bundled into a given direction as they decay into neutrinos and their corresponding charged leptons. Once the leptons have been removed by appropriate absorber materials, a neutrino beam emerges. Since the pions and kaons are produced with their own energy spectra, their two-body decays into a charged lepton and a neutrino lead to a broad neutrino-energy distribution.

In neutrino scatterings, an additional complication comes about because all modern neutrino experiments use nuclear targets, which may introduce model dependence in the analyses identifying the particular reaction mechanism in the nucleus. Even assuming that the neutrino–nucleus interaction can be described as a superposition of quasi-free interactions of the neutrino with individual nucleons, the latter are bound and move with their Fermi motion. As a result, the initial-state neutron moves with a momentum of up to about 225 MeV, smearing the reconstructed neutrino energy around its true value by a few tens of MeV. Furthermore, final-state interactions concerning the hadrons produced—both between themselves and with

## 9.2 Neutrino-Nucleon Interactions

the nuclear environment of the detector—significantly complicate the analyses. A notable example is the fact that final state of a true QE reaction and that of events producing a meson, e.g. a pion, subsequently absorbed inside the nuclear target, are experimentally indistinguishable. Then the data for QE scattering can be heavily affected by the theory used to describe pion production and absorption on nuclei.

### 9.2.1 Charged Current QE Scattering

In a charged current neutrino quasi elastic (CCQE) scattering, the target neutron is converted to a proton

$$\nu_\ell + n \to \ell^- + p \qquad (9.53)$$

In the case of ingoing antineutrinos, the interaction is with protons

$$\bar{\nu}_\ell + p \to \ell^+ + n \qquad (9.54)$$

and a charged lepton and a nucleon are ejected in the final state. A pictorial description of one possible CCQE scattering is given in Fig. 9.4. After the discover of the neutral currents in 1973 (see Sect. 9.3), and up to the 1990s, these reactions were extensively studied with the primary aim to test NC interactions. Today a major interest resides in the fact that CCQE scattering typically gives the largest contribution to the signal samples in many oscillation experiments; besides, it is a two-body reaction, so in principle, one can determine the neutrino energy solely from lepton kinematics.

The process of Eq. (9.54) with $\ell = e$ is the inverse beta decay (IBD)

$$\bar{\nu}_e + p \to e^+ + n. \qquad (9.55)$$

As seen in Sect. 1.3, it has been one of the first reactions to be investigated, and it is typically measured using neutrinos produced from fission in nuclear reactors, in an energy range from the neutrino threshold energy to about 10 MeV. IBD is also important at slightly higher energies of supernova neutrinos (typically around

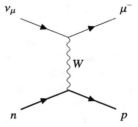

**Fig. 9.4** Sketch of the lowest order muon neutrino $\nu_\mu$ CCQE scattering

20 MeV). One can calculate the IBD energy threshold, following the same steps of Sect. 9.1.2 for neutrino-lepton QE interactions. In the laboratory system, the proton is at rest, and the squared sum $s$ of the antineutrino 4-momenta $p_{\bar{\nu}_e}$ and of the proton $p_p$ is

$$s = (p_{\bar{\nu}_e} + p_p)^2 \simeq m_p^2 + 2m_p E_{\bar{\nu}_e}. \tag{9.56}$$

Here $E_{\bar{\nu}_e}$ is the anti-neutrino energy, $m_p$ the proton mass, and the neutrino mass is neglected. The threshold, neglecting nuclear effects, is found by equating $s$ to the minimum value of the squared sum of the momenta of the final states of (9.55), that is $s = (p_e + p_n)^2_{\min} = m_e^2 + m_n^2 + 2m_e m_n$. The result is

$$E_{\bar{\nu}}^{thr} = \frac{(m_n + m_e)^2 - m_p^2}{2m_p}. \tag{9.57}$$

This value is about 1.8 MeV. A similar formula also applies when $\ell \to \mu$ or $\ell \to \tau$ in Eq. (9.54); different lepton mass values produce different values for the threshold energy, that is about 113 MeV and 3.5 GeV, respectively. Similar reasonings hold for the threshold energy of the process (9.53).

In general, neutrinos interactions with nucleons in targets are very difficult to model because of the complicated nuclear physics involved. The more simple description of CCQE scatterings is obtained by replicating the formalism of neutrino-lepton scattering in the effective framework of Eq. (9.7), where one of the two lepton CC currents is replaced by an hadronic one, which takes into account the effect of strong interactions and the extended (not point-like) dimension of the nucleon. It correspond to a description of the nuclear target as an ensemble of free nucleons. Using the principle of Lorentz covariance, the more general charged hadronic current reads

$$J_\mu^+ = \bar{u}(p') \left( f_1(q^2)\gamma_\mu + if_2(q^2)\sigma_{\mu\nu}\frac{q^\nu}{2M} + f_3(q^2)\frac{q_\mu}{M} + \right.$$
$$\left. + g_1(q^2)\gamma_\mu\gamma_5 + g_2(q^2)\frac{q_\mu}{M}\gamma_5 + ig_3(q^2)\sigma_{\mu\nu}\frac{q^\nu}{2M}\gamma_5 \right) u(p) \tag{9.58}$$

where $u(p)$ and $u(p')$ are the Dirac spinors for the target and final state nucleon of 4-momenta $p$ and $p'$, respectively, and $M = (m_n + m_p)/2 \simeq 938.9$ MeV is the average nucleon mass. We have defined $q = p' - p = p_\nu - p_\ell$, where $p_\nu$ and $p_\ell$ are the 4-momenta of the ingoing and outgoing leptons, respectively. Let us observe that due to finite size of the nucleon, the six dimensionless complex couplings, which are Lorentz invariant and called form factors, are no longer constant. They acquire a dependence on Lorentz invariant terms built with external momenta. Only one independent scalar can be built with the momenta $q$, $p'$, and $p$, because of the 4-momentum conservation and the fact that all interacting particles are on shell, namely $p'^2 = p^2 = M^2$ (neglecting the small mass difference between

## 9.2 Neutrino-Nucleon Interactions

proton and neutron). It is customary to take as the independent scalar the four-momentum transfer squared $t \equiv q^2 = -Q^2$. The form factors $f_1$, $f_2$ and $f_2$ are generally referred to, respectively, as vector, weak magnetism and scalar. The terms including them represent the vector part of the current. The terms including $g_1$, $g_2$ and $g_2$ represent the axial part of the current. Equation (9.58) includes all possible independent vector and axial terms one can build starting from the basis in the four-dimensional Clifford algebra. In the vector current, a term proportional to $p' + p$ does not appear, since it can be re-expressed as a linear combination of the first and second term, by using the Gordon identity.[6] Terms proportional to $\gamma^\mu q_\mu$ are not included since the matrix element is on shell.

It has been suggested [401] to classify the form factors under G-parity, already discussed in Sect. 2.9. Under G-parity, vector and axial vector currents which transform as

$$GV_\mu G^{-1} = V_\mu, \quad GA_\mu G^{-1} = -A_\mu \tag{9.59}$$

are termed as first-class currents. On the contrary, vector and axial vector currents which transform as

$$GV_\mu G^{-1} = -V_\mu, \quad GA_\mu G^{-1} = A_\mu \tag{9.60}$$

are classified as second-class-currents. G-parity is a symmetry for strong interactions, which is broken by mass differences. In the limit of exact flavour $SU(3)$ symmetry, second class current are absent. The vector and axial vector currents of the SM with form factors $f_1$, $f_2$, $g_1$ and $g_2$ transform as in Eq. (9.59) and are first class currents, the vector and axial vector currents with form factors $f_3$ and $g_3$ transform as in Eq. (9.60) and are second class currents. If time reversal holds, second class currents are also absent by assuming charge symmetry.[7] Hence in the limit of equal mass for up and down quarks, a commonly followed and reasonable approximation, two of the form factors in Eq. (9.58) are set to zero

$$f_3(q^2) = g_3(q^2) = 0 \tag{9.61}$$

The differential cross section is given by

$$\frac{d\sigma}{dt} = \frac{G_F^2 |V_{ud}|^2}{64\pi (s - m_p^2)^2} \left[ A(t) \pm (s - u) B(t) + (s - u)^2 C(t) \right] \tag{9.62}$$

---

[6] In case of equal proton and neutron masses, the Gordon identity can be expressed as:

$$\bar{u}(p') i \sigma_{ab} \frac{q^b}{2M} \gamma_5 u(p) = -\bar{u}(p') \frac{p_a + p'_a}{2M} \gamma_5 u(p).$$

[7] For a demonstration, see e.g. Ref. [76] page 607.

in terms of the usual Mandelstam variables $s = (p_\nu + p)^2, t = (p_\nu - p_\ell)^2 = q^2 = -Q^2$, and $u = (p_\ell - p)^2$, the Fermi constant $G_F$ and the CKM matrix element $V_{ud}$.[8] The positive (negative) sign before the $B$ term refers to neutrino (antineutrino) scattering, with incident 4-moment $p_\nu$. Under the approximations (9.61), one has [403, 404]

$$A = (t - m_\ell^2)\bigg[ 8|f_1^2|(4M^2 + t + m_\ell^2) + 8|g_1^2|(-4M^2 + t + m_\ell^2)$$

$$+ 2|f_2^2|(t^2/M^2 + 4t + 4m_\ell^2)$$

$$+ 8m_\ell^2 t |g_2^2|/M^2 + 16\mathrm{Re}[f_1^* f_2](2t + m_\ell^2) + 32m_\ell^2 \mathrm{Re}[g_1^* g_2] \bigg]$$

$$- \Delta^2 \bigg[ (8|f_1^2| + 2t|f_2^2|/M^2)(4M^2 + t - m_\ell^2) + 8|g_1^2|(4M^2 - t + m_\ell^2)$$

$$+ 8m_\ell^2 |g_2^2|(t - m_\ell^2)/M^2$$

$$+ 16\mathrm{Re}[f_1^* f_2](2t - m_\ell^2) + 32m_\ell^2 \mathrm{Re}[g_1^* g_2] \bigg] - 64m_\ell^2 M\Delta \mathrm{Re}[g_1^*(f_1 + f_2)]$$

$$B = 32t\,\mathrm{Re}[g_1^*(f_1 + f_2)] + 8m_\ell^2 \Delta(|f_2^2| + \mathrm{Re}[f_1^* f_2 + 2g_1^* g_2])/M$$

$$C = 8(|f_1^2| + |g_1^2|) - 2t|f_2^2|/M^2 \qquad (9.63)$$

where $\Delta = m_n - m_p \simeq 1.293$ MeV. By including the contribution of second class currents and releasing the approximation (9.61), $A$, $B$ and $C$ include additional terms, which are different if the process is initiated by a neutrino or an antineutrino [297].

In the analyses of these cross sections, it is important not only the assessment of the value, but also the estimate of their uncertainty, which is potentially relevant in cases where the statistical sample is quite large. For instance, the conservative estimate of the uncertainty at low energies on the IBD cross section obtained in [297] is around 2 per mil at the most. The main source for uncertainty on the IBD cross section under a few tens of MeVs is due to the uncertainties of two important constants: the value of the CKM matrix element $V_{ud}$ and that of the axial coupling $g_1(0) = \lim_{q^2 \to 0} g_1(q^2)$. The uncertainty at higher energies, on the other hand, depends also on the behaviour of the form factors as $q^2$ varies. Theory alone says little about the detailed shape of the form factors, which is deduced on the basis of global symmetries of the extended hadrons and checked by measuring the parameters of the assumed shape. In some cases lattice QCD[9] can be used to compute nucleon and nuclear structure functions.

---

[8] For a derivation see e.g. [402] or [399].

[9] Let us just mention that lattice QCD means a QCD gauge theory formulated on a grid or lattice of points in space and time.

## 9.2 Neutrino-Nucleon Interactions

In many early neutrino QE measurements, detectors were bubble chambers (see Sect. 8.2) and target nuclei were hydrogen or deuterium, both of which do not readily involve bound states. In these experiments the final state was clear, and elastic kinematic conditions could be easily verified. The use of heavier nuclear targets and sophisticated detectors in modern neutrino experiments results in more complications, and an increased weight of nuclear effects. In particular, a QE interaction on a nuclear target does not necessarily imply that only a lepton and a single nucleon are ejected in the final state. It is possible to have the ejection of additional particles, especially pions, in the final state. This aspect is essential, because in the two-body dynamics of Eqs. (9.53) and (9.54), the measurement of the energy and angle of the outgoing lepton also determines the incoming energy and the momentum transfer. If the final state is composed of more particles, one has a different kinematics. Therefore recent literature tends to underline which processes involve pion-less (e.g. nucleon-only) final states, referring to them as CC0$\pi$ or QE-like reactions. In the case of neutrino scattering old literature labels as QE both QE and nucleon-only final states processes, while this distinction has always been made in the case of charged lepton scatterings.

### 9.2.2 Single Pion Production

Single pion production becomes possible when the boson exchanged between the neutrino and the target nucleon has the requisite four-momentum to create a pion at the interaction vertex (non-resonant interaction) or to excite the nucleon to a resonance state, which promptly decays to produce a final-state pion (resonant interaction), as shown in Fig. 9.5. In single pion production mediated by a charged

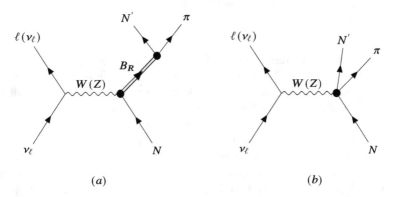

**Fig. 9.5** Representation of (**a**) resonant and (**b**) non-resonant single-pion production induced by neutrino-nucleon interaction

**Fig. 9.6** Example of muon neutrino CC interaction with a neutron $n$, inducing a resonance $\Delta^+$ with single pion production

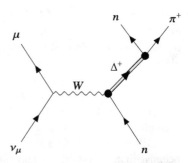

$W^\pm$ boson, the final state is a pion $\pi$ accompanied by a nucleon $N'$ and a charged lepton $\ell^\pm$. The process reads

$$\nu_\ell(\bar{\nu}_\ell) + N \to \ell^-(\ell^+) + \pi + N' \tag{9.64}$$

where $N$ is the nucleon in the initial state. In $Z^0$ mediated production, the charged lepton in the final state is substituted by a neutrino, yielding

$$\nu_\ell(\bar{\nu}_\ell) + N \to \nu_\ell(\bar{\nu}_\ell) + \pi + N'. \tag{9.65}$$

Resonant production generally dominates non-resonant one. In resonant processes, the neutrinos excites the target nucleon to a resonance state. While still in the nuclear medium, the baryon resonance quickly ($\sim 10^{-23}$ s) decays to a variety of possible mesonic final states, producing combinations of nucleons and mesons. In most cases, the final state is composed by a nucleon and single pion final state, that is we have

$$\nu_\ell + N \to \ell^- + B_R$$
$$B_R \to \pi + N' \tag{9.66}$$

where $B_R$ is the baryon resonance (see Fig. 9.5a). This inelastic interaction resolves the nucleon as a whole.[10] Resonant single pion production is the most common mean of single pion production at intermediate energies of a few GeV. The nucleon resonances which are excited in the inelastic reactions are determined by the neutrino energy. The first resonance which can be excited is the $\Delta$ baryon, with zero strangeness ($S = 0$), isospin $I = 3/2$, spin-parity $J^P = 3/2^+$ and mass $\simeq 1232$ MeV. A pictorial example of neutrino CC interaction with a $\Delta$ resonance is given in Fig. 9.6. The next excited resonance is the Roper resonance, indicated with $N(1440)$. It has $S = 0$, $I = 1/2$, $J^P = 1/2^+$ and mass $\simeq 1440$ MeV. In

---

[10] It is to be noted that pion production can also occur in neutrino interaction with a *nucleus* as a whole.

the spectrum of nucleon-like states, i.e. baryons with isospin $I = 1/2$, the Roper resonance lies about 0.4 GeV above the nucleon and 0.15 GeV below the first $J^P = 1/2^-$ state, which has roughly the same width. Other resonances are excited at higher energies. Generic baryon resonances with $S = 0$ are often indicated with $N^*$ when they are nucleon-like states ($I = 1/2$) and $\Delta^*$ when $I = 3/2$ as of the $\Delta$ baryon. To be more specific, one can use the labelling $N(m) J^P$ and $\Delta(m) J^P$, where $m$ is the resonance mass. An older notation reflects the fact that nearly all resonance information used to come from elastic $\pi N$ scattering. The resonances are represented by the symbol $L_{2I2J}(m)$, where $L$ corresponds to the orbital angular momentum of the incoming state and it is indicated by its orbital symbol, that is $L = 0 \to S, L = 1 \to P, L = 2 \to D$ and so on. These notations, for instance, yield $\Delta = \Delta(1232) 3/2^+ = P_{33}(1232)$.

A proper understanding of single pion production is complex, since it requires a realistic model of production at the nucleon level, accompanied by models of additional effects due to the nucleon being bound in the nucleus and to the interactions with the nucleons of the particles produced inside the nucleus on their way out. The latter corrections are commonly referred as final state interaction (FSI) effects.

The knowledge of single pion production is very important in the analysis of neutrino oscillation experiments. For instance, $\pi^0$ production is an important background to experiments that measure $\nu_\mu - \nu_e$ oscillations in the neutrino energy range around 1 GeV. A neutral pion almost always decays through the $\pi^0 \to \gamma\gamma$ channel. Each $\gamma$ initiates an electromagnetic cascade. A similar cascade is created when the electron, which signals the neutrino interaction, travels in the detector medium. The misidentification can occur when the two showers overlap or one of the two photons decay is not detected. The latter might happen when the photon exits the detector before showering or does not have enough energy to initiate an electromagnetic shower. The overlapping of showers might occur when both $\gamma$s travel in similar directions, i.e. the angle between them is small, which is more likely to happen for more energetic pions. In both cases, the result is that a $\pi^0$ event mimics a $\nu_e$ signal event, since the electromagnetic showers instigated by photon and electron are not distinguishable in the usual Cherenkov tanks employed as far detectors. Another example is when a pion is produced and subsequently absorbed; then events where a single pion is produced add to the background in measurements of QE neutrino scattering off nuclear targets.

### 9.2.3 Deep Inelastic Scattering

When a neutrino, with energies well above a few GeV, scatters off a target nucleon exchanging, at tree level, a virtual boson of 4-momentum $q$, such that $|q^2| \gg 1$ GeV$^2$, the wavelength of the virtual boson becomes small enough to resolve the proton constituents. This kinematical region is mostly out of the resonance production region and dominated by deep inelastic scattering processes (DIS).

In DIS reactions the neutrino scatters on an individual quark inside the nucleon. The scattered quark recombines with partons inside the nucleon, in a strong interaction process which also involves emissions of gluons, that in turn gives rise to other gluons or quark–antiquark pairs, and so on. This process is called hadronisation and ends with the production of hadrons in the final state. Hence the basic reaction for the (anti)neutrino DIS process on a free nucleon target $N$ is given by

$$\nu_\ell(\bar{\nu}_\ell) + N \to \ell^-(\ell^+) + X \tag{9.67}$$

for CC interactions, and by

$$\nu_\ell(\bar{\nu}_\ell) + N \to \nu_\ell(\bar{\nu}_\ell) + X \tag{9.68}$$

for NC interactions. All the final hadrons are generically indicated with $X$. The cross sections for these processes are inclusive, that is in the final state only the lepton is observed. No attempt is made to identify any particular hadron in $X$ and all possible outgoing hadronic momenta are included.

Figure 9.7 is a pictorial representation of neutrino DIS processes. The 4-momenta of the particles observed in the inclusive scattering are indicated. They are clearly very similar to DIS of electrons on nucleons. The main differences are that at leading order they are mediated by gauge vector bosons rather than by the photon, and that the coupling at the leptonic vertex includes both vector and axial-vector pieces. The measurements of the cross sections of deep inelastic processes in the 1970s led to the establishment of the quark structure of the nucleon.

As shown in Fig. 9.7, the momentum transferred to virtual gauge bosons is $q = k - k'$, where $k$ and $k'$ are the 4-momenta of the initial and final leptons, respectively. The Lorentz invariant $q^2$ is negative, as it can be easily checked, for instance, in the nucleon rest frame. If $p_{h_i}$, with $i \in \{1 \cdots n\}$, are the 4-momenta of the final hadrons, the scatterings can be also viewed as

$$W/Z(q) + N(p) \to X(p_{h_1}, \ldots, p_{h_n}) \tag{9.69}$$

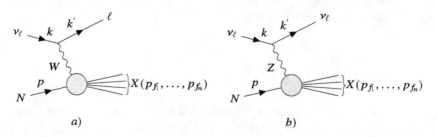

**Fig. 9.7** Representation of neutrino deep inelastic scattering mediated by (**a**) charged and (**b**) neutral currents

## 9.2 Neutrino-Nucleon Interactions

with the neutrino acting as a source of virtual $W$ or $Z$ bosons in CC and NC interactions, respectively.

In DIS we assume, as in the CCQE case, that both the electron vertex and the weak propagator are unaffected by the interactions at the nucleon vertex, and that we can replicate the formalism of neutrino-lepton scattering where one of the two lepton currents is fully specified by the measured electron kinematical variables and the other lepton current is replaced by an hadronic one. For evaluating the cross sections we start directly from the square of the Lorentz invariant amplitude, which can be written at the lowest order in perturbation theory for both CC and NC interactions as

$$\overline{\mathcal{M}^2} = \frac{G_F^2 M_V^4}{f_V} \frac{1}{(q^2 - M_V^2)^2} L_{\mu\nu} W^{\mu\nu}. \qquad (9.70)$$

The subscript $V$ refers to the $W$ or $Z$ gauge bosons, with masses $M_V$, and $\{f_W, f_Z\} = \{2, 8\}$. The first term in Eq. (9.70) comes from the coupling of the gauge bosons to the fermions. For charged $W$ connecting to fermions (V-A electroweak vertexes), it follows from the equality

$$\left(\frac{g}{2\sqrt{2}}\right)^2 = \frac{1}{\sqrt{2}} G_F M_W^2. \qquad (9.71)$$

In case of a neutral $Z$ connecting two neutrino vertexes, the relations

$$\left(\frac{g}{4\cos\theta_W}\right)^2 = \frac{G_F M_W^2}{2\sqrt{2}\cos^2\theta_W} = \frac{1}{2\sqrt{2}} G_F M_Z^2 \qquad (9.72)$$

hold. In DIS one vertex is not a neutrino, hence strictly speaking $f_Z \neq 8$, but factor differences can be taken care of in the tensors definition.

The second term in Eq. (9.70) comes from the gauge boson propagator. $L_{\mu\nu}$ is a leptonic tensor as defined in Eq. (9.32). Since neutrinos are assumed massless at this scales, they are completely longitudinally polarised (100% negative helicity states). Hence, in the definition (9.30) of the averaged squared matrix element $\overline{\mathcal{M}^2}$, the factor of 1/2, that corresponded to averaging over the two spin states, is omitted. The tensor $W^{\mu\nu}$ is an unknown hadronic tensor, containing the complicated dynamics of the strong interactions. In principle, $W^{\mu\nu}$ depends on $p, q$ and all possible outgoing momenta. Since we are measuring inclusive quantities, we sum over all accessible final states $X$ and integrate in the phase space of all possible $p'_{h_1} \ldots p'_{h_n}$ momenta of all final hadrons $h_1, \ldots h_n$. We include this sum and integration in the definition of $W^{\mu\nu}$ appearing in (9.70), that is

$$W_{\mu\nu}(p, q) \equiv \frac{1}{4\pi} \int \ldots \int W'_{\mu\nu} \frac{d^3 \mathbf{p}'_{h_1}}{(2\pi)^3 2 E'_{h_1}} \ldots \frac{d^3 \mathbf{p}'_{h_n}}{(2\pi)^3 2 E'_{h_n}}$$
$$\times (2\pi)^4 \delta^4(q + p - p'_{h_1} - \cdots - p'_{h_n}) \qquad (9.73)$$

where $1/4\pi$ is a conventional factor. As a consequence, $W^{\mu\nu}$ depends only on $p$ and $q$, as in the elastic case.

The Lorentz-invariant cross section can be written symbolically as in Eq. (9.35). The flux factor is $F = 4k \cdot p$, setting to zero the neutrino mass. The Lorentz invariant phase space factor reads

$$\text{dLips} = \frac{d^3\mathbf{k}'}{(2\pi)^3 2E'}. \tag{9.74}$$

It depends only on the final lepton momentum and energy, since the phase spaces of all possible final hadrons have been taken care of in the hadronic tensor. Let us remark that when the scattering is inelastic, the total hadronic moment is no longer constrained by the condition that its squared mass is on the proton mass shell.

It is convenient to express the hadronic tensor $W^{\mu\nu}$ in terms of form factors, by mimicking the same arguments of the elastic case. The form factors depend on two independent scalar factors, instead of one, as in the elastic case. The invariant mass $m_X^2$ of the hadronic state can be defined as

$$m_X^2 \equiv (p+q)^2 = M^2 + q^2 + 2p \cdot q \approx q^2 + 2p \cdot q \tag{9.75}$$

where in the last passage we have neglected the nucleon mass $M$. Since we do not have any constraints on the variable $m_X$, the scalars $q \cdot p$ and $-q^2/2$ are independent.

In the form factors, a common choice of independent scalars is

$$Q^2 \equiv -q^2 > 0 \qquad \nu \equiv \frac{q \cdot p}{M}. \tag{9.76}$$

That is because, in the laboratory frame, $\nu$ is the energy transferred, namely $\nu = E - E'$, where $E$ and $E'$ are the energy of incident and the scattered lepton, respectively. We also have

$$Q^2 = 4E\,E' \sin^2 \theta/2 \tag{9.77}$$

where $\theta$ is the scattering angle of the electron with respect to the direction of incident lepton. In the limit of vanishing lepton masses, a widely used parameterisation, similar to that used in electron-nucleons DIS, is

$$W_{\mu\nu} = \left(-g_{\mu\nu} + \frac{q_\mu q_\nu}{q^2}\right) W_1(Q^2, \nu) + \left(p_\mu - \frac{q \cdot p}{q^2} q_\mu\right)\left(p_\nu - \frac{q \cdot p}{q^2} q_\nu\right)$$
$$\times \frac{W_2(Q^2, \nu)}{M^2} - i\frac{1}{2M^2}\varepsilon_{\mu\nu\lambda\rho} p^\lambda q^\rho\, W_3(Q^2, \nu) + \frac{1}{M^2} q_\mu q_\nu\, W_4(Q^2, \nu)$$
$$+ \frac{1}{M^2}(p_\mu q_\nu + q_\mu p_\nu)\, W_5(Q^2, \nu)$$
$$+ i\frac{1}{M^2}(p_\mu q_\nu - q_\mu p_\nu)\, W_6(Q^2, \nu). \tag{9.78}$$

## 9.2 Neutrino-Nucleon Interactions

The third term arises due to parity violation in weak interactions, and it carries opposite signs for neutrinos and antineutrinos. It is absent when we consider DIS by a charged lepton on a proton at the lowest order, which is a pure QED process, being mediated by a virtual photon. The contribution of the last term vanishes when contracted with the leptonic tensor. When the lepton masses in the final state are negligible, only the first three terms in Eq. (9.78) contribute to the cross sections. At this stage, the form factors $W_i(Q^2, \nu)$ with $i = 1, 2, 3$ are completely arbitrary.

An alternative useful choice of independent scalars is $Q^2$ and $x$ defined as

$$x \equiv \frac{Q^2}{2q \cdot p} = \frac{Q^2}{2M\nu} \qquad 0 \leq x \leq 1. \qquad (9.79)$$

The structure functions can be as well expressed in terms of $Q^2$ and $x$. The elastic case can be regarded as a special case of inelastic scattering, where Eq. (9.75) becomes the constraint

$$(p+q)^2 = M^2 + q^2 + 2p \cdot q \approx -Q^2 + 2p \cdot q = p'^2 = M^2 \approx 0 \qquad (9.80)$$

since the final hadronic state is the same hadron in the initial state, with mass $M$, which is negligible at these high transferred momentum and energy scales. In terms of the variable (9.79), the elastic case correspond to fix $x$ to its maximum value, namely $x = 1$. It is well known that ever since 1968, when experiments at SLAC [405] started firing electrons into protons with beam energies up to 20 GeV, the highest energies then available, DIS processes have been in the front line to reveal the internal composition of the nucleon. In the 1960s, many people regarded quarks simply as a useful book-keeping device to classify the many new "elementary" particles that had been discovered in cosmic rays and bubble-chamber experiments. The big surprise from the SLAC experiments was that the cross section did not depend strongly on $Q^2$, a phenomenon called scaling. From these experimental measurements we started to understand that the nucleon is a complex dynamical system comprised of quarks, gluons and antiquarks, generically indicated as partons.

### 9.2.4 The Bjorken Scaling and the Parton Model

In 1969, James Bjorken [406] analyzed the behaviour of the structure functions in scattering of electrons on protons by using the $SU(3)$ flavour symmetry of the quark model and the methods of current algebra. He predicted the dependence on $Q^2$ to fade away in the so called deep inelastic region, where both $Q^2$ and $\nu$ become much larger than the masses, while their adimensional ratio $x$, defined in Eq. (9.79), stays fixed. More precisely, when

$$Q^2 \to \infty \qquad \nu \to \infty \qquad x \to \text{fixed} \qquad (9.81)$$

the structure functions scale, that is only depend on $x$

$$MW_1(Q^2, \nu) \to F_1(x) \qquad \nu W_2(Q^2, \nu) \to F_2(x). \qquad (9.82)$$

The variable $x$ takes also the name of Bjorken's variable. In other terms, the Bjorken's hypothesis states that, under the previous limits, the functions $F_1(x)$ and $F_2(x)$ exist and are finite, nor infinite or zero. Almost no variation was observed at SLAC,[11] with $Q^2$ going roughly from 1 to 10 GeV, at fixed values of $x$ [407–410].

The constituent model of the nucleon which opened the way for a simple dynamical interpretation of DIS results, included the scaling, was the so-called parton model of the American physicist Richard Feynman [411]. In the parton model, the nucleon is a bound state of point-like constituents, called generically partons. If the virtual boson has sufficiently high $Q^2$, it scatters elastically off a parton of the nucleon; the electroweak probe no more interacts with an extended structure, but with a point-like particle, the parton. The partons constituting a nucleon are strongly bound together as viewed in the rest frame, otherwise the nucleon will fall apart. It seems reasonable to suppose that the binding interactions have a time scale of about 1 fm, the approximate radius of the nucleon in natural units. We can specify the kinematics so that binding effects can be neglected during the large energy transfer to the parton, and partons can be treated as free.

Let us consider for instance the electron-nucleon center-of-mass frame, where an electron and a nucleon arrive from opposite sides. The nucleon is time-dilated and Lorentz contracted with respect to its rest-frame, and it seems reasonable to assume that the binding interactions occur on a time-dilated scale as well (the more, the higher $Q^2$). Equivalently, one can say that the hard interactions of the virtual intermediate boson occur over a scale $1/Q$, which goes to zero as $Q^2 \to \infty$, while the binding strong force acts at fixed distances of order 1 fm. However phrased, this suggests that, during the interaction of the electron with the hadronic system, we can safely assume that the electron interacts with a single fast-moving parton and neglect the strong interactions of the parton with the rest of the nucleon. In other terms, the incoming parton is approximated as a free particle for the purposes of calculating the interaction with the electron.

This framework holds in the Bjorken limiting region (9.81) for deep inelastic scattering from nucleons as viewed from the so-called infinite momentum frame. In this kinematic regime the 4-momentum is also approximately conserved across the interaction vertex of the parton. The term infinite-momentum frame refers to any frame in which the magnitudes of the longitudinal momenta of the particles are very large, also respect to their rest masses. One possible infinite-momentum frame is a center of mass frame which verifies the previous condition. It is different from the laboratory frame, in which the longitudinal momentum of one of the two colliding

---

[11] For their pioneering investigations concerning deep inelastic scattering of electrons on protons and bound neutrons, the US scientists Jerome I. Friedman and Henry W. Kendall, and the Canadian scientist Richard E. Taylor were awarded the Nobel prize in 1990.

## 9.2 Neutrino-Nucleon Interactions

particles is zero. However, it can be obtained from a laboratory frame by a boost of the coordinate system in the longitudinal direction at very high speed. In the infinitum momentum frame we ensure that the transit time across the target is less than the timescale of internal motion. The transferred momentum $Q^2$ is sufficiently high to resolve point-like partons inside the hadron. This justifies the incoherent addition of the probabilities for individual interactions of the virtual intermediate boson with the partons. An equivalent statement is that the parton can only have limited transverse momentum relative to the direction of the parent hadron in the infinitum frame. The experimental results of high energy collisions generally present small average transverse momenta in the decay products, that can be taken as a confirmation of the assumptions of the naïve parton model.

In the parton model, let us assume that the parton carries a charge $q_i$ and a momentum $p_i$ that is a fraction $x'$ of the proton momentum. If the proton momentum is given by $p = (E, p_L, p_T = 0)$, where the subscript $L$ and $T$ indicate the longitudinal and transverse momenta, respectively, then $p_i = (x'E, x'p_L, p_T = 0)$. This implies a mass for the parton $m_i = x'M$, where $M$ is the proton mass. In DIS at the lowest EW order, the virtual photon now interacts with a parton inside the proton. We are then justified in comparing the differential cross sections with the ones obtained in the QED elastic scattering of an electron on a point-like particle, and safely operate the substitutions[12]

$$W_1\left(\nu, Q^2\right) \to e_i^2 \frac{Q^2}{4m_i^2} \delta\left(\nu_i - \frac{Q^2}{2m_i}\right) = e_i^2 \frac{1}{m_i} \frac{Q^2}{4m_i \nu_i} \delta\left(1 - \frac{Q^2}{2m_i \nu_i}\right)$$

$$W_2\left(\nu, Q^2\right) \to e_i^2 \delta\left(\nu_i - \frac{Q^2}{2m_i}\right) = e_i^2 \frac{1}{\nu_i} \delta\left(1 - \frac{Q^2}{2m_i \nu_i}\right). \quad (9.83)$$

Let us observe that $\nu_i = p_i \cdot q / m_i = p \cdot q / M = \nu$. By changing variable one obtains:

$$MW_1\left(\nu, Q^2\right) \to e_i^2 \frac{1}{x'} \frac{Q^2}{4 p_i \cdot q} \delta\left(1 - \frac{Q^2}{2 p_i \cdot q}\right) = \frac{1}{2} e_i^2 \frac{x}{x'^2} \delta\left(1 - \frac{x}{x'}\right)$$

$$= \frac{e_i^2}{2} \frac{x}{x'} \delta\left(x' - x\right)$$

$$\nu W_2\left(\nu, Q^2\right) \to e_i^2 \delta\left(1 - \frac{Q^2}{2 p_i \cdot q}\right) = e_i^2 \delta\left(1 - \frac{x}{x'}\right) = e_i^2 x' \delta\left(x' - x\right). \quad (9.84)$$

Now, let us introduce a function, the parton distribution (or density) function (PDF), $f_i(x')$, to represent the probability that the $i$-th parton carries the fraction $x'$ of the proton momentum, where by definition $x' \in [0, 1]$. The DIS nucleon structure

---

[12] This comparison can be found on most particle books dealing with the parton model, see for instance [412].

functions will be obtained by integrating over $x'$ the structure functions of partons. Then we sum incoherently over all partons inside the hadron, since we look for a sum of probabilities (not amplitudes). We find

$$W_1(x) = \sum_i \int_0^1 dx' \, e_i^2 \frac{1}{2M} \frac{x}{x'} \delta(x' - x) \, f_i(x') = \frac{1}{2M} \sum_i e_i^2 f_i(x)$$

$$W_2(x) = \sum_i \int_0^1 dx' \, e_i^2 \frac{x'}{\nu} \delta(x' - x) \, f_i(x') = \frac{x}{\nu} \sum_i e_i^2 \, f_i(x). \tag{9.85}$$

Note that the fractional momentum of the struck parton $x'$ is identified with the Bjorken variable $x$ through the delta function. We can define two dimensionless structure functions

$$F_1(x) \equiv M W_1(x) = \frac{1}{2} \sum_i e_i^2 \, f_i(x)$$

$$F_2(x) \equiv \nu W_2(x) = x \sum_i e_i^2 \, f_i(x) \tag{9.86}$$

which respect the Bjorken scaling (9.82). In this derivation, the parton mass $x'm$ has been introduced so both the parton energy and longitudinal momentum may be fractions $x'$ of those of the hadron. The transverse components are neglected. This collinear approximation is only really possible in infinitum momentum frame, where all masses and transverse momenta may be neglected. The parton distribution functions are different for different partons, and in the scaling limit, they do not depend on $Q^2$. The same distribution functions appear, in different combinations, for neutron targets. In 1969, Callan and Gross [413] suggested that Bjorken's scaling functions are related

$$F_2(x) = 2x F_1(x). \tag{9.87}$$

We can see directly from the structure functions (9.86) that the Callan Gross relation (9.87) is verified.

The parton picture became even more interesting in the late 1970s and 1980s, when scattering experiments started to use neutrinos and antineutrinos. Since neutrinos and antineutrinos have opposite helicity, their weak interaction with quarks and antiquarks gives different angular distributions. The experimental results indicated the presence of antiquarks within the proton, leading to the well known picture of three valence quarks into a sea of quark–antiquark pairs. They also showed that the total momentum carried by the quarks amounts to only around half of that of the proton. The missing momentum is recovered taking into account the existence of gluons, which bind the quarks together and confine them inside the proton.

## 9.2 Neutrino-Nucleon Interactions

The formalism depicted above can be applied to DIS initiated by neutrinos rather than an electron in an obvious way. By writing the coupling of a the gauge bosons to a quark of type $i$ in the generic form $g_V^i \gamma_\mu - g_A^i \gamma_\mu \gamma_5$, the analogue of Eq. (9.83) reads

$$W_1\left(\nu, Q^2\right) \rightarrow [(g_V^i)^2 + (g_A^i)^2]\frac{Q^2}{4m_i^2 \nu} \delta\left(1 - \frac{Q^2}{2m_i \nu}\right)$$

$$\nu W_2\left(\nu, Q^2\right) \rightarrow [(g_V^i)^2 + (g_A^i)^2]\delta\left(1 - \frac{Q^2}{2m_i \nu}\right)$$

$$\nu W_3\left(\nu, Q^2\right) \rightarrow -2 g_V^i g_A^i \delta\left(1 - \frac{Q^2}{2m_i \nu}\right). \tag{9.88}$$

Than, in analogy to (9.85) and (9.86), we have

$$MW_1(x) \equiv F_1(x) = \frac{1}{2}\sum_i [(g_V^i)^2 + (g_A^i)^2] f_i(x)$$

$$\nu W_2(x) \equiv F_2(x) = x\sum_i [(g_V^i)^2 + (g_A^i)^2] f_i(x)$$

$$\nu W_3(x) \equiv -F_3(x) = -2\sum_i g_V^i g_A^i f_i(x). \tag{9.89}$$

Note that the Callan and Gross relation (9.87) still holds.

The structure of the nucleon is described in terms of PDFs $f_i(x)$, giving the longitudinal, transverse, and spin distributions of quarks within the nucleon. Neutrino scattering plays an important role in the extraction of these fundamental PDFs since only neutrinos via the CC weak interaction can resolve the flavour of the nucleon's constituents. The Bjorken scaling variable $x$ is a key variable in DIS, where the quark can carry a portion of the incoming energy momentum of the struck target. Another useful variable is the inelasticity $y$, related to the energy of the hadronic system, and defined as:

$$y \equiv \frac{q \cdot p}{k \cdot p} \qquad (0 \leq y \leq 1). \tag{9.90}$$

In the laboratory frame, $y$ is the ratio $y = (E - E')/E$, where $E$ and $E'$ are the energy of the incident and the scattered lepton, respectively.

Leaving behind the naïve parton model, one observe that Bjorken scaling is not perfect and the structure functions $F_i$, which, to first approximation, do not vary with $Q^2$, admit instead a logarithmic dependence on $Q^2$ as predicted by QCD. The differential cross section with respect to $x$ and $y$, for DIS initiated by a neutrino or an antineutrino, can be expressed as [412, 414]

$$\frac{d\sigma^{\nu,\bar{\nu}}}{dxdy} = G_F^2 \frac{s}{2\pi} \frac{M_V^4}{(q^2 - M_V^2)^2} \left[ xy^2 F_1^V \left(x, Q^2\right) + \right.$$
$$\left. + \left(1 - y - \frac{xyM^2}{s}\right) F_2^V \left(x, Q^2\right) \pm \left(y - \frac{y^2}{2}\right) x F_3^V \left(x, Q^2\right) \right]. \quad (9.91)$$

The subscript $V$ refers to the $W$ or $Z$ gauge bosons, that is to CC or NC interactions, $s$ is the usual Mandelstam variable and the $+(-)$ sign in the last term refers to neutrino (antineutrino) interactions. Although this expression captures the main features of DIS processes, to achieve an higher level of precision it is necessary to take into account additional effects as the inclusion of lepton masses, nuclear effects, radiative corrections and so on.

## 9.3 The Gargamelle Neutrino Experiment

The parton model was not widely accepted straight away. It had to wait for experimental decisive tests, that came from electron DIS scattering at SLAC and DESY [415] and from neutrino DIS scattering at CERN with the heavy liquid bubble chamber Gargamelle [416], which later discovered the neutral currents.

The discovery of the neutral currents in the Gargamelle neutrino experiment at CERN in 1973 opened a new era in the physics of the weak and electro-magnetic interactions. It confirmed the approach based on the idea of the electroweak unification of the Glashow-Weinberg-Salam model, at the basis of the SM.

Gargamelle was a giant bubble chamber, with a cylindrical body 4.8 m long and 1.85 m wide. It weighed 1000 tonnes and held nearly 12 cubic metres of heavy-liquid freon (CF3Br). Its conceptual design was inspired by the French physicist André Lagarrigue [417]. It was built at the Saclay Laboratory in France and operated at CERN from 1970 to 1976 with a muon-neutrino beam produced by the PS. Later, it was transferred to the neutrino beam at the SPS and equipped with an external muon identifier. In 1979 the chamber ceased operation after cracks had appeared that proved impossible to repair. The name Gargamelle derives from the giantess Gargamelle, who was Gargantua's mother in the novels written in the sixteenth century by François Rabelais.

For neutrino experiments, a heavy liquid bubble chamber has two advantages over an hydrogen bubble chamber: (1) it presents a more dense target yielding a higher interaction rate (2) the distance a neutral particle travels in the liquid before producing charged particles (which leave tracks giving information about the parent neutral particle) is shorter. The disadvantages are that a heavy liquid chamber is less favourable than the hydrogen chamber in the complexity of the target it presents

## 9.3 The Gargamelle Neutrino Experiment

to the incoming beam, and in the accuracy with which the particle tracks can be measured.

Gargamelle was built specially for the study of neutrino processes, although the search for NC induced processes had the eighth priority in its programme, while W search, analyses of DIS, and current algebra sum rules were in pole position [417]. The framework of weak interactions built by Glashow [106], Weimberg [107] and Salam [108] during the 1960s, which had neutral currents at its core, was largely ignored until, in 1971, 't Hooft and Veltman proved the renormalizability of non abelian theories [123]. The question of whether weak neutral currents, in addition to the known charged currents, existed or not, became then a significant issue. By that time two neutrino experiments were running, Gargamelle and the HPWF (Harvard, Pennsylvania, Wisconsin, Fermilab) experiment at what is now Fermilab [418]. Both were suddenly confronted with this challenge.

In December 1972 an event was found by Gargamelle which consisted of an isolated uniquely identified electron and was attributed to antineutrino elastic scattering $\bar{v}_\mu e^- \rightarrow \bar{v}_\mu e^-$. This first candidate for a NC interaction fired the optimism and motivated an intensive search for NC-induced DIS processes $v_\ell(\bar{v}_\ell) + N \rightarrow v_\ell(\bar{v}_\ell) + X$, which have cross sections about two orders of magnitude larger. As in leptonic scattering, the neutrino enters invisibly the bubble chamber, interacts, and then moves on, again invisibly, but for hadronic neutral currents the signal is an event containing only hadrons and no lepton. The main task became then to control events induced by neutrons (produced in CC undetected neutrino interactions), which could simulate neutral currents.

In 1973 the Gargamelle collaboration claimed the observation of events induced by neutral particles and producing hadrons, but no muon or electron, which "behave as expected if they arise from neutral current induced processes" [419]. In the beginning these data were confirmed by the HPWF collaboration, but later the HPWF collaboration modified their apparatus with the net result that the previously observed signal of neutral currents disappeared. This news had a dismaying effect and was a cause for distrust of the Gargamelle result for about one year. By the middle of 1974 the Gargamelle collaboration doubled their statistics and confirmed their original result [420]. In the meantime the HPWF collaboration had elucidated the reason why they lost the signal and also confirmed the Gargamelle finding [421]. The same year other two neutrino experiments gave evidence of weak neutral currents, the 12-foot Argonne bubble chamber at the Zero Gradient Synchrotron (ZGS) located at Argonne National Laboratory (Illinois, USA), reporting single pion production induced by neutral currents, and a new experiment by the California Institute of Technology (Caltech) and Fermilab (CITF) collaboration, reporting the observation of neutral current events in DIS [422]. The discovery of neutral currents was firmly established.

## 9.4 Neutrino Interactions with Nuclei

The process of neutrino and antineutrino scattering off a nucleus with mass number A can be described assuming the Born approximation, i.e. the one-boson exchange approximation, where the exchanged virtual boson is a neutral $Z^0$ boson for the NC process and a charged $W^\pm$ boson for the CC one.

In neutrino scattering experiments, nuclear targets made of the simplest elements, hydrogen or deuterium, have the advantage of almost no dependence on nuclear effects. Nuclear effects represents a major theoretical challenge in passing from neutrino scattering with free particles to neutrino scattering with nucleons bond in the nucleus. The down side is that these targets produce low interaction rates, and they are difficult to build. Moreover, in targets made of hydrogen, the CCQE interactions are only available to antineutrinos (see Eq. (9.54)).

Modern neutrino experiments (from the 1990s) use complex nuclei as targets. Complex, heavier nuclei, such as carbon, oxygen (water) or iron, have higher interaction rates but they may be sizeably affected by nuclear effects. The description of neutrino interactions in terms of scattering off free nucleons is not necessarily reliable. Heavy nuclei also exhibit collective behaviour, which is totally absent in light nuclei. The target with which the neutrino can interact is not necessarily limited to an individual nucleon, but can include correlated nucleon pairs, or any combination of nucleons in a quasi-bound state. An accurate description requires the knowledge of nuclear structure of the initial and final nuclear states, in addition to the knowledge of the neutrino interactions with nucleons *within* the nucleus.

A broad classification of nuclear effects distinguishes initial-state and final-state effects. Initial-state effects are conditions within the nucleus that impact the nucleon prior to and as a part of the neutrino interaction. Final-state effects consist of hadronic interactions that impact the outgoing final-state particles prior to their exit from the nucleus. In the following, we list some of these effects.

- Target nucleons, in their initial state, are subject to strong hadronic interactions inside the nucleus. Changes in their direction and momentum in relation to an incoming neutrino can affect both kinematics and cross section of neutrino-nucleon interactions, especially at low energies and momentum transfer. This effect is often called Fermi smearing.
- Another typical nuclear effect is the so-called Pauli-blocking, which limits the final-state kinematics available to interactions which produce a nucleon. As a fermion, the resulting nucleon is not permitted to be in a state which is already occupied by another nucleon—reducing the available phase space and hence the cross-section.
- After the interaction, the final-state particles propagate out through the nuclear medium, where they undergo strong interactions with the other nucleons inside the nucleus. Because the nucleus is so dense, it is not surprising that the interaction probability in the nucleus is significant. These final-state interactions (FSI) can alter the kinematics of the final-state particles, as well as their type

**Fig. 9.8** A diagram illustrating a charged-current MEC interaction of an electron neutrino with two nucleons, $N_1$ and $N_2$, correlated as through the exchange of a virtual pion $\pi^*$

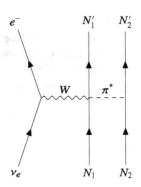

and number. Particles, especially pions, can be absorbed and never escape the nucleus, or their collisions with other nucleons can generate additional particles.

Another help for classification comes from the value of the energy-momentum transferred to the nucleus since, as observed in Sect. 9.2, the various physical mechanisms which dominate the neutrino-nucleus cross section depend on it. At low values neutrino beams probe the whole nucleus, at higher values they interact with individual nucleons or partons.

At the lower end of the energy scale, up to about 1 MeV, one can consider interactions without a threshold, which can be initiated when the neutrino has any low momentum. Such processes include elastic scattering, neutrino capture and coherent scattering. In previous sections we have already discussed the first two processes, and we describe the latter one in Sect. 9.4.1.

As the energy of the neutrino increases, scattering processes allow one to access nucleons individually. In Sect. 9.2 the main reaction mechanisms driving the lepton-nucleon interaction have been identified: quasi-elastic (QE), resonance excitation (RES), and deep inelastic scattering (DIS). In addition, in lepton interactions with the *nucleus*, there are processes where the vector boson from the leptonic current is absorbed by a pair of nucleons (two-body currents). These interactions are generally depicted as in Fig. 9.8, where two correlated nucleons are modelled as interacting through the exchange of a meson. When one of the correlated nucleons interacts with a neutrino through the exchange of a $W$ boson both nucleons can be ejected from the nucleus. The interactions between neutrinos and correlated nucleon pairs are referred to as meson exchange current (MEC) or two-particle-two-hole (2p-2h) processes.[13] In the 1980s, it was suggested [423] that in certain detectors these excitations could be experimentally indistinguishable from true QE events and would thus contribute to the QE cross section. Since last decade, the possible role of 2p-2h effects on the differential QE cross sections has been widely investigated for nuclear targets ranging from carbon to iron.

---

[13] According to this terminology, QE scattering interactions are labelled as one-particle one-hole (1p-1h) interactions.

Determining the nuclear response in neutrino-nucleon basic interaction modes can help one to reduce the systematic uncertainties and increase the precision of the analyses. However, the entire problem of modelling nuclear structure is quite involved. In intermediate energy regimes, about 0.1–10 GeV, where most of the oscillation experiments live, QE neutrino scattering is quite often the main interaction mechanism. In this kinematic region the effect that neutrinos have on nucleons or the nucleus as a whole is often resolved by using simplified theoretical or phenomenological models, each with a different physical assumption. We introduce a few common approaches in Sects. 9.4.2 and 9.4.3.

### 9.4.1 Coherent Nuclear Scattering

The notion of coherent nuclear scattering is well-known in the case of electron-nucleus scattering. The nucleus is probed by the electron as a single coherent object, with coherent scattering across all its nucleons. One adds the amplitudes of scattering on individual nucleons, while in incoherent scattering the probabilities of scattering on individual nucleons are to be added.

In the case of neutrinos, coherent nuclear scattering started to be considered soon after the discovery of weak neutral currents, which implied that neutrinos were capable of coupling to quarks through the exchange of neutral Z bosons. It was suggested that this mechanism should also lead to coherent interactions between neutrinos and all nucleons present in an atomic nucleus [424–427].

The amplitude $\mathscr{A}(\mathbf{k}', \mathbf{k})$ for elastic scattering of a projectile (e.g. a neutrino) off a composite system (e.g. a nucleus) is given by the sum of the contributions from each constituent

$$\mathscr{A}(\mathbf{k}', \mathbf{k}) = \sum_{i=1}^{n} A_i(\mathbf{k}', \mathbf{k}) e^{i\mathbf{q} \cdot \mathbf{x}_i} \qquad (9.92)$$

where $\mathbf{k}'$ and $\mathbf{k}$ are the incoming and outcoming neutrino momenta, $\mathbf{q} = \mathbf{k}' - \mathbf{k}$ is the momentum transfer and $n$ is the number of constituents. In the case of the nucleus and of neutral interaction $n$ equals the mass number $A$, i.e. the total number of nucleons in the nucleus. The differential cross-section is

$$d\sigma = |\mathscr{A}(\mathbf{k}', \mathbf{k})|^2 =$$
$$= \sum_{i=1}^{n} |A_i(\mathbf{k}', \mathbf{k})|^2 + \sum_{i,j}^{i \neq j} A_i(\mathbf{k}', \mathbf{k}) A_j^{\dagger}(\mathbf{k}', \mathbf{k}) e^{i\mathbf{q} \cdot (\mathbf{x}_j - \mathbf{x}_i)}. \qquad (9.93)$$

## 9.4 Neutrino Interactions with Nuclei

If there is only one type of constituent, and only one amplitude $\mathbb{A}(\mathbf{k}', \mathbf{k}) = \mathsf{A}_i(\mathbf{k}', \mathbf{k})$ for any $i$, all non-diagonal terms cancel and the differential cross-section reads

$$d\sigma = \sum_{i=1}^{n} |\mathsf{A}_i(\mathbf{k}', \mathbf{k})|^2 = A\,|\mathbb{A}(\mathbf{k}', \mathbf{k})|^2 \tag{9.94}$$

since $n = A$.

In the general case (9.93), and when the scattering is incoherent, one cannot exclude major cancellations among the $n(n-1)$ terms in the second sum, due to the presence of the phase factors. Then the differential cross section becomes a sum of probabilities

$$d\sigma \simeq \sum_{i=1}^{n} |\mathsf{A}_i(\mathbf{k}', \mathbf{k})|^2. \tag{9.95}$$

The necessary condition for coherent scattering in the nucleus is that all neutrino waves scattered off the different nucleons are in phase with each other. All phase factors may be approximated by unity and the terms add coherently as long as the momentum exchanged remains smaller than the inverse of the nuclear size $R$, namely $|\mathbf{q}|R \ll 1$. This condition effectively restricts the process to neutrino energies at small momentum transfer, below about 50 MeV. By the Heisenberg uncertainty relation a small momentum transfer increases the uncertainty of the coordinate of the scatterer, and it becomes in principle impossible to find out on which nucleon the neutrino has scattered. In the coherent case, the differential cross-section reads

$$d\sigma = |\mathscr{A}(\mathbf{k}', \mathbf{k})|^2 = \left|\sum_{i=1}^{n} \mathsf{A}_i(\mathbf{k}', \mathbf{k})\right|^2 \simeq A^2 |\mathbb{A}(\mathbf{k}', \mathbf{k})|^2 \tag{9.96}$$

in case of only one type of constituent. Evidently, the coherent scattering cross-section grows as the square of the atomic number, $A^2$, and it is enhanced compared to that of a single constituent, an isolated nucleon. In Fig. 9.9 this enhancement is shown by comparing qualitatively the neutrino cross section of the coherent scattering and the one of another threshold-less process, the inverse $\beta$ decay.

In the realistic case of a nucleus with $Z$ protons and $N$ neutrons ($N = A - Z$), and assuming zero nuclear spin, the coherent elastic neutrino-nucleus scattering (CE$\nu$NS) differential cross-section, in the coherence limit $|\mathbf{q}|^2 \to 0$, reads [428, 429]

$$\frac{d\sigma}{d\Omega} = \frac{G_F^2}{16\pi^2} E_\nu^2 (1 + \cos\theta)[(1 - 4\sin^2\theta_W)Z - N]^2 \tag{9.97}$$

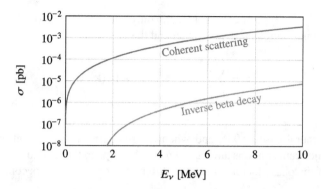

**Fig. 9.9** Qualitative comparison of neutrino cross section in coherent neutrino scattering and IBD processes

where $\theta_W$ is the weak mixing angle, $G_F$ is the Fermi constant and $E_\nu$ is the neutrino initial energy. The angle $\theta$ is the neutrino scattering angle with respect to the incident neutrino direction, in the laboratory frame, and $d\Omega = d\phi\, d(\cos\theta)$ is the phase space factor. We observe that the coefficient of $Z$ nearly vanishes and the cross section essentially scales quadratically with $N$. This process can in principle provide information on the nuclear weak charge $Q_W$, defined as the factor

$$Q_W = N - (1 - 4\sin^2\theta_W)Z. \tag{9.98}$$

For an isoscalar target $Z = N = A/2$; the total cross section is [429]

$$\sigma \simeq 1.68 \times 10^{-42} \sin^4\theta_W A^2 \left(\frac{E_\nu}{10\,\text{MeV}}\right)^2 \text{cm}^2. \tag{9.99}$$

It corresponds to $\sigma \simeq 2.8 \times 10^{-40}$ cm$^2$ for $E_\nu \simeq 10$ MeV and $A = 56$ (Fe).

The characteristic most often associated with neutrinos is a very small probability of interaction which requires large targets (tons to tens of kilotons) to be used for their detection. The enhancement to the scattering cross section, for heavy nuclei and sufficiently intense neutrino sources, can lead to a marked reduction in detector mass, down to a few kilograms. Notwithstanding this advantage, CE$\nu$NS has evaded experimental demonstration for decades since its first theoretical description. The single outcome of the CE$\nu$NS interaction is the low-energy nuclear recoil, ranging from sub-keV to a few tens of keV depending on the nucleus and neutrino source. It is not easy to detect. If we compare with a minimum ionizing particle of the same energy,[14] a recoiling nucleus is less capable to generate measurable scintillation or ionization in common detector materials. This is reflected in the so-called quenching

---

[14] Most relativistic particles (e.g., cosmic-ray muons) have mean energy loss rates close to the minimum, and are said to be minimum ionizing particles.

## 9.4 Neutrino Interactions with Nuclei

factor that measures the ionization efficiency as a function of nuclear recoil energy. Moreover, trying to increase the cross section using larger atomic numbers results in heavier target nuclei, which implies an even smaller maximum recoil energy.

CE$\nu$NS processes were observed for the first time in 2017 by the COHERENT collaboration, which reported a $6.7\sigma$ significance for an excess of events that agreed with the SM prediction to within $1\sigma$. The collaboration used a Cesium iodide (CsI) scintillator exposed to neutrinos produced via pion decays at rest ($\pi$-DAR, pion-decay-at-rest) coming from the Spallation Neutron Source (SNS) at the Oak Ridge National Laboratory, USA [395]. Weighing 14.6 kg, with dimensions comparative to that of a kitchen microwave, the CsI detector was the first ever working handheld neutrino detector. After that first experimental detection, the COHERENT experiment progressed by searching for CE$\nu$NS on different nuclear targets and with different detector technologies. In 2021 the experiment successfully observed CE$\nu$NS with a detector (CENNS-10) containing 24 kg of liquid argon, and equipped with two photo-multipliers tubes (PMTs). The measured cross sections matched the theoretical cross section within error bars [430].

CE$\nu$NS processes are a potential probe to search for physics beyond the SM. Since the SM provides a very direct prediction of the CE$\nu$NS cross section, any possible deviation would immediately represent indications of new physics. Both neutrino magnetic moment and neutrino charge radius affect the CE$\nu$NS cross section. CE$\nu$NS processes are mediated by neutral currents, and in principle they could be affected by possible sterile neutrinos, which do not couple to SM $W^\pm$ mesons. The analyses of CE$\nu$NS processes can also be relevant for cosmology and astrophysics; for instance, in supernova dynamics, it is this scattering process which has the largest cross-section, helping to expel material outward. In addition, solar, diffuse supernovae remnant and atmospheric neutrinos interact via CE$\nu$NS producing an irreducible background ('neutrino floor') in detectors searching for dark matter. Dark matter is a non-baryonic form of matter, whose evidence comes from observations at very different scales in modern cosmology, and whose composition is not known. A precise measurement of the CE$\nu$NS cross section would allow for a better prediction of this background.

Nuclear power plants constitute another important experimental probe for CE$\nu$NS studies, with several experiments currently underway or in preparation. These facilities are intense, well localised, sources of electron antineutrinos with energies below 10 MeV. The expected CE$\nu$NS signal at reactors does not suffer from nuclear physics uncertainties, in contrast to $\pi$-DAR based experiments. On the other hand, although fluxes are generally order of magnitudes larger than at spallation sources, the neutrino energy generated is much lower, hence, the recoil energies are small. There are several experiments, either in data-taking mode, or being planned, at different reactors sites around the world. The first competitive constraint from CE$\nu$NS at reactors for extensions of the SM with light vector and scalar mediators came from the Coherent Neutrino-Nucleus Interaction Experiment (CONNIE), located about 30 m from the core of the 3.8 GW Angra 2 nuclear reactor in Rio de Janeiro, Brazil [431]. As the neutrino emission is isotropic, the flux in the detector strongly depends on its distance from the core of the reactor. Another experiment, the CONUS (COherent Neutrino nUcleus Scattering) experiment, located at 17 m from the core of a 3.9 GW thermal power nuclear plant in Brokdorf, Germany, and operating since 2018, was able to set the best limit for CEνNS from reactor neutrinos using the first data set [432]. In 2021 and 2022 the Dresden-II Collaboration has reported [433, 434] a suggestive piece of evidence pointing to the first ever observation of CEνNS with reactor antineutrinos. The experiment has used a 3 kg germanium detector, with a location, at 10.39 m from the core of the Dresden-II power reactor (2.96 GW thermal power), which has allowed for an unprecedented flux, about $10^{13} \bar{\nu}_e$ cm$^{-2}$ s$^{-1}$.

### 9.4.2 Nuclear Effects

A way to model the nuclear effects in neutrino-nucleus cross sections starts by slightly modifying the formalism of simple currents we used in interactions with free nucleons. The cross section is still expressed by the contraction of the leptonic tensor $L^{\mu\nu}$ (Eq. (9.32)) and an hadronic tensor $W^{\mu\nu}$, namely

$$d\sigma \propto L_{\mu\nu} W^{\mu\nu}. \tag{9.100}$$

As for the leptonic tensor, the hadronic tensor $W^{\mu\nu}$ can be decomposed into two real tensors, which are symmetric and antisymmetric under the exchange of the Lorentz indexes $\mu \leftrightarrow \nu$, namely

$$W^{\mu\nu} = W_s^{\mu\nu} + i W_a^{\mu\nu}. \tag{9.101}$$

The hadronic tensor is given by bilinear products of the matrix elements of the nuclear current operator between the initial and the final states of the target nucleus. It contains all information on nuclear structure and nuclear interactions, i.e. the entire response of the target, and, as such, it is a very complicated object. Its explicit expression depends on the specific process under consideration, and can be constructed from basic symmetry requirements. However, from classical field theory, it is well known that a dispersion relation in terms of the polarization propagator $\Pi^{\mu\nu}$ exists, which reads

$$W_{s,a}^{\mu\nu} = -\frac{1}{\pi} \operatorname{Im} \Pi_{s,a}^{\mu\nu}. \tag{9.102}$$

In general, in an interacting field theory, spectral representations give an expression for the two-point function as a sum or an integral of free propagators.

In the case of QE neutrino interactions with a nuclear target, spectral representations of the polarization propagators $\Pi_{s,a}^{\mu\nu}$ can recast the tensors in a factorizable form that separates the primary weak interaction on a single nucleon in the nucleus from the quantity that arises from modelling the nucleus, and that captures the

## 9.4 Neutrino Interactions with Nuclei

probability that the event occurs, namely, the so-called spectral function (SF). Then we can write

$$\operatorname{Im} \Pi_{s,a}^{\mu\nu} \simeq \int d^4 p \, W_{s,a}^{(\text{nucleon})\,\mu\nu} \otimes S_f(p). \tag{9.103}$$

The $W_{s(a)}^{(\text{nucleon})\mu\nu}$ tensors is the real symmetric (antisymmetric) quasi elastic hadronic tensor for nucleons, that is, for charged currents, the hadronic tensor built from $J_\mu^+$ in Eq. (9.58). The function $S_f(p)$ represents one or more spectral functions, and depends on the momentum $p$ of nucleons in the nucleus. According to (9.103), one can disregard off-shell effects on the nucleon current and adopt the same form factors as for the free nucleon. In a sense, spectral functions account for the interaction of the outgoing nucleon with the medium and describe the momentum distribution of nucleons from nuclei.

It is worth noticing that theoretical calculations using spectral functions [435–441] only model the initial state and do not account for final state interactions. Nuclear environment affects neutrino-nucleus processes not only in the description of the nuclear target but also after the scattering, since hadrons arising in a primary interaction must propagate through nucleus before they can be detected. These effects are collectively indicated with final state interactions (FSI), and they particularly important for NC processes because the outgoing neutrino cannot be detected. The details of the particle spectral functions, and FSI, in general, do not affect the inclusive integrated cross section, but are important for the differential ones, and are crucial to achieve a realistic description of the reaction final state [442].

The most basic, yet widely used, model to describe nucleons bond into a nucleus is the relativistic Fermi-gas (FG) model (see Sect. 9.4.3). In this model, protons and neutrons are considered as moving freely within the nuclear volume, with a constant binding energy. In Sect. 9.4.3 the SFs based on the FG models are indicated; however, it is possible to build different SFs based on more realistic modelling of the nucleus. Indeed, the energy-momentum distribution of the nucleons in the nucleus is quite different from that of a Fermi gas. The Fermi gas is a system of free fermions or fermions bound with no boundary, while the actual nuclei are locally bound systems of the nucleons, and the nucleonic energies are discrete, corresponding to the single-particle orbits. The energy-momentum distribution is modified further by the very strong interactions between the nucleons at short distance, corresponding to short-distance correlations. The range of the correlations is about 1 fm, which affects physics of several hundred MeV/c momentum. The statistical correlations, which are the only correlations included in the Fermi gas, are of the length scale of several times larger, effective mostly in low momenta.

Other approaches have been developed trying to describe nuclear effects not covered by FG models or spectral function approaches, as collective effects and FSI. One of the several examples is the random phase approximation (RPA) [443], which incorporate nucleon-nucleon correlations. In addition to the short-range correlations, RPA describes the excitation of many-body systems, assuming that the

### 9.4.3 The Fermi Gas Model

A simple and common description of the nucleus is provided by the relativistic global Fermi gas model (RgFG) [444, 445]. In this model the nucleus is considered as an ideal gas composed of weakly interacting neutrons and protons, moving practically freely within the nuclear volume. Nucleons are considered distinguishable fermions, obeying Fermi-Dirac statistics, with a constant binding energy $E_B$. Due to the Pauli exclusion principle, each energy state can be occupied at the maximum by two nucleons with different spin projections. The energy of the highest occupied state is given by the Fermi energy $E_F$, and there is a maximum value for the momentum as well, the Fermi momentum $p_F$. Since all states up to the Fermi level are filled, particles cannot be ejected in momentum states lower than this level. This naturally leads to Pauli-blocking, expressed by the requirement that the final-state nucleon's momentum exceed the Fermi momentum. This hard momentum cut-off reduces the cross section.

The number of nucleons that can be contained in a certain volume of space is obtained by dividing that volume by the volume of one state in phase space, namely $(2\pi)^3$

$$dn = \frac{V 4\pi p^2 dp}{(2\pi)^3} \tag{9.104}$$

where $V$ is the nuclear volume and $n$ is the number of protons or neutrons. The total number of protons or neutrons will be then given by

$$n = \frac{V 4\pi \int_0^{p_F} p^2 dp}{(2\pi)^3} = \frac{V p_F^3}{6\pi^2}. \tag{9.105}$$

The momentum-energy distribution of nucleons $P(E, \mathbf{p})$ is given by

$$P(E, \mathbf{p}) = \theta(p_F - |\mathbf{p}|) \delta(E + \sqrt{M_N^2 + |\mathbf{p}|^2} - E_B) \tag{9.106}$$

where $\theta$ is the step function and $M_N$ the nucleon mass. In the RgFG some relevant SF are proportional to $\theta(p_F - |\mathbf{p}|)$, with $p_F$ being a global Fermi momentum. In another approach, often called relativistic local Fermi Gas (RlFG), the Fermi momentum $p_F$ is not considered fixed, but depending on the local density of protons and neutrons [443, 444]. This description introduces space-momentum correlations, that render more realistic the nucleon momentum distribution.

## 9.4 Neutrino Interactions with Nuclei

In the determination of neutrino cross sections, the FG model of the nucleus assumes the impulse approximation (IA) [446], according to which the interaction takes place on single nucleons whose contributions are summed incoherently. When a lepton transfers 3-momentum **q** to the nucleus it interacts with, a region of the nucleus of order $\sim 1/|\mathbf{q}|$ is penetrated. For low values of $|\mathbf{q}|$ the region covers more than one nucleon and, inevitably, a few particles are involved in the scattering. For higher momentum, the region is small enough to treat the nucleus as a set of independent nucleons. It is then justified to describe the interaction as a scattering off a single (bound) nucleon, whose contributions are summed incoherently—namely, the impulse approximation. Around 1 GeV, typical values of momentum transfer are large enough and IA can be used as a reliable approximation. It is worth underlining that ingredients such as nucleon correlations, or meson exchange currents (MEC), are beyond the IA. The influence of the spectator nucleons can only be present in SF or in FSI. However, when $|\mathbf{q}|$ is small and $1/|\mathbf{q}|$ becomes of the size of the distance among nucleons, one should not expect IA to hold. If aside from IA, the nucleon in the final state is further assumed to leave the nucleus after primary interaction without further interaction with the residual nucleus (no FSI), then this nucleon can be described by a plane wave (Plane Wave Impulse Approximation or PWIA). Examples of such an approach for CCQE scattering can be found in Refs. [437, 438, 440, 447].

# Chapter 10
# Theoretical and Experimental Prospects

The study of neutrino interactions has already provided significant physics results, yet several questions remain unsolved and require more analyses. Investigating the scale and nature of neutrino mass addresses fundamental aspects of particle physics. The existence of non-zero neutrino mass is a solid evidence of BSM physics. Neutrino oscillation experiments prove that neutrinos are massive, constrain mixing parameters, put a lower bound to the allowed mass sum (see Sect. 6.5), but, since they are a sort of interference experiment, cannot be expected to measure the absolute value of neutrino masses. For that, we have to rely on other observations and experiments, that we examine in Sect. 10.1.

Detection of neutrinoless double beta decay is to date the only known method with plausible sensitivity to test the Majorana nature of the neutrino. Its observation would provide direct evidence that lepton number is violated. We discuss double beta decay and its neutrinoless version in Sects. 10.2 and 10.3.

A massive neutrino can have non-trivial electromagnetic properties, in particular it can have a dipole magnetic moment. Neutrino magnetic moments can be used to distinguish Dirac and Majorana neutrinos and, in many extensions of the SM, can be large and observable by future experiments, thus representing a powerful probe on BSM physics. We discuss neutrino magnetic moments in Sect. 10.4.

Finally, in Sect. 10.5, we examine theoretical motivations and experimental searches for sterile neutrinos.

## 10.1 Absolute Values of Neutrino Masses

When the SM was developed, neutrinos were introduced as massless leptons in well-defined flavour states, paired in electroweak doublets to massive leptons. The subsequent discovery of neutrino oscillations established that these three flavor states are actually quantum superpositions of three well-defined mass states.

Oscillation experiments are sensitive only to the differences of neutrino squared masses, and not to the individual values of the neutrino masses. The oscillation frequency depends on the splittings between two squared mass values, so the existence of flavour oscillation ensures that there are three distinct mass values in the active neutrino sector.

One can single out three different approaches to probe the absolute mass scale of neutrinos: (1) direct neutrino mass determinations, (2) cosmological observations, (3) analyses of neutrino-less double $\beta$ decay.

The direct neutrino mass determination is pursued by precision investigations of the kinematics of charged particles (leptons, mesons) emitted together with neutrinos in an electroweak process. When the total energy of the initial state is well known and the kinematics of the final state can be measured with precision, it is possible to constrain, using energy and momentum conservation, the neutrino mass. One essentially uses the relativistic energy-momentum relationship $E^2 = p^2 + m^2$, without further assumptions. Therefore this approach is sensitive to the neutrino mass squared. The most sensitive neutrino mass measurement to date is via the kinematics of a single $\beta$-decay. The neutrino masses lead to a reduction of the maximal observed energy of the decay and a small spectral shape distortion close to the kinematic endpoint of the $\beta$-spectrum. As seen in Sect. 1.3, this way to probe neutrino mass was suggested by Francis Perrin [23] and Enrico Fermi [15]. The Karlsruhe Tritium Neutrino (KATRIN) experiment provided the first direct neutrino-mass measurement with sub-eV sensitivity via a high-precision measurement of the tritium $\beta$-decay spectrum close to its endpoint at 18.6 keV [448, 449]. We discuss direct neutrino mass determination in Sect. 10.1.1.

Cosmology also imposes limits on neutrino masses. Within the SM, neutrinos interact exclusively through the electroweak force mediated by the W and Z vector bosons. These interactions, due to their weak nature, fall out of equilibrium early on in the history of the Universe at the freeze-out temperature $T \sim 1$ MeV (see Sect. 7.5.1). After decoupling from other SM particles, neutrinos are thought to free stream throughout the Universe, only interacting with other species through their gravitational interactions. They have large free-streaming lengths that depend on their small masses. Within the standard cosmological paradigm, free-streaming neutrinos smear out fluctuations that are imprinted in the CMB and, in general, affect the large-scale structure of the universe. Roughly speaking, neutrinos free stream out of high density structures in the Universe, diminishing their total mass; hence, they tend to erase them at scales smaller than the distance travelled during a significant fraction of the formation time of such high density structures.

By determining the early fluctuations imprinted on the CMB, and mapping out today's structure of the universe by large galaxy surveys like Sloan Digital Sky Survey (SDSS) [450], conclusions on the sum of the neutrino masses can be drawn. CMB space-based measurements have been carried out first by NASA's Cosmic Background Explorer (COBE) satellite [451], then by the Wilkinson Microwave Anisotropy Probe (WMAP) [452], and finally by the Planck satellite, launched

## 10.1 Absolute Values of Neutrino Masses

in 2009 and deactivated in 2013. To a very good approximation, cosmological observables are mainly sensitive to the sum of neutrino masses, namely

$$\sum m_\nu = \sum_i m_i. \tag{10.1}$$

Up to now, only upper limits on the sum of the neutrino masses have been obtained, which are to some extent model and analysis dependent. The Planck data, in combination with other probes, give the following limit [453]

$$\sum m_\nu < 0.12\,\text{eV} \tag{10.2}$$

at 95% confidence level. This upper limit, combined with the lower limit from oscillation experiments given in Eqs. (6.76) and (6.77), leaves only a narrow window at a value around 0.1 eV, that cries out for explanation in fundamental physics. It also point to disfavouring inverted ordering.

Other options exist for the measurement of the neutrino absolute scale by means of cosmological observations. Next generation galactic surveys will investigate a huge number of galaxies in large portions of the sky and to very far distances, improving sensitivity to neutrino mass through their effect on structure formation. A good example is the Euclid mission,[1] from the European Space Agency (ESA), designed to provide very sharp images of a large fraction of the extragalactic sky and perform near-infrared spectroscopy of hundreds of millions of galaxies and stars over the same sky.[2] If the total sum of neutrino masses $\sum m_\nu$ is larger than 0.1 eV, Euclid should be able to determine the neutrino mass scale independently of the cosmological model assumed. For $\sum m_\nu$ below that value, the sensitivity reaches 0.03 eV in the context of a minimal extension of the Lambda Cold Dark Matter ($\Lambda$CDM) model, a cosmological model which has provided a successful fit for a large part of the astrophysical and cosmological observations carried out over the past decades [454–457]. Similar results can be obtained by the Rubin Observatory Legacy Survey of Space and Time (LSST), before 2019 named Large Synoptic Survey Telescope,[3] still under construction in Chile. It will cover the entire southern sky for a decade. Even better results might be obtained by means of global fits to all available data, including CMB data. It is quite possible that cosmology will give us the neutrino mass scale before direct kinematic experiments during next decade.

Neutrino mass bounds can also be provided by time-of-flight measurements. The idea is that neutrinos travel long distances in a time that depends on their mass as well as on their energy [458]. So there is a delay compared to the time of flight

---

[1] https://www.euclid-ec.org.
[2] Euclid, with a mass in orbit of 2 tonnes, took off on July 1st, 2023, onboard SpaceX Falcon 9 from Cape Canaveral (Florida, USA). In the month after, Euclid has travelled 1.5 million kilometres from Earth, reaching its destination orbit.
[3] https://rubinobservatory.org/.

of a supposedly massless particle that impacts the time spectrum of the neutrino events at the detector. Measuring this effect requires an experiment to have very long baselines and therefore very strong neutrino sources, which only cataclysmic (and rare) astrophysical events like a type II SN bursts could provide. In order to compare arrival times to the detector of a massive neutrino with respect to a massless one, one should know the time at which the neutrino is produced, For supernova neutrinos, this is not known precisely, but one can investigate whether neutrinos with different energies arrive at different times, leading to, for example, a larger-than-expected spread in the neutrino arrival times.

Using SN1987A inverse $\beta$ decay data from Kamiokande, IMB and BUST experiments, an upper limit on neutrino mass of 5.7 eV [459] or 5.8 eV [460] at 95% C.L. was set. In case of a SN event at a distance of 10 kpc, future detectors as JUNO or DUNE should be able to provide sub-eV mass limits [461, 462].

If the neutrino is a Majorana fermion, the observation of the neutrinoless double $\beta$ decay, discussed in Sect. 10.3, would provide indications on its mass value.

### 10.1.1 Direct Neutrino Mass Measurements

The absolute values of the neutrino masses can be directly determined from the kinematics of a weak process without using models or making assumptions on the neutrino mass type (Majorana or Dirac), but using only relativistic energy-momentum relations and 4-momentum conservation. The weak process of election is $\beta$ decay, which allows a direct determination from the precise measurement of the shape of the electron spectrum near the endpoint.

In $\beta$ decays, the energy available from the nuclear mass difference (the $Q$-value) is shared by the electron and the neutrino. The neutrino energy is $E_{\nu_e} = Q - E_k$, where $E_k$ is the kinetic energy of the electron. If neutrinos were massless, $Q$ would coincide with the maximal kinetic energy of the electron. Since the neutrino has mass, the largest possible energy that the electron can receive is lower, hence the endpoint of the electron energy spectrum is reached at lower energy and the shape of a small interval below the endpoint slightly changes. Non-zero neutrino mass can be detected by carefully examining this region.

Since the number of electrons near the endpoint of the spectrum is small, the statistical error is large. It can be reduced by choosing a nucleus with a low $Q$-value, since a larger fraction of the total spectrum then resides within a given interval from the endpoint. The enhancement of source intensity is favoured by choosing a nucleus with a relatively short half-life. A gaseous $\beta$ decay source is also generally advantageous, since it combines low density, so minimal energy loss, which could bias the measurement, with a reasonable number of source atoms to decay.

The $\beta$ decay of the hydrogen isotope tritium

$$^{3}\text{H} \rightarrow {}^{3}\text{He}^{+} + e^{-} + \nu_e \tag{10.3}$$

## 10.1 Absolute Values of Neutrino Masses

is a sensitive choice. It combines a small $Q$-value ($Q = m_{3H} - m_{3He} \sim 18.6$ keV) and a short half-life (12.3 years) with a simple nuclear and atomic structure. When looking for an already rare signal, one wants elements which have the least possible of the nuclear effects which would distort the region near the endpoint in a way difficult to interpret. For general mixing schemes the spectral shape of $\beta$ decay can be rather complex. In the case of three active neutrinos, and in the quasi-degenerate mass regime,[4] the individual mass eigenstates $m_i$ are not resolved, and $\beta$ decay experiments probe the so-called effective electron neutrino mass $m_\beta$[5]

$$m_\beta \equiv \left( \sum_{i=1}^{3} |U_{ei}|^2 m_i^2 \right)^{1/2}. \tag{10.4}$$

It is a real average, with the modules of the PMNS matrix element $|U_{ei}|$ as positive weighting factors, such that no cancellations can occur. The effective mass extracted from the experiment fixes the absolute mass scale in the almost degenerate case, taking into account the small values of $\Delta m^2$ from oscillation experiments. The result (10.4) is qualitatively understandable, if we consider that for each $m_i$ the part of the differential rate associated with the mass effects, $d\Gamma(m_i^2)$, can be expected to be proportional to $m_i^2$, for dimensional reasons. and to $|U_{ei}|^2$, since it is a $\beta$ decay. Summing incoherently one has

$$d\Gamma \sim \sum d\Gamma(m_i^2) \propto \sum |U_{ei}|^2 m_i^2. \tag{10.5}$$

The first experiments of this kind to quantitatively constrain the mass of the neutrino started in 1948 and used gaseous tritium [463]. Building on about 70 years of kinematic searches, the KATRIN (Karlsruhe Tritium Neutrino) experiment began commissioning with tritium in 2018. A neutrino mass has not been extracted yet, but the KATRIN experiment has provided a first experimental constraint on the effective neutrino mass $m_\beta < 0.8$ eV [449], lowered in 2024 to $m_\beta < 0.45$ eV [464], that is by a factor of almost 2.

The KATRIN experiment consists in a 70-m-long beamline located at the Karlsruhe Institute of Technology, Germany. It has the same design as all tritium $\beta$ decay experiments. The beamline is divided into the tritium-containing source and transport section, and the tritium-free section containing the spectrometers and detector system. Electrons are produced in a 10 m long windowless gaseous tritium source, in which a highly purified molecular tritium gas is continuously injected. The gas diffuses towards both ends of the tube where it is pumped out and fed back to the tritium loop system. Tritium decays, releasing an electron and an electron

---

[4] The quasi-degenerate mass regime occurs when the mass values are not exceedingly small and their splittings are negligible with respect to their values.

[5] In literature, also other terms, e.g. $m(\nu_e)$, are used to indicate the effective mass. The effective neutrino mass can also be defined as the squared value of definition (10.4).

antineutrino, which escapes undetected. The $\beta$-emitted electrons are magnetically guided from the source to a large spectrometer known as a MAC-E filter (Magnetic Adiabatic Collimation combined with an Electrostatic filter). The MEC-E filter imposes a retarding electron potential along its length, stopping all but the electrons with high enough energy to surmount the potential barrier, which are then guided into the detector, where they are counted. By changing the value of the height of the potential barrier, the energy threshold changes, and an integrated spectrum of the electrons can be built. The KATRIN experiment is characterized by the size of its MAC-E filter, which has a diameter of 10 m, an order of magnitude larger than those used in previous such experiments.

The goal of the KATRIN experiment is to reach $m_\beta < 0.2$ eV at 90% confidence level. Of course, the real goal is not to produce a smaller limit—it is to find non-zero values. If the electron neutrino mass is above 0.35 (0.30) eV, KATRIN will measure it to a precision of 5 (3) standard deviations. It is a true challenge, and actually disfavoured by cosmological constraints, as seen in Sect. 10.1.

Improvements in sensitivity by experiments like KATRIN are limited by the size of the electrostatic filter, hence new approaches are being developed. The Project 8 experiment employs a technique known as Cyclotron Radiation Emission Spectroscopy (CRES) [465], in pursuit of sensitivity to the effective neutrino mass down to 0.04 eV. This technique relies on the detection and measurement of coherent radiation created from the cyclotron motion of electrons in a strong magnetic field (cyclotron emission). In a constant magnetic field, an electron undergoes cyclotron motion. This occurs at a frequency that depends on the kinetic energy of the electron. Thus, a precise measurement of the frequency is tantamount to a precise measurement of the energy. The experiment is housed at the University of Washington in Seattle, USA. Project 8 has recently applied this technique to the continuous tritium beta spectrum—enabling the first neutrino effective mass limit using CRES, namely $m_\beta < 155$ (152) eV in a Bayesian (frequentist) analysis [466].

Electron capture (see Eq. (1.36))

$$p + e^- \rightarrow n + \nu_e \tag{10.6}$$

has a sensitivity to the neutrino mass entirely analogous to the one of $\beta$-decay. In the electron capture process an atomic electron interacts with a nucleus of charge Z to produce a neutrino, leaving behind a nucleus of charge $Z-1$ and a hole in the orbital of the atom from which the electron was captured. To be (maybe overly) precise, electron capture can determine the electron neutrino mass, while the beta decay of tritium measures the electron antineutrino mass. Currently, three experiments, ECHo [467] (Electron Capture Holmes), HOLMES [468], and NuMECS (Neutrino Mass via Electron Capture Spectroscopy) [469], are investigating the feasibility of the electron capture process

$$^{163}\text{Ho} + e^- \rightarrow {}^{163}\text{Dy}^* + \nu_e. \tag{10.7}$$

## 10.1 Absolute Values of Neutrino Masses

The holmium isotope $^{163}$Ho, which has a half-life of 4570 years, goes into an excited state of the dysprosium $^{163}$Dy* with a low Q-value of 2.8 keV. The basic idea is to place the holmium source inside an absorber material with low heat capacity. X-rays and electrons emitted in the de-excitation of the dysprosium daughter atom create phonons in the absorber material and cause a small temperature increase, which can be detected. Since the $^{163}$Ho source is fully contained in detectors, all of the energy released in the decay of $^{163}$Ho, except that taken away by the electron neutrino, contributes to the signal. Therefore, neutrino mass can be searched for looking at modifications in the end-point of the $^{163}$Ho energy spectrum. A limit on the effective electron neutrino mass has been produced by the ECHo experiment [470], yielding $m_\beta < 150$ eV at 95% confidence level.

Although electron capture or $\beta$ decay allow precise measurements, they are not the only weak processes that can be used to directly measure the neutrino mass. The mass of the muon neutrino can be measured directly by investigating pion decays, namely

$$\pi^+ \rightarrow \mu^+ + \nu_\mu$$
$$\pi^- \rightarrow \mu^- + \bar{\nu}_\mu. \tag{10.8}$$

Given a precise measurement of the pion and muon masses, and the relevant kinematic quantities, the muon neutrino mass can be obtained by energy and momentum conservation. An upper limit on the muon neutrino mass has been made [471], yielding $m_{\nu_\mu} < 170$ keV at 90% confidence level.

The mass of the tau neutrino is measured in the multi-hadronic decay of $\tau$ leptons produced at electron-positron colliders through the process

$$e^+ + e^- \rightarrow \tau^+ + \tau^-. \tag{10.9}$$

The energy of each tauon is just half the centre of mass energy of the beam, so it is well known. Due to its large mass, about 1777 MeV, the tauon can decay into several pions; for example, we can have

$$\tau \rightarrow \nu_\tau + 5\pi^\pm(\pi^0). \tag{10.10}$$

The available phase space for the neutrino is restricted, implying a low branching ratio and a low count rate. Moreover, one has to be sure that the detector has recorded, and reconstructed, all the pions well and with sufficient precision as to be sensitive to the missing energy and momentum carried away by the tau neutrino. The actual upper limit [472] is $m_{\nu_\tau} < 18.2$ MeV at 95% confidence level. It was achieved by the ALEPH experiment at the LEP collider at CERN, using the data collected from 1991 to 1995.

## 10.2 Double $\beta$ Decay

The so-called double beta decay ($2\nu\beta\beta$) is the spontaneous twofold conversion of a neutron into a proton in one nucleus at the same time, accompanied by the emission of two electrons and two anti-neutrinos, which occurs in neutron-rich nuclei:

$$(A, Z) \rightarrow (A, Z + 2) + 2 e^- + 2 \bar{\nu}_e \quad (2\nu\beta^-\beta^- \text{ or } 2\beta^-_{2\nu}) \qquad (10.11)$$

where $A$ and $Z$ are the mass and the atomic number of the nuclei, respectively.

In proton-rich nuclei, the inverse process can be kinematically possible. It involves the conversions of two protons into two neutrons and the emission of two positron and two electron neutrinos:

$$(A, Z) \rightarrow (A, Z - 2) + 2 e^+ + 2 \nu_e \quad (2\nu\beta^+\beta^+ \text{ or } 2\beta^+_{2\nu}). \qquad (10.12)$$

As for the single $\beta$ decay, in the SM also the crossed channels of the mode (10.12) are possible, namely electron capture with an electron in the initial state and only one positron in the final one (EC $\beta^+$), and double electron capture (2EC or EC EC), with two electron in the initial state and no charged leptons in the final one:

$$(A, Z) + e^- \rightarrow (A, Z - 2) + e^+ + 2 \nu_e \quad (\text{EC } \beta^+)$$
$$(A, Z) + 2 e^- \rightarrow (A, Z - 2) + 2 \nu_e \quad (\text{2EC or ECEC or } 2\nu\text{ECEC}). \qquad (10.13)$$

The double beta decay is allowed by the SM and conserves the lepton number. It is mediated by two charged $W$ bosons at the partonic level, and it is equivalent to two simultaneous single $\beta$-decays (see Fig. 10.1, left). The double beta decay occurs regardless of whether neutrinos are their own antiparticles or not, i.e. Majorana or Dirac. Being of second order in the weak interactions, the decay rates are very low and the observation becomes only feasible if single $\beta$ decay is not allowed energetically or at least is kinematically strongly suppressed. Given an isotope $(A, Z)$ of mass $M(A, Z)$, the double beta decay in Eq. (10.11) occurs when the kinematic condition $M(A, Z) > M(A, Z + 2)$ holds. Then, another condition is that single beta decay is kinematically forbidden, i.e. $M(A, Z) < M(A, Z + 1)$, or that it is so much suppressed (e.g. by the angular momentum selection rules) that it does not compete with double beta decay.[6]

The effect that makes searching for double beta decay a viable possibility is the nuclear pairing force. Nucleons in nuclei have magnetic moments and it is usually energetically favourable for them to pair up with spins opposing each-other in equivalent spatial orbitals. This configuration maximises wave-function overlap and the stabilising effect of the attractive spin-opposite spin interaction. The result

---

[6] Let us specify that we are referring to ground-state energies and neglecting the lepton masses.

## 10.2 Double β Decay

is that nuclei with even numbers of protons, or even numbers of neutrons, are nearly always slightly more tightly bound than similar nuclei with odd numbers of both. This effect is particularly evident in the so called even-even nuclei where both proton and neutron numbers are even, and all of the nucleons can pair up. For instance, even-even nuclei $^{136}$Xe and $^{136}$Ba are stabilised by pairing. Odd-odd nuclei $^{136}$I and $^{136}$Cs are less tightly bound. Another example is the case of the nucleus of $^{76}$Ge ($Z=32$) (the highest natural Germanium isotope). Its $\beta$ decay into the nucleus of an isotope next on the periodic table, the Arsenic $^{76}$As ($Z=33$), higher in mass and with a smaller binding energy, is energetically forbidden, while it is allowed its double $\beta$ decay into the nucleus of the stable Selenium isotope $^{76}$Se ($Z=34$), lighter and with a larger binding energy. Both $^{76}$Ge ($Z=32$) and $^{76}$Se ($Z=34$) nuclei are even-even.

The study of double $\beta$ decay was suggested by Maria Goeppert-Mayer in 1935 [473], one year after the Fermi theory of $\beta$ decay [15], but it took more than 50 years to observe it directly in $^{82}$Se [474]. Up to now the double $\beta$ decay has been observed using different techniques in 11 nuclei, with measured half-lives $\tau_{1/2}^{2\nu}$ ranging from approximately $10^{19}$ up to $10^{24}$ years. The two-neutrino double electron capture has been observed only in three nuclei, $^{130}$Ba, $^{78}$Kr and, more recently (2019) in $^{124}$Xe [475].[7]

Three main experimental methods exist for investigating the double beta decay, direct, geochemical, and radiochemical. In direct counting experiments the double $\beta$ decay is analysed in real time by measuring the energies of the decay electrons or their sum from a pure sample containing the isotope of interest. We have two main methods in this contest: one in which a detector is used at the same time as a source of double beta decay events (source = detector), and the one where the source of the double $\beta$ decay is external to the detector (source $\neq$ detector). Experiments of the first type are referred to as active source and have as a major advantage an high registration efficiency. In the second approach there is a possibility to measure several isotopes simultaneously and to get full information about electron tracks. This latter method can also be used to study the gammas that are emitted in addition (or in alternative) to the study of the emitted positrons/electrons.

An active source experiment, which has stopped data taking in 2020, is the Germanium Detector Array (GERDA) experiment. GERDA started at LNGS in 2011 and has operated about 40 kg of $^{76}$Ge detectors directly immersed in a liquid argon (LAr) volume instrumented to detect its scintillation light. GERDA final results for two-neutrino double-$\beta$ decay were obtained with a subset of the entire exposure, that 11.8 kg year, yielding [480]

$$\tau_{1/2}^{2\nu} = (2.022 \pm 0.018_{\text{stat}} \pm 0.038_{\text{sys}}) \times 10^{21} \text{ years.} \tag{10.14}$$

This is one of the most precise measurements of a double $\beta$ decay process.

The indirect methods of observation are the geochemical and the radiochemical ones. In the geochemical experiments one measures the excess of the daughter

---

[7] For a review and average half-lives, see for instance Refs. [476–479].

nuclei $(A, Z + 2)$, accumulated in very large times, in a sample of rocks containing nuclei $(A, Z)$. The sample of the parent rock is dated independently, for instance with geologic means. Its advantage lies in the long geological times during which the daughter nuclei have been accumulated. The parent rock is usually part of an ore sample from which the atom of a daughter isotope is chemically extracted and analysed by mass spectroscopy. The earliest experimental evidence for double beta decay came from geochemical experiments, the first of which was performed in 1949 [481].

Radiochemical experiments use the fact that some double beta decay daughter nuclei are themselves radioactive. An artificial sample of the parent isotope is prepared in a laboratory setting, and the accumulation of the daughter isotope is measured after some years. The daughter nuclei are first chemically isolated; since they are radioactive, they produce a distinctive signal, which is searched for.

## 10.3 Neutrinoless Double $\beta$ Decay

In 1939 the existence of a process similar to the double $\beta$ decay, but without neutrinos, was suggested by Wendell H. Furry [482]. The nuclear transition

$$(A, Z) \rightarrow (A, Z + 2) + 2e^- \qquad (0\nu\beta\beta) \qquad (10.15)$$

is the so-called neutrinoless double $\beta$ decay ($0\nu\beta\beta$ or $2\beta 0\nu$). The existence of this still unobserved process requires one vertex where one virtual antineutrino is emitted and another one where it is absorbed as a virtual neutrino. The $0\nu\beta\beta$ decay can be interpreted as two subsequent steps, the so-called Racah-sequence:

$$(A, Z) \rightarrow (A, Z + 1) + e^- + \bar{\nu}_e$$
$$(A, Z + 1) + \nu_e \rightarrow (A, Z + 2) + e^-. \qquad (10.16)$$

A neutron $\beta$ decays emitting a right-handed $\bar{\nu}_e$, which is absorbed at the second vertex as a left-handed $\nu_e$ (neutrino capture process). In Fig. 10.1 the double $\beta$ decay and its neutrinoless counterpart are depicted. The $0\nu\beta\beta$ decay process in Eq. (10.15) violates the total lepton number of two units: $\Delta L = \pm 2$, and it is therefore forbidden in the SM, where the lepton number is conserved. Besides, in the SM the massless antineutrino is right-handed, while the massless neutrino is left-handed, then the antineutrino has the wrong chirality for being absorbed.

These difficulties are overcome if the SM is extended to include neutrinos which are Majorana particles and have mass. In Sect. 5.3, we have seen that in presence of Majorana neutrinos the total lepton number is not conserved. A neutrino mass is required to allow for the helicity matching. The reason is that a massive neutrino has no fixed helicity and therefore, besides the dominant left-handed contribution, has an admixture of a right-handed component (or vice versa for antineutrinos).

## 10.3 Neutrinoless Double β Decay

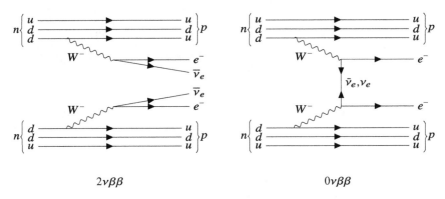

**Fig. 10.1** Pictorial representation of double β decay (left) and neutrinoless double β decay (right), with the proton $p$ and neutron $n$ resolved in their quark components

If neutrinos have mass, one vertex can emit a Majorana neutrino having negative helicity with relative amplitude $\propto m_{\nu_e}/E_{\nu_e}$ (see Sect. 5.3), which is absorbed by the other leptonic vertex with relative amplitude equal to unity.

Let us remark that the process in Eq. (10.15) is not the only process that one can expect to observe once accepted the possibility to have double beta decay processes where an antineutrino is absorbed as a neutrino, or vice versa. In fact, depending on the relative numbers of protons and neutrons in the nucleus, the different mechanisms described in Eqs. (10.12) and (10.13) for double beta decay are also possible *without neutrinos*.

If $0\nu\beta\beta$ decay occurs, we are automatically beyond the SM, and one can envision new mechanisms to induce $0\nu\beta\beta$ decays besides the exchange of Majorana massive neutrinos with SM interactions, as depicted in Fig. 10.1, right. These mechanisms could involve new interactions and/or new particles beyond the SM. For example, the helicity matching condition could be overcome assuming an additional $\gamma_\mu(1 + \gamma_5)$ coupling, rather than the standard EW one $\gamma_\mu(1 - \gamma_5)$. However, it is possible to demonstrate that if neutrinoless double β decay exists, then neutrinos have a massive Majorana nature, irrespective of the mechanism for double β decay [483, 484]. In conclusion, the detection of neutrinoless double β decay would prove the non-conservation of the lepton number and be an evidence that neutrinos are Majorana particles with a non-zero mass.

Both the modes of the double β decay are of second order in weak interactions, hence inherently slow, but neutrinoless double β decay, being forbidden by the lepton number conservation, is expected much slower. A distinctive feature is that, while in the $2\nu\beta\beta$ mode the two neutrons undergoing the transition are uncorrelated (but decay simultaneously), in the $0\nu\beta\beta$ the two neutrons are correlated.

The main experimental advantage of neutrinoless double β decay is easily the very clear experimental signature. The full energy of the decay is the $Q$-value of the β decay $Q = m_i - m_f$, where $m_i$ and $m_f$ are, respectively, the masses of the initial and final nuclei. In single and double beta decay, the neutrinos carry away part of the

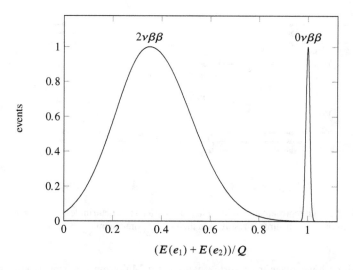

**Fig. 10.2** Schematic (not in scale) plot of summed energy spectra of the two emitted electrons $e_1$ and $e_2$ for double beta decay and neutrinoless double beta decay. The former signal has been convoluted with an arbitrarily chosen experimental Gaussian resolution function

energy released in the decay and the sum of the energy of the two electrons in the final state has a broad energy distribution. In neutrinoless double-beta decay, there are no neutrinos in the final state, so all of the energy is carried by the electrons in the final state (the recoil energy of the final nucleus is considered negligible) and the sum of the energy of the two electrons is practically equal to the $Q$-value of the reaction. The $Q$-value depends on the isotope, ranging between about 1.5 and 5 MeV. In Fig. 10.2 an idealised sum electron spectrum for the double beta decay is presented. In the case of $2\nu 2\beta$ the distribution is continuous, with an endpoint at $Q$. Instead, the $0\nu 2\beta$ decay, if present, would produce a peak at $Q$ with a width determined only by the energy resolution of the detector. The monochromatic peak ranges from 2 to 3 MeV for most of the emitters. This allows one to separate the two double $\beta$ decay modes experimentally by measuring the sum energy of the emitted electrons, even if the decay rate for the $0\nu\beta\beta$ mode is much smaller than for the $2\nu\beta\beta$ mode. Direct detection is the only approach with the capability to distinguish between $2\nu\beta\beta$ and $0\nu\beta\beta$ decays. In order to detect the signal, an experiment must have low background in the region near the $0\nu\beta\beta$ $Q$-value and good energy resolution to avoid losing the signal in the tail of the $2\nu\beta\beta$ spectrum.

## 10.3.1 Total Decay Rate and Effective Majorana Mass

The decay rate (see Eq. (7.13)) of double beta decay can be expressed as a product, with factors which depend on the atomic physics (a phase-space factor

## 10.3 Neutrinoless Double $\beta$ Decay

$G^{0\nu}$) and nuclear structure (a squared modulus of a nuclear matrix elements $M^{0\nu}$), plus a squared modulus of a function $f$ which collects possible particle physics parameters. That is, for $0\nu\beta\beta$ decay we have (see e.g. Ref. [485])

$$\frac{\Gamma}{\ln 2} = \frac{1}{\tau_{1/2}^{0\nu}} = G^{0\nu}|M^{0\nu}|^2|f(m_i, U_{ei})|^2 \tag{10.17}$$

where $\tau_{1/2}^{0\nu}$ is the half-life of $0\nu\beta\beta$ decay. A similar formula holds for the $2\nu\beta\beta$ mode, without the function $f$ and with different phase-space factor and nuclear matrix element ($M^{2\nu}$ and $G^{2\nu}$).

The phase space factors $G^{0\nu}$, which encloses the kinematics of the decay, depend on the $Q$-value and the atomic number $Z$. They are exactly calculable and yield tiny values, around $10^{-26}$ years$^{-1}$ eV$^{-1}$.[8]

The function $f(m_i, U_{ei})$ depends on the neutrino mass eigenvalues $m_i$ and on the matrix elements of the $U_{\text{PMNS}}$ matrix $U_{ei}$ (see Eq. (6.9)). It is rather easy to understand such dependence. If we consider the SM Lagrangian with the addition of the Majorana mass terms, the latter are the only terms that violate the lepton number by two units. This simple consideration motivates the fact that the amplitude of the decay has to be proportional to mass parameters. The dependence on $U_{ei}$ follows from the weak vertices which mediate the decay, involving charged currents as in (6.3). The $f(m_i, U_{ei})$ function cannot be built inside the SM and its precise form depends on the new physics model one uses.

A rather conservative assumption is that the decay can occur through the exchange of a virtual, light but massive, Majorana neutrino between the two nucleons undergoing the transition, and that these neutrinos interact by the standard left-handed weak currents.[9] It leads to [489–492]

$$f(m_i, U_{ei}) = \frac{m_{\beta\beta}}{m_e} \tag{10.18}$$

with

$$m_{\beta\beta} \equiv \left| \sum_{i=1}^{3} U_{ei}^2 m_i \right| \tag{10.19}$$

---

[8] The phase space factors for double beta decay were calculated in the 1980 [486]; a calculation of phase-space factors is reported in the book of Boehm and Vogel [487]. These calculations use an approximate expression for the electron wave functions in the nucleus. Phase space factors have been recently recalculated with exact Dirac electron wave functions and including screening by the electron cloud [488].

[9] Let us underline that this assumption is not the only theoretically possible one, since BSM lepton number violating interactions involving so far unobserved much heavier particles might also lead to neutrinoless double $\beta$ decay of comparable rates.

where the index $i$ runs on the three light neutrinos with given mass $m_i$, and $m_{\beta\beta}$ is the so-called effective Majorana mass.[10] All the physical phases are included in the $U_{ei}^2$. Roughly speaking, the form (10.18) follows from the presence in the calculation (10.17) of the propagator of the left-handed neutrino

$$\sum_{i=1}^{3} U_{ei}^2 P_L \frac{\not{p} - m_i}{p^2 - m_i^2} P_L \qquad (10.20)$$

where $P_L = (1 - \gamma_\mu \gamma_5)/2$. The $U_{ei}^2$ matrix elements appear because an electron is emitted in each of the two vertices of Fig. 10.1 (right). Since $P_L \not{p} P_L = 0$ holds, we are left with a dependence on $m_i$.

In the standard parametrisation (6.8) and (6.9), the effective Majorana mass can be written as

$$m_{\beta\beta} = |c_{13}^2 c_{12}^2 m_1 + c_{13}^2 s_{12}^2 m_2 e^{2i\alpha_2} + s_{13}^2 m_3 e^{2i(\alpha_3 - \delta)}| \qquad (10.21)$$

where $c_{ij} = \cos\theta_{ij}$, $s_{ij} = \sin\theta_{ij}$, $\alpha_{2,3} = [0, 2\pi]$. It depends not only on the mixing angles and Dirac $CP$-violating phase, but also on the Majorana $CP$-violating phases $\alpha_2$ and $\alpha_3$. If $CP$ is conserved $\alpha_2$ and $\alpha_3 - \delta$ are multiple of $\pi$, but generally any value of these phases are possible. Thus, the effective Majorana mass could be complex and cancellations in the sum are possible. One can write (see Sect. 6.5)

$$m_{\beta\beta} = \left| c_{13}^2 c_{12}^2 m_1 + c_{13}^2 s_{12}^2 \sqrt{m_1^2 + \Delta m_{21}^2} e^{2i\alpha_2} + s_{13}^2 \sqrt{m_1^2 + \Delta m_{31}^2} e^{2i(\alpha_3 - \delta)} \right| \qquad (10.22)$$

in the normal order, and

$$m_{\beta\beta} = \left| |c_{13}^2 c_{12}^2 \sqrt{m_3^2 - \Delta m_{31}^2} + c_{13}^2 s_{12}^2 \sqrt{m_3^2 + \Delta m_{21}^2 - \Delta m_{31}^2} e^{2i\alpha_2} + \right.$$
$$\left. + s_{13}^2 m_3 e^{2i(\alpha_3 - \delta)} \right| \qquad (10.23)$$

in the inverted order. We can regard $m_{\beta\beta}$ as a function of several parameters: the mixing matrix elements, the squared mass splittings, the mass of the lightest neutrino, i.e. $m_1$ ($m_3$) in the normal (inverted) order, and the complex phases. If the mixing matrix elements are known with good precision, varying the Majorana phases, we can plot the range of values of $m_{\beta\beta}$ as a function of the absolute mass of the lightest neutrino.

---

[10] In literature, slightly different notations are also used, e.g. $f(m_i, U_{ei}) \equiv m_{\beta\beta}$. The effective mass $m_{\beta\beta}$ is not always defined as an absolute value. It can be indicated as $m_{2\beta}$, $m_{ee}$ or as the average light neutrino mass, namely as $\langle m_\nu \rangle$, or $\langle m_{\beta\beta} \rangle$.

## 10.3 Neutrinoless Double $\beta$ Decay

Through the discovery of neutrinoless double $\beta$ decay it would be possible, in principle, to obtain a non-zero value of $m_{\beta\beta}$—and it would be a pioneering achievement. It would prove the Majorana nature of the neutrinos and show that lepton number is violated. However, a direct extraction of the two Majorana phases from the measurement of the effective Majorana mass is far from obvious, particularly because of the large uncertainty still existing in the calculation of nuclear matrix elements. It is clear from Eqs. (10.17) and (10.18) that information on $m_{\beta\beta}$ can only be extracted from a measurement of the decay rate if the nuclear matrix elements $M^{0\nu}$ is known. Its computation is affected by large theoretical errors, since it requires an accurate nuclear model. Many different techniques exist, requiring extensive computations. The nuclear matrix elements are generally considered the largest of the theoretical uncertainties in predictions of double beta decay rates [479, 493].

A not impossible option is to extract the phases from a large number of independent observations of neutrino-less double beta decays from different nuclei, but there is certainly a long way to go for that. It should be underlined, however, that this is the only known way to probe Majorana phases directly, and we can just hope that the searches in progress with many different nuclei (chiefly $^{136}$Xe, $^{76}$Ge, $^{130}$Te and $^{100}$Mo) may offer a big discovery in the next decade.

### 10.3.2 Experimental Searches

The experimental signature of the neutrinoless double beta decay is a monoenergetic peak at the $Q$-value of the summed energy spectra of the two emitted electrons (see Fig. 10.2). A typical experiment measures the total energy of the two electrons, and perform a signal search over a narrow energy window around $Q$. The width of this region of interest is selected on the basis of the energy resolution of the detector. Let us imagine an ideal experiment with a pure source of double beta decays, a perfect energy resolution and no background. The expected number $N$ of $0\nu\beta\beta$ events is given by

$$N = \frac{M \cdot N_A}{M_A} \cdot \frac{\ln 2}{\tau_{1/2}^{0\nu}} \cdot \varepsilon \cdot t \equiv N_{\beta\beta} \cdot \frac{\ln 2}{\tau_{1/2}^{0\nu}} \cdot \varepsilon \cdot t \qquad (10.24)$$

where $N_A$ is the Avogadro number, $M$ and $M_A$ are the total mass and the atomic mass of the decaying isotope, respectively, $\varepsilon$ is the signal detection efficiency, $\tau_{1/2}^{0\nu}$ is the isotope half-life and $t$ is the data-taking time. The Avogadro number converts the total mass expressed in grams into dalton (mass atomic unit $u$), since

$$1\,g = N_A\,u = 6.023 \times 10^{23} u. \qquad (10.25)$$

Then $N$ is directly proportional to the number $N_{\beta\beta}$ of $\beta\beta$ decaying nuclei and to the decay rate (see Eq. (7.13)).

Different identical experiments running for the same total exposure (defined as $M \cdot t$) would observe different numbers of events, $n$, with a Poisson probability distribution having an expectation value $\mu = N$, namely:

$$P(n; \mu) = \frac{\mu^n}{n!} e^{-\mu}. \tag{10.26}$$

As it is well known, the Poisson probability distribution is appropriate for discrete counts at a fixed rate. The mean of the Poisson distribution coincides with the variance; at high statistic, the Poisson distribution can be approximated by a Gaussian distribution $P(n; \mu) \propto \exp[-(x - \mu)^2/2\mu]$.

The null result $n = 0$ can be observed with a probability

$$P(0; \mu) \equiv \alpha = e^{-\mu} \implies \mu = -\ln \alpha \tag{10.27}$$

when $\mu$ is the true value of the distribution. It means that, for instance, when $\alpha = 0.1$ (namely 90% confidence level), then $\mu \simeq 2.3$ and in a Poisson distribution with this true value one observes 10% of the times $n = 0$.

Let us now consider the effects of the background, by defining a parameter $B$, known as the background index, or level, that is the number of background events in the region of interest (around the $Q$-value) divided by the size of the energy region $\Delta E$, the exposure time $t$ and the active detector mass $M_D$. In most experiments we can make the assumption that the background counts scale linearly with the active detector mass, and that $B$ is measured independently and precisely. Then the number of background events is

$$N_B = M_D \cdot t \cdot B \cdot \Delta E. \tag{10.28}$$

The background events in the region of interest will also follow a Poisson distribution with $\mu = N_B$, that can be approximated at high statistics as a Gaussian distribution ($\sigma = \sqrt{N_B}$). Under this approximation, if no signal is detected by the experiment, one can assume that the number $N$ of signal events is such that

$$N < \sqrt{N_B}. \tag{10.29}$$

Then from Eqs. (10.24) and (10.28), we find the upper limit

$$\tau_{1/2}^{0\nu} > \frac{N_{\beta\beta} \cdot \ln 2 \cdot \varepsilon \cdot t}{\sqrt{M_D \cdot t \cdot B \cdot \Delta E}}. \tag{10.30}$$

## 10.3 Neutrinoless Double $\beta$ Decay

**Table 10.1** Nuclei decays in $0\nu\beta\beta$ processes, their $Q$-values, and limits of the half-life set by experiments listed in the last column

| Decay | $Q$ (KeV) | $\tau_{1/2}^{0\nu}$ (year) | Exp |
|---|---|---|---|
| $^{76}$Ge $\rightarrow$ $^{76}$Se | 2039.0 | $>1.8 \times 10^{26}$ | GERDA [495] |
|  |  | $>8.3 \times 10^{25}$ | Majorana [496] |
| $^{136}$Xe $\rightarrow$ $^{136}$Ba | 2457.8 | $>2.3 \times 10^{26}$ | KamLAND-Zen [497] |
|  |  | $>3.5 \times 10^{25}$ | EXO-200 [498] |
| $^{130}$Te $\rightarrow$ $^{130}$Xe | 2527.5 | $>2.2 \times 10^{25}$ | CUORE [499] |
| $^{128}$Te $\rightarrow$ $^{128}$Xe | 866.7 | $>3.6 \times 10^{24}$ | CUORE [500] |
| $^{82}$Se $\rightarrow$ $^{82}$Kr | 2997.9 | $>4.6 \times 10^{24}$ | CUPID-0 [501] |
|  |  | $>2.5 \times 10^{23}$ | NEMO-3 [502] |
| $^{100}$Mo $\rightarrow$ $^{100}$Ru | 3034.4 | $>1.8 \times 10^{24}$ | CUPID-Mo [503] |
| $^{116}$Cd $\rightarrow$ $^{116}$Sn | 2813.5 | $>2.2 \times 10^{23}$ | AURORA [504] |
| $^{48}$Ca $\rightarrow$ $^{48}$Ti | 4268.0 | $>5.6 \times 10^{22}$ | CANDLES-III [505] |

Experimental searches measure or limit the decay rate of a particular isotope. So far, experiments agree with the null-signal hypothesis,[11] placing lower limits on the isotopes half-life, which translates into upper bounds on the effective Majorana mass for a given nuclear matrix element. The main theoretical uncertainty is coming from nuclear matrix element, which value depends on the nuclear model used for calculation. Double beta decays are energetically allowed for a few tenths of nuclei, but not all of them are suitable as candidate isotopes for direct searches of $0\nu\beta\beta$ decays. In Table 10.1 we list the nuclei decays used in recent experiments, their $Q$-values, and the limits set on the half-life.

Because the counting rate for a signal is very low, neutrinoless double $\beta$ decay experiments require an extremely low background, at least in the region of interest, the one close to the decay energy. Then, a major experimental issue is reduce the background in that region and maintain it low until the end of data taking. For this reason neutrinoless double $\beta$ decay experiments are always located in underground laboratories, that shield them from cosmic rays. They must also be shielded to eliminate background arising from natural radioactivity in the surroundings. For instance, the radon present in trace in the atmosphere in proximity of the experimental site is generally removed by suitable sealed radon removal systems. When all the far sources of background are suppressed, the detector itself can become the main problem. All the materials for the detector construction must be carefully selected, cleaned and stored. The sample of decaying isotope must be extremely pure, since even a very small contamination of a $\beta$-decaying impurity would overwhelm the signal from double-$\beta$ decay. This is a nontrivial requirement—several kilograms of an isotope are generally needed to obtain a

---

[11] With the exception of a controversial result from a subset of collaborators of the Heidelberg-Moscow experiment, who have claimed a measurement of the process in $^{76}$Ge, with 70 kg $\times$ year of data [494].

detectable counting rate, since the $0\nu\beta\beta$ decay rate grows with the exposure (kg × year). In the $2\nu\beta\beta$ decay, the full energy is shared with the two neutrinos in the final state, which leave the detector unregistered, and produce the continuous spectrum whose end point is close to the $Q$-value of the $0\nu\beta\beta$ decay. It follows that the $2\nu\beta\beta$ decay is an unavoidable background source for $0\nu\beta\beta$ decay. Its suppression in measurements requires the use of a detector with a good energy resolution.

As in $2\nu\beta\beta$ decays (see Sect. 10.2), the observation of the $0\nu\beta\beta$ decay can be performed through geochemical, radiochemical and direct techniques. Direct detection experiments have the ability to distinguish between the $0\nu$ and $2\nu$ modes of double beta decay, unlike indirect ones. Direct searches can employ the detector as a source of $0\nu\beta\beta$ decay events (active source or calorimetric or homogeneous approach) or have an external source, a thin film source made of $0\nu\beta\beta$ isotopes which is placed between detectors (tracko-calo or non-homogeneous experiments). The former approach enhances the efficiency for the collection of the electrons emitted in the decay, while in the latter one the loss in efficiency is compensated by the better topological reconstruction of the single electrons, an useful feature in case of discovery. Another distinction can be made between experiments operating with gas or liquid detectors, which are easier to scale to large masses, or with detectors made from crystals, which usually have a better energy resolution.

### *10.3.3 Direct Detection Experiments*

Summarising what we have just discussed in Sect. 10.3.2, we classify double beta decay experiments as:

1. active source experiments

   (a) semiconductors
   (b) bolometers
   (c) gas or liquid detectors

2. non-homogeneous experiments

In 1(a) type, the decaying isotope is a semiconductor, used not only as a sample but also as a solid state detector measuring the energy released in a given decay. In particular, High-Purity Germanium (HPGe) detectors, which use $^{76}$Ge, have been a leading technology for double $\beta$ experiments since the very first beginning. Milestone experiments of this type have been performed by the HMBB [506] and IGEX [507] collaborations. The IGEX [507] collaboration, which stopped data taking in 1999, was located at LSC (Laboratorio Subterraneo de Canfranc) in Spain. The Heidelberg-Moskow Double Beta Decay Experiment (HMBB or HdM) was located at LNGS (Laboratory Nazionali del Gran Sasso) in Italy and it has been taking data from 1990 to 2003. The most sensitive double beta decay searches based on HPGe detectors have been conducted by GERDA, already mentioned in Sect. 10.2, and by the Majorana Demonstrator. The GERDA

## 10.3 Neutrinoless Double $\beta$ Decay

collaboration searched for the $0\nu\beta\beta$ decay of the isotope $^{76}$Ge by operating HPGe detectors isotopically enriched to more than 86% in $^{76}$Ge. It has shown no evidence of neutrinoless beta decay, in contrast with claims made previously by part of the HMBB collaboration [508]. GERDA has given one of the best limit of the half-life of $0\nu\beta\beta$ decay in this isotope, that is [509]

$$\tau_{1/2}^{0\nu} > 1.8 \times 10^{26} \text{ years} \quad [^{76}\text{Ge}] \quad (10.31)$$

at 90% CL.

The Majorana Demonstrator [510], which has stopped data taking in 2023, was located in the Sanford Underground Research Facility (SURF) in South Dakota (US) 1.5 km underground and operated about 30 kg of HPGe detectors in two vacuum cryostats. Based on six years' monitoring, corresponding to an exposure of 64.5 kg × year, the collaboration found [496]

$$\tau_{1/2}^{0\nu} > 8.3 \times 10^{25} \text{ years} \quad [^{76}\text{Ge}] \quad (10.32)$$

at 90% CL. This translates to an upper limit of an effective neutrino mass $m_{\beta\beta}$ of 113-269 meV (90% C.L.), depending on the choice of nuclear matrix elements.

Based on the experience gained with GERDA and Majorana Demonstrator, the next generation experiment will be realised in the framework of the LEGEND project, following two stages named LEGEND-200 and LEGEND-1000. LEGEND-200 is based at Gran Sasso and has started data taking in 2023 with 142 (out of the intended 200) Kg of HPGe detectors in the upgraded GERDA infrastructure. LEGEND-1000 will be realised in a new infrastructure able to host 1000 Kg of active target mass. It is currently under preparation and is expected to come online towards the end of this decade.

At LNGS is also planned the experiment COBRA (Cadmium Zinc Telluride 0-Neutrino Double-Beta Research Apparatus), which uses CdZnTe semiconductor detectors that contain several double beta decay candidate isotopes and operate at room temperature. Since 2013 the COBRA demonstrator setup has been in operation with the goal to investigate the experimental issues and the prospects of this technique. Based on the knowledge gained from the work with this demonstrator, the experiment has been upgraded to COBRA XDEM (short for eXtended DEMonstrator) in 2018 [511].

In 1(b) type experiments, the sample and detector are calorimeters operating at cryogenic temperatures around 10 mK (historically also called bolometers). A calorimeter is a detector measuring, by means of a dedicated sensor, the temperature rise of the material in which a particle interacted, releasing a fraction of its energy. The thermodynamic fluctuations of the internal energy of the absorber, which limit the intrinsic resolution of the measurement, grow with the temperature. At room temperature these kind of detectors would not have the sensitivity to be used as a particle detector and this is the reason why these instruments are generally used at temperatures of the order of mK. In a cryogenic calorimeter, the energy deposited by impinging radiation in the absorber crystal is turned into heat, resulting in a temperature rise, which is subsequently detected. The CUORE (Cryogenic Underground Observatory for Rare Events) experiment [512] consists of 988 tellurium-dioxide (TeO$_2$) crystals serving as bolometers. It became operational in 2017 in the underground LNGS. Among other protections from environmental $\gamma$ background, it has shielding made of ancient Roman lead, recovered from a Roman shipwreck, which has a particularly low level of intrinsic radioactivity. The CUORE cryostat has represented a breakthrough in cryogenic technology, reaching an experimental volume of about 1 m$^3$ and a mass of 1.5 tonne (detectors, holders, shields) to be held just about 10 mK above absolute zero. The CUORE experiment did not found evidence for $0\nu\beta\beta$ decay, yielding a lower bound on the half-life of $^{130}$Te [499]

$$\tau_{1/2}^{0\nu} > 2.2 \times 10^{25} \text{ years} \qquad [^{130}\text{Te}] \qquad (10.33)$$

at 90% CL. This half-life limit converts to a limit on the effective Majorana mass $m_{\beta\beta}$ of 90-305 meV, with the spread induced by different nuclear matrix element calculations. Next generation calorimetric decay searches exploiting these developments are planned. Among these, CUPID (CUORE Upgrade with Particle IDentification) will utilise the same cryogenic infrastructure as CUORE, replacing the TeO$_2$ crystals with scintillating Li$_2^{100}$MoO$_4$ crystals. CUPID-0 [501] at LNGS is the first pilot experiment of CUPID, and it has taken more than one year of data in its first phase (2017–2018) and in its second and last phase (2019–2020). In parallel, the bolometric technique is investigated by the Advanced Molybdenum based Rare process Experiment (AMoRE) [513], which utilises CaMoO$_4$ and Li$_2$MoO$_4$ scintillating crystals, enriched in $^{100}$Mo. The first phase, AMoRE-I, started in 2020, was conducted in a pumped water power plant about 700 m underground at the Yangyang Underground Laboratory (Y2L) in South Korea [514]. The AMoRE-II detector, expected to start data taking soon, is installed in a newly built underground laboratory named Yemilab, in Jeongseon, Korea. Yemilab is located at about 1000 m underground, next to an active iron mine.

Experiments of 1(c) type use liquid/gas detectors. Liquid scintillators are a common choice for large-scale multi-ton rare event searches. As double beta isotopes may be added to the liquid scintillator, such detectors represent a cost-effective way to scale-up in isotope mass, especially when solar/reactor neutrino experiments are repurposed.

Many searches for the $0\nu\beta\beta$ decay are conducted using the noble element xenon.[12] The searched for process is $^{136}$Xe $\rightarrow$ $^{136}$Ba $+ 2e^-$. The $^{136}$Xe isotope has a $Q$-value relatively high, leading to an enhanced decay rate and good spectral separation from many residual radioactive backgrounds. It can be used in gaseous and liquid (cryogenically cooled) forms. Ionizing charged particles generate in liquid xenon free electrons and scintillation light that can be collected

---

[12] Xenon is also commonly used for direct dark matter detection.

## 10.3 Neutrinoless Double β Decay

and, optionally, amplified through various sensors and electronics. Liquid xenon is very dense, and has a large number of electrons per atom, which makes it an excellent gamma-ray absorber providing effective self-shielding against external residual radiation. There are a few inconveniences, like a high cost for procurement and enrichment in the $^{136}$Xe isotope, as its natural abundance is only 8.9%

The KamLAND infrastructure, located in the Kamioka Mine in Japan, was upgraded with a nylon balloon in the active detector volume filled with a liquid scintillator. Enriched xenon was dissolved in the liquid scintillator such that the KamLAND-Zen (Zero Neutrino double beta decay search) experiment could search for $0\nu\beta\beta$ decay of $^{136}$Xe. The nylon balloon was surrounded by another liquid scintillator instrumented with PMTs (see Sect. 7.2.2), whose scintillation light allows the background discrimination and rejection. The first search, referred to as KamLAND-Zen400, took data from 2011 to 2015, using 320–380 kg of enriched xenon gas [515] and yielding the limit

$$\tau^{0\nu}_{1/2} > 1.07 \times 10^{26} \text{ years} \qquad [^{136}\text{Xe}] \qquad (10.34)$$

at 90% C.L. This result sets a strong bound on the effective Majorana mass, $m_{\beta\beta} < 61$–165 meV. The energy range shows the uncertainty in the calculation of the nuclear matrix element. The upgraded experiment, KamLAND-Zen800, accumulated data from 2019 to 2021 using 745 kg of enriched xenon. A combined fit of the KamLAND-Zen400 and KamLAND-Zen800 data sets gives [497]

$$\tau^{0\nu}_{1/2} > 2.3 \times 10^{26} \text{ years} \qquad [^{136}\text{Xe}] \qquad (10.35)$$

at 90% C.L. By using commonly adopted nuclear matrix element calculations, the limit of the half-life yields the upper limits $m_{\beta\beta} < 36$–156 meV.

The SNO+ experiment [249], mentioned at the end of Sect. 7.3.3, repurposes the SNO Canadian facility. The detector re-uses most of the SNO components, with several major upgrades to enable the use of liquid scintillator as target material. During the third phase, or tellurium phase, the scintillator, loaded with about 4 tonnes of natural tellurium, will search for the $0\nu\beta\beta$-decay of $^{130}$Te.

The Enriched Xenon Observatory (EXO) [516] searches for $0\nu\beta\beta$ of the $^{136}$Xe isotope. The EXO program is based on a time projection chamber (TPC) designed to measure both the scintillation of liquid xenon and its ionization signal. Within the EXO program, the EXO-200 represented the first stage. It has taken data from 2011 to 2018 at Waste Isolation Pilot Plant (WIPP) near Carlsbad, in New Mexico. It has used a TPC filled with about 200 kg of liquid xenon, enriched to about 80% in $^{136}$Xe. With its full dataset EXO-200 has set a limit on $^{136}$Xe half-life of $3.5 \times 10^{25}$ year at 90% C.L. [498]. There is a proposed experiment, nEXO (next-generation Enriched Xenon Observatory) [517], which scales up EXO-200 suggesting a TPC with 5 tons of liquid xenon, enriched to about 90% in $^{136}$Xe. Another project searching for $0\nu\beta\beta$ decays of $^{136}$Xe is the NEXT (Neutrino Experiment with a Xenon TPC) experiment at Laboratorio Subterráneo de Canfranc (LSC) in Spain. Its technique, though, is different, since it implements a high-pressure xenon gas TPC with electroluminescent amplification. When a double beta decay occurs, two electrons are emitted and interact with the xenon gas through ionization and excitation. When they excite xenon atoms, they cause the emission of scintillation light detected by an array of PMTs. The electrons produced in the ionization process are drifted by an electric field towards the electroluminescent amplification zone. There, the electrons are accelerated emitting around thousand photo-electrons each, being detected again by the PMTs. Time delay between these two detections allows a better tracking of the two electrons. The experiment has been developing in phases. The NEXT-White detector, taking data from 2016 to 2021 at the LSC, has represented the first large scale demonstrator of the NEXT Experiment, containing approximately an active xenon mass of 5 kg. The NEXT-100 detector started construction in 2021; it was completed and installed in 2024. It scales up NEXT-White by a factor between 2 and 3 in dimensions, increases pressure and incorporates improvements and changes [518].

The most advanced detector of type 2 was NEMO (Neutrino Ettore Majorana Observatory) experiment [519], which was located in Modane Underground Laboratory (LSM) in the Fréjus Tunnel beneath Mont Blanc in the French Alps. The NEMO collaboration has built two prototype detectors (NEMO-1 and NEMO-2) which have been used until 1997. From 1994 to 2001, the NEMO collaboration has designed and constructed the detector NEMO-3: data taking started in January 2003 and ended in January 2011. The physics results motivated the endeavour of designing Super-NEMO [520], that will house 100 kg of isotopes, ten times more than NEMO-3. Its goal is to search for $0\nu\beta\beta$ decays reaching $\tau_{1/2}^{0\nu} > 10^{26}$ year, which corresponds to about $m_{\beta\beta} < 0.1$ eV. The first demonstrator module, which contains about 7 kg of $^{82}$Se, is currently tested at LSM.

### 10.3.4 Majorons

The search for neutrinoless double beta decay could lead to the discovery of a new elementary boson, the majoron, which is one of the candidates for dark matter. Dark matter is a yet unknown form of stable matter whose presence is discerned from its gravitational attraction. It is appropriate to call this matter "dark" because it is detected in no other way; it is not observed to emit or absorb electromagnetic radiation of any wavelength. Evidence for dark matter has been found in the velocity dispersion of galaxies (first noted in the Coma cluster [521]), in rotation curves of galaxies [522, 523] and in gravitational lensing effects [524, 525]. The measurements of the CMB spectrum [526] and of large-scale structures [527] have shown that ordinary (visible baryonic) matter makes up only about 5% of the energy budget of the universe. The larger fraction of about 27% consists of dark matter.[13]

---

[13] The bulk of about 68% of the energy in the universe exists in yet another unknown form, called dark energy.

## 10.3 Neutrinoless Double $\beta$ Decay

Majorana mass terms violate the lepton number $L$ by two units, and therefore also $B - L$ symmetry, which is the only anomaly-free combination of $L$ and the baryon number $B$. The symmetry breaking can be achieved explicitly, by $B - L$ breaking terms in the Lagrangian, or spontaneously. The spontaneous symmetry breaking can be either global or local. As seen in Sect. 3.3.2, the spontaneous breaking of a continuous symmetry implies the existence of massless Goldstone bosons. In case of the $U(1)_{B-L}$ symmetry they are referred to as majorons [528–530]. Let us see a simple realization of how a majoron arises in a one-generation model [528]. We assume that neutrinos have both Majorana and Dirac mass terms and take the form of the Dirac-Majorana mass matrix as in Eq. (5.85):

$$M_{DM} = \begin{pmatrix} 0 & m \\ m & M \end{pmatrix}. \tag{10.36}$$

The mass term in the Lagrangian is (Eq. (5.65))

$$\mathcal{L}_{DM} = -\frac{1}{2} \overline{(n_L)^C} M_{DM} n_L + h.c. \tag{10.37}$$

where

$$n_L \equiv \begin{pmatrix} \nu_L \\ \nu_L^C \end{pmatrix} \tag{10.38}$$

and we have used the convention in Eq. (5.34), namely $\nu_L^C \equiv (\nu_R)^C$. Here $\nu_L$ is part of the leptonic SM doublet $\ell_L$ and $\nu_R$ is a singlet. We can build a Yukawa couplings to the Higgs doublet $\phi$ analogous to the SM one in Eq. (4.86)

$$\mathcal{L}_Y = -Y \, \overline{\ell}_L \, \phi \, \nu_R + h.c. \tag{10.39}$$

with an arbitrary coefficient $Y$, that we can assume real without loss of generality. By adding to the model a singlet Higgs field $\phi_s$, we can build another $SU(2)_L \times U(1)_Y$ invariant Yukawa couplings

$$\mathcal{L}_{Y_s} = -Y^{(s)} \, \phi_s \, \overline{\nu_L^C} \, \nu_R + h.c. \tag{10.40}$$

where $Y^{(s)}$ is again an arbitrary (real) coefficient. A non-zero VEV of the scalar Higgs field $v_s \equiv \langle \phi_s \rangle \neq 0$ induces a spontaneous symmetry breaking which also violates lepton number. In analogy to what has been done for the Higgs doublet in Eq. (4.55), one can parameterise oscillations about the VEV of the scalar Higgs field by a complex, i.e.

$$\phi_s = \frac{1}{\sqrt{2}} (v_s + \sigma + i J). \tag{10.41}$$

The fields $\sigma$ and $J$ are, respectively, a massive and a massless field with zero vacuum expectation value. The field $J$ is the majoron and $\sigma$ is the heavy, scalar majoron partner. The value of the parameter $v_s$ gives the breaking scale of the global symmetry $U(1)_{B-L}$.

As seen in Sect. 5.7, if $M \gg m$ the mass eigenstates $\nu_1$ and $\nu_2$ at the lowest order in $m/M$ have eigenvalues $m^2/M$ and $M$, respectively. According to Eqs. (5.87) and (5.88), at the lowest order in $m/M$ we can rewrite $\mathscr{L}_{Y_s}$ as

$$\mathscr{L}_{Y_s} = -Y^{(s)} \phi_s \overline{(\nu_2)_L} (\nu_2)_R + h.c. \tag{10.42}$$

After SSB, the insertion of $v_s$ in Eq. (10.42) sets the value of the right-handed Majorana mass matrix $M$

$$M = \frac{Y^{(s)} v_s}{\sqrt{2}}. \tag{10.43}$$

The insertion of $iJ$ gives the interaction term

$$\mathscr{L}_{int_J} = -i \frac{Y^{(s)} J}{\sqrt{2}} \bar{\nu}_2 \gamma_5 \nu_2 \tag{10.44}$$

when we take into account the properties of the $\gamma_5$ matrix and the presence of the hermitian conjugate part. Since the Lagrangian must be scalar, this term shows that the majoron is a pseudoscalar field. The coupling of the majoron to light neutrinos is suppressed, and appears at order $m/M$; the couplings to other matter fields are even more suppressed. The only sizeable coupling is the one to the heavy neutrinos, but these particles are unstable because they can rapidly decay into a majoron and a light neutrino. In its original formulation the majoron is so weakly coupled that it cannot produce detectable effects.

If the majoron Yukawa couplings to neutrinos were sufficiently strong, it would have interesting consequences for particle physics, astrophysics, and cosmology. It was clear since early works that the majoron with nontrivial weak isospin ($I$) properties which would have appreciable coupling to neutrinos, would also have a strong coupling to the other leptons. In the 80s, models where the majoron arises from a Higgs triplet ($I = 1$) [529] or a Higgs doublet ($I = 1/2$) [531–533] were suggested. As the first LEP data on the invisible width of the Z boson became available in the early 90s, the number of active light neutrino generations was limited to three, ruling out both triplet and doublet majoron models [147]. Several other models were subsequently proposed, aiming at reconciling the results on the Z decay width with a stronger neutrino-majoron coupling.[14] Assuming the full

---

[14] Let us note that in some models the term majoron is used in a broader sense, departing from its original conception as a Goldstone boson. For example, there are models where the majoron arises as the component of a massive gauge boson [534] or in the context of a brane-bulk scenario [535].

## 10.3 Neutrinoless Double β Decay

Lagrangian (including gravity) contains some small explicit $U(1)_L$-breaking terms, the majoron becomes a pseudo-Goldstone boson with mass different from zero. However, since its interactions remain suppressed, a massive majoron is potentially dark and stable enough to be considered a viable candidate for dark matter. One major signature comes from its eventual decay into SM particles. At tree level only $J \to \nu\nu$ into neutrino mass eigenstates, with coupling $\propto m/M$, is possible. With upcoming neutrino experiments, there are prospects to improve the search for the experimental signature of this channel. Couplings to charged leptons, quarks and photons arise at the one-loop level. The presence of this coupling leads to the possibility of $J \to \bar{\ell}\ell'$, $J \to \bar{q}q$ and $J \to \gamma\gamma$ decays. All these decays are constrained by cosmology at different levels.

One of the most interesting phenomenological consequences of the existence of majorons would be the possibility of their emission in double-$\beta$ decays, giving rise to a final state with two electrons, one or two majorons and no neutrinos:

$$(A, Z) \to (A, Z+2) + 2e^- + (2)J \tag{10.45}$$

In analogy to expression (10.17), the rate of the $\beta\beta$ decay with the emission of one or two majorons can be expressed as:

$$\frac{1}{\tau_{1/2}^{(2)J}} = g_J^{2(4)} |M_{(2)J}|^2 G_{(2)J}. \tag{10.46}$$

Here $g_J$ is the neutrino-majoron coupling, while $M_{(2)J}$ and $G_{(2)J}$ are the nuclear matrix elements and the phase-space factors in the case of one or two emitted majorons. All three terms are model dependent. If one or two majorons are emitted in the double $\beta$ decay, they would escape any detector and carry away part of the decay energy. In analogy with the $2\nu\beta\beta$ decay, the summed electron energy is continuously distributed between 0 and $Q$, and its exact shape depends on the majoron model.

Several experiments designed for the detection of neutrinoless double beta decay have also performed searches for neutrinoless double beta decay decays with majoron emission, e.g. GERDA [536], CUPID-0 [537], EXO-200 [538], KamLAND-Zen [539] and NEMO-3 [540], all mentioned in Sect. 10.3.2. Their results have led to half-life constraints in the range $10^{22}$–$10^{24}$ years, depending on the models and isotopes being investigated.

Although they rely upon additional assumptions and model dependence, limits on the neutrino-majoron coupling are also available from astrophysics, by studying the role of majorons in a supernova explosion [541, 542].

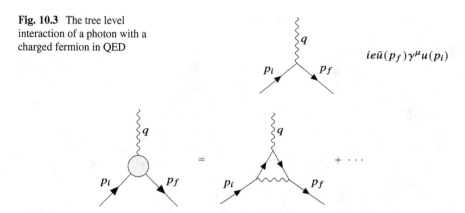

**Fig. 10.3** The tree level interaction of a photon with a charged fermion in QED

**Fig. 10.4** Loop interactions of a photon with a charged fermion in QED. The blob (left) indicates the perturbative series. The Feynman diagram of the lowest radiative correction is indicated on the right

## 10.4 Neutrino Magnetic Moments

In Sect. 1.1.1 we have seen that the magnetic moment stems from the term $\sigma_{\mu\nu}F^{\mu\nu}$, which connects the spin of the charged lepton to the magnetic field. The value of the g-factor, $g = 2$, is contained in the coefficient of this term. Therefore, when calculating the magnetic moment in quantum field theory, we look for terms proportional to $\sigma_{\mu\nu}F^{\mu\nu}$, which can induce corrections to the g-factor. Since the electromagnetic tensor $F^{\mu\nu}$ contains derivatives of the photon field, in momentum space this term translates into elements $\propto \sigma_{\mu\nu}q^\mu$ where $q^\mu$, the 4-momentum of the photon, is equal to the difference between the final and initial momentum of the fermion, i.e. $q^\mu = p_f^\mu - p_i^\mu$.

At the lowest order in QED, the amplitude for a process where a photon is coupled to a cherged fermion of charge $e$, and the corresponding Feynman rule, are given in Fig. 10.3. The photon corresponds to an external electromagnetic field and is off-shell. The Gordon identity applied to this electromagnetic current yields

$$ie\bar{u}(p_f)\gamma^\mu u(p_i) = \bar{u}(p_f)\left(i\frac{e}{2m}(p_f^\mu + p_i^\mu) - \frac{e}{2m}\sigma^{\mu\nu}q_\nu\right)u(p_i). \qquad (10.47)$$

This identity allows to single out the term $C\sigma^{\mu\nu}q_\nu$, with $C = e/(2m)$. It gives the magnetic moment with coefficient $g = 2$, that we can identify with $4mC/e$.[15] In order to the calculate corrections to the g-factor, one has to calculate the corrections to the coefficient $C$ coming from loop diagrams that contribute to the process in Fig. 10.3. They are indicated in Fig. 10.4. The amplitude for this process can be

---

[15] For more detailed derivations, see quantum field theory textbooks, e.g. Refs. [91, 543].

## 10.4 Neutrino Magnetic Moments

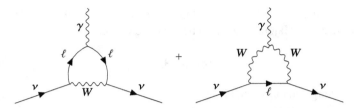

**Fig. 10.5** One loop interactions of a photon with a neutrino in the SM, mediated by a virtual $W$ boson and a virtual charged lepton $\ell$

obtained directly in effective field theory with an effective Lagrangian including the term

$$L_{eff} \propto \bar{\psi}\sigma_{\mu\nu}\psi F^{\mu\nu}. \qquad (10.48)$$

In the effective formalism, the electroweak properties of a spin 1/2 Dirac particle are described in terms of an effective Lagrangian having two vector form factors (the electric and magnetic form factors), which in the static limit define the charge and magnetic moment, and two axial vector form factors (axial and tensor form factors), which in the static limit define the axial charge and electric dipole moment.

Let us now consider neutrino magnetic moments. Obviously, since neutrinos are electrically neutral, they cannot couple to the photon as in Fig. 10.3. However, in principle, they can couple to the photon through loop diagrams generated by weak interactions, as the ones in Fig. 10.5. By decomposing the fermion field in the effective Lagrangian $L_{eff}$ in terms of left- and right-chiral parts, one obtains

$$(\bar{\psi}_R + \bar{\psi}_L)\sigma_{\mu\nu}(\psi_R + \psi_L)F^{\mu\nu} = (\bar{\psi}_R\sigma_{\mu\nu}\psi_L + \bar{\psi}_L\sigma_{\mu\nu}\psi_R)F^{\mu\nu} \qquad (10.49)$$

which shows that the interaction induces a chirality flip. Thus, to have magnetic moments, both the right and left-chiral components of the fermion need to be present. That certainly happens for massive neutrinos of Dirac nature. If neutrinos are massless, they have only one chirality state (Weyl neutrinos) and do not possess magnetic moment. That also implies that neutrino magnetic moments are, in first approximation, proportional to the value $m_\nu$ of the neutrino masses, and therefore heavily suppressed. Since the Feynman diagrams in Fig. 10.5 contain two weak vertices and one electromagnetic vertex, we expect the order of magnitude of the neutrino magnetic moment $\mu_\nu$ to be

$$\mu_\nu \simeq eG_F m_\nu = 2m_e\mu_B G_F m_\nu \sim 10^{-18}\mu_B\left(\frac{m_\nu}{10^{-1}\,\text{eV}}\right) \qquad (10.50)$$

where $e$ is the electric charge, $m_e \simeq 0.5$ MeV the electron mass, $G_F \simeq 10^{-5}$ GeV$^{-2}$ the Fermi constant and the Bohr magneton $\mu_B = e/2m_e \simeq 6 \times 10^{-15}$ MeV/Gauss.

If neutrino fields have Majorana nature, different considerations are in order. As seen in Sect. 5.4, a Majorana neutrino field can be written as

$$\psi = \psi_L + \psi_R^C \tag{10.51}$$

where the charge conjugate field $\psi_R^C = (\psi_L)^C = P_R \psi$ plays the part of the right-chiral field in the magnetic moment interaction.[16] In the generic case of initial and final Majorana neutrinos $\psi_i$ and $\psi_f$, we can rewrite the effective Lagrangian as

$$L_{eff} \propto \mu_{fi} F^{\mu\nu} \left( \bar{\psi}_f \sigma_{\mu\nu} \psi_i + \overline{(\psi_f)^C} \sigma_{\mu\nu} (\psi_i)^C \right). \tag{10.52}$$

By exploiting the properties (2.124), (2.127), and (2.122), and remembering that $\sigma_{\mu\nu}$ is antisymmetric, and a minus sign is gained when exchanging spinors, we obtain

$$\overline{(\psi_f)^C} \sigma_{\mu\nu} (\psi_i)^C = -\bar{\psi}_i \sigma_{\mu\nu} \psi_f. \tag{10.53}$$

Inserting into the effective Lagrangian, we have

$$L_{eff} \propto \mu_{fi} F^{\mu\nu} \left( \bar{\psi}_f \sigma_{\mu\nu} \psi_i - \bar{\psi}_i \sigma_{\mu\nu} \psi_f \right). \tag{10.54}$$

If the initial and final state are the same the effective Lagrangian vanishes—Majorana fermions can only have the so-called *transition* magnetic moments. While only the Dirac neutrino can have a magnetic moment, the transition magnetic moment, which is relevant to the process $\nu_i \to \nu_j + \gamma$, may exist for both Dirac and Majorana neutrinos. Since the effective Lagrangian (10.54) must be symmetric under the exchange $i \leftrightarrow f$, and the quantity in the brackets is antisymmetric, we find that the matrix of transition magnetic moments for Majorana fermions is antisymmetric, namely

$$\mu_{fi} = -\mu_{if}. \tag{10.55}$$

In Sect. 2.3 we have seen that in the effective Lagrangian having a term

$$L_{eff} \propto \bar{\psi} \gamma_5 \sigma_{\mu\nu} \psi F^{\mu\nu} \tag{10.56}$$

produces the interaction $\propto d_E \mathbf{s} \cdot \mathbf{E}$ in the non-relativistic limit, where $d_E$ is the $CP$-violating electric dipole moment (EDM). Thus, the Lorentz-invariant effective Lagrangian for the $\nu_i \to \nu_j + \gamma$ process is in general described by

$$L_{eff} \propto \bar{\psi}_f \left( \mu_{ij} + i \varepsilon_{ij} \gamma_5 \right) \sigma_{\mu\nu} \psi_i F^{\mu\nu} \tag{10.57}$$

---

[16] The following reasonings can be adapted also to the Majorana field defined in Eq. (5.35).

## 10.4 Neutrino Magnetic Moments

where $\mu_{ij}$ and $\varepsilon_{ij}$ are the magnetic and electric transition dipole moments [544]. This result is valid for both Dirac and Majorana neutrinos, taking care of adding the charged conjugate fields in the Majorana case, as done in Eq. (10.52). The Feynman diagrams in Fig. 10.5 do not give contributions to the axial term, so neutrinos have no electric dipole moments at one loop when $i = j$, independently of their nature. One can compute the transition magnetic moments in the framework of the SM with three massive Dirac (or Majorana) neutrinos using the one-loop Feynman diagrams (and their charge-conjugate ones) shown in Fig. 10.5 and the appropriate PMNS unitary mixing matrix [544–548]. In the Majorana case, the transition magnetic and electric moments are suppressed because of the unitarity of the mixing matrix (GIM mechanism).

One way to detect neutrino magnetic moments is by deviations from expectations of the SM in $\nu_e$, $\nu_\mu$ and $\nu_\tau$ scattering off electrons in accelerator experiments (see e.g. [549–551]). The neutrino-electron scattering cross section depends on an effective neutrino magnetic moment, which is a function of the neutrino magnetic and electric moments, and of elements of the lepton mixing matrix. This happens since the neutrino-photon interaction in Fig. 10.6 receives a contribution from the electric dipole moment in addition to the magnetic moment. Moreover, the neutrinos which participate in the neutrino-electron scattering in the detector are produced in flavour eigenstates, which are superpositions of mass eigenstates. Solar $\nu_e$ and reactor $\bar{\nu}_e$ are also used by laboratory experiments to look for neutrino magnetic moments. Analyses of neutrino scattering off electrons at low energies allow to establish bounds on the neutrino magnetic moment [552–558]. A further way to determine the existence of a magnetic moment is the observation of coherent interactions of neutrinos in a material (i.e. a crystal) through the magnetic moment interaction (see e.g. Ref. [559]). In addition to terrestrial experiments, one can use astrophysical and cosmological data to set constraints on neutrino magnetic moments.

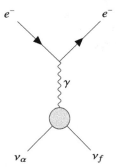

**Fig. 10.6** The neutrino electromagnetic interaction which contributes to neutrino-electron scattering, where $\nu_\alpha$ is an active neutrino flavour state and $\nu_f$ is a mass eigenstate. The gray blob represents the neutrino electromagnetic interaction, including the magnetic moment, whose perturbative series was also represented with a blob in Fig. 10.4 (left). In this diagram, the blob also includes the mixing and oscillation of the initial neutrino

Following Eq. (10.50) and considering the current limits on the neutrino mass, the prediction of the minimal extension of the Standard Model with right-handed neutrinos is $\mu_\nu \approx 10^{-18} \mu_B$, 6–7 orders of magnitude below the current experimental limits on $\mu_\nu$. Up to now, no effect of neutrino electromagnetic properties has been found. However, the experimental studies are stimulated by the hope that new physics beyond the minimally extended SM with right-handed neutrinos might give stronger, and measurable, contributions.

## 10.5 Sterile Neutrinos

The existence of three active neutrinos has been confirmed by the precision measurement of the partial decay width of the $Z^0$ boson, but this result does not exclude the possibility of additional neutrinos with no SM interactions. In many of the previous chapters we have already mentioned such neutrinos, said sterile, which were introduced by Bruno Pontecorvo [34, 35, 560]. They are singlets of the SM gauge group and interact with ordinary matter only through mixing with the usual neutrinos. In general terms, we can define a sterile neutrino as a neutral lepton with no ordinary weak interactions except those induced by mixing.

The impact of sterile neutrinos could manifest in several ways, and below, we provide a tentative list.

1. An admixture of light (eV-scale) sterile neutrinos with the three active species in the SM produces observable effects in the direct mass experiments discussed in Sect. 10.1.1.
   
   In the process of single $\beta$ decay, the addition of a fourth sterile neutrino changes the electron spectrum $\Gamma$, which has to include a decay branch $\propto |U_{e4}|^2 d\Gamma(m_4, \ldots)$, where $U_{e4}$ is the matrix element responsible of the mixing with the sterile fourth neutrino of mass $m_4$. There is a change also in the endpoint energy, which passes from a value $E_0$ to a value $E_{04} = E_0 - m_4$, corresponding to the endpoint energy for decays into an electron plus a sterile neutrino final state. The sterile branch emerges at electron energies $E \leq E_{04}$, which results in a kink-like spectra distortion. Such kinks have been searched for in the $\beta$ decay spectra of numerous isotopes, and several constraints have been imposed on the $|U_{e4}|^2$ parameter (see e.g. Ref. [561, 562]).
   
   Nuclear beta decay constraints are less important at very low $m_4$, less than 10 eV, where the kink in the beta decay spectrum moves too close to the endpoint to be discernible. At the other extreme, for $m_4$ above $\sim$MeV, sterile neutrino production in nuclear decays becomes kinematically forbidden. In this case, one can investigate the lepton spectra from meson decays, e.g. $\pi \to e\nu$ [563] and $K^+ \to e\nu$ [564]. Electron capture experiments searching for the effects of neutrino masses are also sensitive to an additional fourth neutrino [565].

2. Sterile neutrinos can affect neutrino oscillations experiments provided there is non-negligible mixing between active and sterile states of the same chirality.

## 10.5 Sterile Neutrinos

By definition, the mixing can only originate because of the mass terms in the Lagrangian (see Eq. (5.60)). It should be noted that it does not occur for sterile Majorana neutrinos with only Majorana mass. They can have arbitrary mass, but they have no SM interactions, and they decouple (unless there are suitable BSM interactions). There is no mixing also in the case of Dirac neutrinos, which only have Dirac mass. They have right handed components with no SM interactions except for the Yukawa coupling to the Higgs. The right and left-handed states combine to form a Dirac neutrino with a conserved lepton number, and do not mix. Active-sterile mixing requires that at least some Dirac $m_D$ and Majorana masses ($m_{M_R}$ and/or $m_{M_L}$) are simultaneously non-zero, as we discuss in Sect. 10.5.1.

The seesaw mechanism, seen in Sect. 5.7, interferes negatively with sterile-active mixing. In fact, when $m_{M_R} \gg m_D$ and $m_{M_L} = 0$, we have naturally 3 light neutrinos, that can be identified with the ordinary neutrinos, and very heavy sterile neutrinos, right-handed at all effects, whose mass is generally considered proportional to a possible higher symmetry breaking scale. The mixing angle (5.86) $\theta \simeq m_D/m_{M_R}$ is very small; sterile neutrinos are too heavy to be relevant to oscillations. A sizeable sterile-active mixing should have the $m_D$ value comparable to the $m_{M_R}$ and/or $m_{M_L}$ values. In the case of several families, though, there may be cancellations in the neutrino mass matrix that allow individual entries of the mixing matrix to be much bigger than this $\theta$ estimate.

It is to be noted that another signal of the presence of active-sterile mixing would be the non-unitarity of the 3 × 3 lepton mixing matrix. Besides, in presence of sterile neutrinos, oscillation effects are expected for both charged and neutral currents. This is not what happens in neutrino oscillations with three generations, where only charged currents display an oscillatory pattern and the event rates in neutral currents remain unchanged.

3. Light and heavy sterile neutrinos, with masses in the eV–GeV and even TeV range, can impact the rate of neutrinoless beta decay. If the mass of sterile neutrinos is intermediate, and especially when it is of the order of the Fermi momentum in the nucleus, about 100 MeV, the factorization (10.17) of the inverse of the half-life into phase space factor, nuclear matrix element and $f(m_i, U_{ei})$ function is no more allowed since nuclear effects take a dominant role [566]. The presence of light sterile neutrinos changes the constraints on the effective Majorana mass. For instance, a fourth light neutrino turns the effective Majorana mass $m_{\beta\beta}$ (10.19) into

$$m_{\beta\beta} = \left| \sum_{i=1}^{3} U_{ei}^2 m_i + |U_{e4}|^2 e^{i\alpha_4} m_4 \right| \quad (10.58)$$

where $\alpha_4$ is an unknown phase. The additional contribution could increase the spread of the allowed values of $m_{\beta\beta}$ and render neutrinoless double beta decay within reach of next generation experiments.

4. By definition, only gravitational interactions affect sterile neutrinos. Hence, even in the absence of significant active-sterile oscillations, the existence of sterile neutrino mass states can have observable effects in astrophysics and in cosmology. Most BSM models involve one or more sterile neutrinos, with model-dependent masses which can have almost any value. Sterile neutrinos at the eV scale or lower can contribute to the number of relativistic degrees of freedom (effective number of neutrino species $N_{eff}$), see e.g. Ref. [567]. The latest Planck data [568] show a standard value $N_{eff} \simeq 3$ at the time of recombination, disfavouring the existence of a thermal population of light sterile neutrinos which would have altered this value. Increasing neutrino masses, a large variety of signatures is still possible; some examples are the search of decay products of these sterile neutrinos in accelerator experiments, their possible role in supernova explosions, cosmological signatures, once one assumes their production in the early Universe and their decay into lighter species. Astrophysical and cosmological limits are generally stronger than the ones obtained in laboratories which perform direct searches or look for indirect signatures. Yet, they have the disadvantage to be strongly model depending.

5. The keV mass range is especially interesting because sterile neutrinos could be dark matter (DM) candidates. In 1983 the hot, warm, cold DM terminology was introduced [569, 570]. Roughly speaking, cold DM would be composed by particles moving non-relativistically at the time that galaxies started forming, hot DM by particles that were relativistically moving at that time and warm DM would be in-between. They have drastically different influences on the formation of galaxies, galaxy clusters and larger structures (structure formation), see e.g. Refs. [247, 571, 572]. Another class, cool DM, has been introduced in 1999 [573]. Neutrinos are the only viable candidates for DM within the SM, since they are stable, with no electromagnetic or strong interaction, and have mass, necessary to explain gravitational effects. However, active neutrinos fall into the category of hot DM, whose impact on structure formation seems to result in a scenario different from what it is observed. Another reason to exclude the known neutrinos follows from the study of phase space density of DM dominated objects, which sets a lower bound on the mass of DM particles [574]. Broadly speaking, SM neutrinos cannot compose all of the observed DM because of the smallness of their mass and the magnitude of their coupling to other particles. Hence, one obvious solution is to candidate sterile neutrinos. Depending on their production mechanism in the early universe, sterile neutrinos can act as effectively cold, cool, or even warm dark matter (see e.g. Ref. [575]). In 2014 there was a tentative evidence for a 3.5 keV line in the X-ray spectra of galaxy clusters or galaxies [576–578]. One possible interpretation was that it stemmed from the decay of a (DM) sterile neutrino. Not all subsequent observations have confirmed the presence of the line, and explanations such as statistical fluctuations, systematic effects or astrophysical emission line have not been ruled out.

6. A straightforward approach to building a DM model without altering established SM physics is to posit the existence of a hidden (dark) sector, composed of

## 10.5 Sterile Neutrinos

particles which do not interact with SM forces. The dark sector could be relatively minimal, comprising just one or a few SM gauge singlet states, or it could be as complex as the SM, featuring new dark interactions and symmetries. Apart from providing the DM candidate, it could offer solutions to current problems or anomalies.

By definition, it would seem that the hidden sector could have no interactions with the visible one aside from gravity. However, there are only a few interactions permitted by SM symmetries that can act as a "portal" from the visible sector into the dark one. To exemplify the concept of a portal, let's delve into the interpretation of beta decays in the 1930s (see Sect. 1.2). It was observed that a neutron $n$ decaying into a proton $p$ and an electron $e$ produced an "anomalous" continuous electron spectrum. In this scenario, we could interpret the undetected neutrino $\nu$ as a dark particle. The "anomaly" is resolved when we postulate the interaction of the hidden neutrino with the visible proton and neutron through a four Fermi interaction $(\bar{p}\gamma^\mu n)(\bar{e}\gamma_\mu \nu)$, which acts as a portal.

Let's return to the present day. In Sect. 5.5, we have observed that it is possible to construct a SM gauge-invariant Yukawa coupling involving the weak isospin doublet state[17] $\ell_L = (\nu_{\ell L}, \ell_L)^T$ of hypercharge $Y = -1/2$, and a sterile neutrino $\nu_s$, whose both components are a weak isospin singlet state with zero hypercharge. As $\nu_s$ is a singlet of the SM gauge group, it can act as a mediator between the dark sector and the SM. In the case of more generations, Eq. (5.49) reads

$$-\mathscr{L}_D = Y_D^{\alpha\beta} \, \overline{\ell}_L^\alpha \, \tilde{\phi} \, \nu_s^\beta + h.c. \tag{10.59}$$

where $\alpha$ and $\beta$ are generation indices, $\tilde{\phi}$ is the field of hypercharge $Y = -1/2$ defined as $\tilde{\phi} = i\tau_2 \phi^*$, and $\phi$ is the SM Higgs doublet. The coupling (10.59) represents the sole viable renormalizable coupling between the SM doublets and the singlet neutral fermions, which could additionally interact in a non-SM manner with particles in the dark sector. This type of operator is termed the neutrino portal.

After SSB, $\mathcal{L}_D$ converts to the usual Dirac term

$$\mathscr{L}_D = -M_D^{\alpha\beta} \bar{\nu}^\alpha \nu_s^\beta + h.c. \tag{10.60}$$

where $M_D^{\alpha\beta} \equiv Y_D^{\alpha\beta} v/\sqrt{2}$ and $v/\sqrt{2}$ is the non-zero VEV of the SM Higgs field.

For completeness, we observe that, at the renormalizable level, there are only two other possible portal interactions: the kinetic mixing portal $\lambda_b F'_{\mu\nu} F^{\mu\nu}$ for gauge bosons, where $F'_{\mu\nu}$ is the field strength tensor of a new dark sector $U(1)'$ symmetry, and the Higgs portal for scalars $\lambda_s \, \phi^\dagger \phi \, \phi_s^\dagger \phi_s$, where $\phi_s$ is a dark sector scalar field; $\lambda_b$ and $\lambda_s$ are dimensionless coupling constants.

---

[17] $T$ stays for transpose.

Although the vast majority of the oscillation data can be explained by three neutrinos, there are intriguing anomalies that cannot be accounted for within this framework, hinting to BSM physics. Data from a variety of short-baseline experiments have produced results that are either inconsistent with the 3-neutrino picture, or inconsistent with another observable, such as the reactor antineutrino flux. Laboratory experiments suggesting oscillations of active into eV-scale sterile neutrinos are discussed in Sect. 10.5.2.

### 10.5.1 Mixing in the Dirac-Majorana Case

In Chap. 5 we observed that the addition of a Dirac or a Majorana ($M_{M_L}$) neutrino mass term in the SM Lagrangian leads to the field redefinition

$$\tilde{\nu}_L = U_L^{\nu\dagger} \hat{\nu}_L \tag{10.61}$$

where $\hat{\nu} = (\nu_e, \nu_\mu, \nu_\tau)$ are the neutrino states whose left handed components enter the weak Lagrangian and $\tilde{\nu} = (\nu_1, \nu_2, \nu_3)$ are the mass eigenstates. The matrix $U_L^\nu$ is a $3 \times 3$ unitary matrix, whether the neutrino has a Dirac or Majorana nature. In the charged lepton sector, we have the $3 \times 3$ unitary matrix $U_L^\ell$ which plays the same role and diagonalizes the charged lepton mass matrix. As seen in Sect. 6.1, the leptonic weak charged current written in terms of massive fields contains the lepton mixing (or PMNS) unitary matrix

$$U \equiv U_L^{\ell\dagger} U_L^\nu. \tag{10.62}$$

For example, we have (see Eq. (6.3))

$$J_W^{\ell-\mu} = \frac{1}{\sqrt{2}} \overline{\hat{\ell}_L} \gamma^\mu \hat{\nu}_L = \frac{1}{\sqrt{2}} \overline{\tilde{\ell}_L} \gamma^\mu U \tilde{\nu}_L \tag{10.63}$$

where $\tilde{\ell} \equiv (e, \mu, \tau)$ indicates the charged lepton states of definite mass and $\hat{\ell}$ the corresponding flavour eigenstates. To understand how this framework changes in the presence of sterile neutrinos, we start from the one generation case.

A right-handed sterile neutrino is needed to build a Dirac mass term; in the general case, in which also Majorana mass terms are present, there is mixing among active and sterile neutrino fields. We have seen in Eq. (5.66) that, even in the case of one generation, diagonalizing the mass matrix is equivalent to introduce a unitary $2 \times 2$ matrix $U^\nu$ such that

$$\begin{pmatrix} \nu_L \\ (\nu_R)^c \end{pmatrix} = U^\nu \begin{pmatrix} \nu_{1L} \\ \nu_{2L} \end{pmatrix}. \tag{10.64}$$

## 10.5 Sterile Neutrinos

The active neutrino field $\nu_L$ and the sterile field $\nu_R$ are linear combinations of the massive neutrino fields $\nu_{1L}$ and $\nu_{2L}$. As a consequence, mixing between active and sterile states is possible. In a way analogous to what described in Sect. 6.3, the fact that massive neutrinos of different mass evolve differently in time implies that the probability to detect the active neutrino $\nu_L$ oscillates with time. The oscillation between the active and the sterile state causes the survival probability of the active state to be smaller than 1. The sterile field cannot be directly detected since it does not participate to the SM interactions.

The active-sterile neutrino mixing affects the charged currents, as the one in Eq. (10.63), which in terms of massive neutrinos becomes

$$J_W^{\ell-\mu} = \frac{1}{\sqrt{2}} \overline{\ell_L} \gamma^\mu \nu_L = \frac{1}{\sqrt{2}} \overline{\ell_L} \gamma^\mu (U_{\ell 1}^\nu \nu_{1L} + U_{\ell 2}^\nu \nu_{2L}) = \frac{1}{\sqrt{2}} \sum_{i=1,2} U_{\ell i}^\nu \overline{\ell_L} \gamma^\mu \nu_{iL} \tag{10.65}$$

where $\ell_L$ and $\nu_L$ are the charged lepton and the neutrino participating to the interaction in the one generation case. The neutrino neutral current is reported in Eq. (4.43). We can express it in terms of massive states

$$J_Z^\mu = g_L^\nu \bar{\nu}_L \gamma^\mu \nu_L = g_L^\nu \sum_{i,j=1,2} U_{\ell i}^{\nu *} U_{\ell j}^\nu \overline{\nu_{iL}} \gamma^\mu \nu_{jL} \tag{10.66}$$

where $g_L^\nu = 1/(2\cos\theta_W)$. A significant feature is that the neutrino neutral current is no more diagonal in the massive fields and it is possible to have neutral-current transitions among the two massive neutrinos. An example of processes induced by a flavour changing neutral current at tree level are $Z \to \bar{\nu}_1 \nu_2$ and $Z \to \bar{\nu}_2 \nu_1$ decays. This may be confusing, since when dealing with the 1-generation coupling, e.g. $Z\bar{\nu}_e \nu_e$, we expect that the corresponding decay width, of the Z boson into the electron neutrino, is proportional to one. In order to understand how this happens, we observe that the $2 \times 2$ matrix $U^\nu$ can be assumed real and expressed as a generic orthogonal matrix in terms of an arbitrary mixing angle $\theta$ without loss of generality

$$U^\nu = \begin{pmatrix} \cos\theta & \sin\theta \\ -\sin\theta & \cos\theta \end{pmatrix}. \tag{10.67}$$

Keeping only the essential factors, the Z vertex reads

$$\bar{\nu}_e \nu_e = (\cos\theta \bar{\nu}_1 + \sin\theta \bar{\nu}_2)(\cos\theta \nu_1 + \sin\theta \nu_2) = $$
$$= \cos^2\theta \, \bar{\nu}_1 \nu_1 + \cos\theta \sin\theta (\bar{\nu}_1 \nu_2 + \bar{\nu}_2 \nu_1) + \sin^2\theta \, \bar{\nu}_2 \nu_2. \tag{10.68}$$

The width of Z boson decaying into any of $\bar{\nu}_i \nu_j$ ($i,j = 1,2$) is proportional to the sum of the squared amplitudes, namely

$$\Gamma \propto \cos^4\theta + \sin^4\theta + 2\cos^2\theta \sin^2\theta = (\cos^2\theta + \sin^2\theta)^2 = 1. \tag{10.69}$$

Thus, we find that the decay width is proportional to one, as expected, even if there are two massive neutrinos both interacting with Z. The original coupling $Z\bar{\nu}_e \nu_e$ is redistributed over four combinations of two fields which preserve the rate of one active flavour.

In reality, the number of left-handed fields is greater than one; experimental constraints fix the number of active generations to 3. This limit does not apply to sterile neutrinos. In this general case Eq. (10.64) still holds, but the column matrix of 2 elements on the left is substituted by the $3 + n$ column matrix $n_L$ (5.62) of the left-handed fields in the flavour basis, with 3 components of $\nu_L$ and $n$ components of $(\nu_R)^C$. The column matrix of 2 elements on the right is substituted by the $3 + n$ column matrix of the $3 + n$ Majorana eigenstates $\nu_1 \ldots \nu_{3+n}$, obtained by diagonalizing the Dirac-Majorana mass matrix. The matrix $U^\nu$ becomes a $(3 + n) \times (3 + n)$ unitary matrix. It can be considered the equivalent of the matrix $U_L^\nu$ in Eq. (10.61) extended to all, active and sterile, neutrino fields. However, only a sub-matrix of $U^\nu$ affects SM interactions. Indeed, for the charged current, we have again

$$J_W^{\ell-\mu} = \frac{1}{\sqrt{2}} \overline{\hat{\ell}_L} \gamma^\mu \hat{\nu}_L = \frac{1}{\sqrt{2}} \overline{\tilde{\ell}_L} \gamma^\mu U \tilde{\nu}_L \qquad (10.70)$$

but now the mixing is determined by the $3 \times (3 + n)$ rectangular sub-matrix of elements

$$U_{\alpha i} = \sum_\beta (U_L^{\ell\dagger})_{\alpha\beta} (U^\nu)_{\beta i} \qquad \alpha, \beta \in (e, \mu, \tau), \quad i \in (1, \ldots, 3+n). \qquad (10.71)$$

This matrix is obviously not unitary. Its parameterizations are different than the ones derived in Sect. 6.1.1.[18] As in the one generation case, oscillations between active and sterile states are possible, since active and sterile neutrino fields are linear combinations of the same massive neutrino fields. Indeed, the non-unitarity of the mixing matrix implies that the total probability of active flavors is not conserved.

Let us remark that it is generally accepted to use the convention discussed in Sect. 6.1, namely to choose the basis where the mass matrix of the charged leptons is diagonal. Within this convention, in the charged currents the neutrino flavour couples to the corresponding charged lepton only, and $U$ does not appear.

As in the one generation case, the neutrino neutral current is not diagonal in the massive fields:

$$J_Z^\mu = g_L^\nu \overline{\hat{\nu}_L} \gamma^\mu \hat{\nu}_L = g_L^\nu \overline{\tilde{\nu}_L} U^{\nu\dagger} \gamma^\mu U^\nu \tilde{\nu}_L. \qquad (10.72)$$

---

[18] The mixing matrix can be parameterized in terms of $3 + 3n$ mixing angles and $3 + 3n$ phases, divided into $1 + 2n$ Dirac phases and $2 + n$ Majorana phases. For a derivation, see e.g. Sect. 6.7.3 of [400].

where $g_L^\nu = 1/(2\cos\theta_W)$. It is possible to have neutral-current transitions among the two massive neutrinos. For instance, in the case of Z decays, also the decays $Z \to \bar{\nu}_i \nu_j$, with $i, j \in (\nu_1 \ldots \nu_{3+n})$, are allowed. It is possible to repeat a discussion analogous to what has been done in the one generation case, based this time on the unitarity of the matrix $U^\nu$. One finds that the expected width is determined by three generations of leptons, avoiding inconsistency with the observed decay width of the Z boson.

### 10.5.2 In Search of eV-Scale Sterile Neutrino Oscillations

The standard three-flavor framework of neutrino oscillations is well established, with two independent squared-mass differences, namely $\Delta m_{21}^2$ and $\Delta m_{31}^2$, whose ratio is $\Delta m_{21}^2/\Delta m_{31}^2 \simeq 34$, as discussed in Sect. 6.5. However, there are some oscillation experiments whose results do not fit in this framework. These results are often referred to as neutrino oscillation anomalies. Should they stand up to further scrutiny, their explanation might require the introduction of sterile neutrinos.

For simplicity, let us consider only one scheme, the one sometimes indicated as 3+1, where one assumes the existence of one sterile massive neutrino $\nu_4$, heavier than the three usual massive neutrinos, and with a large mass gap $\Delta m$. In this scheme we do not expect significative changes in the oscillations of solar and atmospheric neutrinos, and neutrinos produced at long-baseline experiments, in agreement with current data. That may not be the case with short-baseline (SBL) neutrino oscillation experiments. As seen in Sect. 6.3, oscillations are maximized when the ratio of the baseline over the neutrino beam energy is $L/E = 1.2/\Delta m^2$, so SBL experiments are particularly apt to probe relatively large mass gaps. Since most neutrino beams have primarily $E < 5$ GeV, studies of this fourth mass state generally call for a much shorter baseline than the ones needed for studies of atmospheric and solar oscillations.

By assuming a large $\Delta m^2$, the effects of the smaller $\Delta m_{21}^2$ and $\Delta m_{31}^2$ can be neglected, and one is left with the effective probabilities as in the case of two-neutrino mixing, namely

$$P_{\nu_\alpha \to \nu_\beta} = \delta_{\alpha\beta} + (-1)^{\delta_{\alpha\beta}} \sin^2 2\theta \sin^2 \frac{\Delta m^2 L}{4E} \tag{10.73}$$

in agreement with formula (6.40). We have $\Delta m^2 \equiv \Delta m_{41}^2$ and $\sin^2 2\theta \equiv 4|U_{\alpha 4}|^2 |U_{\beta 4}|^2$. The amplitudes are expressed in terms of the effective mixing angle $\theta$, that depends only on the elements of the mixing matrix $U$ connecting the flavour neutrinos $\nu_\alpha$ and $\nu_\beta$ to the massive $\nu_4$.

### 10.5.2.1 LSND

The experiment LSND (Liquid Scintillator Neutrino Detector) [340] was the first to observe a surplus of positron events in a neutrino beam from $\mu^+$ decaying at rest (DAR, see Sect. 8.2.1) at a short baseline (about 30 m). The statistical significance of the excess was about three standard deviations. LSND was an experiment searching for a $\bar{\nu}_\mu \rightarrow \bar{\nu}_e$ appearance signal, operating from 1993 to 1998. It used the accelerator located at a major LANL facility, the Los Alamos Neutron Science Center (LANSCE), or Los Alamos Meson Physics Facility (LAMPF), as it was called until 1995. The pions were produced by a proton current of 1 mA at energy of 798 MeV hitting a target followed by a Cu beam dump, where the pions came at rest before decaying. Negatively charged pions were rapidly absorbed by the target without producing neutrinos. Thus most of the $\pi^-$ were absorbed by the target, and DAR neutrinos were dominantly produced by $\pi^+ \rightarrow \mu^+ + \nu_\mu$ and $\mu^+ \rightarrow e^+ + \nu_e + \bar{\nu}_\mu$. No electron antineutrinos are produced in these decays, making convenient the search for $\bar{\nu}_\mu - \bar{\nu}_e$ oscillations. The mean neutrino energy was about between 30 and 60 MeV, averaging to $L/E \sim 1$ m/MeV. The detector consisted of an approximately cylindrical tank filled with 167 tons of mineral oil, lightly doped with liquid scintillator, viewed by 1220 photomultiplier tubes. This low scintillator concentration allowed the detection of both Cherenkov light and scintillation light. The signature of $\bar{\nu}_e$ appearance events was the inverse beta decay $\bar{\nu}_e + p = n + e^+$. The $\bar{\nu}_e$ interactions were tagged via the coincident observation of the prompt positron signal and the delayed detection of the 2.2 MeV photon from neutron capture on a free proton.

The most frequently entertained explanation of the LSND excess is in terms of neutrino flavour transitions to a sterile neutrino with a large mass square difference [340]

$$|\Delta m^2| \gtrsim 0.1 \, \text{eV}^2. \tag{10.74}$$

However, this explanation is very difficult to reconcile with results from other experiments, and the LSND data still represent an anomaly more than 20 years after the data taking stopped.

### 10.5.2.2 KARMEN

Another SBL experiment operating in a similar setup, but with overall lower statistics, was the KARMEN (KArlsruhe Rutherford Medium Energy Neutrino) experiment [579]. KARMEN was located at the spallation neutron source ISIS of the Rutherford Appleton Laboratory (RAL) in UK and was powered by an 800 MeV proton accelerators of 0.2 mA intensity. The final data set recorded with the full experimental setup was taken from 1997 until 2001. Its baseline was about 18 m and the detector was a liquid scintillator with a total mass of 56 tons. KARMEN searched for $\bar{\nu}_\mu \rightarrow \bar{\nu}_e$ oscillations in the same way as in LSND, by looking for the inverse beta decay, $\bar{\nu}_e + p = n + e^+$, which gives a twofold signature of a positron followed by one photon from neutron capture. KARMEN did not observe an excess as LSND [387], although a joint analysis of the two experiments showed that their data sets are compatible with oscillations occurring either in a band with $|\Delta m^2|$ from 0.2 to 1 eV$^2$ or in a region around $|\Delta m^2| \sim 7$ eV$^2$ [580].

### 10.5.2.3 MiniBooNE

The anomaly at LSND largely motivated the Booster Neutrino Experiment (MiniBooNE) [581], collecting data from 2002 to 2019.[19] It uses an high-intensity proton source, the Fermilab booster, a synchrotron accelerator with a circumference of 474 m located at FNAL. Instead of a stopped pion source this experiment employs a neutrino beam produced by pions decaying in flight (DIF), see Sect. 8.2. The neutrinos are produced by the 8 GeV proton beam from the Fermilab booster interacting on a beryllium target inside a magnetic focusing horn, followed by meson decay in a 50 m decay pipe. The sign of the pions that are focused towards the detector is determined by the polarity of a focusing horn. The MiniBooNE experiment has operated in neutrino mode, with a focused beam of positive pions whose decays produces an almost pure flux of $\nu_\mu$ peaking at $E \sim 600$ MeV, and in anti-neutrino mode, where the flux of $\bar{\nu}_\mu$ produced by negative pions peaks at approximately 400 MeV. The detector, placed at a distance of 541 m from the target, consists of a tank filled with 818 tons of pure mineral oil (CH$_2$). The array of 1520 photomultiplier tubes lining the MiniBooNE detector records Cherenkov and scintillation photons from the charged particles produced in neutrino interactions. The MiniBooNE experiment was designed with approximately the same $L/E \simeq 1$ km/GeV value as LSND. Whereas the LSND neutrino beam travels about 30 m with a typical energy of 30 MeV, the MiniBooNE neutrino beam travels a distance of about 500 m and has a typical energy of 500 MeV. With neutrino energies an order of magnitude higher, the MiniBooNE backgrounds and systematic errors are completely different from those of LSND. MiniBooNE, therefore, constitutes an independent check of the LSND evidence for neutrino oscillations at the $\sim 1$ eV$^2$ mass scale.

In 2007 the first MiniBooNE results, in neutrino mode, observed no excess over the background, contradicting LSND data [582]. Completely different results came in 2018, when MiniBooNE reported an excess of electron neutrino events in the energy range $200 < E < 1250$ MeV in both neutrino and antineutrino running modes [560], excess which was confirmed, at a significance of 4.8 $\sigma$, with the full data set [584]. Parameter values are consistent with the ones from LSND and also data interpretation, which in terms of two-neutrino oscillations requires at least four

---

[19] The prefix Mini in the name refers to fact that this is a downscaled version of a proposed experiment BooNE, which would have consisted of two similar detectors placed at different baselines, and which was never built.

neutrino types. The range of the ratio $L/E$ (m/MeV) for the LSND excess was about $0.5 < L_{\text{LSND}}/E_{\text{LSND}} < 1.5$. To compare with MiniBooNE one sets

$$L_{\text{LSND}}/E_{\text{LSND}} \simeq L_{\text{MiniBooNE}}/E_{\text{MiniBooNE}} \tag{10.75}$$

Therefore the LSND signal, which corresponds to a maximum of 1.5 m/MeV, should be seen in MiniBooNE, which has a longer baseline, at neutrino energies about $E_{\text{MiniBooNE}} > 360$ MeV. Since part of the energy range where the excess is observed corresponds to values of L/E outside the LSND range, the low-energy excess is often considered an effect different from the LSND anomaly, and referred to as the MiniBooNE low-energy anomaly.

#### 10.5.2.4 Short-Baseline Neutrino Program at Fermilab

Fermilab has currently a neutrino program [585] based on SBL accelerators. It relies on three detectors located on the same neutrino beamline than MiniBooNE, at different baselines. The three detectors are liquid argon time projection chambers (LAr-TPC), which benefit of significant advantages.[20] In the neutrino-Argon interaction the charged particles produced ionize (or excite) the argon as they move through the volume. When the argon de-excites or relaxes it emits scintillation light. Measuring the amount of scintillation in addition to the amount of ionization improves the measurements of the energy deposited along the track ($dE/dx$) as a function of distance, which allow to perform particle identification in the TPC.

The first of the three detectors, known as MicroBooNE [586], operated from 2015 to 2021. Its detector was a 170 ton total mass LAr-TPC along with an array of photomultiplier tubes (PMT) for scintillation light detection. The wavelength of the scintillation light was shifted (from 128 nm to 425–450 nm) by wavelength shifting material to be directly detected by the PMTs. MicroBooNE was designed to address the nature of the MiniBooNE low-energy anomaly. It was located in an open pit approximately 6 m below ground level and at 470 m from the target, in order to yield a similar $L/E \sim 1$ m/MeV as the MiniBooNE detector.

The key characteristic of the anomalous excess in the MiniBooNE Cherenkov detector was the presence of an electromagnetic shower in each event, presumably from the outgoing electron in a $\nu_e$ charged-current quasi-elastic (CCQE) interaction (see Sect. 9.2.1). The same signature, however, is shared by a single photon that could be produced in a $\nu_\mu$ neutral-current interaction, inducing a $\pi^0$ decay where only one of the two decay photons is visible.[21] This background event cannot be distinguished from the signal in the MiniBooNE detector, at a variance with the Lar-TPC detector of the MicroBooNE experiment, which provided uniquely distinguishable signals for photons and electrons by analysing the $dE/dx$ for the

---

[20] Some of them were delineated while discussing the LBL experiment ICARUS in Sect. 8.2.2.3.
[21] See Sect. 9.2.2 for pion production.

## 10.5 Sterile Neutrinos

start of the electromagnetic shower. An electron produces a track with a $dE/dx$ near the minimum until it begins to produce an electromagnetic shower, while an high energy photon does not leave a trace until it pair converts or Compton scatters. Other advantages of MicroBooNE were improved electron neutrino efficiency relative to MiniBooNE (a factor of two better) and the fact that the MicroBooNE detector had the sensitivity to reconstructed neutrino energies much lower than the 200 MeV threshold of MiniBooNE.

As of today, no excess of $\nu_e$ was found [587] and no evidence of light sterile neutrino oscillations in the $3+1$ oscillation hypothesis [588]. Further MicroBooNE physics data analysis is ongoing, incorporating additional statistics (nearly doubling the data set analyzed so far), and expanding beyond the simplest interpretations of the MiniBooNE low energy excess.

ICARUS [589] is another detector of the SBL neutrino program at Fermilab. It is the first large scale operating Lar-TPC, containing 760 tons of ultra-pure liquid Argon. It began its scientific life at LNGS in 2010, recording data from a neutrino beam sent by CERN.[22] The detector was shipped to CERN in 2014, where it was upgraded and prepared to be transported to Fermilab. In June 2017, it left CERN in two parts, loaded onto trucks, and it boarded a boat on the Rhine to a port in Antwerp (Belgium). From there, it crossed the Atlantic, headed to Fermilab via the Great Lakes. It arrived in July 2017—the journey had lasted in total six weeks. At Fermilab, the detector is located at a baseline of 600 m, at ground level with a limited overburden, and in the SBN program it operates as the far detector, being the farthest from the target on the Booster neutrino beamline. In addition, its position is such that it can also be exposed ($6^0$ off-axis) to neutrinos of the NuMI beam, a higher energy neutrino beam provided by Fermilab. ICARUS finished commissioning and started the first physics data run in 2022.

The last detector of the program, the Short Baseline Near Detector (SBND), currently starting data taking, uses about 260 tons of liquid argon. It is located in proximity of the neutrino source (about 110 m), in order to record a large number (over two million) of neutrino interactions per year. Being at ground level, it will be also susceptible to a sizeable backgrounds induced by cosmic interactions.

Together these three detectors provide a powerful accelerator probe for sterile neutrino searches. The use of multiple detectors at different baselines aims at reducing systematic uncertainties, caused by the challenge of predicting absolute neutrino fluxes in accelerator beam experiments and the large uncertainties associated with neutrino-nucleus interactions. Besides, the physics of these interactions is an important element of future neutrino experiments that will employ the LAr-TPC technology, such as the long-baseline experiment DUNE.[23]

---

[22] See Sect. 8.2.2.

[23] See Sect. 8.2.4.

### 10.5.2.5 JSNS$^2$

In 2021 the JSNS$^2$ experiment at J-PARC (J-PARC Sterile Neutrino Search at J-PARC Spallation Neutron Source) has started the search for neutrino oscillations with $\Delta m^2 \sim 1$ eV$^2$ with a single detector [590]. JSNS$^2$ is aiming to provide a direct and ultimate test of the LSND observations. In analogy with LSND, it searches for a $\bar{\nu}_\mu \to \bar{\nu}_e$ appearance signal, using DAR neutrinos, via inverse beta decay reaction. The $\bar{\nu}_\mu$ beam is produced by injecting 3 GeV protons (1 MW beam) from a rapid cycling synchrotron on a mercury spallation target. The detector, placed at 24 m from the target, consists of 50 tons of liquid scintillator, including 17 tons of Gadolinium loaded liquid scintillator inside an inner acrylic tank. In the second phase, JSNS$^2$-II, under preparation, a second detector, at a baseline of 48 m, will be added as the far detector.

### 10.5.2.6 Gallium and Reactor Anti-neutrino Anomalies

Besides the LSND anomaly, there are two other anomalies which can be interpreted as due to a possible mixing with sterile neutrinos. One is the anomaly in two experiments that use Gallium, namely SAGE and GALLEX (described in Sect. 7.2.4), where the measured flux of solar electron neutrinos is consistently lower than expected ($\nu_e$ disappearance). Such a deficit, confirmed by the experiments GNO and BEST, is known as the Gallium anomaly [591]. The discrepancy is commonly expressed through the ratio $R = N_{\text{meas}}/N_{\text{pred}}$ of the number of measured to predicted events. The current estimate, considering the measurements of SAGE, Gallex, and GNO, is $R = 0.87 \pm 0.05$ [592], namely a deviation of about $2.5\sigma$ from 1. With the inclusion of the BEST results the significance grows to about $4\sigma$ [593].

Another anomaly was discovered in as a consequence of more detailed calculations of fluxes of $\bar{\nu}_e$'s produced in reactors in 2011 [300], as already mentioned in Sect. 8.1.2. Comparison of the new theoretical predictions with data resulted in the observation of a deficit of $\bar{\nu}_e$ from very short baseline experiments near nuclear reactors ($10 < L < 100$ m), where no oscillations are expected in the standard three flavor oscillation framework. One explanation of this discrepancy, dubbed the reactor anti-neutrino anomaly (RAA), is the assumption of $\bar{\nu}_e$ disappearance caused by oscillations with $\Delta m^2 \sim 1$ eV$^2$ [594].[24] Another possible explanation for the RAA anomaly lies in the deficiencies of the 2011 theoretical flux model, as some recent ones show significantly better agreement with data [307, 308].

The 2011 flux predictions have not been supported by the reactor antineutrino flux measurements at Daya Bay [305], an experiment discussed in Sect. 8.1.3. NEOS is another experiment which has investigated the mixing with sterile neutrino and the RAA. The NEOS detector is a Gd-loaded liquid scintillator located just 24 m from the core of the 2.8GW Hamit nuclear power plant in South Korea.

---

[24] For a review see e.g. Refs. [562, 595, 596].

## 10.5 Sterile Neutrinos

No strong evidence of an active-to-sterile neutrino oscillation was found in 2017 when the spectral shape was compared with the Daya Bay spectrum [597]. More recently, NEOS has teamed up with RENO, leveraging their shared location at the Hanbit Nuclear Power Plant complex, to minimize uncertainties related to reactor operations [317]. The RENO experiment has been already mentioned in Sect. 8.1.3. Their combined search gives a 95% C.L. excluded region of $0.1 < |\Delta m_{41}^2| < 7$ eV$^2$, but also a 68% C.L allowed region with the best fit of $|\Delta m_{41}^2| = 2.41$ eV$^2$ and $\sin^2 2\theta_{14} = 0.08$ [317].

### 10.5.2.7 Unstable Sterile Neutrinos

One of the strongest constraints on sterile neutrino induced $\nu_\mu$ and $\bar{\nu}_\mu$ disappearance can be provided from the study of high-energy atmospheric neutrinos ($\gtrsim$500 GeV) observed by the IceCube neutrino observatory (discussed in Sect. 7.7.1) [598]. Their results after eight years of searches have not given evidence for eV-scale sterile neutrinos in the 3+1 framework [599, 600]. No evidence of deviations from the standard atmospheric neutrino oscillations due to the sterile neutrino mixing came either from the analyses using three years of the lower energy data of DeepCore, the denser part of IceCube, in the energy range approximately between 10 and 60 GeV [601]. IceCube has also performed a search for nonstandard neutrino interactions from TeV-scale $\nu_\mu$ disappearance, reporting strong constraints on any nonstandard parameter from any oscillation channel [602].

Another model that could offer an explanation for the observed anomalies is the so-called 3+1+decay model, wherein the fourth additional sterile neutrino is unstable. IceCube has the unique capability to test this model through entirely different processes than vacuum oscillations. The presence of a sterile neutrino at the eV-scale can be revealed as a resonant, matter-enhanced flavor transition for either muon antineutrinos or muon neutrinos passing through the core of the Earth. This results in a shortage of upgoing muon (anti)neutrinos at TeV-scale energies. IceCube cannot distinguish between neutrinos and antineutrinos. Consequently, the only signature is a deficit in the combined distribution of muon neutrino and muon antineutrino charged-current events at TeV energies. The assumption that the 3+1+decay model provides a more accurate description of the data than the 3+1 model was investigated using an eight-year database, yielding no significant evidence in support of this hypothesis [603].

# References

1. W. Roentgen. Ueber eine neue Art von Strahlen. Vorläufige Mitteilung. *Aus den Sitzungsberichten der Würzburger Physik.-medic. Gesellschaft Würzburg*, 137–147, 1895.
2. H. Becquerel. On the rays emitted by phosphorescence. *Compt. Rend. Hebd. Seances Acad. Sci.*, 122(8):420–421, 1896.
3. J. J. Thomson. Cathode rays. *Phil. Mag. Ser. 5*, 44:293–316, 1897.
4. A. I. Akhiezer and V. B. Berestetskii. *Quantum electrodinamica*. Interscience Publishers, 1965.
5. P. Kusch and H. M. Foley. The magnetic moment of the electron. *Phys. Rev.*, 74(3):250, 1948.
6. J. S. Schwinger. On Quantum electrodynamics and the magnetic moment of the electron. *Phys. Rev.*, 73:416–417, 1948.
7. J. Chadwick. Intensitätsverteilung im magnetischen Spectrum der $\beta$-Strahlen von radium B + C. *Verhandl. Dtsc. Phys. Ges.*, 16:383, 1914.
8. J. Chadwick and C. D. Ellis. A preliminary investigation of the intensity distribution in the $\beta$-ray spectra of radium B and C. *Proceedings of the Cambridge Philosophical Society*, 21, 1922.
9. C. D. Ellis and W. A. Wooster. The average energy of disintegration of radium E. *Proceedings of the Royal Society (London)*, A117:109–123, 1927.
10. N. Bohr. Atomic stability and conservation laws. *Reale Accademia d'Italia*, 119–130, 1932.
11. L. M. Brown. The idea of the neutrino. *Phys. Today*, 31N9:23–28, 1978.
12. J. Chadwick. Possible existence of a neutron. *Nature*, 129:312, 1932.
13. G. Amaldi. *Materia e antimateria*. Mondadori, 1961.
14. E. Fermi. Trends to a theory of beta radiation. (In Italian). *Nuovo Cim.*, 11:1–19, 1934.
15. E. Fermi. An attempt of a theory of beta radiation. 1. *Z. Phys.*, 88:161–177, 1934.
16. E. Rutherford. Collision of $\alpha$ particles with light atoms. IV. An anomalous effect in nitrogen. *Phil. Mag.*, 90(sup1):31–37, 1919.
17. J. C. Street and E. C. Stevenson. New evidence for the existence of a particle of mass intermediate between the proton and electron. *Phys. Rev.*, 52:1003–1004, 1937.
18. S. H. Neddermeyer and C. D. Anderson. Note on the nature of cosmic ray particles. *Phys. Rev.*, 51:884–886, 1937.
19. B. Pontecorvo. Nuclear capture of mesons and the meson decay. *Phys. Rev.*, 72:246, 1947.
20. E. Majorana. Theory of the symmetry of electrons and positrons. *Nuovo Cim.*, 14:171–184, 1937.
21. G. Racah. On the symmetry of particle and antiparticle. *Nuovo Cim.*, 14:322–328, 1937.
22. C. S. Wu, E. Ambler, R. W. Hayward, D. D. Hoppes, and R. P. Hudson. Experimental test of parity conservation in beta decay. *Phys. Rev.*, 105:1413–1414, 1957.

23. F. Perrin. Possibilité d'émission de particules neutres de masse intrinsèque nulle dans les radioactivités beta. *Comptes-Rendus*, 197:1625, 1933.
24. Ch. Kraus, B. Bornschein, L. Bornschein, J. Bonn, B. Flatt, et al. Final results from phase II of the Mainz neutrino mass search in tritium beta decay. *Eur. Phys. J.*, C40:447–468, 2005.
25. V. N. Aseev et al. An upper limit on electron antineutrino mass from Troitsk experiment. *Phys. Rev.*, D84:112003, 2011.
26. H. Bethe and R. Peierls. The 'neutrino'. *Nature*, 133:532, 1934.
27. B. Pontecorvo. Inverse beta processes. *Chalk River Report*, 6:429, 1946.
28. F. Reines and C. L. Cowan. The neutrino. *Nature*, 178:446–449, 1956.
29. F. Reines and C. L. Cowan. Free anti-neutrino absorption cross-section. 1: Measurement of the free anti-neutrino absorption cross-section by protons. *Phys. Rev.*, 113:273–279, 1959.
30. F. Reines and C. Cowan. The Reines-Cowan experiments: Detecting the Poltergeist. *Los Alamos Sci.*, 25:4–27, 1997.
31. B. Pontecorvo. Mesonium and anti-mesonium. *Sov. Phys. JETP*, 6:429, 1957.
32. M. Goldhaber, L. Grodzins, and A. W. Sunyar. Helicity of neutrinos. *Phys. Rev.*, 109:1015–1017, 1958.
33. B. Pontecorvo. Inverse beta processes and nonconservation of lepton charge. *Sov. Phys. JETP*, 7:172–173, 1958.
34. B. Pontecorvo. Neutrino experiments and the problem of conservation of leptonic charge. *Zh. Eksp. Teor. Fiz.*, 53:1717–1725, 1967.
35. B. Pontecorvo. Neutrino experiments and the problem of conservation of leptonic charge. *Sov. Phys. JETP*, 26:984–988, 1968.
36. G. Danby, J. M. Gaillard, K. A. Goulianos, L. M. Lederman, N. B. Mistry, et al. Observation of high-energy neutrino reactions and the existence of two kinds of neutrinos. *Phys. Rev. Lett.*, 9:36–44, 1962.
37. Z. Maki, M. Nakagawa, and S. Sakata. Remarks on the unified model of elementary particles. *Prog. Theor. Phys.*, 28:870–880, 1962.
38. C. V. Achar et al. Detection of muons produced by cosmic ray neutrinos deep underground. *Phys. Lett.*, 18:196–199, 1965.
39. F. Reines, M. F. Crouch, T. L. Jenkins, W. R. Kropp, H. S. Gurr, G. R. Smith, J. P. F. Sellschop, and B. Meyer. Evidence for high-energy cosmic ray neutrino interactions. *Phys. Rev. Lett.*, 15:429–433, 1965.
40. R. Davis Jr., D. S. Harmer, and K. C. Hoffman. Search for neutrinos from the sun. *Phys. Rev. Lett.*, 20:1205–1209, 1968.
41. V. N. Gribov and B. Pontecorvo. Neutrino astronomy and lepton charge. *Phys. Lett.*, B28:493, 1969.
42. L. Wolfenstein. Neutrino oscillations in matter. *Phys. Rev.*, D17:2369–2374, 1978.
43. S. P. Mikheyev and A. Yu. Smirnov. Resonance amplification of oscillations in matter and spectroscopy of solar neutrinos. *Sov. J. Nucl. Phys.*, 42:913–917, 1985.
44. S. P. Mikheev and A. Yu. Smirnov. Resonant amplification of neutrino oscillations in matter and solar neutrino spectroscopy. *Nuovo Cim. C*, 9:17–26, 1986.
45. I. Shelton. *IAU (International Astronomical Union) Circular*, page 4316, 1987.
46. K. S. Hirata et al. Observation in the Kamiokande-II detector of the neutrino burst from supernova SN 1987a. *Phys. Rev. D*, 38:448–458, 1988.
47. R. M. Bionta et al. Observation of a neutrino burst in coincidence with supernova SN 1987a in the large magellanic cloud. *Phys. Rev. Lett.*, 58:1494, 1987.
48. E. N. Alekseev, L. N. Alekseeva, V. I. Volchenko, and I. V. Krivosheina. Possible detection of a neutrino signal on 23 February 1987 at the Baksan Underground Scintillation Telescope of the Institute of Nuclear Research. *JETP Lett.*, 45:589–592, 1987.
49. K. S. Hirata et al. Experimental study of the atmospheric neutrino flux. *Phys. Lett. B*, 205:416, 1988.
50. D. Casper et al. Measurement of atmospheric neutrino composition with IMB-3. *Phys. Rev. Lett.*, 66:2561–2564, 1991.

51. D. Denegri, B. Sadoulet, and M. Spiro. The number of neutrino species. *Rev. Mod. Phys.*, 62:1–42, 1990.
52. K. Kodama et al. Observation of tau neutrino interactions. *Phys. Lett. B*, 504:218–224, 2001.
53. Y. Fukuda et al. Measurement of the flux and zenith angle distribution of upward through going muons by Super-Kamiokande. *Phys. Rev. Lett.*, 82:2644–2648, 1999.
54. Q. R. Ahmad et al. Direct evidence for neutrino flavor transformation from neutral current interactions in the Sudbury Neutrino Observatory. *Phys. Rev. Lett.*, 89:011301, 2002.
55. K. Eguchi et al. First results from KamLAND: Evidence for reactor anti-neutrino disappearance. *Phys. Rev. Lett.*, 90:021802, 2003.
56. M. H. Ahn et al. Indications of neutrino oscillation in a 250 km long baseline experiment. *Phys. Rev. Lett.*, 90:041801, 2003.
57. D. G. Michael et al. Observation of muon neutrino disappearance with the MINOS detectors and the NuMI neutrino beam. *Phys. Rev. Lett.*, 97:191801, 2006.
58. K. Abe et al. Measurements of neutrino oscillation in appearance and disappearance channels by the T2K experiment with $6.6 \times 10^{20}$ protons on target. *Phys. Rev.*, D91(7):072010, 2015.
59. F. P. An et al. New measurement of antineutrino oscillation with the full detector configuration at Daya Bay. *Phys. Rev. Lett.*, 115(11):111802, 2015.
60. S. B. Kim. New results from RENO and prospects with RENO-50. *Nucl. Part. Phys. Proc.*, 265–266:93–98, 2015.
61. T. Abrahão et al. Measurement of $\theta_{13}$ in Double Chooz using neutron captures on hydrogen with novel background rejection techniques. *JHEP*, 01:163, 2016. [JHEP01,163(2016)].
62. T. Araki et al. Experimental investigation of geologically produced antineutrinos with KamLAND. *Nature*, 436:499–503, 2005.
63. G. Bellini et al. Observation of geo-neutrinos. *Phys. Lett. B*, 687:299–304, 2010.
64. V. S. Berezinsky and G. T. Zatsepin. Cosmic neutrinos of superhigh energy. *Yad. Fiz.*, 11:200–205, 1970.
65. M. G. Aartsen et al. First observation of PeV-energy neutrinos with IceCube. *Phys. Rev. Lett.*, 111:021103, 2013.
66. M. G. Aartsen et al. Evidence for high-energy extraterrestrial neutrinos at the IceCube detector. *Science*, 342:1242856, 2013.
67. M. G. Aartsen et al. Multimessenger observations of a flaring blazar coincident with high-energy neutrino IceCube-170922A. *Science*, 361(6398):eaat1378, 2018.
68. W. Heisenberg. On the structure of atomic nuclei. *Z. Phys.*, 77:1–11, 1932.
69. B. Hall. *Lei Groups, Lie Algebras, and Representations*. Springer International Publishing Switzerland, 2003.
70. N. Goldstein, C. Poole, and J. Safko. *Classical Mechanics*. Addison Wesley, 2000.
71. I. I. Bigi, G. Ricciardi, and M. Pallavicini. *New Era for CP Asymmetries*. World Scientific, 7 2021.
72. J. Schwichtenberg. *Physics from Symmetry*. Undergraduate Lecture Notes in Physics. Springer International Publishing, Cham, 2018.
73. S. Weinberg. *The Quantum Theory of Fields, Vol. 1 Foundations*. Cambridge University Press, 1996.
74. Emmy Noether. Invariante Variationsprobleme Nachrichten der Königlichen Gesellschaft der Wissenschaften zu Göttingen. *Math. Phys. Kl.*, 235, 1918.
75. C. D. Anderson. The apparent existence of easily deflectable positives. *Science*, 76:238–239, 1932.
76. T. D. Lee and C. N. Yang. Question of parity conservation in weak interactions. *Phys.Rev.*, 104:254–258, 1956.
77. R. Feynam. *The Character of Physical Law*. The MIT Press, 1964.
78. R. L. Garwin, L. M. Lederman, and M. Weinrich. Observations of the failure of conservation of parity and charge conjugation in meson decays: the magnetic moment of the free muon. *Phys. Rev.*, 105:1415–1417, 1957.
79. J. I. Friedman and V. L. Telegdi. Nuclear emulsion evidence for parity nonconservation in the decay chain $\pi^+ \to \mu^+ \to e^+$. *Phys. Rev.*, 106:1290–1293, 1957.

80. M. Gell-Mann and A. Pais. Behavior of neutral particles under charge conjugation. *Phys. Rev.*, 97:1387–1389, 1955.
81. K. Lande, E. T. Booth, J. Impeduglia, L. M. Lederman, and W. Chinowsky. Observation of long-lived neutral V particles. *Phys. Rev.*, 103:1901–1904, 1956.
82. J. H. Christenson, J. W. Cronin, V. L. Fitch, and R. Turlay. Evidence for the $2\pi$ decay of the $K_2^0$ meson. *Phys. Rev. Lett.*, 13:138–140, 1964.
83. K. Abe et al. Observation of large CP violation in the neutral $B$ meson system. *Phys. Rev. Lett.*, 87:091802, 2001.
84. B. Aubert et al. Observation of CP violation in the $B^0$ meson system. *Phys. Rev. Lett.*, 87:091801, 2001.
85. R. Aaij et al. Observation of CP violation in charm decays. *Phys. Rev. Lett.*, 122(21):211803, 2019.
86. A. Angelopoulos et al. First direct observation of time reversal noninvariance in the neutral kaon system. *Phys. Lett.*, B444:43–51, 1998.
87. L. Wolfenstein. The search for direct evidence for time reversal violation. *Int. J. Mod. Phys.*, E8:501–511, 1999.
88. L. Alvarez-Gaume, C. Kounnas, S. Lola, and P. Pavlopoulos. Violation of time reversal invariance and CPLEAR measurements. *Phys. Lett.*, B458:347–354, 1999.
89. J. P. Lees et al. Observation of time reversal violation in the $B^0$ meson system. *Phys. Rev. Lett.*, 109:211801, 2012.
90. J. P. Lees et al. Observation of time reversal violation in the $B^0$ meson system. *Phys. Rev. Lett.*, 109:211801, 2012.
91. M. D. Schwartz. *Quantum Field Theory and the Standard Model*. Cambridge University Press, 3 2014.
92. M. S. Sozzi. *Discrete Symmetries and CP Violation: From Experiment to Theory*. Oxford University Press, 2008.
93. G. Luders. On the equivalence of invariance under time reversal and under particle-antiparticle conjugation for relativistic field theories. *Kong. Dan. Vid. Sel. Mat. Fys. Med.*, 28N5(5):1–17, 1954.
94. R. F. Streater and A. S. Wightman. *PCT, Spin and Statistics, and All That*. Addison-Wesley, 1989.
95. R. Jost. A remark on the C.T.P. theorem. *Helv. Phys. Acta*, 30:409–416, 1957.
96. O. W. Greenberg. Why is CPT fundamental? *Found. Phys.*, 36:1535–1553, 2006.
97. N. N. Bogolyubov, A. A. Logunov, and I. T. Todorov. *Introduction to Axiomatic Quantum Field Theory*. Benjamin, 1975.
98. O. W. Greenberg. CPT violation implies violation of Lorentz invariance. *Phys. Rev. Lett.*, 89:231602, 2002.
99. S. W. Hawking. Breakdown of predictability in gravitational collapse. *Phys. Rev. D*, 14:2460–2473, 1976.
100. G. Luders and B. Zumino. Some consequences of TCP-invariance. *Phys. Rev.*, 106:385–386, 1957.
101. J. C. Maxwell. A dynamical theory of electromagnetic field. *Philosophical Transactions of the Royal Society of London*, 155:459–512, 1865.
102. V. Fock. On the invariant form of the wave equation and the equations of motion for a charged point mass. (In German and English). *Z. Phys.*, 39:226–232, 1926.
103. H. Weyl. Electron and gravitation. 1. (In German). *Z. Phys.*, 56:330–352, 1929.
104. J. D. Jackson and L. B. Okun. Historical roots of gauge invariance. *Rev. Mod. Phys.*, 73:663–680, 2001.
105. R. Utiyama. Invariant theoretical interpretation of interaction. *Phys. Rev.*, 101:1597–1607, 1956.
106. S. L. Glashow. Partial symmetries of weak interactions. *Nucl. Phys.*, 22:579–588, 1961.
107. S. Weinberg. A model of leptons. *Phys. Rev. Lett.*, 19:1264–1266, 1967.
108. A. Salam. Weak and electromagnetic interactions. *Conf. Proc.*, C680519:367–377, 1968.

# References

109. G. 't Hooft. Renormalizable lagrangians for massive Yang-Mills fields. *Nucl. Phys.*, B35:167–188, 1971.
110. G. 't Hooft. Renormalization of massless Yang-Mills fields. *Nucl. Phys.*, B33:173–199, 1971.
111. R. P. Feynman. Quantum theory of gravitation. *Acta Phys. Polon.*, 24:697–722, 1963. [,272(1963)].
112. L. D. Faddeev and V. N. Popov. Feynman diagrams for the Yang-Mills field. *Phys. Lett.*, B25:29–30, 1967. [,325(1967)].
113. Y. Aharonov and D. Bohm. Significance of electromagnetic potentials in the quantum theory. *Phys. Rev.*, 115:485–491, 1959. [,95(1959)].
114. J. Goldstone. Field theories with superconductor solutions. *Nuovo Cim.*, 19:154–164, 1961.
115. J. Goldstone, A. Salam, and S. Weinberg. Broken symmetries. *Phys. Rev.*, 127:965–970, 1962.
116. Y. Nambu and G. Jona-Lasinio. Dynamical model of elementary particles based on an analogy with superconductivity. 1. *Phys. Rev.*, 122:345–358, 1961.
117. Y. Nambu and G. Jona-Lasinio. Dynamical model of elementary particles based on an analogy with superconductivity. II. *Phys. Rev.*, 124:246–254, 1961.
118. Y. Nambu. Axial vector current conservation in weak interactions. *Phys. Rev. Lett.*, 4:380–382, 1960.
119. P. W. Higgs. Broken symmetries and the masses of gauge bosons. *Phys. Rev. Lett.*, 13:508–509, 1964. [,160(1964)].
120. P. W. Higgs. Broken symmetries, massless particles and gauge fields. *Phys. Lett.*, 12:132–133, 1964.
121. F. Englert and R. Brout. Broken symmetry and the mass of gauge vector mesons. *Phys. Rev. Lett.*, 13:321–323, 1964. [,157(1964)].
122. G. S. Guralnik, C. R. Hagen, and T. W. B. Kibble. Global conservation laws and massless particles. *Phys. Rev. Lett.*, 13:585–587, 1964. [,162(1964)].
123. G. 't Hooft. Renormalizable lagrangians for massive Yang-Mills fields. *Nucl. Phys.*, B35:167–188, 1971. [,201(1971)].
124. I. Aitchison. *Supersymmetry in Particle Physics-An Elementary Introduction*. Cambridge University Press, 2007.
125. H. Goldstein, C.P. Poole, and J. L. Safko. *Classical Mechanics*. Addison Wesley, 2000.
126. S. Coleman. *Notes from Sidney Coleman's Physics 253a: Quantum Field Theory*. 2011.
127. P. A. Zyla et al. Review of particle physics. *PTEP*, 2020(8):083C01, 2020.
128. T. van Ritbergen and R. G. Stuart. On the precise determination of the Fermi coupling constant from the muon lifetime. *Nucl. Phys.*, B564:343–390, 2000.
129. S. Weinberg. Current algebra and gauge theories. 1. *Phys. Rev.*, D8:605–625, 1973. especially footnote 8.
130. G. Feinberg, P. Kabir, and S. Weinberg. Transformation of muons into electrons. *Phys. Rev. Lett.*, 3:527–530, 1959. especially footnote 9.
131. T. P. Cheng and L. F. Li. *Gauge Theory of Elementary Particle Physics*. Oxford University Press, Oxford, UK, 1984.
132. R. N. Mohapatra and P. B. Pal. Massive neutrinos in physics and astrophysics. Third edition. *World Sci. Lect. Notes Phys.*, 72, 2004.
133. N. Cabibbo. Unitary symmetry and leptonic decays. *Phys. Rev. Lett.*, 10:531–533, 1963. [,648(1963)].
134. S. L. Glashow, J. Iliopoulos, and L. Maiani. Weak interactions with Lepton-Hadron symmetry. *Phys. Rev.*, D2:1285–1292, 1970.
135. J. E. Augustin, A. M. Boyarski, M. Breidenbach, F. Bulos, J. T. Dakin, G. J. Feldman, G. E. Fischer, D. Fryberger, G. Hanson, B. Jean-Marie, R. R. Larsen, V. Lüth, H. L. Lynch, D. Lyon, C. C. Morehouse, J. M. Paterson, M. L. Perl, B. Richter, P. Rapidis, R. F. Schwitters, W. M. Tanenbaum, F. Vannucci, G. S. Abrams, D. Briggs, W. Chinowsky, C. E. Friedberg, G. Goldhaber, R. J. Hollebeek, J. A. Kadyk, B. Lulu, F. Pierre, G. H. Trilling, J. S. Whitaker, J. Wiss, and J. E. Zipse. Discovery of a narrow resonance in $e^+e^-$ annihilation. *Phys. Rev. Lett.*, 33:1406–1408, Dec 1974.

136. J. J. Aubert, U. Becker, P. J. Biggs, J. Burger, M. Chen, G. Everhart, P. Goldhagen, J. Leong, T. McCorriston, T. G. Rhoades, M. Rohde, S. C. C. Ting, S. L. Wu, and Y. Y. Lee. Experimental observation of a heavy particle j. *Phys. Rev. Lett.*, 33:1404–1406, Dec 1974.
137. S. W. Herb et al. Observation of a dimuon resonance at 9.5-GeV in 400-GeV proton-nucleus collisions. *Phys. Rev. Lett.*, 39:252–255, 1977.
138. F. Abe et al. Observation of top quark production in $\bar{p}p$ collisions. *Phys. Rev. Lett.*, 74:2626–2631, 1995.
139. S. Abachi et al. Observation of the top quark. *Phys. Rev. Lett.*, 74:2632–2637, 1995.
140. M. Kobayashi and T. Maskawa. CP violation in the renormalizable theory of weak interaction. *Prog. Theor. Phys.*, 49:652–657, 1973.
141. L. Wolfenstein. Parametrization of the Kobayashi-Maskawa matrix. *Phys. Rev. Lett.*, 51:1945, 1983.
142. A. J. Buras, M. E. Lautenbacher, and G. Ostermaier. Waiting for the top quark mass, $K^+ \to \pi^+ \nu \bar{\nu}$, $B_s^0 - \bar{B}_s^0$ mixing and CP asymmetries in B decays. *Phys. Rev.*, D50:3433–3446, 1994.
143. R. S. Chivukula and H. Georgi. Composite technicolor standard model. *Phys. Lett.*, B188:99–104, 1987.
144. I. I. Bigi and A. I. Sanda. *CP Violation*, volume 9. Cambridge University Press, 9 2009.
145. D. Brandt, H. Burkhardt, M. Lamont, S. Myers, and J. Wenninger. Accelerator physics at LEP. *Rept. Prog. Phys.*, 63:939–1000, 2000.
146. SLAC Linear Collider Conceptual Design Report. 1980.
147. S. Schael et al. Precision electroweak measurements on the Z resonance. *Phys. Rept.*, 427:257–454, 2006.
148. M. Aaboud et al. Precision measurement and interpretation of inclusive $W^+$, $W^-$ and $Z/\gamma^*$ production cross sections with the ATLAS detector. *Eur. Phys. J.*, C77(6):367, 2017.
149. S. Schael et al. Electroweak measurements in electron-positron collisions at W-boson-pair energies at LEP. *Phys. Rept.*, 532:119–244, 2013.
150. R. Aaij et al. Measurement of forward $W \to e\nu$ production in $pp$ collisions at $\sqrt{s} = 8$ TeV. *JHEP*, 10:030, 2016.
151. A. Abulencia et al. Measurements of inclusive W and Z cross sections in p anti-p collisions at s**(1/2) = 1.96-TeV. *J. Phys.*, G34:2457–2544, 2007.
152. A. Pich. Precision tau physics. *Prog. Part. Nucl. Phys.*, 75:41–85, 2014.
153. A. Aguilar-Arevalo et al. Improved measurement of the $\pi \to e\nu$ branching ratio. *Phys. Rev. Lett.*, 115(7):071801, 2015.
154. C. Lazzeroni et al. Precision measurement of the ratio of the charged kaon leptonic decay rates. *Phys. Lett.*, B719:326–336, 2013.
155. D. M. Asner et al. Physics at BES-III. *Int. J. Mod. Phys.*, A24:S1–794, 2009.
156. M. Ablikim et al. Precision measurements of $B[\psi(3686) \to \pi^+\pi^- J/\psi]$ and $B[J/\psi \to l^+l^-]$. *Phys. Rev.*, D88(3):032007, 2013.
157. Z. Li et al. Measurement of the branching fractions for $J/\psi \to l^+l^-$. *Phys. Rev.*, D71:111103, 2005.
158. V. V. Anashin et al. Measurement of $\Gamma_{ee}(J/\psi)$ with KEDR detector. *JHEP*, 05:119, 2018.
159. R. Aaij et al. Test of lepton universality in beauty-quark decays. *Nature Phys.*, 18(3):277–282, 2022.
160. R. Aaij et al. Test of lepton universality with $B^0 \to K^{*0}\ell^+\ell^-$ decays. *JHEP*, 08:055, 2017.
161. Y. S. Amhis et al. Averages of b-hadron, c-hadron, and $\tau$-lepton properties as of 2018. *Eur. Phys. J.*, C81:226, 2021. updated results and plots available at https://hflav.web.cern.ch/.
162. Measurement of the ratios of branching fractions $\mathcal{R}(D^*)$ and $\mathcal{R}(D^0)$. *Phys. Rev. Lett.*, 131:111802, 2023.
163. Y. S. Amhis et al. Averages of b-hadron, c-hadron, and $\tau$-lepton properties as of 2021. *Phys. Rev. D*, 107(5):052008, 2023.
164. J. W. F. Valle and J. C. Romao. *Neutrinos in High Energy and Astroparticle Physics*. Physics textbook. Wiley-VCH, Weinheim, 2015.

165. A. Tumasyan et al. Precision measurement of the Z boson invisible width in pp collisions at s = 13 TeV. *Phys. Lett. B*, 842:137563, 2023.
166. L. D. Landau. On the conservation laws for weak interactions. *Nucl. Phys.*, 3:127–131, 1957.
167. T. D. Lee and C. N. Yang. Parity nonconservation and a two component theory of the neutrino. *Phys. Rev.*, 105:1671–1675, 1957. [,245(1957)].
168. A. Salam. On parity conservation and neutrino mass. *Nuovo Cim.*, 5:299–301, 1957.
169. K. A. Olive et al. Review of particle physics. *Chin. Phys.*, C38:090001, 2014.
170. A. S. Goldhaber and M. Goldhaber. The neutrinos elusive helicity reversal. *Phys. Today*, 64:40–43, 2011.
171. T. Takagi. *Japan J. Math.*, 1:83, 1925.
172. M. Tanabashi et al. Review of particle physics. *Phys. Rev.*, D98(3):030001, 2018.
173. P. Minkowski. $\mu \to e\gamma$ at a rate of one out of $10^9$ muon decays? *Phys. Lett.*, B67:421–428, 1977.
174. M. Gell-Mann, P. Ramond, and R. Slansky. *Supergravity*. F. van Nieuwenhuizen and D. Freedman, North Holland, Amsterdam, 1979.
175. T. Yanagida. *Proc. of the Workshop on Unified Theory and the Baryon Number of the Universe*. KEK, Japan, 1979.
176. R. N. Mohapatra and G. Senjanovic. Neutrino mass and spontaneous parity violation. *Phys. Rev. Lett.*, 44:912, 1980.
177. J. Schechter and J. W. F. Valle. Neutrino masses in SU(2) x U(1) theories. *Phys. Rev.*, D22:2227, 1980.
178. S. Glashow. *Proc. of 1979 Cargese Workshop Quarks and Leptons, Eds M. Levy et al.* Plenum, New York, 1980.
179. J. Schechter and J. W. F. Valle. Neutrino decay and spontaneous violation of lepton number. *Phys. Rev.*, D25:774, 1982.
180. Y. Cai, J. Herrero-García, M. A. Schmidt, A. Vicente, and R. R. Volkas. From the trees to the forest: a review of radiative neutrino mass models. *Front. in Phys.*, 5:63, 2017.
181. P. Ramond. The family group in grand unified theories. In *International Symposium on Fundamentals of Quantum Theory and Quantum Field Theory*, pages 265–280, 2 1979.
182. M. Gell-Mann, P. Ramond, and R. Slansky. Complex spinors and unified theories. *Conf. Proc. C*, 790927:315–321, 1979.
183. S. Weinberg. Baryon and lepton nonconserving processes. *Phys. Rev. Lett.*, 43:1566–1570, 1979.
184. M. Magg and C. Wetterich. Neutrino mass problem and gauge hierarchy. *Phys. Lett.*, 94B:61–64, 1980.
185. G. Lazarides, Q. Shafi, and C. Wetterich. Proton lifetime and fermion masses in an SO(10) model. *Nucl. Phys.*, B181:287–300, 1981.
186. R. N. Mohapatra and G. Senjanovic. Neutrino masses and mixings in gauge models with spontaneous parity violation. *Phys. Rev.*, D23:165, 1981.
187. E. Ma and U. Sarkar. Neutrino masses and leptogenesis with heavy Higgs triplets. *Phys. Rev. Lett.*, 80:5716–5719, 1998.
188. A. Arhrib, R. Benbrik, M. Chabab, G. Moultaka, M. C. Peyranere, L. Rahili, and J. Ramadan. The Higgs potential in the type II seesaw model. *Phys. Rev. D*, 84:095005, 2011.
189. A. Melfo, M. Nemevsek, F. Nesti, G. Senjanovic, and Y. Zhang. Type II seesaw at LHC: the roadmap. *Phys. Rev. D*, 85:055018, 2012.
190. T. P. Cheng and L. F. Li. Neutrino masses, mixings and oscillations in SU(2) x U(1) models of electroweak interactions. *Phys. Rev.*, D22:2860, 1980.
191. S. M. Bilenky, J. Hosek, and S. T. Petcov. On oscillations of neutrinos with Dirac and Majorana masses. *Phys. Lett.*, 94B:495–498, 1980.
192. I. Yu. Kobzarev, B. V. Martemyanov, L. B. Okun, and M. G. Shchepkin. The phenomenology of neutrino oscillations. *Sov. J. Nucl. Phys.*, 32:823, 1980. [Yad. Fiz. 32,1590(1980)].
193. S. Ashanujjaman and K. Ghosh. Revisiting type-II see-saw: present limits and future prospects at LHC. *JHEP*, 03:195, 2022.

194. R. Foot, H. Lew, X. G. He, and G. C. Joshi. Seesaw neutrino masses induced by a triplet of leptons. *Z. Phys.*, C44:441, 1989.
195. P. B. Denton, M. Friend, M. D. Messier, H. A. Tanaka, S. Böser, J. A. B. Coelho, M. Perrin-Terrin, and T. Stuttard. Snowmass neutrino frontier: NF01 topical group report on three-flavor neutrino oscillations. 12 2022.
196. C. Giunti. Theory of neutrino oscillations. In *Proceedings, 15th Conference on High Energy Physics (IFAE 2003): Lecce, Italy, April 23-26, 2003*, 2003.
197. L. B. Okun, M. V. Rotaev, M. G. Schepkin, and I. S. Tsukerman. Plane waves and wave packets in particle oscillations. 2003.
198. C. Giunti. Reply to "A Remark on the 'Theory of neutrino oscillations". 2003.
199. C. Jarlskog. Commutator of the quark mass matrices in the standard electroweak model and a measure of maximal CP violation. *Phys. Rev. Lett.*, 55:1039, 1985.
200. K. Abe et al. Constraint on the matter–antimatter symmetry-violating phase in neutrino oscillations. *Nature*, 580(7803):339–344, 2020.
201. K. Abe et al. Improved constraints on neutrino mixing from the T2K experiment with **3.13 × $10^{21}$** protons on target. *Phys. Rev. D*, 103(11):112008, 2021.
202. M. A. Acero et al. Improved measurement of neutrino oscillation parameters by the NOvA experiment. *Phys. Rev. D*, 106(3):032004, 2022.
203. R. L. Workman et al. Review of particle physics. *PTEP*, 2022:083C01, 2022.
204. M. C. Gonzalez-Garcia and M. Yokoyama. Neutrino masses, mixing, and oscillations. In *Review of Particle Physics*, 2022.
205. A. Y. Smirnov. The Mikheyev-Smirnov-Wolfenstein (MSW) effect. In *International Conference on History of the Neutrino: 1930-2018*, 1 2019.
206. S. P. Mikheyev and A. Yu. Smirnov. Resonance amplification of oscillations in matter and spectroscopy of solar neutrinos. *Sov. J. Nucl. Phys.*, 42:913–917, 1985.
207. S. P. Mikheyev and A. Yu. Smirnov. Resonant neutrino oscillations in matter. *Prog. Part. Nucl. Phys.*, 23:41–136, 1989.
208. L. B. Okun. *Leptons and quarks*. North-Holland, Amsterdam, Netherlands, 1982.
209. S. J. Parke. Nonadiabatic level crossing in resonant neutrino oscillations. *Phys. Rev. Lett.*, 57:1275–1278, 1986.
210. J. N. Bahcall, N. A. Bahcall, and G. Shaviv. Present status of the theoretical predictions for the Cl-36 solar neutrino experiment. *Phys. Rev. Lett.*, 20:1209–1212, 1968. [,45(1968)].
211. W. C. Haxton, R. G. Hamish Robertson, and A. M. Serenelli. Solar neutrinos: status and prospects. *Ann. Rev. Astron. Astrophys.*, 51:21–61, 2013.
212. G. Bellini et al. First evidence of pep solar neutrinos by direct detection in Borexino. *Phys. Rev. Lett.*, 108:051302, 2012.
213. M. Agostini et al. Experimental evidence of neutrinos produced in the CNO fusion cycle in the Sun. *Nature*, 587:577–582, 2020.
214. R. Davis. Solar neutrinos. II: Experimental. *Phys. Rev. Lett.*, 12:303–305, 1964.
215. W. Hampel et al. GALLEX solar neutrino observations: Results for GALLEX IV. *Phys. Lett. B*, 447:127–133, 1999.
216. J. N. Abdurashitov et al. Results from SAGE. *Phys. Lett. B*, 328:234–248, 1994.
217. K. S. Hirata et al. Observation of B-8 solar neutrinos in the Kamiokande-II detector. *Phys. Rev. Lett.*, 63:16, 1989.
218. Y. Fukuda et al. The super-kamiokande detector. *RNuclear Instruments and Methods in Physics Research A*, 2003.
219. G. Aardsma et al. A heavy water detector to resolve the solar neutrino problem. *Phys. Lett. B*, 194:321–325, 1987.
220. A. Bellerive, J. R. Klein, A. B. McDonald, A. J. Noble, and A. W. P. Poon. The Sudbury neutrino observatory. *Nucl. Phys. B*, 908:30–51, 2016.
221. A. B. McDonald. Nobel lecture: The Sudbury neutrino observatory: observation of flavor change for solar neutrinos. *Rev. Mod. Phys.*, 88(3):030502, 2016.
222. T. Kajita. Nobel lecture: Discovery of atmospheric neutrino oscillations. *Rev. Mod. Phys.*, 88(3):030501, 2016.

223. R. Davis. A review of the Homestake solar neutrino experiment. *Prog. Part. Nucl. Phys.*, 32:13–32, 1994.
224. B. Pontecorvo. *Chalk River Report PD*, 205, 1946.
225. P. A. Cerenkov. Visible radiation produced by electrons moving in a medium with velocities exceeding that of light. *Phys. Rev.*, 52:378–379, 1937.
226. I. M. Frank and I. E. Tamm. Coherent visible radiation of fast electrons passing through matter. *Compt. Rend. Acad. Sci. URSS*, 14(3):109–114, 1937.
227. W. Hampel et al. Final results of the Cr-51 neutrino source experiments in GALLEX. *Phys. Lett. B*, 420:114–126, 1998.
228. J. N. Abdurashitov et al. Measurement of the solar neutrino capture rate with gallium metal. *Phys. Rev. C*, 60:055801, 1999.
229. P. Anselmann et al. GALLEX results from the first 30 solar neutrino runs. *Phys. Lett. B*, 327:377–385, 1994.
230. N. Jelley, A. B. McDonald, and R. G. Hamish Robertson. The Sudbury neutrino observatory. *Ann. Rev. Nucl. Part. Sci.*, 59:431–465, 2009.
231. S. N. Ahmed et al. Measurement of the total active B-8 solar neutrino flux at the Sudbury Neutrino Observatory with enhanced neutral current sensitivity. *Phys. Rev. Lett.*, 92:181301, 2004.
232. B. Aharmim et al. An independent measurement of the total active B-8 solar neutrino flux using an array of He-3 proportional counters at the Sudbury Neutrino Observatory. *Phys. Rev. Lett.*, 101:111301, 2008.
233. A. Gando et al. Limit on neutrinoless $\beta\beta$ decay of Xe-136 from the first phase of KamLAND-Zen and comparison with the positive claim in Ge-76. *Phys. Rev. Lett.*, 110(6):062502, 2013.
234. M. Agostini et al. Comprehensive measurement of $pp$-chain solar neutrinos. *Nature*, 562(7728):505–510, 2018.
235. M. Agostini et al. First simultaneous precision spectroscopy of $pp$, $^7$Be, and $pep$ solar neutrinos with Borexino phase-II. *Phys. Rev. D*, 100(8):082004, 2019.
236. M. Agostini et al. Experimental evidence of neutrinos produced in the CNO fusion cycle in the Sun. *Nature*, 587:577–582, 2020.
237. J. Guillochon, J. Parrent, L. Z. Kelley, and R. Margutti. An open catalog for supernova data. *Astrophys. J.*, 835(1):64, 2017.
238. K. Hirata, T. Kajita, M. Koshiba, M. Nakahata, Y. Oyama, N. Sato, A. Suzuki, M. Takita, Y. Totsuka, T. Kifune, T. Suda, K. Takahashi, T. Tanimori, K. Miyano, M. Yamada, E. W. Beier, L. R. Feldscher, S. B. Kim, A. K. Mann, F. M. Newcomer, R. Van, W. Zhang, and B. G. Cortez. Observation of a neutrino burst from the supernova sn1987a. *Phys. Rev. Lett.*, 58:1490–1493, Apr 1987.
239. R. M. Bionta, G. Blewitt, C. B. Bratton, D. Casper, A. Ciocio, R. Claus, B. Cortez, M. Crouch, S. T. Dye, S. Errede, G. W. Foster, W. Gajewski, K. S. Ganezer, M. Goldhaber, T. J. Haines, T. W. Jones, D. Kielczewska, W. R. Kropp, J. G. Learned, J. M. LoSecco, J. Matthews, R. Miller, M. S. Mudan, H. S. Park, L. R. Price, F. Reines, J. Schultz, S. Seidel, E. Shumard, D. Sinclair, H. W. Sobel, J. L. Stone, L. R. Sulak, R. Svoboda, G. Thornton, J. C. van der Velde, and C. Wuest. Observation of a neutrino burst in coincidence with supernova 1987a in the large magellanic cloud. *Phys. Rev. Lett.*, 58:1494–1496, Apr 1987.
240. E. N. Alekseev, L. N. Alekseeva, I. V. Krivosheina, and V. I. Volchenko. Detection of the neutrino signal from SN1987A in the LMC using the Inr Baksan underground scintillation telescope. *Phys. Lett. B*, 205:209–214, 1988.
241. W. Hillebrandt and J. C. Niemeyer. Type Ia supernova explosion models. *Ann. Rev. Astron. Astrophys.*, 38:191–230, 2000.
242. D. A. Howell et al. The effect of progenitor age and metallicity on luminosity and 56Ni yield in Type Ia supernovae. *Astrophys. J.*, 691:661–671, 2009.
243. H. A. Bethe and J. R. Wilson. Revival of a stalled supernova shock by neutrino heating. *Astrophys. J.*, 295:14–23, 1985.
244. K. Rozwadowska, F. Vissani, and E. Cappellaro. On the rate of core collapse supernovae in the milky way. *New Astron.*, 83:101498, 2021.

245. J. F. Beacom. The diffuse supernova neutrino background. *Ann. Rev. Nucl. Part. Sci.*, 60:439–462, 2010.
246. M. Aglietta et al. On the event observed in the Mont Blanc Underground Neutrino observatory during the occurrence of Supernova 1987a. *EPL*, 3:1315–1320, 1987.
247. A. D. Dolgov. Neutrinos in cosmology. *Phys. Rept.*, 370:333–535, 2002.
248. K. Hirata et al. Observation of a neutrino burst from the supernova SN 1987a. *Phys. Rev. Lett.*, 58:1490–1493, 1987.
249. V. Albanese et al. The SNO+ experiment. *JINST*, 16(08):P08059, 2021.
250. A. Coleman et al. Ultra high energy cosmic rays The intersection of the Cosmic and Energy Frontiers. *Astropart. Phys.*, 149:102819, 2023.
251. R. Abbasi et al. The IceCube high-energy starting event sample: Description and flux characterization with 7.5 years of data. *Phys. Rev. D*, 104:022002, 2021.
252. Y. Oyama et al. Experimental study of upward going muons in Kamiokande. *Phys. Rev. D*, 39:1481, 1989.
253. C. B. Bratton et al. Irvine-Michigan-Brookhaven (I. M. B.) nucleon decay search status report. In *2nd Workshop on Grand Unification*, 1981.
254. J. M. LoSecco. The history of "anomalous" atmospheric neutrino events: a first person account. *Phys. Perspect.*, 18(2):209–241, 2016.
255. Y. Fukuda et al. Evidence for oscillation of atmospheric neutrinos. *Phys. Rev. Lett.*, 81:1562–1567, 1998.
256. Z. Li et al. Measurement of the tau neutrino cross section in atmospheric neutrino oscillations with Super-Kamiokande. *Phys. Rev. D*, 98(5):052006, 2018.
257. M. Apollonio et al. Limits on neutrino oscillations from the CHOOZ experiment. *Phys. Lett. B*, 466:415–430, 1999.
258. F. Boehm et al. Final results from the Palo Verde neutrino oscillation experiment. *Phys. Rev. D*, 64:112001, 2001.
259. W. Winter. Atmospheric neutrino oscillations for Earth tomography. *Nucl. Phys. B*, 908:250–267, 2016.
260. R. Gandhi, C. Quigg, M. H. Reno, and I. Sarcevic. Ultrahigh-energy neutrino interactions. *Astropart. Phys.*, 5:81–110, 1996.
261. A. Donini, S. Palomares-Ruiz, and J. Salvado. Neutrino tomography of Earth. *Nature Phys.*, 15(1):37–40, 2019.
262. L. Maderer, E. Kaminski, J. A. B. Coelho, S. Bourret, and V. Van Elewyck. Unveiling the outer core composition with neutrino oscillation tomography. *Front. Earth Sci.*, 11:1008396, 2023.
263. G. Gamow. Expanding universe and the origin of elements. *Phys. Rev.*, 70:572–573, 1946.
264. R. A. Alpher, H. Bethe, and G. Gamow. The origin of chemical elements. *Phys. Rev.*, 73:803–804, 1948.
265. R. A. Alpher and R. C. Herman. On the relative abundance of the elements. *Phys. Rev.*, 74(12):1737–1742, 1948.
266. R. A. Alpher and R. Herman. Evolution of the Universe. *Nature*, 162(4124):774–775, 1948.
267. A. A. Penzias and R. W. Wilson. A Measurement of excess antenna temperature at 4080-Mc/s. *Astrophys. J.*, 142:419–421, 1965.
268. E. Baracchini et al. PTOLEMY: a proposal for thermal relic detection of massive neutrinos and directional detection of MeV dark matter. 8 2018.
269. Y. Cheipesh, V. Cheianov, and A. Boyarsky. Navigating the pitfalls of relic neutrino detection. *Phys. Rev. D*, 104(11):116004, 2021.
270. A. G. Cocco, G. Mangano, and M. Messina. Probing low energy neutrino backgrounds with neutrino capture on beta decaying nuclei. *JCAP*, 06:015, 2007.
271. E. Roulet and F. Vissani. *Neutrinos in Physics and Astrophysics*. World Scientific, 10 2022.
272. A. Albert et al. All-flavor search for a diffuse flux of cosmic neutrinos with nine years of ANTARES data. *Astrophys. J. Lett.*, 853(1):L7, 2018.
273. R. Abbasi et al. Evidence for neutrino emission from the nearby active galaxy NGC 1068. *Science*, 378(6619):538–543, 2022.

274. M. G. Aartsen et al. Neutrino emission from the direction of the blazar TXS 0506+056 prior to the IceCube-170922A alert. *Science*, 361(6398):147–151, 2018.
275. T. Fujii. Rapporteur talk: CRI. *PoS*, ICRC2023:031, 2024.
276. K. Greisen. End to the cosmic ray spectrum? *Phys. Rev. Lett.*, 16:748–750, 1966.
277. G. T. Zatsepin and V. A. Kuzmin. Upper limit of the spectrum of cosmic rays. *JETP Lett.*, 4:78–80, 1966.
278. I. Zheleznykh. The Soviet DUMAND program (1980-1991) and development of alternative large-scale neutrino telescopes. In *International Conference on History of the Neutrino: 1930-2018*, 2019.
279. V. A. Balkanov et al. The BAIKAL neutrino project: Status report. *Nucl. Phys. B Proc. Suppl.*, 118:363–370, 2003.
280. V. Ayutdinov et al. Results from the BAIKAL neutrino telescope. In *28th International Cosmic Ray Conference*, 5 2003.
281. M. Ackermann et al. New results from the AMANDA neutrino telescope. *Nucl. Phys. B Proc. Suppl.*, 145:319–322, 2005.
282. P. K. F. Grieder. The NESTOR neutrino telescope. *Nuovo Cim. C*, 24:771–776, 2001.
283. S. Adrián-Martínez et al. Long term monitoring of the optical background in the Capo Passero deep-sea site with the NEMO tower prototype. *Eur. Phys. J. C*, 76(2):68, 2016.
284. A. Margiotta. The ANTARES neutrino telescope. *PoS*, EPS-HEP2021:093, 2022.
285. A. Margiotta. The KM3NeT infrastructure: status and first results. In *21st International Symposium on Very High Energy Cosmic Ray Interactions*, 8 2022.
286. A. Sinopoulou, R. Coniglione, C. Markou, R. Muller, and E. Tzamariudaki. Atmospheric neutrinos with the first detection units of KM3NeT-ARCA. *PoS*, ICRC2021:1134, 2021.
287. D. Zaborov et al. Recent results from the Baikal-GVD neutrino telescope. *PoS*, ICHEP2022:083, 11 2022.
288. F. Halzen and A. Kheirandish. *IceCube and High-Energy Cosmic Neutrinos*, chapter Neutrino physics and astrophysics. World Scientific, 2 2022.
289. S. L. Glashow. Resonant scattering of antineutrinos. *Phys. Rev.*, 118:316–317, 1960.
290. M. G. Aartsen et al. Detection of a particle shower at the Glashow resonance with IceCube. *Nature*, 591(7849):220–224, 2021. [Erratum: Nature 592, E11 (2021)].
291. M. Bustamante and A. Connolly. Extracting the energy-dependent neutrino-nucleon cross section above 10 TeV using IceCube showers. *Phys. Rev. Lett.*, 122(4):041101, 2019.
292. M. G. Aartsen et al. Measurement of atmospheric neutrino oscillations at 6–56 GeV with IceCube DeepCore. *Phys. Rev. Lett.*, 120(7):071801, 2018.
293. G. Fiorentini, M. Lissia, and F. Mantovani. Geo-neutrinos and Earth's interior. *Phys. Rept.*, 453:117–172, 2007.
294. S. Abe et al. Abundances of uranium and thorium elements in Earth estimated by geoneutrino spectroscopy. 5 2022.
295. G. Eder. Terrestrial neutrinos. *Nucl. Phys*, 78:657–662, 1966.
296. G. Marx. Geophysics by neutrinos. *Czechoslovak Journal of Physics B*, 19:1471–1479, 1969.
297. G. Ricciardi, N. Vignaroli, and F. Vissani. An accurate evaluation of electron (anti-)neutrino scattering on nucleons. *JHEP*, 08:212, 2022.
298. S. Kumaran, L. Ludhova, Ö. Penek, G. Settanta. Borexino results on neutrinos from the Sun and Earth. *Universe*, 7(7):231, 2021.
299. M. Agostini et al. Comprehensive geoneutrino analysis with Borexino. *Phys. Rev. D*, 101(1):012009, 2020.
300. Th. A. Mueller et al. Improved predictions of reactor antineutrino spectra. *Phys. Rev. C*, 83:054615, 2011.
301. O. Hahn and F. Strassmann. Über den Nachweis und das Verhalten der bei der Bestrahlung des Urans mittels Neutronen entstehenden Erdalkalimetalle. *Naturwiss*, 27:11, 1939.
302. O. Hahn and F. Strassmann. Nachweis der Entstehung activer Bariumisotope aus Uran und Thorium durch Neutronenbestrahlung; Nachweis weiterer aktiver Bruchtucke bei der Uranspaltung. *Naturwiss*, 27:89, 1939.

303. L. Meitner and O. Frisch. Disintegration of uranium by neutrons: a new type of nuclear reaction. *Nature*, 143:239, 1939.
304. D. Adey et al. Improved measurement of the reactor antineutrino flux at Daya Bay. *Phys. Rev. D*, 100(5):052004, 2019.
305. F. P. An et al. Improved measurement of the evolution of the reactor antineutrino flux and spectrum at Daya Bay. *Phys. Rev. Lett.*, 130(21):211801, 2023.
306. H. de Kerret et al. Double Chooz $\theta_{13}$ measurement via total neutron capture detection. *Nature Phys.*, 16(5):558–564, 2020.
307. M. Estienne et al. Updated summation model: an improved agreement with the Daya Bay antineutrino fluxes. *Phys. Rev. Lett.*, 123(2): 022502 (2019).
308. C. Zhang, X. Qian, and M. Fallot. Reactor antineutrino flux and anomaly. *Prog. Part. Nucl. Phys.*, 136: 104106 (2024).
309. F. P. An et al. Measurement of the reactor antineutrino flux and spectrum at Daya Bay. *Phys. Rev. Lett.*, 116(6):061801, 2016. [Erratum: Phys. Rev. Lett. 118, 099902 (2017)].
310. A. G. Beda, V. B. Brudanin, V. G. Egorov, D. V. Medvedev, V. S. Pogosov, E. A. Shevchik, M. V. Shirchenko, A. S. Starostin, and I. V. Zhitnikov. Gemma experiment: The results of neutrino magnetic moment search. *Phys. Part. Nucl. Lett.*, 10:139–143, 2013.
311. F. Reines and C. L. Cowan. Detection of the free neutrino. *Phys. Rev.*, 92:830–831, 1953.
312. C. L. Cowan. Anatomy of an experiment: an account of the discovery of the neutrino. *Smithsonian Institution Annual Report*, 409:830–831, 1964.
313. F. Boehm, J. F. Cavaignac, F. v. Feilitzsch, A. A. Hahn, H. E. Henrikson, D. H. Koang, H. Kwon, R. L. Mössbauer, B. Vignon, and J. L. Vuilleumier. Experimental study of neutrino oscillations at a fission reactor. *Phys. Lett. B*, 97:310–314, 1980.
314. H. Kwon, F. Boehm, A. A. Hahn, H. E. Henrikson, J. L. Vuilleumier, J. F. Cavaignac, D. H. Koang, B. Vignon, F. Von Feilitzsch, and R. L. Mossbauer. Search for neutrino oscillations at a fission reactor. *Phys. Rev. D*, 24:1097–1111, 1981.
315. P. Huber. On the determination of anti-neutrino spectra from nuclear reactors. *Phys. Rev. C*, 84:024617, 2011. [Erratum: Phys. Rev. C 85, 029901 (2012)].
316. I. Alekseev et al. Search for sterile neutrinos at the DANSS experiment. *Phys. Lett. B*, 787:56–63, 2018.
317. Z. Atif et al. Search for sterile neutrino oscillations using RENO and NEOS data. *Phys. Rev. D*, 105(11):L111101, 2022.
318. M. Andriamirado et al. Improved short-baseline neutrino oscillation search and energy spectrum measurement with the PROSPECT experiment at HFIR. *Phys. Rev. D*, 103(3):032001, 2021.
319. H. Almazán et al. STEREO neutrino spectrum of $^{235}$U fission rejects sterile neutrino hypothesis. *Nature*, 613(7943):257–261, 2023.
320. A. P. Serebrov et al. Search for sterile neutrinos with the Neutrino-4 experiment and measurement results. *Phys. Rev. D*, 104(3):032003, 2021.
321. M. Apollonio et al. Search for neutrino oscillations on a long baseline at the CHOOZ nuclear power station. *Eur. Phys. J. C*, 27:331–374, 2003.
322. P.F. Harrison, D.H. Perkins, and W.G. Scott. Tri-bimaximal mixing and the neutrino oscillation data. *Phys. Lett.*, B530:167, 2002.
323. G. Altarelli and F. Feruglio. Discrete flavor symmetries and models of neutrino mixing. *Rev. Mod. Phys.*, 82:2701–2729, 2010.
324. Y. Abe et al. Indication of reactor $\bar{\nu}_e$ disappearance in the double Chooz experiment. *Phys. Rev. Lett.*, 108:131801, 2012.
325. J. K. Ahn et al. Observation of reactor electron antineutrino disappearance in the RENO experiment. *Phys. Rev. Lett.*, 108:191802, 2012.
326. R. Leitner. Recent results of Daya Bay reactor neutrino experiment. *Nucl. Part. Phys. Proc.*, 285–286:32–37, 2017.
327. Y. Abe et al. Indication for the disappearance of reactor electron antineutrinos in the Double Chooz experiment. *Phys. Rev. Lett.*, 108:131801, 2012.

328. C. D. Shin et al. Observation of reactor antineutrino disappearance using delayed neutron capture on hydrogen at RENO. *JHEP*, 04:029, 2020.
329. F. P. An et al. Precision measurement of reactor antineutrino oscillation at kilometer-scale baselines by Daya Bay. *Phys. Rev. Lett.*, 130(16):161802, 2023.
330. A. Abusleme et al. JUNO physics and detector. *Prog. Part. Nucl. Phys.*, 123:103927, 2022.
331. F. An et al. Neutrino physics with JUNO. *J. Phys. G*, 43(3):030401, 2016.
332. B. Pontecorvo. Electron and muon neutrinos. *Zh. Eksp. Teor. Fiz.*, 37:1751–1757, 1959.
333. B. Pontecorvo. Electron and muon neutrinos. *Sov. Phys. JETP*, 10:1236–1240, 1960.
334. M. Schwartz. Feasibility of using high-energy neutrinos to study the weak interactions. *Phys. Rev. Lett.*, 4:306–307, 1960.
335. S. Fukui and S. Miyamoto. A new type of particle detector: the discharge chamber. *Nuovo Cim.*, 11:113, 1959.
336. S. van der Meer. A directive device for charged particles and its use in an enhanced neutrino beam. 2 1961.
337. J. K. Bienlein et al. Spark chamber study of high-energy neutrino interactions. *Phys. Lett.*, 13:80–86, 1964.
338. G. Bernardini et al. Search for intermediate boson production in high-energy neutrino interactions. *Phys. Lett.*, 13:86–91, 1964.
339. D. Beavis et al. Long baseline neutrino oscillation experiment at the AGS approved by the HENPAC as AGS experiment 889. 4 1995.
340. A. Aguilar-Arevalo et al. Evidence for neutrino oscillations from the observation of antineutrino(electron) appearance in a anti-neutrino(muon) beam. *Phys. Rev.*, D64:112007, 2001.
341. K. Eitel. The KARMEN search for appearance of anti-nu/e's. *Prog. Part. Nucl. Phys.*, 48:89–98, 2002.
342. A. E. Asratian et al. Search for prompt neutrinos in 70-GeV $pN$ collisions. *Phys. Lett. B*, 79:497, 1978.
343. H. Abramowicz et al. Prompt neutrino production in a proton beam dump experiment. *Z. Phys. C*, 13:179, 1982.
344. P. Fritze et al. Further study of the prompt neutrino flux from 400-GeV proton - nucleus collisions using BEBC. *Phys. Lett. B*, 96:427–434, 1980.
345. M. Jonker et al. Experimental study of X distributions in semileptonic neutral current neutrino reactions. *Phys. Lett. B*, 128:117, 1983.
346. D. Neuffer. Design of muon storage rings for neutrino oscillations experiments. *IEEE Trans. Nucl. Sci.*, 28:2034–2036, 1981.
347. P. Zucchelli. A novel concept for a anti-nu/e / nu/e neutrino factory: The beta beam. *Phys. Lett. B*, 532:166–172, 2002.
348. K. S. Hirata et al. Results from one thousand days of real time directional solar neutrino data. *Phys. Rev. Lett.*, 65:1297–1300, 1990.
349. M. H. Ahn et al. Measurement of neutrino oscillation by the K2K experiment. *Phys. Rev.*, D74:072003, 2006.
350. T. Nakaya and K. Nishikawa. Long baseline neutrino oscillation experiments with accelerators in Japan: From K2K to T2K. *Eur. Phys. J. C*, 80(4):344, 2020.
351. M. H. Ahn et al. Measurement of neutrino oscillation by the K2K experiment. *Phys. Rev. D*, 74:072003, 2006.
352. S. Yamamoto et al. An improved search for nu(mu) —> nu(e) oscillation in a long-baseline accelerator experiment. *Phys. Rev. Lett.*, 96:181801, 2006.
353. P. Adamson et al. Measurement of neutrino oscillations with the MINOS detectors in the NuMI beam. *Phys. Rev. Lett.*, 101:131802, 2008.
354. L. H. Whitehead. Neutrino oscillations with MINOS and MINOS+. *Nucl. Phys. B*, 908:130–150, 2016.
355. D. G. Michael et al. Observation of muon neutrino disappearance with the MINOS detectors and the NuMI neutrino beam. *Phys. Rev. Lett.*, 97:191801, 2006.
356. P. Adamson et al. Measurement of neutrino oscillations with the MINOS detectors in the NuMI beam. *Phys. Rev. Lett.*, 101:131802, 2008.

357. P. Adamson et al. Measurement of the neutrino mass splitting and flavor mixing by MINOS. *Phys. Rev. Lett.*, 106:181801, 2011.
358. P. Adamson et al. An improved measurement of muon antineutrino disappearance in MINOS. *Phys. Rev. Lett.*, 108:191801, 2012.
359. P. Adamson et al. Measurement of neutrino and antineutrino oscillations using beam and atmospheric data in MINOS. *Phys. Rev. Lett.*, 110(25):251801, 2013.
360. K. Abe et al. Indication of electron neutrino appearance from an accelerator-produced off-axis muon neutrino beam. *Phys. Rev. Lett.*, 107:041801, 2011.
361. P. Adamson et al. Precision constraints for three-flavor neutrino oscillations from the full MINOS+ and MINOS dataset. *Phys. Rev. Lett.*, 125(13):131802, 2020.
362. E. Gschwendtner, K. Cornelis, I. Efthymiopoulos, A. Pardons, H. Vincke, J. Wenninger, and I. Krätschmer. CNGS, CERN neutrinos to Gran Sasso, five years of running a 500 kilowatt neutrino beam facility at CERN. In *4th International Particle Accelerator Conference*, page MOPEA058, 2013.
363. C. Farnese. The ICARUS experiment. *Universe*, 5(2):49, 2019.
364. N. Agafonova et al. New results on $\nu_\mu \to \nu_\tau$ appearance with the OPERA experiment in the CNGS beam. *JHEP*, 1311:036, 2013.
365. T. Adam et al. Measurement of the neutrino velocity with the OPERA detector in the CNGS beam. *JHEP*, 10:093, 2012.
366. M. Dracos. Measurement of the neutrino velocity in OPERA experiment. *Nucl. Phys. B Proc. Suppl.*, 235–236:283–288, 2013.
367. M. Antonello et al. Measurement of the neutrino velocity with the ICARUS detector at the CNGS beam. *Phys. Lett. B*, 713:17–22, 2012.
368. N. Agafonova et al. Final results of the search for $\nu_\mu \to \nu_e$ oscillations with the OPERA detector in the CNGS beam. *JHEP*, 06:151, 2018.
369. N. Agafonova et al. Observation of a first $\nu_\tau$ candidate in the OPERA experiment in the CNGS beam. *Phys. Lett.*, B691:138–145, 2010.
370. M. Tenti. Final results from the OPERA experiment in the CNGS neutrino beam. *PoS*, NuFACT2018:062, 2019.
371. N. Agafonova et al. Final results of the OPERA experiment on $\nu_\tau$ appearance in the CNGS neutrino beam. *Phys. Rev. Lett.*, 120(21):211801, 2018. [Erratum: Phys. Rev. Lett. 121, 139901 (2018)].
372. K. Abe et al. Indication of electron neutrino appearance from an accelerator-produced off-axis muon neutrino beam. *Phys. Rev. Lett.*, 107:041801, 2011.
373. D. S. Ayres et al. NOvA: proposal to build a 30 kiloton off-axis detector to study $\nu_\mu \to \nu_e$ oscillations in the NuMI beamline. 2004.
374. K. Abe et al. Observation of electron neutrino appearance in a muon neutrino beam. *Phys. Rev. Lett.*, 112:061802, 2014.
375. K. Abe et al. Search for CP violation in neutrino and antineutrino oscillations by the T2K experiment with $2.2 \times 10^{21}$ protons on target. *Phys. Rev. Lett.*, 121(17):171802, 2018.
376. J. G. Walsh. CP-violation search with T2K data. In *20th Conference on Flavor Physics and CP Violation*, 8 2022.
377. K. Abe et al. T2K measurements of muon neutrino and antineutrino disappearance using $3.13 \times 10^{21}$ protons on target. *Phys. Rev. D*, 103(1):L011101, 2021.
378. M. A. Ramírez et al. Updated T2K measurements of muon neutrino and antineutrino disappearance using $3.6 \times 10^{21}$ protons on target. 5 2023.
379. M. A. Acero et al. New constraints on oscillation parameters from $\nu_e$ appearance and $\nu_\mu$ disappearance in the NOvA experiment. *Phys. Rev. D*, 98:032012, 2018.
380. M. A. Acero et al. First measurement of neutrino oscillation parameters using neutrinos and antineutrinos by NOvA. *Phys. Rev. Lett.*, 123(15):151803, 2019.
381. M. Yokoyama. The hyper-Kamiokande experiment. In *Prospects in Neutrino Physics*, 4 2017.
382. K. Abe et al. Physics potential of a long-baseline neutrino oscillation experiment using a J-PARC neutrino beam and Hyper-Kamiokande. *PTEP*, 2015:053C02, 2015.

383. B. Abi et al. *The DUNE Far Detector Interim Design Report Volume 1: Physics, Technology and Strategies.* 2018.
384. B. Abi et al. Deep Underground Neutrino Experiment (DUNE), far detector technical design Report, Volume I Introduction to DUNE. *JINST*, 15(08):T08008, 2020.
385. E. Eskut et al. Final results on nu(mu) - nu(tau) oscillation from the CHORUS experiment. *Nucl. Phys.*, B793:326–343, 2008.
386. C.T. Kullenberg et al. A search for single photon events in neutrino interactions in NOMAD. *Phys. Lett.*, B706:268–275, 2012.
387. B. Armbruster et al. Upper limits for neutrino oscillations muon-anti-neutrino —> electron-anti-neutrino from muon decay at rest. *Phys. Rev. D*, 65:112001, 2002.
388. A. A. Aguilar-Arevalo et al. A combined $\nu_\mu \to \nu_e$ and $\bar{\nu}_\mu \to \bar{\nu}_e$ oscillation analysis of the MiniBooNE excesses. 7 2012.
389. S. Hatakeyama et al. Measurement of the flux and zenith angle distribution of upward through going muons in Kamiokande II + III. *Phys. Rev. Lett.*, 81:2016–2019, 1998.
390. P. Adamson et al. Improved search for muon-neutrino to electron-neutrino oscillations in MINOS. *Phys. Rev. Lett.*, 107:181802, 2011.
391. F.P. An et al. Observation of electron-antineutrino disappearance at Daya Bay. *Phys. Rev. Lett.*, 108:171803, 2012.
392. J. K. Ahn et al. Observation of reactor electron antineutrino disappearance in the RENO experiment. *Phys. Rev. Lett.*, 108:191802, 2012.
393. F. P. An et al. Improved measurement of electron antineutrino disappearance at Daya Bay. *Chin. Phys.*, C37:011001, 2013.
394. I. Esteban, M. C. Gonzalez-Garcia, M. Maltoni, T. Schwetz, A. Zhou. The fate of hints: updated global analysis of three-flavor neutrino oscillations (with data available in November 2022, NuFIT 5.2, www.nu-fit.org.). *JHEP*, 09:178, 2020.
395. D. Akimov et al. Observation of coherent elastic neutrino-nucleus scattering. *Science*, 357(6356):1123–1126, 2017.
396. F. Halzen and A. D. Martin. *Quarks and Leptons: An Introductory Course in Modern Particle Physics.* John Wiley and Sons, 1984.
397. A. Bettini. *Introduction to Elementary Particle Physics.* Cambridge University Press, 2008.
398. A. Branca, G. Brunetti, A. Longhin, M. Martini, F. Pupilli, and F. Terranova. A new generation of neutrino cross section experiments: challenges and opportunities. *Symmetry*, 13(9):1625, 2021.
399. M. Sajjad Athar and S. K. Singh. *The Physics of Neutrino Interactions.* Cambridge University Press, 5 2020.
400. C. Giunti and C. W. Kim. *Fundamentals of Neutrino Physics and Astrophysics.* Oxford University Press, 2007.
401. S. Weinberg. Charge symmetry of weak interactions. *Phys. Rev.*, 112:1375–1379, 1958.
402. C. H. Llewellyn Smith. Neutrino reactions at accelerator energies. *Phys. Rept.*, 3:261–379, 1972.
403. P. Vogel and J. F. Beacom. Angular distribution of neutron inverse beta decay, anti-neutrino(e) + p —> e+ + n. *Phys. Rev. D*, 60:053003, 1999.
404. A. Strumia and F. Vissani. Precise quasielastic neutrino/nucleon cross-section. *Phys. Lett. B*, 564:42–54, 2003.
405. E. D. Bloom, D. H. Coward, H. DeStaebler, J. Drees, G. Miller, L. W. Mo, R. E. Taylor, M. Breidenbach, J. I. Friedman, G. C. Hartmann, and H. W. Kendall. High-energy inelastic $e - p$ scattering at 6° and 10°. *Phys. Rev. Lett.*, 23:930–934, Oct 1969.
406. J. D. Bjorken. Asymptotic sum rules at infinite momentum. *Phys. Rev.*, 179:1547–1553, 1969.
407. D. H. Coward, H. C. DeStaebler, R. A. Early, J. Litt, A. Minten, et al. Electron - proton elastic scattering at high momentum transfers. *Phys. Rev. Lett.*, 20:292–295, 1968.
408. E. D. Bloom et al. High-energy inelastic e p scattering at 6-degrees and 10-degrees. *Phys. Rev. Lett.*, 23:930–934, 1969.

409. M. Breidenbach, J. I. Friedman, H. W. Kendall, E. D. Bloom, D. H. Coward, H. C. DeStaebler, J. Drees, L. W. Mo, and R. E. Taylor. Observed behavior of highly inelastic electron-proton scattering. *Phys. Rev. Lett.*, 23:935–939, 1969.
410. E. D. Bloom, G. Buschhorn, R. Leslie Cottrell, D. H. Coward, H. C. DeStaebler, et al. *Recent Results in Inelastic Electron Scattering.* 1970.
411. R. P. Feynman. Very high-energy collisions of hadrons. *Phys. Rev. Lett.*, 23:1415–1417, 1969.
412. E. Leader and E. Predazzi. *An Introduction to Gauge Theories and Modern Particle Physics. Vol. 1: Electroweak Interactions, the New Particles and the Parton Model.* Cambridge University Press, 4 2011.
413. C. G. Callan Jr. and D. J. Gross. High-energy electroproduction and the constitution of the electric current. *Phys. Rev. Lett.*, 22:156–159, 1969.
414. J. A. Formaggio and G. P. Zeller. From eV to EeV: neutrino cross sections across energy scales. *Rev. Mod. Phys.*, 84:1307–1341, 2012.
415. R. E. Taylor. Inelastic electron - proton scattering in the deep continuum region. *Conf. Proc. C*, 690914:251–260, 1969.
416. H. Deden et al. Experimental study of structure functions and sum rules in charge changing interactions of neutrinos and anti-neutrinos on nucleons. *Nucl. Phys. B*, 85:269–288, 1975.
417. D. Haidt. The discovery of weak neutral currents. In *International Conference on History of the Neutrino: 1930-2018*, 2019.
418. F. Sciulli. An experimenter's history of neutral currents. *Prog. Part. Nucl. Phys.*, 2:41, 1979.
419. F. J. Hasert et al. Observation of neutrino like interactions without muon or electron in the Gargamelle neutrino experiment. *Phys. Lett. B*, 46:138–140, 1973.
420. F. J. Hasert et al. Observation of neutrino like interactions without muon or electron in the Gargamelle neutrino experiment. *Nucl. Phys. B*, 73:1–22, 1974.
421. B. Aubert et al. Further observation of muonless neutrino induced inelastic interactions. *Phys. Rev. Lett.*, 32:1454–1457, 1974.
422. B. C. Barish. Recent high-energy electron, muon, and neutrino experiments. *AIP Conf. Proc.*, 23:1–21, 1975.
423. J. Delorme and M. Ericson. Exploration of the spin - isospin nuclear response function by neutrinos. *Phys. Lett. B*, 156:263–266, 1985.
424. J. Bernabeu. Neutral currents in semi-leptonic processes delta-t=0. *Lett. Nuovo Cim.*, 10S2:329–332, 1974.
425. J. Bernabeu. Low-energy elastic neutrino-nucleon and nuclear scattering and its relevance for supernovae. *Astron. Astrophys.*, 47:375–379, 1976.
426. D. Z. Freedman. Coherent neutrino nucleus scattering as a probe of the weak neutral current. *Phys. Rev. D*, 9:1389–1392, 1974.
427. D. Z. Freedman, D. N. Schramm, and D. L. Tubbs. The weak neutral current and its effects in Stellar collapse. *Ann. Rev. Nucl. Part. Sci.*, 27:167–207, 1977.
428. M. Abdullah et al. Coherent elastic neutrino-nucleus scattering: Terrestrial and astrophysical applications. 3 2022.
429. M. Fukugita and T. Yanagida. *Physics of Neutrinos and Applications to Astrophysics.* Theoretical and Mathematical Physics. Springer-Verlag, Berlin, Germany, 2003.
430. D. Akimov et al. First measurement of coherent elastic neutrino-nucleus scattering on argon. *Phys. Rev. Lett.*, 126(1):012002, 2021.
431. I. Nasteva. Low-energy reactor neutrino physics with the CONNIE experiment. *J. Phys. Conf. Ser.*, 2156(1):012115, 2021.
432. H. Bonet et al. Constraints on elastic neutrino nucleus scattering in the fully coherent regime from the CONUS experiment. *Phys. Rev. Lett.*, 126(4):041804, 2021.
433. J. Colaresi, J. I. Collar, T. W. Hossbach, A. R. L. Kavner, C. M. Lewis, A. E. Robinson, and K. M. Yocum. First results from a search for coherent elastic neutrino-nucleus scattering at a reactor site. *Phys. Rev. D*, 104(7):072003, 2021.
434. J. Colaresi, J. I. Collar, T. W. Hossbach, C. M. Lewis, and K. M. Yocum. Measurement of coherent elastic neutrino-nucleus scattering from reactor antineutrinos. *Phys. Rev. Lett.*, 129(21):211802, 2022.

435. O. Benhar, A. Fabrocini, S. Fantoni, and I. Sick. Spectral function of finite nuclei and scattering of GeV electrons. *Nucl. Phys. A*, 579:493–517, 1994.
436. H. Nakamura and R. Seki. Quasielastic neutrino nucleus scattering and spectral function. *Nucl. Phys. B Proc. Suppl.*, 112:197–202, 2002.
437. O. Benhar, N. Farina, H. Nakamura, M. Sakuda, and R. Seki. Electron- and neutrino-nucleus scattering in the impulse approximation regime. *Phys. Rev. D*, 72:053005, 2005.
438. O. Benhar and D. Meloni. Total neutrino and antineutrino nuclear cross-sections around 1-GeV. *Nucl. Phys. A*, 789:379–402, 2007.
439. O. Benhar and D. Meloni. Impact of nuclear effects on the determination of the nucleon axial mass. *Phys. Rev. D*, 80:073003, 2009.
440. A. M. Ankowski and J. T. Sobczyk. Argon spectral function and neutrino interactions. *Phys. Rev. C*, 74:054316, 2006.
441. A. M. Ankowski and J. T. Sobczyk. Construction of spectral functions for medium-mass nuclei. *Phys. Rev. C*, 77:044311, 2008.
442. L. Alvarez-Ruso, Y. Hayato, and J. Nieves. Progress and open questions in the physics of neutrino cross sections at intermediate energies. *New J. Phys.*, 16:075015, 2014.
443. J. Nieves, J. E. Amaro, and M. Valverde. Inclusive quasi-elastic neutrino reactions. *Phys. Rev.*, C70:055503, 2004. [Erratum: Phys. Rev. C72, 019902 (2005)].
444. R. A. Smith and E. J. Moniz. Neutrino reactions on nuclear targets. *Nucl. Phys. B*, 43:605, 1972. [Erratum: Nucl. Phys. B 101, 547 (1975)].
445. E. J. Moniz, I. Sick, R. R. Whitney, J. R. Ficenec, R. D. Kephart, and W. P. Trower. Nuclear Fermi momenta from quasielastic electron scattering. *Phys. Rev. Lett.*, 26:445–448, 1971.
446. S. Frullani and J. Mougey. Single particle properties of nuclei through (e, e' p) reactions. *Adv. Nucl. Phys.*, 14:1–283, 1984.
447. A. V. Butkevich. Quasi-elastic neutrino charged-current scattering off medium-heavy nuclei: 40Ca and 40Ar. *Phys. Rev. C*, 85:065501, 2012.
448. M. Aker et al. Improved upper limit on the neutrino mass from a direct kinematic method by KATRIN. *Phys. Rev. Lett.*, 123(22):221802, 2019.
449. M. Aker et al. Direct neutrino-mass measurement with sub-electronvolt sensitivity. *Nature Phys.*, 18(2):160–166, 2022.
450. Abdurro'uf and others. The seventeenth data release of the Sloan Digital Sky Surveys: complete release of MaNGA, MaStar, and APOGEE-2 data. *Astrophysical Journal*, 259(2):35, April 2022.
451. N. W. Boggess et al. *Astrophysical Journal*, 397:420, October 1992.
452. C.L. Bennett, et al. Nine-year Wilkinson Microwave Anisotropy Probe (WMAP) observations: final maps and results. *The Astrophysical Journal Supplement Series*, 208(2):20, 2013.
453. N. Aghanim et al. Planck 2018 results. I. Overview and the cosmological legacy of Planck. *Astron. Astrophys.*, 641:A1, 2020.
454. S. M. Carroll. The cosmological constant. *Living Rev. Rel.*, 4:1, 2001.
455. P. J. E. Peebles. Tests of cosmological models constrained by inflation. *Astrophys. J.*, 284:439–444, 1984.
456. P. J. E. Peebles and B. Ratra. The cosmological constant and dark energy. *Rev. Mod. Phys.*, 75:559–606, 2003.
457. L. Perivolaropoulos and F. Skara. Challenges for $\Lambda$CDM: An update. *New Astron. Rev.*, 95:101659, 2022.
458. S. Pakvasa and K. Tennakone. Neutrinos of non-zero rest mass. *Phys. Rev. Lett.*, 28:1415, 1972.
459. T. J. Loredo and D. Q. Lamb. Bayesian analysis of neutrinos observed from supernova SN-1987A. *Phys. Rev. D*, 65:063002, 2002.
460. G. Pagliaroli, F. Rossi-Torres, and F. Vissani. Neutrino mass bound in the standard scenario for supernova electronic antineutrino emission. *Astropart. Phys.*, 33:287–291, 2010.
461. R. S. L. Hansen, M. Lindner, and O. Scholer. Timing the neutrino signal of a Galactic supernova. *Phys. Rev. D*, 101(12):123018, 2020.

462. B. Abi et al. Supernova neutrino burst detection with the Deep Underground Neutrino Experiment. *Eur. Phys. J. C*, 81(5):423, 2021.
463. S. Curran et al. Beta spectrum of tritium. *Nature*, 162:302, 1948.
464. M. Aker et al. Direct neutrino-mass measurement based on 259 days of KATRIN data. arXiv eprint 2406.13516, 2024.
465. B. Monreal and J. A. Formaggio. Relativistic cyclotron radiation detection of tritium decay electrons as a new technique for measuring the neutrino mass. *Phys. Rev. D*, 80:051301, 2009.
466. A. Ashtari Esfahani et al. Tritium beta spectrum measurement and neutrino mass limit from cyclotron radiation emission spectroscopy. *Phys. Rev. Lett.*, 131(10):102502, 2023.
467. L. Gastaldo et al. The electron capture $^{163}$Ho experiment ECHo: an overview. *J. Low Temp. Phys.*, 176(5–6):876–884, 2014.
468. B. Alpert et al. HOLMES - The electron capture decay of $^{163}$Ho to measure the electron neutrino mass with sub-eV sensitivity. *Eur. Phys. J. C*, 75(3):112, 2015.
469. M. P. Croce et al. Development of holmium-163 electron-capture spectroscopy with transition-edge sensors. *J. Low Temp. Phys.*, 184(3–4):958–968, 2016.
470. C. Velte et al. High-resolution and low-background $^{163}$Ho spectrum: interpretation of the resonance tails. *Eur. Phys. J. C*, 79(12):1026, 2019.
471. K. Assamagan et al. Upper limit of the muon-neutrino mass and charged pion mass from momentum analysis of a surface muon beam. *Phys. Rev. D*, 53:6065–6077, 1996.
472. R. Barate et al. An Upper limit on the tau-neutrino mass from three-prong and five-prong tau decays. *Eur. Phys. J. C*, 2:395–406, 1998.
473. M. Goeppert-Mayer. Double beta-disintegration. *Phys. Rev.*, 48:512–516, 1935.
474. S. R. Elliott, A. A. Hahn, and M. K. Moe. Direct evidence for two neutrino double beta decay in $^{82}$Se. *Phys. Rev. Lett.*, 59:2020–2023, 1987.
475. E. Aprile et al. Observation of two-neutrino double electron capture in $^{124}$Xe with XENON1T. *Nature*, 568(7753):532–535, 2019.
476. R. Saakyan. Two-neutrino double-beta decay. *Ann. Rev. Nucl. Part. Sci.*, 63:503–529, 2013.
477. A. S. Barabash. Average and recommended half-life values for two-neutrino double beta decay: upgrade-2019. *AIP Conf. Proc.*, 2165(1):020002, 2019.
478. A. S. Barabash. Precise half-life values for two neutrino double beta decay. *Phys. Rev. C*, 81:035501, 2010.
479. A. Barabash. Double beta decay experiments: recent achievements and future prospects. *Universe*, 9(6):290, 2023.
480. M. Agostini et al. Final results of GERDA on the two-neutrino double-$\beta$ decay half-life of Ge76. *Phys. Rev. Lett.*, 131(14):142501, 2023.
481. M. G. Inghram and J. H. Reynolds. Double beta-decay of Te-130. *Phys. Rev.*, 78:822–823, 1950.
482. W. H. Furry. On transition probabilities in double beta-disintegration. *Phys. Rev.*, 56:1184–1193, 1939.
483. J. Schechter and J. W. F. Valle. Neutrinoless double beta decay in SU(2) x U(1) theories. *Phys. Rev.*, D25:2951, 1982.
484. E. Takasugi. Can the neutrinoless double beta decay take place in the case of dirac neutrinos? *Phys. Lett.*, B149:372, 1984.
485. F. T. Avignone III, S. R. Elliott, and J. Engel. Double beta decay, majorana neutrinos, and neutrino mass. *Rev. Mod. Phys.*, 80:481–516, 2008.
486. M. Doi, T. Kotani, H. Nishiura, K. Okuda, and E. Takasugi. Neutrino mass, the right-handed interaction and the double beta decay. 1. Formalism. *Prog. Theor. Phys.*, 66:1739, 1981. [Erratum: Prog. Theor. Phys. 68, 347 (1982)].
487. P. Vogel, F. Boehm. *Physics of Massive Neutrinos, Second edition*. Cambridge University Press, 2 edition, 1992.
488. J. Kotila and F. Iachello. Phase space factors for double-$\beta$ decay. *Phys. Rev.*, C85:034316, 2012.
489. W. C. Haxton and G. J. Stephenson. Double beta decay. *Prog. Part. Nucl. Phys.*, 12:409–479, 1984.

490. M. Doi, T. Kotani, and E. Takasugi. Double beta decay and majorana neutrino. *Prog. Theor. Phys. Suppl.*, 83:1, 1985.
491. J. Suhonen and O. Civitarese. Weak-interaction and nuclear-structure aspects of nuclear double beta decay. *Phys. Rept.*, 300:123–214, 1998.
492. S. M. Bilenky. Neutrinoless double beta-decay. *Phys. Part. Nucl.*, 41:690–715, 2010.
493. F. Pompa, T. Schwetz, and J. Y. Zhu. Impact of nuclear matrix element calculations for current and future neutrinoless double beta decay searches. *JHEP*, 06:104, 2023.
494. H. V. Klapdor-Kleingrothaus, A. Dietz, H. L. Harney, and I. V. Krivosheina. Evidence for neutrinoless double beta decay. *Mod. Phys. Lett. A*, 16:2409–2420, 2001.
495. M. Agostini et al. Final results of GERDA on the search for neutrinoless double-$\beta$ decay. *Phys. Rev. Lett.*, 125(25):252502, 2020.
496. I. J. Arnquist et al. Final result of the majorana demonstrator's search for neutrinoless double-$\beta$ decay in Ge76. *Phys. Rev. Lett.*, 130(6):062501, 2023.
497. S. Abe et al. Search for the majorana nature of neutrinos in the inverted mass ordering region with KamLAND-Zen. *Phys. Rev. Lett.*, 130(5):051801, 2023.
498. G. Anton et al. Search for neutrinoless double-$\beta$ decay with the complete EXO-200 dataset. *Phys. Rev. Lett.*, 123(16):161802, 2019.
499. D. Q. Adams et al. Search for Majorana neutrinos exploiting millikelvin cryogenics with CUORE. *Nature*, 604(7904):53–58, 2022.
500. D. Q. Adams et al. New direct limit on neutrinoless double beta decay half-life of Te128 with CUORE. *Phys. Rev. Lett.*, 129(22):222501, 2022.
501. O. Azzolini et al. Final result on the neutrinoless double beta decay of $^{82}$Se with CUPID-0. *Phys. Rev. Lett.*, 129(11):111801, 2022.
502. R. Arnold et al. Final results on $^{82}Se$ double beta decay to the ground state of $^{82}Kr$ from the NEMO-3 experiment. *Eur. Phys. J. C*, 78(10):821, 2018.
503. C. Augier et al. Final results on the $0\nu\beta\beta$ decay half-life limit of $^{100}$Mo from the CUPID-Mo experiment. *Eur. Phys. J. C*, 82(11):1033, 2022.
504. A. S. Barabash et al. Final results of the Aurora experiment to study $2\beta$ decay of $^{116}$Cd with enriched $^{116}$CdWO$_4$ crystal scintillators. *Phys. Rev. D*, 98(9):092007, 2018.
505. S. Ajimura et al. Low background measurement in CANDLES-III for studying the neutrinoless double beta decay of $^{48}$Ca. *Phys. Rev. D*, 103(9):092008, 2021.
506. H. V. Klapdor-Kleingrothaus, A. Dietz, L. Baudis, G. Heusser, I. V. Krivosheina, et al. Latest results from the Heidelberg-Moscow double beta decay experiment. *Eur. Phys. J.*, A12:147–154, 2001.
507. C. E. Aalseth et al. The IGEX Ge-76 neutrinoless double beta decay experiment: Prospects for next generation experiments. *Phys. Rev.*, D65:092007, 2002.
508. H. V. Klapdor-Kleingrothaus and I. V. Krivosheina. The evidence for the observation of 0nu beta beta decay: The identification of 0nu beta beta events from the full spectra. *Mod. Phys. Lett.*, A21:1547–1566, 2006.
509. M. Agostini et al. Final results of GERDA on the search for neutrinoless double-$\beta$ decay. *Phys. Rev. Lett.*, 125(25):252502, 2020.
510. R. Gaitskell et al. The Majorana zero neutrino double beta decay experiment. 2003.
511. J. H. Arling et al. Commissioning of the COBRA extended demonstrator at the LNGS. *Nucl. Instrum. Meth. A*, 1010:165524, 2021.
512. D. R. Artusa et al. Searching for neutrinoless double-beta decay of $^{130}$Te with CUORE. *Adv. High Energy Phys.*, 2015:879871, 2015.
513. V. Alenkov et al. Technical Design Report for the AMoRE $0\nu\beta\beta$ Decay Search Experiment. 12 2015.
514. Y. Oh. Status of AMoRE. *J. Phys. Conf. Ser.*, 2156:012146, 2021.
515. A. Gando et al. Search for Majorana neutrinos near the inverted mass hierarchy region with KamLAND-Zen. *Phys. Rev. Lett.*, 117(8):082503, 2016. [Addendum: Phys. Rev. Lett. 117, 109903 (2016)].
516. M. Auger et al. The EXO-200 detector, part I: Detector design and construction. *JINST*, 7:P05010, 2012.

517. J. B. Albert et al. Sensitivity and discovery potential of nEXO to neutrinoless double beta decay. *Phys. Rev. C*, 97(6):065503, 2018.
518. J. Martín-Albo et al. Sensitivity of NEXT-100 to neutrinoless double beta decay. *JHEP*, 05:159, 2016.
519. J. Argyriades et al. Measurement of the double beta decay half-life of Nd-150 and search for neutrinoless decay modes with the NEMO-3 detector. *Phys. Rev.*, C80:032501, 2009.
520. R. Arnold et al. Probing new physics models of neutrinoless double beta decay with SuperNEMO. *Eur. Phys. J.*, C70:927–943, 2010.
521. F. Zwicky. Die Rotverschiebung von extragalaktischen Nebeln. *Helv. Phys. Acta*, 6:110–127, 1933.
522. V. C. Rubin and W. K. Ford, Jr. Rotation of the andromeda nebula from a spectroscopic survey of emission regions. *Astrophys. J.*, 159:379–403, 1970.
523. K. G. Begeman, A. H. Broeils, and R. H. Sanders. Extended rotation curves of spiral galaxies: Dark haloes and modified dynamics. *Mon. Not. Roy. Astron. Soc.*, 249:523, 1991.
524. L. V. E. Koopmans and T. Treu. The structure and dynamics of luminous and dark matter in the early-type lens galaxy of 0047-281 at z=0.485. *Astrophys. J.*, 583:606–615, 2003.
525. D. Clowe, A. Gonzalez, and M. Markevitch. Weak lensing mass reconstruction of the interacting cluster 1E0657-558: Direct evidence for the existence of dark matter. *Astrophys. J.*, 604:596–603, 2004.
526. P. A. R. Ade et al. Planck 2013 results. XVI. Cosmological parameters. *Astron. Astrophys.*, 571:A16, 2014.
527. M. Tegmark et al. Cosmological parameters from SDSS and WMAP. *Phys. Rev. D*, 69:103501, 2004.
528. Y. Chikashige, R. N. Mohapatra, and R. D. Peccei. Spontaneously broken lepton number and cosmological constraints on the neutrino mass spectrum. *Phys. Rev. Lett.*, 45:1926, 1980.
529. G. B. Gelmini and M. Roncadelli. Left-handed neutrino mass scale and spontaneously broken lepton number. *Phys. Lett.*, B99:411, 1981.
530. H. M. Georgi, S. L. Glashow, and S. Nussinov. Unconventional model of neutrino masses. *Nucl. Phys.*, B193:297, 1981.
531. A. Santamaria and J. W. F. Valle. Spontaneous R-parity violation in supersymmetry: a model for solar neutrino oscillations. *Phys. Lett. B*, 195:423–428, 1987.
532. A. Santamaria and J. W. F. Valle. Supersymmetric Majoron signatures and solar neutrino oscillations. *Phys. Rev. Lett.*, 60:397–400, 1988.
533. A. Santamaria and J. W. F. Valle. Solar neutrino oscillation parameters and the broken R parity Majoron. *Phys. Rev. D*, 39:1780–1783, 1989.
534. C. D. Carone. Double beta decay with vector majorons. *Phys. Lett. B*, 308:85–88, 1993.
535. R. N. Mohapatra, A. Perez-Lorenzana, and C. A. de S Pires. Neutrino mass, bulk Majoron and neutrinoless double beta decay. *Phys. Lett. B*, 491:143–147, 2000.
536. M. Agostini et al. Search for exotic physics in double-$\beta$ decays with GERDA Phase II. *JCAP*, 12:012, 2022.
537. O. Azzolini et al. Search for Majoron-like particles with CUPID-0. *Phys. Rev. D*, 107(3):032006, 2023.
538. S. Al Kharusi et al. Search for Majoron-emitting modes of $^{136}$Xe double beta decay with the complete EXO-200 dataset. *Phys. Rev. D*, 104(11):112002, 2021.
539. A. Gando et al. Limits on Majoron-emitting double-beta decays of Xe-136 in the KamLAND-Zen experiment. *Phys. Rev. C*, 86:021601, 2012.
540. R. Arnold et al. Detailed studies of $^{100}$Mo two-neutrino double beta decay in NEMO-3. *Eur. Phys. J. C*, 79(5):440, 2019.
541. M. Kachelriess, R. Tomas, and J. W. F. Valle. Supernova bounds on Majoron emitting decays of light neutrinos. *Phys. Rev. D*, 62:023004, 2000.
542. Y. Farzan. Bounds on the coupling of the Majoron to light neutrinos from supernova cooling. *Phys. Rev. D*, 67:073015, 2003.
543. M. E. Peskin and D. V. Schroeder. *An Introduction to Quantum Field Theory*. Addison-Wesley, Reading, USA, 1995.

544. R. E. Shrock. Electromagnetic properties and decays of dirac and Majorana neutrinos in a general class of gauge theories. *Nucl. Phys. B*, 206:359–379, 1982.
545. B. W. Lee and R. E. Shrock. Natural suppression of symmetry violation in gauge theories: muon - lepton and electron lepton number nonconservation. *Phys. Rev. D*, 16:1444, 1977.
546. W. J. Marciano and A. I. Sanda. Exotic decays of the muon and heavy leptons in gauge theories. *Phys. Lett. B*, 67:303–305, 1977.
547. P. B. Pal and L. Wolfenstein. Radiative decays of massive neutrinos. *Phys. Rev. D*, 25:766, 1982.
548. J. Schechter and J. W. F. Valle. Majorana neutrinos and magnetic fields. *Phys. Rev. D*, 24:1883–1889, 1981. [Erratum: Phys. Rev. D 25, 283 (1982)].
549. R. C. Allen et al. Study of electron-neutrino electron elastic scattering at LAMPF. *Phys. Rev. D*, 47:11–28, 1993.
550. L. B. Auerbach et al. Measurement of electron - neutrino - electron elastic scattering. *Phys. Rev. D*, 63:112001, 2001.
551. R. Schwienhorst et al. A new upper limit for the tau - neutrino magnetic moment. *Phys. Lett. B*, 513:23–29, 2001.
552. E. Aprile et al. Excess electronic recoil events in XENON1T. *Phys. Rev. D*, 102(7):072004, 2020.
553. M. Agostini et al. Limiting neutrino magnetic moments with Borexino Phase-II solar neutrino data. *Phys. Rev. D*, 96(9):091103, 2017.
554. L. Ludhova. Limiting the effective magnetic moment of solar neutrinos with the Borexino detector. *J. Phys. Conf. Ser.*, 1342(1):012033, 2020.
555. H. T. Wong et al. A search of neutrino magnetic moments with a high-purity germanium detector at the Kuo-Sheng nuclear power station. *Phys. Rev. D*, 75:012001, 2007.
556. A. G. Beda, V. B. Brudanin, V. G. Egorov, D. V. Medvedev, V. S. Pogosov, M. V. Shirchenko, and A. S. Starostin. The results of search for the neutrino magnetic moment in GEMMA experiment. *Adv. High Energy Phys.*, 2012:350150, 2012.
557. D. W. Liu et al. Limits on the neutrino magnetic moment using 1496 days of Super-Kamiokande-I solar neutrino data. *Phys. Rev. Lett.*, 93:021802, 2004.
558. A. Gal. Neutrino effects in two-body electron-capture measurements at GSI. *Nucl. Phys. A*, 842:102–112, 2010.
559. M. Atzori Corona, M. Cadeddu, N. Cargioli, F. Dordei, C. Giunti, Y. F. Li, C. A. Ternes, and Y. Y. Zhang. Impact of the Dresden-II and COHERENT neutrino scattering data on neutrino electromagnetic properties and electroweak physics. *JHEP*, 09:164, 2022.
560. B. Pontecorvo. Superweak interactions and double beta decay. *Phys. Lett.*, B26:630–632, 1968.
561. D. A. Bryman and R. Shrock. Improved constraints on sterile neutrinos in the MeV to GeV mass range. *Phys. Rev. D*, 100(5):053006, 2019.
562. B. Dasgupta and J. Kopp. Sterile neutrinos. *Phys. Rept.*, 928:1–63, 2021.
563. A. Aguilar-Arevalo et al. Improved search for heavy neutrinos in the decay $\pi \to e\nu$. *Phys. Rev. D*, 97(7):072012, 2018.
564. A. Abada, D. Bečirević, O. Sumensari, C. Weiland, and R. Zukanovich Funchal. Sterile neutrinos facing kaon physics experiments. *Phys. Rev. D*, 95(7):075023, 2017.
565. L. Gastaldo, C. Giunti, and E. M. Zavanin. Light sterile neutrino sensitivity of 163Ho experiments. *JHEP*, 06:061, 2016.
566. J. Barea, J. Kotila, and F. Iachello. Limits on sterile neutrino contributions to neutrinoless double beta decay. *Phys. Rev. D*, 92:093001, 2015.
567. J. Lesgourgues, G. Mangano, G. Miele, and S. Pastor. *Neutrino Cosmology*. Cambridge University Press, 2 2013.
568. N. Aghanim et al. Planck 2018 results. VI. Cosmological parameters. *Astron. Astrophys.*, 641:A6, 2020. [Erratum: Astron. Astrophys. 652, C4 (2021)].
569. J. R. Primack and G. R. Blumenthal. What is the dark matter? Implications for galaxy formation and particle physics. In *3rd Moriond Astrophysics Meeting: Galaxies and the Early Universe*, pages 445–464, 1983.

570. J. R. Bond, A. S. Szalay, J. Centrella, and J. R. Wilson. Dark matter and shocked pancakes. In *3rd Moriond Astrophysics Meeting: Galaxies and the Early Universe*, pages 87–99, 1983.
571. K. N. Abazajian. Sterile neutrinos in cosmology. *Phys. Rept.*, 711–712:1–28, 2017.
572. M. Drewes et al. A white paper on keV sterile neutrino dark matter. *JCAP*, 01:025, 2017.
573. X. D. Shi and G. M. Fuller. A new dark matter candidate: Nonthermal sterile neutrinos. *Phys. Rev. Lett.*, 82:2832–2835, 1999.
574. A. Boyarsky, O. Ruchayskiy, and D. Iakubovskyi. A lower bound on the mass of dark matter particles. *JCAP*, 03:005, 2009.
575. A. Boyarsky, M. Drewes, T. Lasserre, S. Mertens, and O. Ruchayskiy. Sterile neutrino dark matter. *Prog. Part. Nucl. Phys.*, 104:1–45, 2019.
576. E. Bulbul, M. Markevitch, A. Foster, R. K. Smith, M. Loewenstein, and S. W. Randall. Detection of an unidentified emission line in the stacked x-ray spectrum of galaxy clusters. *Astrophys. J.*, 789:13, 2014.
577. A. Boyarsky, O. Ruchayskiy, D. Iakubovskyi, and J. Franse. Unidentified line in X-ray spectra of the Andromeda galaxy and Perseus galaxy cluster. *Phys. Rev. Lett.*, 113:251301, 2014.
578. A. Boyarsky, J. Franse, D. Iakubovskyi, and O. Ruchayskiy. Checking the dark matter origin of a 3.53 keV line with the Milky Way Center. *Phys. Rev. Lett.*, 115:161301, 2015.
579. H. Gemmeke et al. The High resolution neutrino calorimeter KARMEN. *Nucl. Instrum. Meth. A*, 289:490–495, 1990.
580. E. D. Church, K. Eitel, G. B. Mills, and M. Steidl. Statistical analysis of different muon-anti-neutrino —> electron-anti-neutrino searches. *Phys. Rev. D*, 66:013001, 2002.
581. A. A. Aguilar-Arevalo et al. The MiniBooNE detector. *Nucl. Instrum. Meth. A*, 599:28–46, 2009.
582. A. A. Aguilar-Arevalo et al. A search for electron neutrino appearance at the $\Delta m^2 \sim 1eV^2$ scale. *Phys. Rev. Lett.*, 98:231801, 2007.
583. A. A. Aguilar-Arevalo et al. Significant excess of ElectronLike events in the MiniBooNE short-baseline neutrino experiment. *Phys. Rev. Lett.*, 121(22):221801, 2018.
584. A. A. Aguilar-Arevalo et al. Updated MiniBooNE neutrino oscillation results with increased data and new background studies. *Phys. Rev. D*, 103(5): 052002, 2021.
585. R. Acciarri et al. a proposal for a three detector short-baseline neutrino oscillation program in the Fermilab booster neutrino beam. 3 2015.
586. B. Fleming. The MicroBooNE Technical Design Report. 2 2012.
587. P. Abratenko et al. Search for an excess of electron neutrino interactions in MicroBooNE using multiple final-state topologies. *Phys. Rev. Lett.*, 128(24):241801, 2022.
588. P. Abratenko et al. First constraints on light sterile neutrino oscillations from combined appearance and disappearance searches with the MicroBooNE detector. *Phys. Rev. Lett.*, 130(1):011801, 2023.
589. M. Antonello, B. Baibussinov, P. Benetti, E. Calligarich, N. Canci, et al. Experimental search for the "LSND anomaly" with the ICARUS detector in the CNGS neutrino beam. *Eur. Phys. J.*, C73(3):2345, 2013.
590. S. Ajimura et al. The JSNS2 detector. *Nucl. Instrum. Meth. A*, 1014:165742, 2021.
591. C. Giunti and M. Laveder. Statistical significance of the gallium anomaly. *Phys. Rev.*, C83:065504, 2011.
592. J. N. Abdurashitov et al. Measurement of the solar neutrino capture rate with gallium metal. III: Results for the 2002–2007 data-taking period. *Phys. Rev. C*, 80:015807, 2009.
593. S. R. Elliott, V. Gavrin, and W. Haxton. The gallium anomaly. *Prog. Part. Nucl. Phys.*, 134:104082, 2024.
594. G. Mention, M. Fechner, Th. Lasserre, Th. A. Mueller, D. Lhuillier, M. Cribier, and A. Letourneau. The reactor antineutrino anomaly. *Phys. Rev.*, D83:073006, 2011.
595. C. Giunti, Y. F. Li, C. A. Ternes, and Z. Xin. Reactor antineutrino anomaly in light of recent flux model refinements. *Phys. Lett. B*, 829:137054, 2022.
596. C. Giunti and T. Lasserre. eV-scale sterile neutrinos. *Ann. Rev. Nucl. Part. Sci.*, 69:163–190, 2019.

597. F. P. An et al. Improved measurement of the reactor antineutrino flux and spectrum at Daya Bay. *Chin. Phys. C*, 41(1):013002, 2017.
598. M. G. Aartsen et al. Searches for sterile neutrinos with the IceCube detector. *Phys. Rev. Lett.*, 117(7):071801, 2016.
599. M. G. Aartsen et al. eV-Scale sterile neutrino search using eight years of atmospheric muon neutrino data from the IceCube neutrino observatory. *Phys. Rev. Lett.*, 125(14):141801, 2020.
600. M. G. Aartsen et al. Searching for eV-scale sterile neutrinos with eight years of atmospheric neutrinos at the IceCube neutrino telescope. *Phys. Rev. D*, 102(5):052009, 2020.
601. M. G. Aartsen et al. Search for sterile neutrino mixing using three years of IceCube DeepCore data. *Phys. Rev. D*, 95(11):112002, 2017.
602. R. Abbasi et al. Strong constraints on neutrino nonstandard interactions from TeV-Scale $\nu_\mu$ disappearance at IceCube. *Phys. Rev. Lett.*, 129(1):011804, 2022.
603. R. Abbasi et al. Search for unstable sterile neutrinos with the IceCube neutrino observatory. *Phys. Rev. Lett.*, 129(15):151801, 2022.
604. K. Abe et al. First joint oscillation analysis of Super-Kamiokande atmospheric and T2K accelerator neutrino data, 2024.

# Index

**A**
Accidental symmetries, 135, 137, 149
Active Galactic Nuclei (AGN), 17, 247, 258
Adjoint representation, 23
ALEPH, 145, 351
Alternating Gradient Synchrotron (AGS), 13, 16, 277
AMANDA, 260, 261
Anomalies, 138
Appearance experiments, 267
ARCA, 261
Astronomy with a Neutrino Telescope and Abyss environmental Research (ANTARES), 257, 258, 261
Astrophysical neutrinos, 252
    high energy, 256
Atmospheric neutrino, 246
    anomaly, 15, 250

**B**
BaBar, 53, 141
Baikal Deep Underwater Neutrino Telescope (BDUNT), 245
Baksan Experiment on Sterile Transitions (BEST), 232, 386
Baksan Neutrino Observatory (BNO), 244
Baksan Underground Scintillator Telescope (BUST), 14, 244, 245, 348
Baryon number, 11, 37–39, 127, 136–138
Baseline
    long, 267
    short, 267, 381
Beam dump, 282
Belle, 53, 141
Beryllium neutrinos, 221, 236
BES-III, 141
Beta decay, 6, 9, 14, 15, 131, 237, 263, 374
    allowed, 131
    double, 352, 362
    Fermi, 10
    Gamow-Teller, 10
    inverse, 12, 15, 225, 232, 235, 236, 244, 245, 264, 269, 271, 272, 300, 317, 337, 338, 382, 383, 386
    neutrinoless double, 354, 359
    positive, 10
    superallowed, 131
$B$ Factories, 56
Big Bang, 219, 252, 253
    nucleosynthesis, 253
Bjorken scaling, 316, 327, 331
Blazar, 17
Borexino, 17, 222, 235, 236, 245, 265
Boron neutrinos, 222, 234
Bottomness, 37
Bubble chamber, 279, 286, 321
    Argonne, 333

**C**
Cabibbo angle, 128
Cabibbo-Kobayashi-Maskawa (CKM) matrix, 126, 129, 134, 139
Cadmium Zinc Telluride 0-Neutrino Double-Beta Research Apparatus (COBRA), 363
Carbon cycle, 220
Carbon-nitrogen-oxygen (CNO) cycle, 221, 223, 236

CENNS-10, 339
CERN, 118, 141, 145, 278, 279, 282, 283, 286, 298, 332, 351, 385
CERN Hybrid Oscillation Search Apparatus (CHORUS), 298
CERN Neutrinos to Gran Sasso (CNGS), 286
Charge conjugation, 38, 46, 61
   symmetry, 50, 133
Charged current quasi elastic (CCQE), 308, 315–318, 325, 334, 343, 384
Charge symmetry, 50, 319
Charm quantum number, 37
Cherenkov, 225, 226, 228–231, 233, 234, 236, 244, 245, 250, 251, 258, 260, 261, 276, 283, 293, 296, 323
Chirality, 69
CHOOZ, 251, 274, 276
CITF, 333
COHERENT, 339
Coherent elastic neutrino-nucleus scattering (CE$\nu$NS), 337
Coherent Neutrino-Nucleus Interaction Experiment (CONNIE), 339
Coherent nuclear scattering, 335, 336
Cosmic Microwave Background (CMB), 17, 219, 253–255, 259, 346, 347
Cosmic Neutrino Background (CNB), 252, 254, 255
Cosmic neutrinos, 252
Cosmic rays, 246
Covariant derivative, 102
Cowan and Reines experiment, 12, 267, 272, 273
*CP*
   invariance, 51
   symmetry, 133
   violation, 53, 197, 201, 205, 300, 301, 358
CPLEAR, 55
*CPT*
   invariance, 206, 301
   transformation, 75, 150
   violation, 285
Crossing symmetry, 12
Cross section, 12, 128, 231, 233, 241, 242, 251, 269, 271, 273, 275, 281, 303, 304, 310, 312
Cryogenic Underground Observatory for Rare Events (CUORE), 364
CUORE Upgrade with Particle IDentification (CUPID), 369
Cyclotron Radiation Emission Spectroscopy (CRES), 350

**D**

Dark matter, 339, 347, 364, 366, 369, 376
Dark sector, 376
Daya Bay, 16, 270, 275, 386
Decay at rest of pions (DAR), 281, 298, 339, 382, 386
Decay constant, 227
Decay in flight (DIF), 280, 281, 383
DeepCore, 261, 387
Deep inelastic scattering, 304, 316, 323, 324, 327, 332, 333, 335
Deep Underwater Muon and Neutrino Detector (DUMAND), 260
Deep Underground Neutrino Experiment (DUNE), 295–297, 348, 385
DELPHI, 145
Desy, 332
Diffuse supernova neutrino background, 243, 277
Dirac adjoint, 60, 73, 153
Dirac conjugate, 60
Dirac equation, 2–5, 19, 34, 38, 59, 63, 65, 67, 70, 73, 145
Dirac-Majorana mass term, 159, 160, 162, 164, 367, 378
Dirac mass terms, 146, 159, 162
Dirac matrices, 60
Dirac spinors, 71
Disappearance experiments, 267
DOM, 261
DONUT, 16, 282, 298
Double Chooz, 16, 270, 275, 276, 300
Double electron capture, 352
Dresden-II, 340

**E**

ECHo, 350
Effective Lagrangian, 169, 172
Electric dipole moment, 28, 29, 372
Electron capture, 12, 15, 350, 352
Enriched Xenon Observatory (EXO), 365, 369
Euclid, 347

**F**

Fermi constant, 8, 118, 170, 209, 320, 371
Fermi gas model, 342
Fermi interaction, 8, 68
Fermi Large Area Telescope (Fermi-LAT), 258
Fermi National Accelerator Laboratory (FNAL), 130, 383

Fermi smearing, 334
Final state interactions, 323, 334, 341
Flavour
  anomalies, 141
  changing neutral currents, 128
  symmetry, 136

## G

GALLEX, 232, 386
Gallium anomaly, 386
Gamma-ray bursts, 219, 247
Gargamelle, 304, 332
Gauge
  principle, 83
  symmetry
    electroweak, 102
    Standard Model, 153, 155, 157
  transformations, 79, 81
Gell-Mann-Nishijima formula, 38
Geoneutrinos, 263
GERDA, 353, 362, 363, 369
Gigaton Volume Detector (GVD), 261
GIM mechanism, 128, 129, 373
Global symmetries, 19
GNO, 232, 386
Goldhaber, Grodzins and A.W. Sunyar experiment, 13, 133, 145
Goldstone boson, 95, 97
G-parity, 39, 48
Grand Unified Theories (GUT), 138, 143, 165, 171, 250, 296
Group
  representation, 22
  theory, 21

## H

Half-life, 227, 232, 239, 254, 264, 349, 353, 359
Harvard, Pennsylvania, Wisconsin, Fermilab (HPWF), 333
Heat-producing elements, 264
Heidelberg-Moskow Double Beta Decay Experiment (HMBB), 362
Heisenberg ferromagnet, 88
Helicity, 69
Hep-chain, 222, 223
Hierarchy, 202
  inverted, 203
  normal, 202
Higgs mechanism, 90, 98, 100, 102, 114, 118, 138, 165
Homestake, 225–228, 232

Hyper-Kamiokande, 226, 245, 295–297

## I

IceCube, 17, 245, 252, 256, 261, 387
IGEX, 362
Imaging Cosmic and Rare Underground Signals (ICARUS), 251, 286, 288, 385
Impulse approximation, 343
Incoherent transition, 192, 195
Interactions
  electroweak, 133
  neutrino-electron, 305
    elastic scattering, 306
    quasi elastic scattering, 306
  neutrino-nucleon, 315
Invisible width, 144
Irvine-Michigan-Brookhaven (IMB), 14, 244, 250, 348

## J

Jiangmen Underground Neutrino Observatory (JUNO), 245, 265, 277, 348
J-PARK, 293, 295, 297, 386
JSNS$^2$, 386

## K

Kamiokande, 14, 225, 231, 234, 244, 250, 276, 348
KamLAND, 16, 17, 234, 235, 245, 264, 276, 365
KamLAND-Zen, 365, 369
Kaons
  neutral, 51, 55
  oscillations, 13
KArlsruhe Rutherford Medium Energy Neutrino (KARMEN), 281, 299, 382
Karlsruhe Tritium Neutrino (KATRIN), 346, 349, 350
KEDR, 141
KEK, 53, 56, 283
KEK to Kamioka (K2K), 16, 251, 283, 299
Klein-Gordon equation, 2, 5, 65, 91
KM3Net, 245, 261

## L

L3, 145
Laboratori Nazionali del Gran Sasso (LNGS), 225, 283, 286, 353, 362–364, 385

Lambda Cold Dark Matter ($\Lambda$CDM) model, 347
Lattice QCD, 320
Legacy Survey of Space and Time (LSST), 347
LEGEND, 363
LEP, 16, 139, 145, 351, 368
Lepton family number, 137, 149
Lepton number, 38, 137, 149, 155
Lepton universality, 138
Levi-Civita tensor, 61, 103, 110, 175
LHCb, 141
Lie algebras, 22, 23
Lie groups, 22, 23, 31, 32, 175
Lifetime, 227
Linear sigma model, 96
Liquid argon time projection chamber (LAr-TPC), 286, 297, 384, 385
Liquid Scintillator Neutrino Detector (LSND), 281, 298, 382, 386
Lorentz
    group, 29, 32, 145
    invariance, 116
    transformation, 59, 290
Lorenz
    gauge, 80
Los Alamos Meson Physics Facility (LAMPF), 382
Los Alamos National Laboratory (LANL), 271, 281, 382
Los Alamos Neutron Science Center (LANSCE), 382
Los Alamos Scientific Laboratory (LASL), 271
Luminosity, 221
    apparent, 224

## M
Magnetars, 247
Magnetic horn, 279, 280, 284, 383
Magnetic moment, 6, 9, 28, 45, 352
    dipole, 28
    neutrino, 245, 339, 345, 370
Main sequence, 220, 237, 238
Majorana
    effective mass, 356
    mass, 152, 157, 159, 162, 172
    neutrino, 149, 152, 162, 354, 359
    particle, 9
Majorana Demonstrator, 363
Major Atmospheric Gamma Imaging Cherenkov (MAGIC), 258
Majoron, 366
Matrix elements of hadronic currents, 134

Maxwell equation, 81
Meson exchange current (MEC), 335
MicroBooNE, 384
Mikheyev-Smirnov-Wolfenstein (MSW) mechanism, 14, 212, 236
MiniBooNE, 299, 383, 384
MINOS, 16, 263, 284, 299
Multi-messanger astronomy, 258

## N
NA62, 140
Natural units, 26, 29, 195, 253, 312
NEOS, 386
NESTOR, 260
Neutrino
    astronomy, 18
    discovery
        muon neutrino, 13, 278
        tau neutrino, 16, 282
    generations, 145
    magnetic moment, 271
    mass, 10, 143, 146, 148, 162, 345, 348, 349
        effective, 349
    maximal mixing, 163
    mixing, 163
    oscillations, 13, 183, 188, 190
        matter, 207, 212, 236, 252, 285
    portal, 377
    sterile, 13, 17, 147, 151, 152, 154, 157, 159, 160, 163–166, 168, 186, 273, 274, 286, 298, 299, 339, 374, 378, 387
Neutrino beams
    beam dump, 282
    conventional, 280
    stopped pions, 281
Neutrino capture, 227, 228, 256, 335
    on $\beta$-decay nuclei, 255
Neutrino Experiment with a Xenon TPC (NEXT), 365
Neutrino floor, 339
NEutrino Mediterranean Observatory (NEMO), 369
Neutrino Oscillation MAgnetic Detector (NOMAD), 298
Next-generation Enriched Xenon Observatory (neXO), 365
NGC 1068, 258
Nöther's theorem, 36
NO$\nu$A, 201, 251, 263, 292, 294, 299
Nuclear fission, 12, 268–272, 317
Nuclear weak charge, 338
NuMI, 284, 294, 385

# Index

## O
Off-axis, 281, 288, 293
OPAL, 145
ORCA, 261
Oscillation
  length, 195, 213, 235, 251
  parameters, 299
Oscillation Project with Emulsion Racking Apparatus (OPERA), 251, 263, 286, 287, 298

## P
Palo Verde, 251, 274
Parity, 10, 25, 28, 29, 39–41, 61, 134
  symmetry, 133
Parton model, 316, 327, 332
  PDF, 329, 331
Pauli matrices, 3, 49, 50, 105, 106, 111, 175
Pep-chain, 222, 236
Photomultiplier tube (PMT), 230, 231, 245, 275, 339, 365, 384
PIENU, 140
PMNS (or lepton mixing) matrix, 184, 188, 190, 194, 197, 200, 207, 263, 298–300, 349, 373
Poincaré
  group, 34
  transformation, 34
Pp-chain, 220–222, 232, 236
Probability
  survival, 190, 195, 205, 217, 218, 236, 267, 274, 299
  transition, 189, 194, 196
    average, 194
Project 8, 350
Prompt neutrinos, 282
Proton decay, 171, 226, 231, 250, 296
Protons on target (POT), 280, 284
Proton synchrotron (PS), 278, 332
Ptolemy, 255, 256
Pulsar wind nebula, 247

## Q
Quantum electrodynamics (QED), 81
Quantum numbers, 36, 37
Q-value, 10, 11, 220, 348, 349, 351, 355, 356, 359

## R
Random phase approximation (RPA), 341

Reactor
  anti-neutrino anomaly, 386
Relic neutrinos, 17
RENO, 16, 275, 387
Roper resonance, 323
Rutherford Appleton Laboratory (RAL), 281, 382

## S
Sanford Underground Research Facility (SURF), 297
Scintillator, 224, 234–236, 244, 245, 272
Seesaw mechanism, 164, 166, 173
  type I, 173
  type II, 173–175, 179
  type III, 173, 174, 179
Short Baseline Near Detector (SBND), 385
Single pion production, 321
SLAC, 16, 53, 56, 59, 129, 327, 328, 332
SLC, 16, 139
Sloan Digital Sky Survey (SDSS), 346
SN1987A, 14, 225, 231, 237, 244
SNO+, 245, 265, 365
Solar neutrino, 14, 207, 219, 224, 225, 227, 228, 231–235, 274, 276, 277, 296, 299
  flux, 224
  problem, 14, 16, 224–226, 228, 231, 232, 234
Soviet–American Gallium Experiment (SAGE), 232, 386
Spallation source, 281, 339
  ISIS, 382
  J-PARC, 386
  SNS, 339
Spark chamber, 278
Spectral function, 341
Spontaneous symmetry breaking (SSB), 87, 90, 93, 96, 157, 159, 165, 377
Standard Model, 9, 14, 80
Standard solar model (SSM), 219, 231, 232, 234
Strangeness, 37
Sudbury Neutrino Observatory (SNO), 16, 196, 225, 226, 233–235, 245, 251, 365
Sun fusion reactions, 219
Super Proton Synchrotron (SPS), 298, 332
Super-Kamiokande, 16, 225, 226, 231, 245, 250, 251, 262, 283, 293, 296, 299
Supernova, 236, 240
  neutrino, 14, 236
  remnant, 247

## T

Tachyon, 92, 93
Tau, 14
　decays, 140
Tevatron, 130, 139, 282
Time-projection chamber (TPC), 286, 365, 384
Time reversal, 25, 27–29, 54, 55, 61, 62
Tokai to Hyper-Kamiokande (T2HK), 292, 295, 297
Tokai to Kamioka (T2K), 16, 201, 251, 263, 285, 292, 293, 299
Topness, 37
Two-body currents, 335
Two-particle-two-hole (2p-2h) process, 335
TXS 0506056, 258

## W

Weak (or Weinberg) angle, 103, 305, 338
Weinberg operator, 171, 172, 174, 179, 181
Weyl
　fields, 72, 73
　spinors, 145

## Y

Yang-Mills, 80, 86

## Z

Zero Gradient Synchrotron (ZGS), 333